Strawberry

Growth, Development and Diseases

Strawberry

Growth, Development and Diseases

Edited by

Amjad M. Husaini and Davide Neri

CABI

CABI is a trading name of CAB International

CABI
Nosworthy Way
Wallingford
Oxfordshire OX10 8DE
UK

Tel: +44 (0)1491 832111
Fax: +44 (0)1491 833508
E-mail: info@cabi.org
Website: www.cabi.org

CABI
745 Atlantic Avenue
8th Floor
Boston, MA 02111
USA

Tel: +1 (617) 682 9015
E-mail: cabi-nao@cabi.org

A catalogue record for this book is available from the British Library, London, UK.

Library of Congress Cataloging-in-Publication Data

Names: Husaini, Amjad M., editor. | Neri, Davide, editor.
Title: Strawberry : growth, development and diseases / [edited by] Dr. Amjad M. Husaini and Davide Neri.
Description: Boston, MA : CABI, [2016] | Includes bibliographical references and index.
Identifiers: LCCN 2016022767 (print) | LCCN 2016023629 (ebook) | ISBN 9781780646633 (hbk : alk. paper) | ISBN 9781780646640 (ePDF) | ISBN 9781780646657 (ePub)
Subjects: LCSH: Strawberries.
Classification: LCC SB385 .S743 2016 (print) | LCC SB385 (ebook) | DDC 634/.75--dc23
LC record available at https://lccn.loc.gov/2016022767

ISBN-13: 978 1 78064 663 3

Commissioning editor: Rachael Russell
Editorial assistant: Emma McCann
Production editor: Shankari Wilford

Typeset by SPi, Pondicherry, India
Printed and bound in the UK by CPI Group (UK) Ltd, Croydon, CR0 4YY, UK

Contents

About the Editors

Dr Amjad M. Husaini

Dr Husaini, Centre for Plant Biotechnology, Sher-e-Kashmir University of Agricultural Sciences & Technology of Kashmir (India), has an M.Sc. from Dr Y.S. Parmar University of Horticulture and Forestry, Solan, India. He holds a Ph.D. in Biotechnology and a PG Diploma in Bioinformatics (Jamia Hamdard, New Delhi), as well as certificates in Intellectual Property Rights (Indian Law Institute, New Delhi) and Remote Sensing Applications in Agriculture (Indian Agricultural Research Institute-Indian Space Research Organization). He has post-doctoral experience of working at the Sustainable Agricultural Systems Laboratory, United States Department of Agriculture, Beltsville, Maryland, USA; and the Plant Chemetics Laboratory, Department of Plant Sciences, University of Oxford, Oxford, UK. He has edited/authored four books, and published more than 40 research/review papers in journals of international repute. He has several awards such as the CREST Award-2013 (DBT, India), Young Scientist Award in Agriculture 2009 (J&K, India), Jawahar Lal Nehru Award for Agricultural Research-2008 (ICAR, India), Junior Scientist of the Year Award-2007 (NESA, India), and Junior and Senior Research Fellowships of CSIR, India. His biography was included in 27th edition of the Marquis Who's Who. His research areas include plant tissue culture, genetic engineering and plant chematics, as well as conventional, molecular and participatory plant breeding.

Dr Davide Neri

Dr Neri has published about 130 scientific, technical and extension papers in international refereed journals, refereed proceedings, national technical and scientific journals, national refereed proceedings, and chapters in scientific and technical books; one technical book; seven edited books and proceedings; extension papers; and multimedia extension items. He is the editor of 23 of the 53 episodes of the agricultural TV series *Le Mille Agricolture* by TVRS. He was Chairman of the strawberry physiology working group of the COST 836 'Integrated berry production' and of the bilateral project Italy-Japan 'Strawberry fruit quality: genetic and physiological background' 2000–2003. He has contributed to the organization of several international and national symposia: COST 836 Final Workshop – Euro Berry Symposium; V International Strawberry Symposium in Australia 2004; and the National Scientific Meeting of Horticultural Society SOI 2013.

Contributors

Ana Aguado, Centro IFAPA Las Torres-Tomejil. CAPDER-Junta de Andalucía. Alcalá del Río, Sevilla, Spain.

Gianluca Baruzzi, Consiglio per la Ricerca in Agricoltura e L'Analisi dell'economia agraria, Unità di Ricerca per la Frutticoltura, Forlì, Italy.

Jesús M. Cantoral, Laboratory of Microbiology, Department of Biochemistry and Biotechnology, Environmental and Marine Sciences Faculty. University of Cádiz, 11510, Puerto Real, Spain.

Nieves Capote, Centro IFAPA Las Torres-Tomejil, CAPDER-Junta de Andalucía, Alcalá del Río, Sevilla, Spain.

María Carbú, Laboratory of Microbiology, Department of Biochemistry and Biotechnology, Environmental and Marine Sciences Faculty, University of Cádiz, 11510, Puerto Real, Spain.

Nadia R. Chalfoun, Instituto Superior de Investigaciones Biológicas (INSIBIO), CONICET-UNT; and Instituto de Química Biológica 'Dr. Bernabé Bloj', Facultad de Bioquímica, Química y Farmacia, Universidad Nacional de Tucumán, Chacabuco 461, T4000ILI – San Miguel de Tucumán, Argentina.

Manuel Chamorro, Centro IFAPA Las Torres-Tomejil, CAPDER-Junta de Andalucía, Alcalá del Río, Sevilla, Spain.

Berta de los Santos, Centro IFAPA Las Torres-Tomejil. CAPDER-Junta de Andalucía. Alcalá del Río, Sevilla, Spain.

Juan C. Díaz Ricci, Instituto Superior de Investigaciones Biológicas (INSIBIO), CONICET-UNT; and Instituto de Química Biológica 'Dr. Bernabé Bloj', Facultad de Bioquímica, Química y Farmacia, Universidad Nacional de Tucumán, Chacabuco 461, T4000ILI – San Miguel de Tucumán, Argentina.

Atilla Eris, Istanbul Bilgi University, Faculty of Engineering and Natural Sciences, Department of Genetics and Bioengineering, Istanbul, Turkey.

Walther Faedi, Consiglio per la Ricerca in Agricoltura e L'Analisi dell'economia agraria, Unità di Ricerca per la Frutticoltura – Forlì, Italy.

Marina Gambardella, Departamento de Fruticultura y Enología – Pontificia Universidad Católica de Chile, Santiago, Chile.

Carlos Garrido, Laboratory of Microbiology, Department of Biochemistry and Biotechnology, Environmental and Marine Sciences Faculty. University of Cádiz, 11510, Puerto Real, Spain.

Victoria E. González-Rodríguez, Laboratory of Microbiology, Department of Biochemistry and Biotechnology, Environmental and Marine Sciences Faculty. University of Cádiz, 11510, Puerto Real, Spain.

Carlos F. Grellet-Bournonville, Instituto Superior de Investigaciones Biológicas (INSIBIO), CONICET-UNT; and Instituto de Química Biológica 'Dr. Bernabé Bloj', Facultad de Bioquímica, Química y Farmacia, Universidad Nacional de Tucumán, Chacabuco 461, T4000ILI – San Miguel de Tucumán, Argentina.

Hatice Gulen, Istanbul Bilgi University, Faculty of Engineering and Natural Sciences, Department of Genetics and Bioengineering, Istanbul, Turkey.

Verónica Hael-Conrad, Instituto Superior de Investigaciones Biológicas (INSIBIO), CONICET-UNT; and Instituto de Química Biológica 'Dr. Bernabé Bloj', Facultad de Bioquímica, Química y Farmacia, Universidad Nacional de Tucumán, Chacabuco 461, T4000ILI – San Miguel de Tucumán, Argentina.

Amjad M. Husaini, Centre for Plant Biotechnology, Sher-e-Kashmir University of Agricultural Sciences and Technology of Kashmir, Srinagar, Shalimar, Jammu and Kashmir-190025, India.

Gustavo G. Martos, Instituto Superior de Investigaciones Biológicas (INSIBIO), CONICET-UNT; and Instituto de Química Biológica 'Dr. Bernabé Bloj', Facultad de Bioquímica, Química y Farmacia, Universidad Nacional de Tucumán, Chacabuco 461, T4000ILI – San Miguel de Tucumán, Argentina.

Francesca Massetani, HORT Soc. Coop. (Horticulture Oriented to Recreation and Technique), part of Università Politecnica delle Marche, 60131 Ancona, Italy; and Dipartimento di Scienze Agrarie, Alimentari e Ambientali, Università Politecnica delle Marche, Ancona, Italy.

Yoh-ichi Matsubara, Faculty of Applied Biological Sciences, Gifu University, Gifu 501-1193, Japan.

Juan J. Medina-Mínguez, Centro IFAPA Huelva, CAPDER-Junta de Andalucía. Huelva, Spain.

Rudy Montenegro, Facultad de Agronomía, Universidad Austral de Chile, Casilla 567, Campus Isla Teja, Valdivia, 5090000, Chile.

Davide Neri, Centro di Ricerca per la Frutticoltura – Consiglio per la Ricerca in Agricoltura e L'analisi Dell'economia Agraria, Rome, Italy. E-mail davide.neri@crea.gov.it, and Dipartimento di Scienze Agrarie, Alimentari e Ambientali – Università Politecnica delle Marche, 60131 Ancona, Italy.

Marisa Perato, Instituto Superior de Investigaciones Biológicas (INSIBIO), CONICET-UNT; and Instituto de Química Biológica 'Dr. Bernabé Bloj', Facultad de Bioquímica, Química y Farmacia, Universidad Nacional de Tucumán, Chacabuco 461, T4000ILI – San Miguel de Tucumán, Argentina.

Cristina Quispe, Laboratorio de Química de Productos Naturales, Instituto de Química de Recursos Naturales, Universidad de Talca, Casilla 747, Talca, 3460000, Chile.

Fernando Romero, Centro IFAPA Las Torres-Tomejil. CAPDER-Junta de Andalucía. Alcalá del Río, Sevilla, Spain.

S. Sánchez, Departamento de Fruticultura y Enología, Pontificia Universidad Católica de Chile. Santiago, Chile.

Guillermo Schmeda-Hirschmann, Laboratorio de Química de Productos Naturales, Instituto de Química de Recursos Naturales, Universidad de Talca, Casilla 747, Talca, 3460000, Chile.

Mario Simirgiotis, Laboratorio de Productos Naturales, Departamento de Química, Facultad de Ciencias Básicas, Universidad de Antofagasta, Casilla 170, Antofagasta 1240000, Chile.

Ursula Tonello, Instituto Superior de Investigaciones Biológicas (INSIBIO), CONICET-UNT; and Instituto de Química Biológica 'Dr. Bernabé Bloj', Facultad de Bioquímica, Química y Farmacia, Universidad Nacional de Tucumán, Chacabuco 461, T4000ILI – San Miguel de Tucumán, Argentina.

Ece Turhan, Eskisehir Osmangazi University, Faculty of Agriculture, Department of Agricultural Biotechnology, Eskisehir, Turkey.

Ezequiel Viveros-Valdez, Laboratorio de Química de Productos Naturales, Instituto de Química de Recursos Naturales, Universidad de Talca, Casilla 747, Talca, 3460000, Chile; and Universidad Autónoma de Nuevo León, Facultad de Ciencias Biológicas, Pedro de Alba s/n, Cd. Universitaria, San Nicolás de los Garza, Nuevo León, 66450, México.

Y. Wen Xu, Pharmaceutical Botany, Sichuan Agricultural University, Ya'an 625014, PR China.

Farooq A. Zaki, Faculty of Horticulture, Sher-e-Kashmir University of Agricultural Sciences and Technology of Kashmir, Shalimar, Jammu and Kashmir-190025, India.

Gustavo Martínez Zamora, Instituto Superior de Investigaciones Biológicas (INSIBIO), CONICET- UNT; and Instituto de Química Biológica 'Dr. Bernabé Bloj', Facultad de Bioquímica, Química y Farmacia, Universidad Nacional de Tucumán, Chacabuco 461, T4000ILI – San Miguel de Tucumán, Argentina.

Foreword by Farooq A. Zaki

Strawberry is a delicious fruit with a special economic position in the fruit industry because of its post-harvest processing and antioxidant functional value. There are a large number of cultivars grown worldwide, encompassing major regions distributed across all six continents. Despite its importance, I have come across only a few books dedicated exclusively to the subject of the strawberry. This book provides a broad, well-structured review of strawberries and their cultivation under current environmental conditions.

Methods of strawberry cultivation have undergone many improvements, and this book covers aspects from plant propagation, architecture, genetic resources, breeding, abiotic stresses and climate change to evolving diseases and their control. The book is aimed at those involved in strawberry research and development, and also to those who are interested in cultivation of strawberry for commercial gain.

The first chapter gives a general introduction to strawberry with some important statistics. It is followed by Section 1 on Genetics, Breeding and Omics, which contains five chapters. Chapters 2 and 3 discuss modern aspects related to available genetic resources and breeding for yield, quality and disease traits. The scope of metabolomics is discussed with special reference to white strawberry in Chapter 4. Chapter 5 is an interesting study on the transcriptomic profile of some key genes in relation to systemic acquired resistance, while Chapter 6 discusses different aspects of genetic transformation in strawberry. Section 2 on Cultivation Systems and Propagation contains three chapters. These chapters discuss at length the plant growth, flowering, root growth and architecture, replant problems and plant propagation techniques. Section 3 on Disease and Stress Management contains six chapters dealing with biotic and abiotic stresses of strawberry. The first four chapters discuss traditional and emerging fungal diseases, their diagnosis and modern biointensive management strategies. Chapters 14 and 15 deal with the emerging challenges posed due to climate change and its impact on the changing magnitude and dimensions of abiotic stresses on strawberry.

This book is different on many accounts from the available compilations. It is up to date, containing the latest information available on the subject and is quite comprehensive. Each chapter is on a specialized theme, contributed by leading researchers across the globe. Figures and tables are included to make the subject comprehensible and informative. The book also provides an insight into the different aspects of emerging challenges.

Overall, this is a job well done by the authors and editors. I am confident that researchers, teachers, students and commercial growers will find this book useful, interesting and informative.

I congratulate the authors on their endeavour and wish them success.

Dr Farooq Ahmad Zaki
Dean
Faculty of Horticulture
Sher-e-Kashmir University of Agricultural Sciences and Technology of Kashmir
India

Foreword by Nazeer Ahmed

Strawberry is a fruit of immense significance in temperate and subtropical regions of the world. It generally fetches a high price in the market if supplied fresh. It is in demand in large quantities, both fresh as well as its products like juices, jams, jellies, ice creams, chocolates, pies, syrups and milkshakes. Artificial strawberry flavourings and aromas are also widely used in many products like candies, sanitizers, perfumes, lip sticks etc. Strawberries are rich in nutrients like vitamin C, potassium, folic acid and fibre. These are of medicinal importance. Quercetin, a flavonoid contained in them, has been shown to reduce the risk of atherosclerosis and to provide protection against the damage caused by low-density lipoprotein cholesterol. The high polyphenol content in them may be helpful in reducing the risk of cardiovascular diseases. Due to their high potassium content, strawberries are recommended to people with high blood pressure as it helps to counter the effects of sodium in the body. Furthermore, their low glycaemic index and high fibre content may help in regulating blood sugar and hence they are a smart choice for diabetics.

There are some challenges to strawberry cultivation. Scientists have been successful in making headway in addressing some of these challenges. The most common issues pertain to strawberry improvement for traits that help improve shelf-life as well as increased tolerance against diseases and pests.

This book is an excellent contribution by a group of international strawberry experts. I congratulate CABI and Drs Amjad M. Husaini and Davide Neri for their painstaking efforts in bringing forth such an up-to-date book relevant to strawberry researchers, academics, growers and industry. I am happy to recommend this book to all those interested in strawberry and believe they would find it useful.

Prof. Nazeer Ahmed
Vice-Chancellor
Sher-e-Kashmir University of Agricultural Sciences and Technology of Kashmir
India

Acknowledgements

It is difficult for me to find the words to acknowledge the support given by contributors of this book, especially because it took us 3 years from conception to completion. The patience with which each contributor worked in the lengthy editing process is particularly commendable. I would also like to thank Professor Davide Neri (co-editor), Dr Sreepat Jain (former Commissioning Editor, CABI) and Dr Rachael Russell (Commissioning Editor, CABI) for supporting this project at different stages of the process. I need to acknowledge the immense support that my friend, Wen Xu, provided in formatting and updating all the references correctly, which was rather a painstaking task. Lastly, I express my gratitude to Professors Farooq A. Zaki and Riaz A. Shah for their support and encouragement.

I would be totally dishonest if I ignore the support of my wife, Maroofa, and my lovely parents. They had to bear my absence during long hours of work on this book, and much of the time that I should have been sharing with them had to be sacrificed for this project.

Amjad M. Husaini

Abbreviations

1,3-D	1,3-dichloropropene
2,4-D	2,4-dichlorophenoxyacetic acid
ABA	abscisic acid
AFLP	amplified fragment length polymorphism
AMF	arbuscular mycorrhizae fungi
APS	American Phytopathological Society
APX	ascorbate peroxidase
ASA	ascorbate
ASD	anaerobic soil disinfestation
Avr	avirulence
BA	benzyladenine
BCA	biological control agent
BTH	benzothiadiazole S-methyl ester
CAT	catalase
CGIAR	Consultative Group on International Agricultural Research
CML	calmodulin-like
DAMP	damage-associated molecular pattern
DGGE	denaturing gradient gel electrophoresis
DMDS	dimethyl disulphide
DPPH	1,1-diphenyl-2-picrylhydrazyl
ELISA	enzyme-linked immunosorbent assay
EMBL	European Molecular Biology Laboratory
EPPO	European and Mediterranean Plant Protection Organization
ESI	electrospray ionization
EST	expressed sequence tag
ETI	effector-triggered immunity
EU	European Union
FAO	United Nations Food and Agriculture Organization
GC	glucosinolate
GMO	genetically modified organism
GR	glutathione reductase
GSH	glutathione
GST	glutathione *S*-transferase

H_2DFFDA	2′,7′-dichlorodihydrofluorescein diacetate
HPLC-DAD	high-performance liquid chromatography with diode array detection
HR	hypersensitive response
HSP	heat-shock protein
HST	heat stress tolerance
IAA	indole-3-acetic acid
IBA	indole-3-butyric acid
IFAD	International Fund for Agricultural Development
ISR	induced systemic resistance
ITC	isothiocyanate
ITS	internal transcribed spacer
JA	jasmonic acid
Kan	kanamycin
MeBr	methyl bromide
MS/MS	tandem mass spectrometry
NAA	1-naphthaleneacetic acid
NAM	non-AMF-inoculated
$O_2^{\bullet-}$	superoxide anion
p.t.	post-treatment
PA	phytic acid
PAL	phenylalanine ammonia-lyase
PAMP	pathogen-associated molecular pattern
PCR	polymerase chain reaction
PCR-DGGE	polymerase chain reaction with denaturing gradient gel electrophoresis
PEG	polyethylene glycol
PGPB	plant growth-promoting bacteria
PGPR	plant growth-promoting rhizobacteria
PR	pathogenesis-related
PRD	partial root zone drying
ProCa	prohexadione-calcium
PRR	pattern-recognition receptor
PRX	peroxidase
PTI	PAMP-triggered immunity
Q-3-Gluc	quercetin 3-*O*-glucuronide
RAPD	random amplified polymorphic DNA
Rf	retardation factor
RFLP	restriction fragment length polymorphism
ROS	reactive oxygen species
rRNA	ribosomal RNA
RT	reverse transcription
SA	salicylic acid
SAR	systemic acquired resistance
SCAR	sequence characterized amplified region
SOD	superoxide dismutase
SSR	simple sequence repeat
TDZ	thidiazuron
TIF	totally impermeable film
USDA	US Department of Agriculture
VAM	vesicular–arbuscular mycorrhiza
VCG	vegetative compatibility group
VIF	virtually impermeable film
VOC	volatile organic compound
WFP	United Nations World Food Programme

1 Strawberries: a General Account

Amjad M. Husaini[1]* and Farooq A. Zaki[2]
[1]Centre for Plant Biotechnology, Sher-e-Kashmir University of Agricultural Sciences and Technology of Kashmir, Shalimar, Jammu and Kashmir-190025, India; [2]Faculty of Horticulture, Sher-e-Kashmir University of Agricultural Sciences and Technology of Kashmir, Shalimar, Jammu and Kashmir-190025, India

1.1. Origin and History of Strawberry Cultivation

The genus *Fragaria* belongs to the family Rosaceae. Recorded history of the *Fragas* dates back to 23–79 AD in the writings of Pliny (Darrow, 1966). Early colonists in North America cultivated the native strawberry, *Fragaria virginiana*, which was a hardy plant with the ability to withstand cold temperature and drought. In the early 1600s, *F. virginiana* was imported to Europe from North America. In the 1700s, explorers found a wild strawberry in Chile, *Fragaria chiloensis*, which grew large fruit but was not well suited to a wide range of climates. Northern Europe, including France, cultivated the woodland strawberry, *Fragaria vesca* (L.), as early as 1300. It was appreciated as much for its flowers as for the fruit. Additionally, musky strawberries, *Fragaria moschata*, were also cultivated in Europe and Russia for centuries. Musky strawberries are light red to purple, and have a strong vinous flavour like Muscat grapes.

In 1714, the most important event in the history of the modern strawberry took place (Darrow, 1966). Amédée-Francois Frézier, a member of the French army, returned from duty in Peru and Chile with some plants of *F. chiloensis*. When he arrived, he distributed his plants. One of them was interplanted alongside *F. virginiana* in Brest, France (Hancock and Luby, 1995). A natural hybrid comprising a hardy plant with large fruit developed by natural crossing and was therefore noticed. This natural hybrid was called *Fragaria* × *ananassa* Duch., and many former species have been supplanted by its cultivation ever since.

* amjadhusaini@skuastkashmir.ac.in

1.2. Taxonomy and Biology

The French botanist Antoine Nicolas Duchesne is credited with identification of the natural hybrid *Fragaria* × *ananassa*. The cultivated strawberry *F.* × *ananassa* Duch. is a member of the family Rosaceae, subfamily Rosoideae, along with blackberries and raspberries. There are about 34 species of *Fragaria* found in Asia, America (North and South) and Europe, of which two are cultivated commercially for their fruit: *F. moschata*, the musky or Hautboy strawberry, and *F. vesca*, the woodland or alpine strawberry. These species were cultivated for centuries, but there is very little production of them today, due to the success of *F.* × *ananassa*.

The *F.* × *ananassa* is a perennial and arises from a crown of meristematic tissue or compressed stem tissue. The leaves, stems, runners, axillary crowns, inflorescences and roots all arise from the crown. The plant has trifoliate leaves that spiral around the crown, with buds in the leaf axils giving rise to the runners. The runners have two nodes, with a plant produced at the distal node. Strawberry blossoms contain many pistils, each with its own style and stigma attached to the receptacle. Botanically, the strawberry fruit is an 'accessory fruit' and is not a true berry. When fertilization occurs, the receptacle develops into a fleshy fruit. The flesh consists of the greatly enlarged flower receptacle and is embedded with the many true fruits, or achenes, which are popularly called seeds. These seeds are arranged on the outside of the receptacle tissue. The growth of the receptacle is dependent on successful fertilization of the ovules, with its size and shape dependent on the number of achenes formed (Darnell, 2003).

Strawberries can be diploid, tetraploid, hexaploid, octoploid and even decaploid. The woodland strawberry, *F. vesca*, and most of the native species around the world are diploid. They range from dioecious to hermaphroditic and self-fertile to self-incompatible. Three known tetraploids are *Fragaria moupinensis*, *Fragaria orientalis*, and *Fragaria corymbosa*. *F. moschata* is a hexaploid strawberry and is known for its musky flavour. *F. chiloensis* and *F. virginiana* are both octoploid, with their flowers mostly being dioecious although some are hermaphroditic (Hancock *et al.*, 1996). This polyploidy of the *Fragaria* spp. makes selection of desirable traits via traditional breeding using cross-pollination of the flowering plants tedious and time consuming (Husaini *et al.*, 2011).

1.3. Area, Production and Yield

Strawberry is a highly popular crop and is in great demand for fresh markets as well as in the fruit processing industry for preparing jams and other products (Husaini and Abdin, 2008). Its popularity can be judged from the fact that the production of strawberries has increased considerably in recent years (Table 1.1, Figs 1.1–1.3).

The figures clearly show that worldwide strawberry production has shown a remarkable increase of about 53.5% and an expansion in area of about 12% in the period between 2003 and 2013. The steepest increase has been observed in Africa, where the production increased by 125.9% and area increased by 70.7%

Table 1.1. Total area and production of strawberry across major regions.

	Production (t)		Area (ha)	
Region	2003	2013	2003	2013
World	5,041,331	7,739,622	320,990	361,662
Europe	1,224,692	1,484,987	162,543	162,315
Asia	2,334,869	3,845,553	113,121	143,036
USA	977,945	1,360,869	19,587	23,549
Africa	184,582	417,135	6,250	10,671

Source: http://faostat.fao.org/.

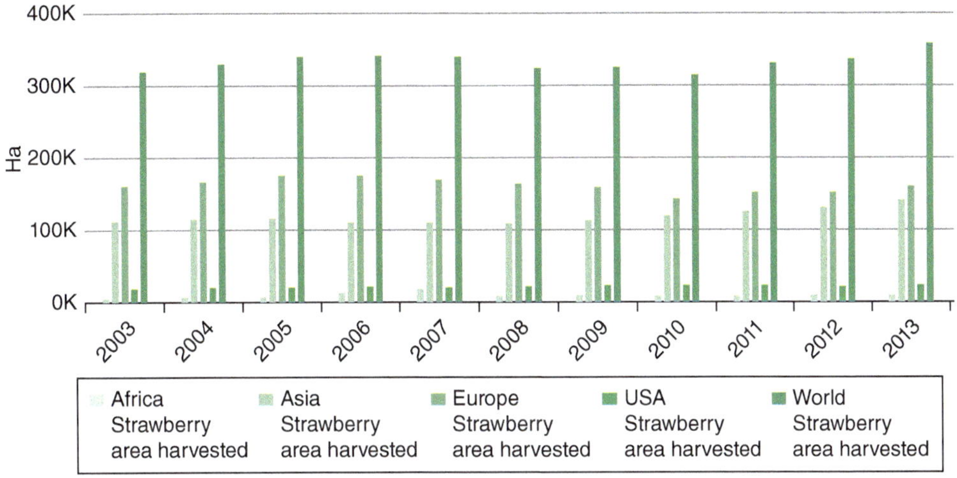

Fig. 1.1. Trend in strawberry area harvested across major regions. K, thousand.

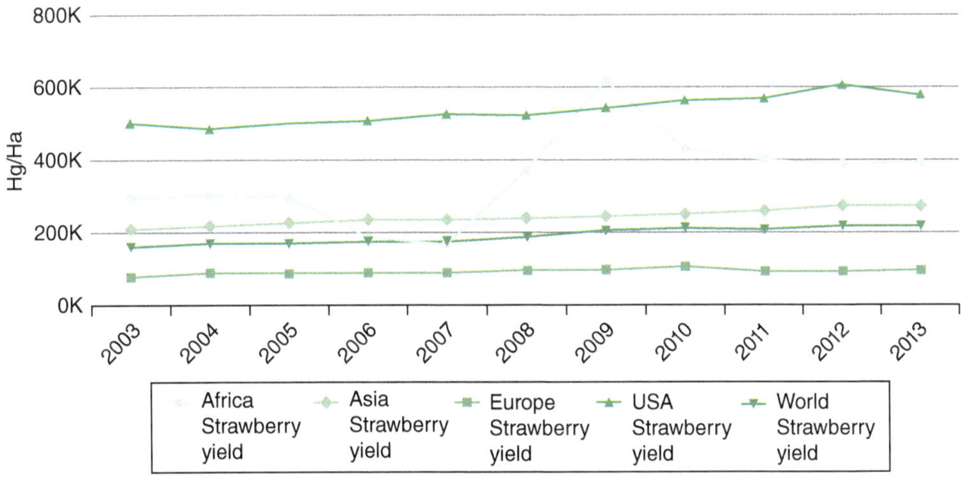

Fig. 1.2. Trend in strawberry yield across major regions. K, thousand.

in this decade. Next in rank comes Asia, where there has been 64.7% increase in production and 26.4% increase in area. The production and area in USA have increased by 39.1 and 20.2%, respectively. Europe has recorded an increase of 21.2% in production, despite a small decrease of 0.1% in the area under cultivation. Overall, the figures are encouraging, revealing the profitability and popularity of this glamour fruit across all major regions of the world.

There are hundreds of different strawberry cultivars. These have been produced by plant breeders to fit particular environmental or marketing niches, and generally no single cultivar is grown worldwide or even nationwide. Each cultivar performs differently, depending on the climate and conditions in which it is grown. Octoploid strawberry accessions are extremely variable in morphology, photoperiod sensitivity and fruit quality (Husaini, 2010). To maximize strawberry production, it is important to choose a suitable strawberry cultivar that is well suited to a growing region. A good source of this information can be found on websites such as

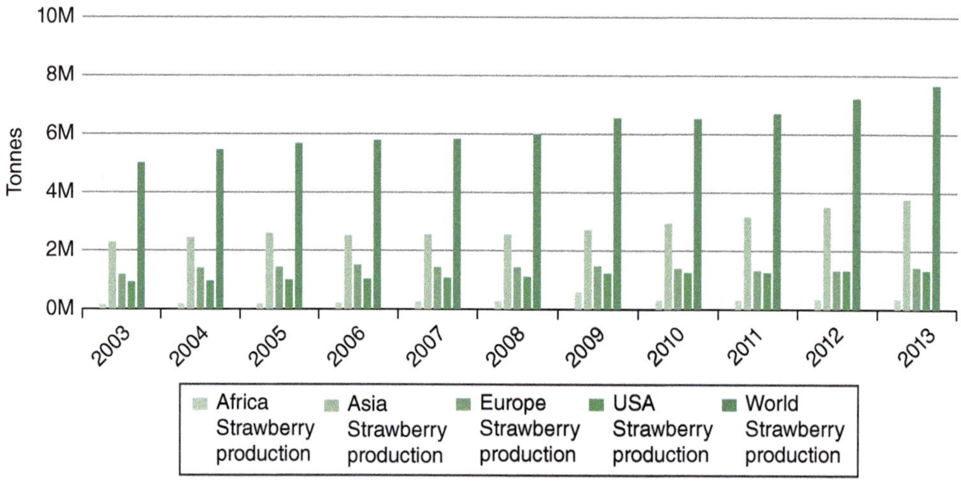

Fig. 1.3. Trend in strawberry production across major regions. M, million.

Strawberry Plants.org (http://strawberryplants.org/2010/05/strawberry-varieties/). Due to the difficulties imposed by the complicated octoploid genome on conventional breeding strategies, manipulation through recombinant DNA technology, Golden Gate cloning and CRISPR (clustered regularly interspaced short palindromic repeats)/Cas systems are favourable options in strawberry improvement. The problem of strawberry fruit softening is a classic example of this kind of intervention by biotechnological tools. Genetic transformation has also improved strawberries for many traits that confer adaptive advantage to these plants such as the challenges imposed by climate change (Husaini *et al.*, 2012) (see Chapter 14, this volume).

1.4. Health-promoting Properties

In the past few years, the antioxidant power of fruit has been considered an indicator of the beneficial bioactive compounds present in foodstuffs and therefore of their healthfulness. Indeed, strawberry phenolics are best known for their antioxidant and anti-inflammatory action, and possess direct and indirect antimicrobial, anti-allergy, and anti-hypertensive properties, as well as the capacity for inhibiting the activities of some physiological enzymes and receptors, preventing oxidative stress-related diseases (Wang *et al.*, 1996). The major class of strawberry polyphenols is flavonoids, mainly anthocyanins. The most quantitatively important phenolic compounds present in strawberries are in the form of pelargonidin and cyanidin derivatives (Giampieri *et al.*, 2012, 2013, 2014).

There is consolidated evidence to classify strawberries as a functional food with several preventive and therapeutic health benefits (Basu *et al.*, 2014). Strawberries possess anticarcinogenic, antioxidative and genoprotective properties against multiple human and mouse cancer cell types both *in vitro* (Wang *et al.*, 2005; Zhang *et al.*, 2008) and *in vivo* in animal models (Carlton *et al.*, 2001; Stoner *et al.*, 2007), but human studies are still rare, and investigations particularly focused on patients with precancerous conditions are strongly advisable. Strawberry phenolics are able to: (i) detoxify free radicals, blocking their production; (ii) modulate the expression of genes involved in metabolism, cell proliferation and antioxidant defence; and (iii) protect and repair DNA damage. Several polyphenolic compounds such as anthocyanins, kaempferol, quercetin, fisetin, ellagitannins and ellagic acid have been reported in strawberries (Giampieri *et al.*, 2012, 2013, 2014). Fisetin possesses antioxidant, anti-inflammatory and anti-proliferative effects in a wide variety of cancers (Ravichandran *et al.*, 2011). Most of the studies have been performed *in vivo*, in particular in lung cancer (Ravichandran *et al.*, 2011; Touil *et al.*, 2011), prostate cancer (Khan *et al.*, 2008), teratocarcinoma (Tripathi *et al.*, 2011) and skin cancer (Syed *et al.*, 2011).

A highly prevalent problem affecting nearly 21% of the world population is depression, and its prevalence has increased significantly by 6% during the past two decades. According to the World Health Organization, depression will become the second leading cause of disease-related disability by the year 2020. The antidepressant potential of fisetin has been investigated in two classical mouse models of despair tasks: tail suspension and forced swimming tests. Fisetin application (10 and 20 mg kg^{-1}, *per os*) inhibited the immobility time in both behavioural tests in a dose-dependent way, while the doses that affected the immobile response did not affect locomotor activity. In addition, neurochemical assays showed that fisetin produced an increase in serotonin and noradrenaline levels in the frontal cortex and hippocampus (Zhen *et al.*, 2012). These findings indicate that fisetin could serve as a novel natural antidepressant agent.

Anticarcinogenic effects of strawberries are mediated mainly through the detoxification of carcinogens, scavenging of reactive oxygen species, the decrease in oxidative DNA damage (Xue *et al.*, 2001; Stoner *et al.*, 2008), the reduction of cancer cell proliferation through apoptosis (Seeram *et al.*, 2006) and cell-cycle arrest (Stoner *et al.*, 2007), downregulation of activator protein 1 and nuclear factor-κB, inhibition of Wnt signalling, tumour necrosis factor-α (Zhang *et al.*, 2008) and angiogenesis (Atalay *et al.*, 2003; Duthie, 2007).

Strawberries (*F.* × *ananassa* Duch.) are a rich source of a wide variety of nutritive compounds such as sugars, vitamins and minerals, as well as non-nutritive, bioactive compounds such as flavonoids, anthocyanins and phenolic acids. The most abundant class of phytochemicals in strawberries is ellagitannins (i.e. sanguiin-H-6), followed by flavonols (i.e. quercetin and kaempferol-3-malonyl glucoside), flavanols (i.e. catechins and procyanidins), and phenolic acids (i.e. caffeic and hydroxybenzoic acid derivatives) (Wang *et al.*, 1996; Giampieri *et al.*, 2012, 2013, 2014). All of these compounds exert a synergistic and cumulative effect on human health promotion and in disease prevention. Of its many positive characteristics, the nutritional value of strawberries is nearly perfect (Table 1.2). Eight medium strawberries contain more vitamin C than an orange, 20% of the recommended daily allowance for folic acid, no fat and no cholesterol, and are considered high in fibre. Another significant nutritional feature is the concentration of folate (24 μg per 100 g of fresh fruit): among fruit, strawberries are one of the richest natural sources of this indispensable micronutrient, which represents an essential factor in health promotion and disease prevention (Tulipani *et al.*, 2008, 2009). Strawberries are also a notable source of manganese, and a good source of iodine, magnesium, copper, iron and phosphorus. Moreover, both their dietary fibre and fructose contents may contribute to regulating blood sugar levels by slowing digestion, while the fibre content may control calorie intake by its satiating effect.

Table 1.2. Nutritional composition of strawberry (*Fragaria* × *ananassa* Duch.). (From US Department of Agriculture: http://ndb.nal.usda.gov/ndb/search/list?qlookup=09316&format=Full.)

Component	Per 100 g	Standard error	Component	Per 100 g	Standard error
Nutrient			Lipids		
Water (g)	90.95	0.214	Fatty acids, total saturated (g)	0.015	–
Energy (kcal)	32	–	16:00 (g)	0.012	–
Energy (kJ)	136	–	18:00 (g)	0.003	–
Protein (g)	0.67	0.026	Fatty acids, total monounsaturated (g)	0.043	–
Total lipid (fat) (g)	0.3	0.047	16:1 undifferentiated (g)	0.001	–
Ash (g)	0.4	0.021	18:1 undifferentiated (g)	0.042	–
Carbohydrate, by difference (g)	7.68	–	Fatty acids, total polyunsaturated (g)	0.155	–
Fibre, total dietary (g)	2	0.152	18:2 undifferentiated (g)	0.09	–
Sugars, total (g)	4.89	–	18:3 undifferentiated (g)	0.065	–
Sucrose (g)	0.47	0.328	18:4 (g)	0	–
Glucose (dextrose) (g)	1.99	0.194	20:4 undifferentiated (g)	0	–
Fructose (g)	2.44	0.198	20:5 *n*-3 (EPA) (g)	0	–
Lactose (g)	0	0	22:5 *n*-3 (DPA) (g)	0	–
Maltose (g)	0	0	22:6 *n*-3 (DHA) (g)	0	–
Galactose (g)	0	0	Cholesterol (mg)	0	–
Starch (g)	0.04	0.029	Phytosterols (mg)	12	–
Vitamins			Amino acids		
Vitamin C, total ascorbic acid (mg)	58.8	2.473	Tryptophan (g)	0.008	–
Thiamin (mg)	0.024	0.003	Threonine (g)	0.02	–
Riboflavin (mg)	0.022	0.008	Isoleucine (g)	0.016	–
Niacin (mg)	0.386	0.037	Leucine (g)	0.034	–
Pantothenic acid (mg)	0.125	0.003	Lysine (g)	0.026	–
Vitamin B-6 (mg)	0.047	0.012	Methionine (g)	0.002	–
Folate, total (μg)	24	5.465	Cystine (g)	0.006	–
Folic acid (μg)	0	–	Phenylalanine (g)	0.019	–
Folate, food (μg)	24	5.465	Tyrosine (g)	0.022	–
Folate, DFE (μg)	24	–	Valine (g)	0.019	–
Choline, total (mg)	5.7	–	Arginine (g)	0.028	–
Betaine (mg)	0.2	–	Histidine (g)	0.012	–
Vitamin B-12 (μg)	0	–	Alanine (g)	0.033	–
Vitamin B-12, added (μg)	0	–	Aspartic acid (g)	0.149	–
Vitamin A, RAE (μg)	1	0.031	Glutamic acid (g)	0.098	–

Retinol (µg)	0	–	Glycine (g)	0.026	–
Carotene, β (µg)	7	0.22	Proline (g)	0.02	–
Carotene, α (µg)	0	0	Serine (g)	0.025	–
Cryptoxanthin, β (µg)	0	0	Minerals		
Vitamin A, (IU)	12	0.625	Calcium (Ca) (mg)	16	0.562
Lycopene (µg)	0	0	Iron (Fe) (mg)	0.41	0.026
Lutein + zeaxanthin (µg)	26	8.04	Magnesium (Mg) (mg)	13	0.222
Vitamin E (α-tocopherol) (µg)	0.29	0.024	Phosphorus (P) (mg)	24	0.72
Vitamin E, added (mg)	0	–	Potassium (K) (mg)	153	4.073
Tocopherol, β (mg)	0.01	0.002	Sodium (Na) (mg)	1	0.1
Tocopherol, γ (mg)	0.08	0.01	Zinc (Zn) (mg)	0.14	0.013
Tocopherol, δ (mg)	0.01	0.005	Copper (Cu) (mg)	0.048	0.004
Vitamin D (D2 + D3) (µg)	0	–	Manganese (Mn) (mg)	0.386	0.018
Vitamin D (IU)	0	–	Selenium (Se) (µg)	0.4	–
Vitamin K (phylloquinone) (µg)	2.2	0.29	Fluoride (F) (µg)	4.4	0.4
Anthocyanidins			Flavonols		
Petunidin (mg)	0.1	0.1	Isorhamnetin (mg)	0	–
Delphinidin (mg)	0.3	0.28	Kaempferol (mg)	0.5	0.01
Malvidin (mg)	0	0.01	Myricetin (mg)	0	0.04
Pelargonidin (mg)	24.8	0.69	Quercetin (mg)	1.1	0.04
Peonidin (mg)	0	0.05	Isoflavones		
Cyanidin (mg)	1.7	0.05	Daidzein (mg)	0	0
Flavan-3-ols			Genistein (mg)	0	0
(+)-Catechin (mg)	3.1	0.19	Glycitein (mg)	0	–
(–)-Epigallocatechin (mg)	0.8	0.35	Total isoflavones (mg)	0	0.005
(–)-Epicatechin (mg)	0.4	0.13	Formononetin (mg)	0	–
(–)-Epicatechin 3-gallate (mg)	0.2	0.02	Coumestrol (mg)	0	–
(–)-Epigallocatechin 3-gallate (mg)	0.1	0.06	Proanthocyanidin		
(+)-Gallocatechin (mg)	0	0.005	Proanthocyanidin monomers (mg)	3.7	0.8
Flavanones			Proanthocyanidin dimers (mg)	5.3	1.89
Hesperetin (mg)	0	0	Proanthocyanidin trimers (mg)	4.9	2.27
Naringenin (mg)	0.2	0.25	Proanthocyanidin 4–6mers (mg)	28.1	6.47
Flavones			Proanthocyanidin 7–10mers (mg)	23.9	3.47
Apigenin (mg)	0	0	Proanthocyanidin polymers (>10mers) (mg)	75.8	13.36
Luteolin (mg)	0	0.001			

References

Atalay, M., Gordillo, G., Roy, S., Rovin, B., Bagchi, D., Bagchi, M. and Sen, C.K. (2003) Anti-angiogenic property of edible berry in a model of hemangioma. *FEBS Letters* 544, 252–257.

Basu, A., Nguyen, A., Betts, N.M. and Lyons, T.J. (2014) Strawberry as a functional food: an evidence-based review. *Critical Reviews in Food Science and Nutrition* 54, 790–806.

Carlton, P.S., Kresty, L.A., Siglin, J.C., Morse, M.A., Lu, J., Morgan, C. and Stoner, G.D. (2001) Inhibition of *N*-nitrosomethylbenzylamine-induced tumorigenesis in the rat esophagus by dietary freeze-dried strawberries. *Carcinogenesis* 22, 441–446.

Darnell, R.L. (2003) Strawberry growth and development. In: Childers, N.F. (ed.) *The Strawberry: A Book for Growers, Others*. Gainesville, FL: Dr Norman F. Childers.

Darrow, G.M. (1966) *The Strawberry: History, Breeding and Physiology*. Holt, Rinehart and Winston, New York.

Duthie, S.J. (2007) Berry phytochemicals, genomic stability and cancer: evidence for chemoprotection at several stages in the carcinogenic process. *Molecular Nutrition and Food Research* 51, 665–674.

Giampieri, F., Tulipani, S., Alvarez-Suarez, J.M., Quiles, J.L., Mezzetti, B. and Battino, M. (2012) The strawberry: composition, nutritional quality, and impact on human health. *Nutrition* 28, 9–19.

Giampieri, F., Alvarez-Suarez, J.M., Mazzoni, L., Romandini, S., Bompadre, S., Diamanti, J., Capocasa, F., Mezzetti, B., Quiles, J.L., Ferreiro, M.S., Tulipani, S. and Battino, M. (2013) The potential impact of strawberry on human health. *Natural Product Research* 27, 448–455.

Giampieri, F., Alvarez-Suarez, J.M. and Battino, M. (2014) Strawberry and human health: effects beyond antioxidant activity. *Journal of Agricultural and Food Chemistry* 62, 3867–3876.

Hancock, J.F. and Luby, J.J. (1995) Adaptive zones and ancestry of the most important can strawberry cultivars. *Fruit Varieties Journal* 49, 85–89.

Hancock, J.F., Scott, D.H. and Lawrence, F.J. (1996) Strawberries. In: Janick and J.N. Moore (eds) *Fruit Breeding*, Vol. II. *Vine and Small Fruits*. John Wiley & Sons. New York, pp. 419–470.

Husaini, A.M. (2010) Pre- and post-agroinfection strategies for efficient leaf disk transformation and regeneration of transgenic strawberry plants. *Plant Cell Reports* 29, 97–110.

Husaini, A.M. and Abdin, M.Z. (2008) Development of transgenic strawberry (*Fragaria*×*ananassa* Duch.) plants tolerant to salt stress. *Plant Science* 174, 446–455.

Husaini, A.M., Mercado, J.A., Schaart, J.G. and Teixeira da Silva, J.A. (2011) Review of factors affecting organogenesis, somatic embryogenesis and *Agrobacterium tumefaciens*-mediated transformation of strawberry. In: Husaini, A.M. and Mercado, J.A. (eds) *Genomics, Transgenics, Molecular Breeding and Biotechnology of Strawberry*. Global Science Books, UK, pp. 1–11.

Husaini, A.M., Abdin, M.Z., Khan, S., Xu, Y.W., Aquil, S. and Anis, M. (2012) Modifying strawberry for better adaptability to adverse impact of climate change. *Current Science* 102, 1660–1673.

Khan, N., Asim, M., Afaq, F., Abu Zaid, M. and Mukhtar, H. (2008) A novel dietary flavonoid fisetin inhibits androgen receptor signaling and tumor growth in athymic nude mice. *Cancer Research* 68, 8555–8563.

Ravichandran, N., Suresh, G., Ramesh, B. and Vijaiyan Siva, G. (2011) Fisetin, a novel flavonol attenuates benzo(a)pyrene-induced lung carcinogenesis in Swiss albino mice. *Food and Chemical Toxicology* 49, 1141–1147.

Seeram, N.P., Adams, L.S., Zhang, Y., Lee, R., Sand, D., Scheuller, H.S. and Heber, D. (2006) Blackberry, black raspberry, blueberry, cranberry, red raspberry, and strawberry extracts inhibit growth and stimulate apoptosis of human cancer cells in vitro. *Journal of Agricultural and Food Chemistry* 54, 9329–9339.

Stoner, G.D., Wang, L.-S., Zikri, N., Chen, T., Hecht, S.S., Huang, C., Sardo, C. and Lechner, J.F. (2007) Cancer prevention with freeze-dried berries and berry components. *Seminars in Cancer Biology* 17, 403–410.

Stoner, G.D., Wang, L.-S. and Casto, B.C. (2008) Laboratory and clinical studies of cancer chemoprevention by antioxidants in berries. *Carcinogenesis* 29, 1665–1674.

Syed, D.N., Afaq, F., Maddodi, N., Johnson, J.J., Sarfaraz, S., Ahmad, A., Setaluri, V. and Mukhtar, H. (2011) Inhibition of human melanoma cell growth by the dietary flavonoid fisetin is associated with disruption of Wnt/β-catenin signaling and decreased Mitf levels. *Journal of Investigative Dermatology* 131, 1291–1299.

Touil, Y.S., Seguin, J., Scherman, D. and Chabot, G.G. (2011) Improved antiangiogenic and antitumour activity of the combination of the natural flavonoid fisetin and cyclophosphamide in Lewis lung carcinoma-bearing mice. *Cancer Chemotherapy and Pharmacology* 68, 445–455.

Tripathi, R., Samadder, T., Gupta, S., Surolia, A. and Shaha, C. (2011) Anticancer activity of a combination of cisplatin and fisetin in embryonal carcinoma cells and xenograft tumors. *Molecular Cancer Therapeutics* 10, 255–268.

Tulipani, S., Romandini, S., Suarez, J.M.A., Capocasa, F., Mezzetti, B., Battino, M., Busco, F., Bamonti, F. and Novembrino, C. (2008) Folate content in different strawberry genotypes and folate status in healthy subjects after strawberry consumption. *Biofactors* 34, 47–55.

Tulipani, S., Mezzetti, B. and Battino, M. (2009) Impact of strawberries on human health: insight into marginally discussed bioactive compounds for the Mediterranean diet. *Public Health Nutrition* 12, 1656–1662.

Wang, H., Cao, G. and Prior, R.L. (1996) Total antioxidant capacity of fruits. *Journal of Agricultural and Food Chemistry* 44, 701–705.

Wang, S.Y., Feng, R., Lu, Y., Bowman, L. and Ding, M. (2005) Inhibitory effect on activator protein-1, nuclear factor-κB, and cell transformation by extracts of strawberries (*Fragaria* × *ananassa* Duch.). *Journal of Agricultural and Food Chemistry* 53, 4187–4193.

Xue, H., Aziz, R.M., Sun, N., Cassady, J.M., Kamendulis, L.M., Xu, Y., Stoner, G.D. and Klaunig, J.E. (2001) Inhibition of cellular transformation by berry extracts. *Carcinogenesis* 22, 351–356.

Zhang, Y., Seeram, N.P., Lee, R., Feng, L. and Heber, D. (2008) Isolation and identification of strawberry phenolics with antioxidant and human cancer cell antiproliferative properties. *Journal of Agricultural and Food Chemistry* 56, 670–675.

Zhen, L., Zhu, J., Zhao, X., Huang, W., An, Y., Li, S., Du, X., Lin, M., Wang, Q., Xu, Y. and Pan, J. (2012) The antidepressant-like effect of fisetin involves the serotonergic and noradrenergic system. *Behavioural Brain Research* 228, 359–366.

2 Genetic Resources of the Strawberry

M. Gambardella* and S. Sánchez
Pontificia Universidad Católica de Chile

2.1. Historical Background and Botanical Classification of the Genus *Fragaria*

The history of the strawberry dates back to Roman times, and perhaps even to Greek times; however, it is difficult to find ancient references to this species. Darrow (1966), in his book *The Strawberry: History, Breeding, and Physiology*, discusses a large part of the early history of the strawberry, through which it is possible to establish the origins of this species. For example, Darrow refers to *Natural History*, Book 21, of Pliny (23–79 AD), which mentions the use of the strawberry as a natural product in Italy. There are other references, especially in medical documents, as therapeutic properties were attributed to this plant. However, the strawberry was not grown in Europe until 1300 AD. The first references mention the use of *Fragaria vesca* as ground cover in French gardens. Initially, the strawberry was used as an ornamental plant, although interest in its fruit increased over time. Wilhelm and Sagen (1974) mention the species description in the *Latin Herbarius*, published in Mainz, Germany, in 1484, in which one of the first drawings of the plant appeared under its botanical name, *Fragaria*, which comes from the Latin *fragans*, meaning fragrance.

In 1500, three European species of strawberry were described: *F. vesca*, *Fragaria*

* mgambardella@uc.cl

moschata and *Fragaria viridis*. The most common was *F. vesca*, of which two subspecies were cited, one with white fruits and the other with red fruits. The species *Fragaria sylvestris semperflorens*, which blossoms and fructifies throughout the entire growing season, was later described (Darrow, 1966). Historically, the most significant event regarding the cultivated strawberry occurred in 1714, when the French explorer Amédée-François Frézier, commissioned by Louis XIV, collected *Fragaria chiloensis* plants on the shores of Concepción, in southern Chile. Although Frézier was an engineer, and the expedition's objective was to study the Spanish fortifications, he had a great interest in botany and natural science. Frézier was attracted by a strawberry species (*F. chiloensis*) that had unusually large fruits and was cultivated by local communities. On his trip back to France, he carefully kept five plants on the ship's deck during the 6-month voyage. When Frézier arrived in France, he gave one of the plants as a present to the Director of the Royal Garden in Paris, who placed it together with plants of *Fragaria virginiana*. Frézier's plant had only female flowers, whereas the *Fragaria virginiana* specimens had only male flowers, which favoured the spontaneous cross between both species. The progeny of this cross exhibited exceptional characteristics in terms of fruit size, shape and colour. This was the beginning of the dynamic and fruitful improvement process of the cultivated strawberry, *Fragaria* × *ananassa* Duch. (Darrow, 1966).

In contrast, the background of the introduction of *F. virginiana* into Europe is not known precisely. In 1623, nearly a century before Frézier's expedition, in a book about native and exotic plants grown in Parisian gardens, J. Robin and V. Robin mention an American *Fragaria*, which would correspond to *F. virginiana* (Staudt, 1999). This species was supposedly collected in the early 16th century in eastern North America by colonists and explorers, and was soon introduced into Europe. In 1629, Perkinson referred to 'Virginia strawberry', noting that, even though its plants had abundant flowers each year, it had not been possible to harvest even a single fruit in seven seasons (Staudt, 1999). In all likelihood, Perkinson's plants comprised only one sex, as we now know that *F. virginiana* is a trioecious species.

In 1766, Duchesne was the first to provide a complete description of American strawberries in a monograph on the genus *Fragaria*, although the nomenclature codes he used did not follow accepted taxonomic standards. In 1768, Miller published the correct names of the species *F. chiloensis* and *F. virginiana* in the *Gardener's Dictionary* (Staudt, 1999). Although, as mentioned above, most of the strawberry plants cultivated today correspond to the hybrid *Fragaria* × *ananassa*, whose parents are of American origin, the genus *Fragaria* includes more than 150 species widely distributed in cold, temperate and subtropical regions. If we take into account only the most important species, this number can be reduced to 20 (Hancock, 1999; Staudt, 1999), which are grouped according to the number of chromosomes into diploid, tetraploid, hexaploid and octoploid species, with a basic haploid number of seven chromosomes (Table 2.1).

These 20 species and their geographical distribution are represented in the map shown in Fig. 2.1, as proposed by Rousseau-Gueutin *et al.* (2009) and adapted according to information compiled from other publications (Hancock and Luby, 1993; Staudt, 1999).

As shown in Fig. 2.1, diploid species are distributed throughout Eurasia, although one species, *F. vesca*, has a wider distribution, and may also be found in America. Tetraploid species, on the other hand, are restricted to East Asia, while the only hexaploid species, *F. moschata*, is found in Europe. *F. chiloensis* grows along the entire Pacific coast of America, from southern Chile to Alaska (Staudt, 1999). *Fragaria iturupensis* was described by Staudt (1973) as the only Asian octoploid species. However, Hummer *et al.* (2009) postulated it is a decaploid species. It should be noted that *F. iturupensis* ploidy is not completely resolved, so it is generally considered as a species with varying ploidy (octoploid and decaploid). *F. iturupensis* can be found in north Pacific islands, specifically the Kuril Islands.

Table 2.1. Main species of the genus *Fragaria* and their ploidy.

Ploidy	Species	Ploidy	Species
Diploids (2x)[a]	*F. vesca*	Tetraploids (4x)	*F. corymbosa*
	F. viridis		*F. orientalis*
	F. nilgerrensis		*F. moupinensis*
	F. daltoniana	Penta and hexaploids (5x and 6x)	*F. bringhurstii*
	F. nubicola	Hexaploids (6x)	*F. moschata*
	F. innumae	Octoploids (8x)	*F. chiloensis*
	F. yesoensis		*F. virginiana*
	F. mandshurica		*F. iturupensis*[b]
	F. nipponica		*F.* × *ananassa*
	F. gracilis	Decaploids (10x)	*F. iturupensis*[b]
	F. pentaphylla		

[a]x, Basic chromosome number in the haploid genome (x=7).
[b]There is no complete clarity on the ploidy of *F. iturupensis*; apparently, some accessions are octoploid, while others are decaploid.

The possible origins and evolutionary processes of this genus can be understood by analysing the distribution map of species from the genus *Fragaria*. These are complex processes that involve mainly the species that currently grow spontaneously in America.

2.2. Evolution and Origin of the Genome

There is no certainty yet about the origin of the genome of *Fragaria*, nor about the key species that are present in the hybrid *F.* × *ananassa*. Compatibility studies have been carried out through interspecific crosses, chromosomal analysis and, more recently, molecular studies with the aim of clarifying genome evolution in the genus *Fragaria*.

2.2.1. Diploid species

Most diploid species of the genus *Fragaria* can cross normally, and meiosis of hybrids occurs regularly. Nevertheless, in some cases sterile hybrids are produced, suggesting that among these species there are genomes with hidden structural differences. According to compatibility characteristics and molecular analyses, there would be three affinity groups of diploid species (Bors and Sullivan, 1998; Potter *et al.*, 2000; Rousseau-Gueutin *et al.*, 2009; Njuguna *et al.*, 2013). Table 2.2 shows the composition of these three groups according to analyses conducted by various authors, using different methodologies.

In a first approach, and as a result of interspecific crosses between nine diploid species of the genus *Fragaria*, Bors and Sullivan (1998) proposed three affinity groups, which overlapped with each other (Table 2.2a). The authors also pointed out that *F. vesca* would be the common ancestor of all diploid species, as there is a strong affinity between this species and most diploid species described, even with *F. nilgerrensis*, which is sexually isolated from the rest of the species studied. Geographical distribution areas also overlap with each other (Fig. 2.1).

Thanks to the development of various molecular techniques, in the last decade there has been great progress in the genetic study of the genus *Fragaria*. Potter *et al.* (2000) conducted a phylogenetic study that included 43 accessions, based on variations of DNA non-coding region sequences in chloroplasts and in the nucleus (Table 2.2b). They determined that diploid species could be classified into three groups. The species *F. iinumae*, from western Japan, would form a monophyletic group, independent of the other diploid species studied. Furthermore,

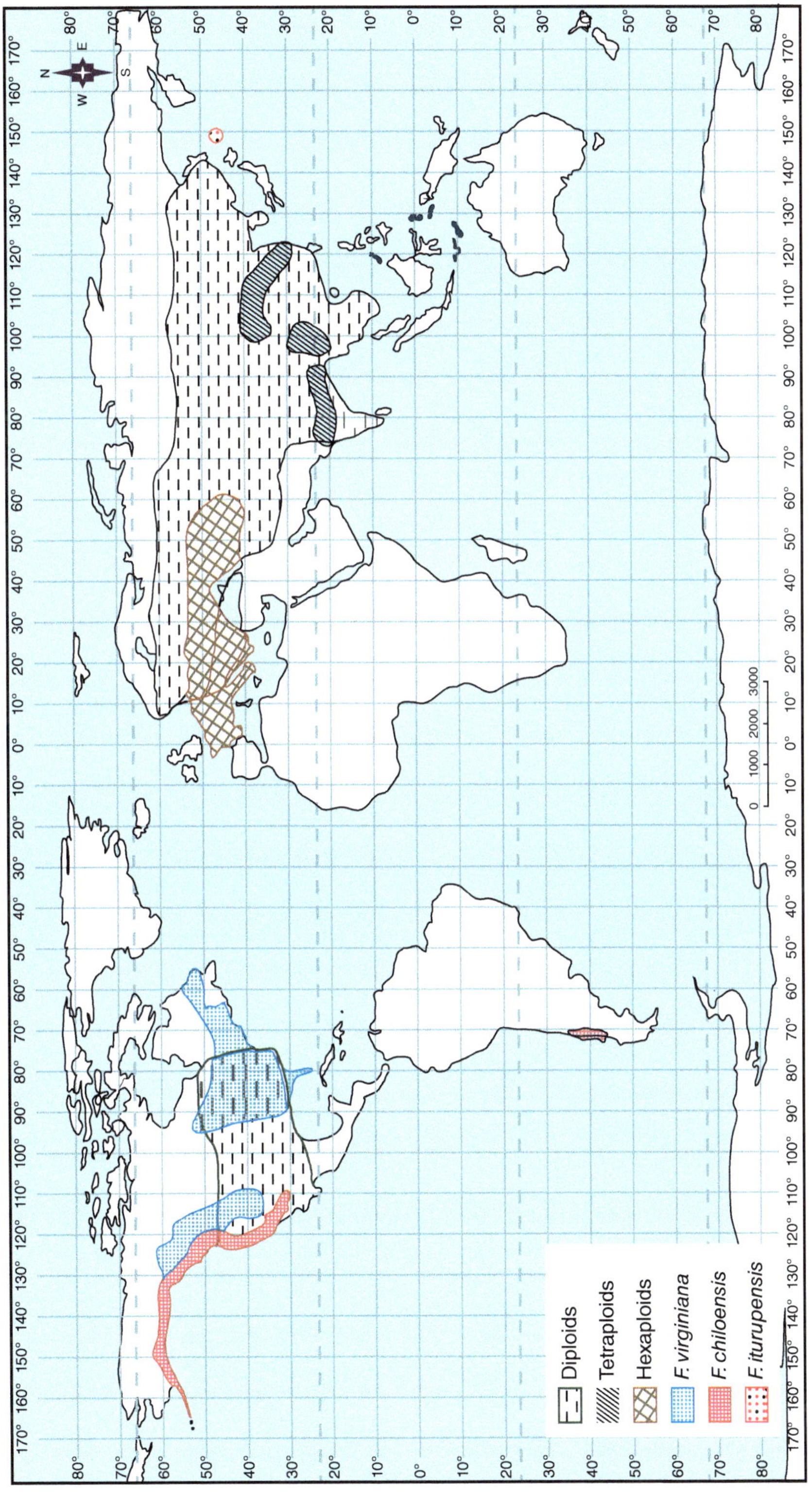

Fig. 2.1. World distribution of *Fragaria* spp.

Table 2.2. Classification of diploid species of the genus *Fragaria*.

Group	Methodology			
	(a) Interspecific crosses (Bors and Sullivan, 1998)	(b) Phylogenetic (Potter *et al*., 2000)[a]	(c) Phylogenetic (Rousseau-Gueutin *et al*., 2009)[b]	(d) Phylogenetic (Njuguna *et al*., 2013)[c]
I	*F. vesca*, *F. pentaphylla*, *F. viridis*, *F. nubicola*	*F. iinumae*	*F. iinumae*	*F. iinumae*
II	*F. vesca*, *F. pentaphylla*, *F. daltoniana*, *F. nilgerrensis*	*F. vesca*, *F. nubicola*	*F. vesca*, *F. mandshurica*, *F. viridis*	*F. vesca*, *F. mandshurica*, *F. bucharica*
III	*F. pentaphylla*, *F. nipponica*, *F. iinumae*, *F. gracilis*	*F. pentaphylla*, *F. nipponica*, *F. daltoniana*, *F. gracilis*, *F. nilgerrensis*	*F. pentaphylla*, *F. nipponica*, *F. daltoniana*, *F. nubicola*, *F. nilgerrensis*, *F. yezoensis*	*F. pentaphylla*, *F. nipponica*, *F. daltoniana*, *F. chinensis*

[a]Nuclear internal transcribed spacer region and a region of the chloroplast genome.
[b]Nuclear genes *GBSSI-2* and *DHAR*.
[c]Chloroplast genome sequencing.

they established that *F. vesca* and *F. nubicola*, which belong to a second group, would be the closest diploid species to the octoploid *F. chiloensis* and *F. virginiana*. *F. pentaphylla*, *F. nipponica*, *F. daltonian*, *F. gracilis* and *F. nilgerrensis* would form a third group of diploid species. Later, Rousseau-Gueutin *et al.* (2009) studied the evolution of ploidy in the species of the genus through phylogenetic analyses of nuclear genes, also giving rise to three affinity groups (Table 2.2c). They proposed that *F. iinumae* stands apart from the other diploid species studied. In a second group, *F. vesca* and *F. mandshurica* would have greater genetic proximity, in contrast to *F. viridis*. In this case, the third group would be composed of six diploid species, including *F. nilgerrensis*, although this species had a higher genetic distance with respect to the rest of group III.

In a recent study conducted by Njuguna *et al.* (2013), in which phylogenetic analysis of chloroplast genome sequences was carried out, it was determined that the diploid species studied could be classified in three groups similar to those described by Rousseau-Gueutin *et al.* (2009) (Table 2.2d). In this study, the location of *F. viridis* and *F. nilgerrensis* remains uncertain; however, there is a strong relationship between *F. viridis* and group II, in accordance with what was suggested by Rousseau-Gueutin *et al.* (2009).

To sum up, the information available to date indicates there are three affinity groups among diploid species. Different authors have made progress regarding the composition of each group, although some contradictions remain. New studies are needed to move forward, which should include various accessions of the different diploid species, in conjunction with species from other ploidies in order to establish the genomic components of octoploid species with greater certainty.

2.2.2. Tetraploid species

Among tetraploid species, *F. orientalis* has been one of the most studied, and two hypotheses have been proposed about its origin. The first states that this species is an autotetraploid, with *F. vesca* or *F. mandshurica* being the parental species. According to the second hypothesis, *F. orientalis* has an allotetraploid origin, where the species *F. vesca* and *F. mandshurica* correspond to the parentals (Rousseau-Gueutin *et al.*, 2009). With respect to other tetraploid species, the same authors established that *F. corymbosa*, *F. gracilis*, *F. moupinensis* and *F. tibetica* would be genetically related to the diploid species *F. nipponica*, *F. nubicola*, *F. pentaphylla* and *F. yezoensis*, but were unable to clearly determine the parental relationships between them.

2.2.3. Hexaploid species

In the case of *F. moschata*, although the chromosomes have high affinity during meiosis, they have been suggested to have different origins. Staudt (1959) suggested that possible diploid ancestors of this species would be *F. nubicola*, *F. viridis* or *F. vesca*. Results obtained by Potter *et al.* (2000) are consistent with the hypothesis proposed by Staudt (1959), suggesting that *F. moschata* would be a hybrid between *F. vesca* and *F. nubicola*. Meanwhile, Rousseau-Gueutin *et al.* (2009) suggested that *F. moschata* would come from a natural cross between the diploid species *F. vesca* and *F. viridis*, with the latter probably being the maternal donor. They also proposed that unreduced gametes from one of the parents would provide two sets of haploid chromosomes and that duplication of the resulting triploids would follow.

2.2.4. The genome of octoploid species

Compatibility studies based on crosses between species and data obtained through molecular techniques both confirm the participation of different genomes in configuring the ancestors of *F.* × *ananassa*.

Polyploidy in the genus *Fragaria* is probably due to the union of 2*n* gametes, as several authors have pointed out that non-disjunction is very common in the different strawberry species (Hancock, 1999). A study on natural populations of *F. chiloensis* and *F. vesca* showed that approximately 1% of pollen grains corresponded to unreduced gametes, and that more than 10% of natural hybrids between these species arose from the union of these gametes (Bringhurst and Senanayake, 1966).

According to analyses performed by Rousseau-Gueutin *et al.* (2009), there is evidence of the allopolyploid origin of the octoploid species *F. chiloensis*, *F. iturupensis* and *F. virginiana*. It is proposed that these three species are the result of hybridization events that combined the genomes of the group formed by the diploid species *F. vesca* and *F. mandshurica*, and the group represented by the diploid species *F. iinumae*. It is suggested that the first group is the maternal genome donor and the second is the paternal genome donor. In all likelihood, the octoploid *F. iturupensis*, the common ancestor, migrated from East Asia into North America through the Bering Strait. Subsequently, the two octoploid species differentiated themselves, and extended their distribution range towards the south as *F. chiloensis*, and towards the east as *F. virginiana* (Potter *et al.*, 2000; Rousseau-Gueutin *et al.*, 2009).

Inheritance studies have shown that octoploid species would have a diploidized genome, as well as the presence of a large number of bivalents in the meiosis of *F.* × *ananassa* (Bringhurst, 1990). Molecular markers confirmed the genome's diploid behaviour in the octoploid *F.* × *ananassa*. Using CAPS (cleavage amplified polymorphic sequence) markers, it was determined that most loci of the cultivated strawberry are transmitted to the progeny according to Mendelian laws of inheritance, and present a disomic inheritance (Kunihisa *et al.*, 2005). Afterwards, through comparative genetic mapping, a prevalence of linkage groups in the coupling/repulsion phase was observed in the progeny from crosses of one diploid and one octoploid species (*F. vesca* × *F. bucharica*), also demonstrating disomic behaviour during meiosis of cultivated strawberry (Rousseau-Gueutin *et al.*, 2008).

Despite the advanced molecular techniques currently available, the composition of the octoploid genome of *F.* × *ananassa* is still under discussion. Initially, and based on cytological comparison studies, it was thought that the octoploid genome of *F.* × *ananassa* was composed of the structure AA A′A′ BB BB (Senanayake and Bringhurst, 1967). However, the most widely accepted theory is that proposed by Bringhurst (1990), who suggested that the genomic structure of *F.* × *ananassa* corresponds to the polyploid composition AA A′A′ BB B′B′. While both hypotheses indicate that cultivated strawberry corresponds to a polyploid species with multiple independent genomes, Bringhurst's approach is consistent with the thesis that the octoploid genome of *F.* × *ananassa* has undergone diploidization, and that its inheritance behaves in a predominantly disomic manner. Figure 2.2 shows an adaptation of the diagram proposed by Bringhurst (1990) with respect to the origin of octoploid genotypes of the genus *Fragaria*, generated by combining Bringhurst's hypothesis with the theory proposed by Rousseau-Gueutin *et al.* (2009), who suggested that the genome of octoploid species would be composed of species of group I (A and A′) and group II (B), while the B genome would come from a single diploid, *F. iinumae.*

2.3. Breeding Programmes Involving Native Germplasm

Over the past half century, strawberry breeding programmes have expanded rapidly. This is one of the fruit species in which many more varieties have been registered: up to 92 registrations per year (UPOV, 2013). Nevertheless, it has been determined that the genetic basis of modern varieties is surprisingly narrow. Until 1990, most varieties had originated just from ten parental genotypes (Dale and Sjulin, 1990). Despite this, harmful effects of this situation are not particularly evident, due to the fact that this species is an octoploid, which has allowed preservation of a large percentage of variability in the genome copies. RAPD (randomly amplified polymorphic DNA) markers applied to a large number of commercial cultivars have shown that there is sufficient variability among cultivated octoploid genotypes (Gambardella *et al.*, 2005). Other authors have corroborated these results, noting that 200 years of breeding have produced a slight reduction in the genetic variability of the cultivated strawberry (Gil-Ariza *et al.*, 2009).

Another factor that contributes to maintaining genetic variability is the use of wild germplasm. Bringhurst and Voth (1984) found that only three generations of backcrosses were required to recover the size of the fruit in varieties of *F.* × *ananassa* after the cross with *F. virginiana*, and that autofertility was strongly restored in the same three generations. In general, the introduction of wild genes is a strategy used

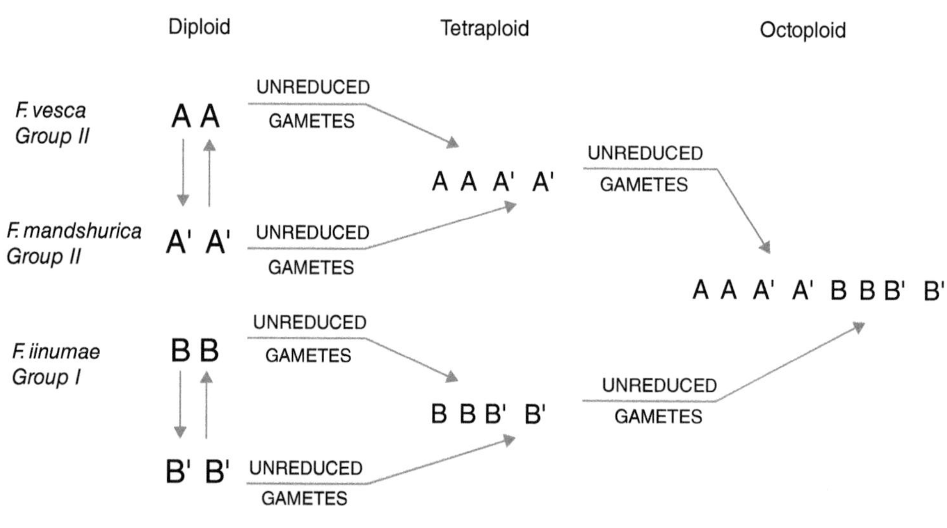

Fig. 2.2. Diagram of the origin of polyploidy in *Fragaria* spp. (Adapted from Bringhurst, 1990.)

by very few breeders. In most cases, it is in public programmes where introgression with wild genes has been carried out, as these programmes normally have greater access to germplasm banks with long-term stable funding, ensuring long periods of crosses, backcrosses and selection. However, it must be considered that, in recent years, strawberry genetic breeding activity has shifted to the private sector. Fruit marketing companies and plant nurseries increasingly develop their own programmes, associating their varieties with business strategies. In the coming years, therefore, there will be a greater probability of narrowing the genetic basis among commercial varieties of strawberries. Such a narrow genetic base should be a cause for concern due to the possibility of harmful inbreeding effects.

The cultivation of strawberries worldwide also faces multiple challenges that could not be addressed if it were not for the search for new genes among wild populations. Fruit quality characteristics, such as better flavour and aroma, as well as a higher content of dietary substances that benefit human health (e.g. high levels of antioxidants), are the qualities required by the modern consumer. The need to minimize the use of pesticides in order to make agriculture environmentally sustainable requires the continual introduction of plant pest- and disease-resistant genes in new varieties. Furthermore, the need to expand cultivation areas and to face changing climates require the presence of abiotic stress resistance genes. These are some of the goals of modern genetic breeding for this species.

Octoploid species, especially *F. chiloensis* and *F. virginiana*, have been the species most used as sources of genetic variability becasue, as species of the same ploidy level, they can easily be crossed. These species also grow in a wide geographical distribution area and are subject to selection pressures in extreme environments, where biotic and abiotic stress resistance genes are present. When analysing the history of the main varieties developed in the USA, it is worth noting the work performed by Albert Etter in the early 20th century (1903–1920) in California, who obtained more than 50 strawberry cultivars by crossing plants of *F.* × *ananassa* with wild accessions of *F. chiloensis*. Later, C.L. Powers and A.C. Hildreth, from the Department of Agriculture (USDA), used *F. virginiana* subsp. *glauca* in the programmes they conducted between 1930 and 1940. R.S. Bringhurst and V. Voth, from the University of California, used *F. virginiana* to produce day-neutral varieties. They also used *F. chiloensis*, primarily to increase fruit size in short-day cultivars (Hancock and Luby, 1993).

Recent studies have shown that an interesting breeding strategy is the independent selection of wild genotypes of *F. virginiana* and *F. chiloensis*, which are then used to rebuild the hybrid *F.* × *annanassa* from outstanding clones. In this way, it is possible to reduce the presence of unfavourable genes, which are often closely linked to genes of agronomic interest (Hancock *et al.*, 2001a, 2003, 2010). Some interesting characteristics have also been found in other species, such as *F. vesca*, *F. moschata* and *F. viridis*. However, given the wide distribution range of the species *F. chiloensis* and *F. virginiana*, it is likely that there is still a large amount of genetic resources in natural populations of these two parental species, as well as in natural populations of the hybrid between the two.

2.3.1. *Fragaria chiloensis*

From the point of view of its botanical classification, Staudt (1962, 1999) identified two forms of this species: *F. chiloensis* subsp. *chiloensis*, *chiloensis* form, and F. *chiloensis* subsp. *chiloensis*, *patagonica* form.

The *chiloensis* form corresponds to the cultivated Chilean strawberry, which was domesticated by pre-Columbian inhabitants in southern Chile. It is distinguished by its vigorous growth habit, thick stems, strong runners and thick, greyish-green, densely hairy leaves. The calyx of the flower is large, with female and hermaphrodite flowers, which always have more than five (five to nine) white petals. Fruits are large (about 35 mm long) and are pale red, pink or white in colour, with large, dark-coloured achenes. These are the most salient characteristics, giving it an exotic feature. The fruits are also very aromatic and with great sweetness, albeit with low firmness (Fig. 2.3a, b).

The *patagonica* form, on the other hand, designated as the wild Chilean strawberry, is characterized by being smaller than the *chiloensis* form, being about 21 cm high. Its leaves are thick, coriaceous, dark green and shiny. The plants are usually dioecious or trioecious (female, male and hermaphrodite) with more than five flower petals (five to seven) in most cases. The red fruits are significantly smaller, 22 mm long on average, with a rounded conical shape (Fig. 2.3c, d).

In the south of Chile, *F. chiloensis* grows spontaneously under diverse environmental conditions, from coastal areas, directly on the sand near the breakwater, to foothill areas, often associated with native undergrowth in volcanic soils. This species is also found under cultivation in small family gardens where propagation material has been preserved from generation to generation.

In surveys conducted in southern Chile by a group of breeders between 1996, 1998 and 1999 (Gambardella *et al.*, 2000a,b, 2005), different phenotypic forms were observed, which could not be restricted to only two botanical forms as proposed by Staudt. Characterization of material collected in terms of growth habit and the morphological characteristics of leaves, flowers and fruits enabled researchers to distinguish four different types, which varied according to the type of habitat in which they were collected. The four morphological types of *F. chiloensis* were characterized morphologically and molecularly.

Another important collection was made in the southern region of the North American range from California to British Columbia. This collection was compared for morphological characteristics, yield component and isozyme traits (Hancock and Bringhurst, 1988).

Fig. 2.3. Morphological types of *F. chiloensis*. (a, b) *F. chiloensis* subsp. *chiloensis* f. *chiloensis*. (c, d) *F. chiloensis* subsp. *chiloensis* f. *patagonica*.

2.3.2. *Fragaria virginiana*

This species also grows under a wide range of ecological conditions. It can be found in open forests and wetland meadows, as well as on dry rocky slopes. The plants are thin and very tall, with many stolons, and their leaves are bushy and dark green, with highly toothed margins. They are dioecious plants, with large imperfect flowers. Staminate flowers are larger than pistillate ones. The fruit is soft, with many seeds, rounded, bright red, and with a white pulp and an acid and aromatic flavour (Staudt, 1999).

Four subspecies of *F. virginiana* have been described: subsp. *virginiana*, *glauca*, *platypetala* and *grayana*. However, the most studied are *F. virginiana* subsp. *virginiana* and *F. virginiana* subsp. *glauca*. *F. virginiana* subsp. *virginiana* is found throughout the eastern area of North America, from the boreal forests in Ontario, Quebec and Newfoundland, bounded in the north by the subarctic open forest, to the deciduous forests in the Appalachian Mountains and the Piedmont Plateau in the south (Staudt, 1999).

F. virginiana subsp. *glauca* is characterized by its macroscopically glabrous petioles, peduncles, pedicels and runners. Its habits are similar to those of *F. virginiana* subsp. *virginiana*, and it grows from Alaska to New Mexico, Iowa and New York. It is common in the Yukon Territory and the Rocky Mountains in British Columbia. Unlike subsp. *virginiana*, subsp. *glauca* grows on the Pacific coast, at the Fraser River's mouth (Staudt, 1999).

Exploration of *F. virginiana* germplasm has intensified in recent years, in part stimulated by the successful incorporation of the day-neutrality gene. Two collections sponsored by the USDA have concentrated on *F. virginiana* subsp. *glauca*. These collections are composed of seed and clonal accessions, collected from 23 sites in the Cascade, Olympic, Siskiyou and Coast mountains in Washington State and Oregon, in 1985. Luby and Hancock collected almost 1000 clones from the northern Rocky Mountains in 1989 (Luby *et al.*, 1991). These collections were variable in terms of fruit size, shape, firmness, skin colour and flavour. The species occurred over a broad range of habitats, inclusding dry ponderosa and lodgepole pine forests, mesic subalpine forests, high wet mountain meadows, and bogs and openings in dense rain forests. The plants appeared to tolerate well the drought prevailing at the time of collection.

2.4. Sources of Genes of Agronomic Interest in Native Germplasm

Regarding genetic improvement, it is not always possible to find the desired characteristics in commercial varieties of the hybrid *F.* × *ananassa*; therefore, searches and characterization of wild genotypes are required. While interspecific crosses have been carried out in some breeding programmes, overall the introduction of wild germplasm is unusual and there is a risk of excessively reducing the genetic base of commercial varieties.

It is possible to find works in the literature aimed at finding genes of agronomic interest in wild plants, mainly from *F. chiloensis*, *F. virginiana* and *F. vesca*. However, more information is still needed regarding a detailed characterization of collections in germplasm banks and the study of hereditary mechanisms involved in each characteristic, as well as the interactions between them. There is also a need to know the capacity of wild genotypes to transfer favourable characteristics to commercial varieties.

One of the most interesting examples is the introduction of the day-neutral characteristic, led by Bringhurst and Voth (1984). These breeders used the genotype *F. virginiana* subsp. *glauca*, collected from the Wasatch Mountains in Utah, as the genetic source of this characteristic (Hancock and Luby, 1993). The mode of inheritance has not been fully elucidated; most studies indicate that it is governed by a dominant locus and that it is also affected by some minor genes (Ahmadi *et al.*, 1990; Shaw and Famula, 2005). Other authors suggest it would be a quantitative character, i.e. polygenic inheritance (Serçe and Hancock, 2005b; Weebadde *et al.*, 2008). Nevertheless, the diversity of responses to photoperiod, and interaction with other environmental and epigenetic factors make the analysis difficult. Moreover, classification normally used to describe the response of flowering to photoperiod – consisting of short-day, long-day, infra-day and day-neutral cultivars – seems too rigid, and it is not always possible to clearly identify the expression of the genotype. Depending on the objectives of the study and the amplitude of the response observed in the progeny, classification as remontant and not remontant is often preferred.

Another aspect indicated by Hancock *et al.* (2002) is that there would be different remontancy genes coming from various sources, mainly from natural mutations in clones of *F.* × *ananassa* and *F. virginiana*. The same authors noted that it is relatively easy to use germplasm of *F. virginiana* to introduce the day-neutral character through breeding. Recently, wild accessions of *F. virginiana* have been described, with varying degrees of photoperiod insensitivity or continuous flowering, although more information and crosses are needed to incorporate these new sources into commercial varieties (Hancock *et al.*, 2001a; Serçe and Hancock, 2005a,b).

Thanks to the availability of *F. vesca* genomic information, and to the amenability of this species to genetic manipulation techniques, it has been possible to identify and characterize the gene that inhibits photoperiod sensitivity, *FvTFL1*, as well as to develop molecular markers for assisted selection (Koskela *et al.*, 2012). Results obtained through this molecular approach reinforce the suggestion that it is a mainly monogenic character or that only a few genes are involved. However, it will be necessary to make further progress in the study of this complex characteristic, which is becoming increasingly important in the modern breeding of this species.

It is worth noting, moreover, that the flowering habit of the strawberry is directly related to temperature, and strongly interacts with photoperiod. This factor affects induction, initiation and differentiation of flower buds. It has been shown that cool summer temperatures (17°C) allow induction to occur under long photoperiods, even in short-day varieties. This means that some cultivars considered as short-day types behave as remontants in cool climates. High temperatures inhibit flowering under any conditions regarding photoperiod or variety, although it has been observed that the critical temperature is higher in day-neutral cultivars (Manakasem and Goodwin, 2001; Stewart and Folta, 2010). It would be desirable to find genotypes that are able to bloom under high-temperature conditions, although apparently there is no information on wild material with this characteristic.

In relation to pest and disease resistance genes, various authors agree that *F. vesca* would be a natural source of resistance to important diseases affecting the crop, such as powdery mildew, *Verticillium* wilt, and root and crown rot (Gooding *et al.*, 1981; Hancock and Luby, 1993; Korbin, 2011). Powdery mildew immunity in *F. moschata*, and red stele, powdery mildew and leaf spot resistance in clones of *F. chiloensis* have also been described (Hancock *et al.*, 1989).

Furthermore, a collection of native germplasm from the species *F. virginiana* and *F. chiloensis*, which is kept at the US Clonal Germplasm Repository in Corvallis, Oregon, has been characterized with respect to the response to several foliar diseases affecting the crop, resistance to black root rot and resistance to northern root-knot nematode (*Meloidogyne hapla*) and root-lesion nematode (*Pratylenchus penetrans*). These studies were able to identify various genotypes resistant to a number of important of pathogens (Hancock *et al.*, 2003).

Extensive studies carried out within the USDA programme at Beltsville, Maryland, have found various sources of resistance to different pathogens. For example, they found resistance to *Xanthomonas fragariae* in a clone of *F. virginiana* from Minnesota, and in a hybrid between *F. virginiana* and *F.* × *ananassa* (Maas *et al.*, 2000). This programme deals with the selection of germplasm tolerant or resistant to the main fungal diseases: red stele (*Phytophthora fragariae*), *Verticillium* wilt, leaf spot (*Mycosphaerella fragariae*), leaf scorch (*Diplocarpon earlianum*

(Ellis & Everh) F.A. Wolf), leaf blight (*Phomopsis obscurans* (Ellis & Everh) Sutton), powdery mildew (*Sphaerotheca macularis* f.sp. *fragariae*), fruit rot or 'grey mould' (*Botrytis cinerea*) and crown rot. This is one of the most complete programmes for searching disease resistance (Galletta *et al.*, 1997).

The chilling requirement, i.e. the accumulation of chilling hours between 0 and 7°C, is another factor affecting reproductive and vegetative growth in strawberry plants. This vernalization period is required to break bud dormancy and is highly dependent on genotype. It is a mechanism intended to prevent plants from developing (budding and flourishing) early in the season, when spring frost probability is still high. Therefore, in areas with springs that are too cold, cultivars with a high chilling requirement should be chosen. In Norway, in a population of *F. vesca* called 'Alta', a much delayed budding and flowering was observed, attributable to a high winter chilling requirement, compared with other studied populations of the same species. However, most information on this characteristic is reported for cultivars of the hybrid *F.* × *ananassa* (Heide and Sønsteby, 2007).

In most cases, genotypes collected in cold environments tend to show greater hardiness and are usually more tolerant to spring frost damage during flowering. In a study that compared accessions of native American octoploid genotypes, it was determined that those of *F. virginiana*, regardless of their origin, had a greater resistance to cold weather than those of *F. chiloensis*. Within the latter, clones collected in North America were more resistant than those collected in South America (Hancock *et al.*, 2001b). In another study, which described a wider collection of germplasm of *F. chiloensis* and *F. virginiana*, a high degree of cold resistance was found in clone PI 552091 of *F. chiloensis*, collected in Chile at an elevation of 1900 m (Hancock *et al.*, 2001a).

Another factor determining the degree of a plant's hardiness is its ability to adapt to extreme drought or salinity conditions. In this context, the harmful effect of salinity on growth, yield, and fruit quality is widely known, mainly because it affects photosynthetic capacity. Several studies have aimed to detect significant differences between genotypes related to the ability of accumulating chloride ions (Cl^-) in different organs. An individual that accumulates this ion in its roots and crowns has a better performance against salinity than one that accumulates Cl^- in its petioles and leaves (Saied *et al.*, 2005). Hancock *et al.* (2001a) pointed out that individuals of *F. chiloensis* have been collected in extremely arid zones, even growing on sand in coastal regions of Chile or the USA (clones PI 602567, PI 612317, PI 551728 and PI 612490); thus, they would be highly tolerant to salt and drought, and would be an important source of tolerance to water stress for modern cultivars.

Through characterization of various genotypes of *F. chiloensis* and *F. virginiana* under five different environmental conditions, it was determined that *F. chiloensis* is usually superior to *F. virginiana* in parameters such as the number of crowns, fruit size, soluble solid content and number of seeds. In turn, *F. virginiana* was superior to *F. chiloensis* in the production of stolons, peduncle length, the number of fruits, and external and internal colour. *F. chiloensis* has an earlier budding (Hancock *et al.*, 2003). Some of these differences may relate to the fact that *F. chiloensis* was subjected to a process of domestication by the pre-Columbian population of southern Chile, and therefore yield and fruit size parameters were improved empirically. In recent years, the trend has been to consume fruit with beneficial health features, especially with anticancer properties and the ability to prevent cardiovascular diseases (Wang *et al.*, 2007; Zafra-Stone *et al.*, 2007). These benefits are attributed mainly to the total antioxidant capacity, which is determined by a set of bioactive compounds present in the fruit. In general, the strawberry is considered a species with a high level of antioxidants, and it has been observed that this characteristic is strongly determined by both the genotype and its interaction with environmental conditions (Capocasa *et al.*, 2008). Regarding this characteristic, the strategy of introducing greater genetic variability through interspecific crosses of *Fragaria* wild material seems to be the most reasonable alternative; therefore, studies aimed at characterizing available germplasm are being developed.

In a comparison of 19 strawberry accessions of the species *F. virginiana*, *F. chiloensis* and *F.* × *ananassa*, it was determined that *F. virginiana* had higher antioxidant activity compared with the rest of the genotypes studied (Wang and Lewers, 2007). Other studies have characterized the different antioxidant compounds present in *F. chiloensis* genotypes native to South America, and have compared them with those of *F. vesca* and *F.* × *ananassa* 'Chandler'. Significant differences were found between species in the types of antioxidant compounds and their location in various fruit tissues (Cheel *et al.*, 2007; Simirgiotis *et al.*, 2009).

Finally, it is necessary to mention an interesting study carried out by Luby *et al.* (1991), who aimed to raise awareness of the *Fragaria* germplasm base in North America in the 21st century and the strategies to increase genetic diversity of the genus. They outlined the geographical distribution and the economically important traits of 15 wild *Fragaria* spp. (Table 2.3).

2.5. Germplasm Collections

Great part of strawberry germplasm is maintained in the collections belonging to the major breeding programs. Often

Table 2.3. Agronomic interest traits and geographic distribution of wild *Fragaria* species. (Data from Luby *et al.*, 1991; Hancock, 1999.)

Fragaria species	Distribution	Traits
Diploid		
F. vesca L.	Northern hemisphere	Wide adaptation, aroma
F. viridis Duch.	Europe, Asia	Firmness, high pH, cold tolerance
F. nilgerensis Schlect.	South-east Asia	Fruit size
F. daltoniana J. Gay	Himalayas	Fruit size, cold tolerance
F. nubicola Lind1. ex Lac.	Himalayas	Unknown
F. iinumae Makino	Japan	Complete dormancy
F. yezoensis Hara	Japan	Complete dormancy
F. nipponica Makino	Japan	Cold tolerance
F. mandshurica Staudt	Manchuria	Cold tolerance
Tetraploid		
F. orientalis Losinsk	Western Asia	Cold and drought tolerance
F. moupinensis (Fr.) Card.	South-west China	Cold tolerance
Hexaploid		
F. moschata Duch.	Northern Europe	Aroma, uniform ripening, cold tolerance
Octoploid		
F. iterupensis Staudt	Juril Island	
F. virginiana Duch.	North America	Disease resistance (red stele, *Verticillium* wilt, powdery mildew, leaf spot, scorch, leaf blight, root knot nematode, root-lesion nematode), stress tolerance (waterlogged soil, heat, drought, high soil pH, cold, frost), aroma, photosynthetic efficiency at high temperatures
F. chiloensis (L.) Duch.	North and South America	Disease resistance (red stele, Verticillium wilt, powdery mildew, leaf blight, viruses, root-lesion nematode), pest resistance (two-spotted mite, strawberry weevil, black vine weevil), stress tolerance (drought, high soil pH, salinity, cold), low chilling, fruit firmness, fruit size, high photosynthesis level

Table 2.4. Major germplasm collections. (Data from FAO, 2010; Jiajun *et al.*, 2012; USDA 2015.)

Institution	Country	Accessions (*n*)
National Germplasm Repository, Corvallis	USA	1807
Vavilov Research Institute of Plant Industry	Russia	940
National Institute of Agrobiological Sciences	Japan	912
Julius Kühn Institute, Federal Research Centre for Cultivated Plants	Germany	622
National Strawberry Germplasm Repository, Beijing	China	580
Centro Regional de Investigación, Instituto de Investigaciones Agropecuarias, Quilamapu	Chile	500
East Malling Research	United Kingdom	329
Consiglio per la Ricerca e la Sperimentazione in Agricoltura, Unità di Ricerca per la Frutticoltura	Italy	220
Fruit Growing Research Institute Maracineni-Arges	Romania	201

resources to maintain and characterize collections are not available, since the work is slow, requires a high degree of specialization and high costs are involved. Thus, much of the genetic diversity is lost or not available. Part of the variability present in wild species has been incorporated into modern cultivars, but there are many traits of interest in wild germplasm that could be used to improve crop characteristics. In order to accomplish this, it is necessary to describe, collect and maintain the genotypes of interest and it is the job of breeders to take advantage of available genetic resources. Usually there are very few examples of official germplasm collections, centralized and dependent on public institutions.

FAO made a compilation of such gene banks worldwide, this compilation along with information obtained from the USDA and the latest strawberry symposium, has combined to give rise to a list of the largest public germplasm banks (Table 2.4).

References

Ahmadi, H., Bringhurst, R.S. and Voth, V. (1990) Modes of inheritance of photoperiodism in *Fragaria*. *Journal of the American Society for Horticultural Science* 115, 146–152.

Bors, B. and Sullivan, J.A. (1998) Interspecific crossability of nine diploid *Fragaria* species. *HortScience* 33, 483.

Bringhurst, R.S. (1990) Cytogenetics and evolution in American *Fragaria*. *HortScience* 25, 879–881.

Bringhurst, R. and Senanayake, Y. (1966) The evolutionary significance of natural *Fragaria chiloensis* × *F. vesca* hybrids resulting from unreduced gametes. *American Journal of Botany* 53, 1000–1006.

Bringhurst, R.S. and Voth, V. (1984) Breeding octoploid strawberries. *Iowa State Journal of Research* 58, 371–381.

Capocasa, F., Diamanti, J., Mezzetti, B., Tulipani, S. and Battino, M. (2008) Breeding strawberry (*Fragaria × ananassa* Duch) to increase fruit nutritional quality. *Biofactors* 34, 67–72.

Cheel, J., Theoduloz, C., Rodríguez, J.A., Caligari, P.D.S. and Schmeda-Hirschmann, G. (2007) Free radical scavenging activity and phenolic content in achenes and thalamus from *Fragaria chiloensis* ssp. chiloensis, *F. vesca* and *F. × ananassa* cv. Chandler. *Food Chemistry* 102, 36–44.

Dale, A. and Sjulin, T.M. (1990) Few cytoplasm contribute to North American strawberry cultivars. *HortScience* 25, 1341–1342.

Darrow, G.M. (1966) *The Strawberry: History, Breeding and Physiology*. Holt, Rinehart and Winston, New York.

FAO (2010) *Second Report on the State of the World's Plant Genetic Resources for Food and Agriculture*. Food and Agriculture Organization of the United Nations (FAO), Rome, Italy.

Galletta, G.J., Maas, J.L., Finn, C.E., Smith, B.J. and Gupton, C.L. (1997) The United States Department of Agriculture strawberry breeding program. *Fruit Varieties Journal* 51, 204–210.

Gambardella, M., Infante, R., López-Aranda, J., Faedi, W. and Roudeillac, P. (2000a) Collection of wild and cultivated native *Fragaria* in southern Chile. *Acta Horticulturae* 567, 61–63.

Gambardella, M., Cadavid-Labrada, A. and Diaz, V. (2000b) Isozyme and RAPD characterization of wild and cultivated native *Fragaria* in Southern Chile. *Acta Horticulturae* 567, 81–84.

Gambardella, M., Cadavid, A., Díaz, V. and Pertuzé, R. (2005) Molecular and morphological characterization of wild and cultivated native fragaria in Southern Chile. *HortScience* 40, 1640–1641.

Gil-Ariza, D.J., Amaya, I., López-Aranda, J.M., Sánchez-Sevilla, J.F., Ángel Botella, M. and Valpuesta, V. (2009) Impact of plant breeding on the genetic diversity of cultivated strawberry as revealed by expressed sequence tag-derived simple sequence repeat markers. *Journal of the American Society for Horticultural Science* 134, 337–347.

Gooding, H.J., McNicol, R.J. and MacIntyre, D. (1981) Methods of screening strawberries for resistance to *Sphaerotheca macularis* (Wall ex Frier) and *Phytophthora cactorum* (Leb. and Cohn). *Journal of Horticultural Science* 56, 239–245.

Hancock, J.F. (1999) *Strawberries*. Crop Production Science in Horticulture Series no. 11. CABI Publishing, Wallingford, UK.

Hancock, J.F. and Bringhurst, R.S. (1988) Yield component interactions in wild populations of California *Fragaria*. *HortScience* 23, 889–890.

Hancock, J.F. and Luby, J.J. (1993) Genetic resources at our doorstep: the wild strawberries. *Bioscience* 43, 141–147.

Hancock, J., Flore, J. and Galletta, G. (1989) Variation in leaf photosynthetic rates and yield in strawberries. *Journal of Horticultural Science* 64, 449–454.

Hancock, J.F., Callow, P.W., Dale, A., Luby, J.J., Finn, C.E., Hokanson, S.C. and Hummer, K.E. (2001a) From the Andes to the Rockies: native strawberry collection and utilization. *HortScience* 36, 221–225.

Hancock, J.F., Finn, C.A., Hokanson, S.C., Luby, J.J., Goulart, B.L., Demchak, K., Callow, P.W., Serce, S., Schilder, A.M.C. and Hummer, K.E. (2001b) A multistate comparison of native octoploid strawberries from North and South America. *Journal of the American Society for Horticultural Science* 126, 579–586.

Hancock, J.F., Luby, J.J., Dale, A., Callow, P.W., Serçe, S. and El-Shiek, A. (2002) Utilizing wild *Fragaria virginiana* in strawberry cultivar development: inheritance of photoperiod sensitivity, fruit size, gender, female fertility and disease resistance. *Euphytica* 126, 177–184.

Hancock, J.F., Callow, P.W., Serçe, S. and Son, P.Q. (2003) Variation in the horticultural characteristics of native *Fragaria virginiana* and *F. chiloensis* from North and South America. *Journal of the American Society for Horticultural Science* 128, 201–208.

Hancock, J.F., Finn, C.E., Luby, J.J., Dale, A., Callow, P.W. and Serçe, S. (2010) Reconstruction of the strawberry, *Fragaria × ananassa*, using genotypes of *F. virginiana* and *F. chiloensis*. *HortScience* 45, 1006–1013.

Heide, O.M. and Sønsteby, A. (2007) Interactions of temperature and photoperiod in the control of flowering of latitudinal and altitudinal populations of wild strawberry (*Fragaria vesca*). *Physiologia Plantarum* 130, 280–289.

Hummer, K.E., Nathewet, P. and Yanagi, T. (2009) Decaploidy in *Fragaria iturupensis* (Rosaceae). *American Journal of Botany* 96, 713–716.

Jiajun, L., Yuntao, Z. and Mizhen, Z. (2012) Strawberries in China. In: *7th International Strawberry Symposium*. 18–22 February, Beijing, China, p. 107.

Korbin, M. (2011) Molecular approaches to disease resistance in *Fragaria* spp. *Journal of Plant Protection Research* 51, 60–65.

Koskela, E.A., Mouhu, K., Albani, M.C., Kurokura, T., Rantanen, M., Sargent, D.J., Battey, N.H., Coupland, G., Elomaa, P. and Hytönen, T. (2012) Mutation in *TERMINAL FLOWER1* reverses the photoperiodic requirement for flowering in the wild strawberry *Fragaria vesca*. *Plant Physiology* 159, 1043–1054.

Kunihisa, M., Fukino, N. and Matsumoto, S. (2005) CAPS markers improved by cluster-specific amplification for identification of octoploid strawberry (*Fragaria × ananassa* Duch.) cultivars, and their disomic inheritance. *Theoretical and Applied Genetics* 110, 1410–1418.

Luby, J.J., Hancock, J.F. and Cameron, J.S. (1991) Expansion of the strawberry germplasm base in North America. In: Dale, A. and Luby, J.J. (eds) *The Strawberry into the 21st Century*. Timber Press, Portland, pp. 66–75.

Maas, J.L., Hartung, J.S., Gouin-Behe, C. and Hokanson, S.C. (2000) Sources of resistance for two differentially pathogenic strains of *Xanthomonas fragariae* in *Fragaria* genotypes. *HortScience* 35, 128–131.

Manakasem, Y. and Goodwin, P. (2001) Responses of dayneutral and Junebearing strawberries to temperature and daylength. *Journal of Horticultural Science and Biotechnology* 76, 629–635.

Njuguna, W., Liston, A., Cronn, R., Ashman, T.-L. and Bassil, N. (2013) Insights into phylogeny, sex function and age of *Fragaria* based on whole chloroplast genome sequencing. *Molecular Phylogenetics and Evolution* 66, 17–29.

Potter, D., Luby, J.J. and Harrison, R.E. (2000) Phylogenetic relationships among species of *Fragaria* (Rosaceae) inferred from non-coding nuclear and chloroplast DNA sequences. *Systematic Botany* 25, 337–348.
Rousseau-Gueutin, M., Lerceteau-Köhler, E., Barrot, L., Sargent, D.J., Monfort, A., Simpson, D., Arús, P., Guérin, G. and Denoyes-Rothan, B. (2008) Comparative genetic mapping between octoploid and diploid *Fragaria* species reveals a high level of colinearity between their genomes and the essentially disomic behavior of the cultivated octoploid strawberry. *Genetics* 179, 2045–2060.
Rousseau-Gueutin, M., Gaston, A., Aïnouche, A., Aïnouche, M.L., Olbricht, K., Staudt, G., Richard, L. and Denoyes-Rothan, B. (2009) Tracking the evolutionary history of polyploidy in *Fragaria* L. (strawberry): new insights from phylogenetic analyses of low-copy nuclear genes. *Molecular Phylogenetics and Evolution* 51, 515–530.
Saied, A.S., Keutgen, A.J. and Noga, G. (2005) The influence of NaCl salinity on growth, yield and fruit quality of strawberry cvs. 'Elsanta' and 'Korona'. *Scientia Horticulturae* 103, 289–303.
Senanayake, Y.D.A. and Bringhurst, R.S. (1967) Origin of *Fragaria* polyploids. I. Cytological analysis. *American Journal of Botany* 54, 221–228.
Serçe, S. and Hancock, J.F. (2005a) The temperature and photoperiod regulation of flowering and runnering in the strawberries, *Fragaria chiloensis, F. virginiana,* and *F. × ananassa. Scientia Horticulturae* 103, 167–177.
Serçe, S. and Hancock, J.F. (2005b) Inheritance of day-neutrality in octoploid species of *Fragaria. Journal of the American Society for Horticultural Science* 130, 580–584.
Shaw, D.V. and Famula, T.R. (2005) Complex segregation analysis of day-neutrality in domestic strawberry (*Fragaria × ananassa* Duch.). *Euphytica* 145, 331–338.
Simirgiotis, M.J., Theoduloz, C., Caligari, P.D.S. and Schmeda-Hirschmann, G. (2009) Comparison of phenolic composition and antioxidant properties of two native Chilean and one domestic strawberry genotypes. *Food Chemistry* 113, 377–385.
Staudt, G. (1959) Cytotaxonomy and phyllogenetic relationship in the genus *Fragaria*. In: *Proceedings of the 9th International Congress of Botany*, 19–29 August, Montreal, Quebec, p. 377.
Staudt, G. (1962) Taxonomic studies in the genus *Fragaria*. Typification of *Fragaria* species known at the time of Linnaeus. *Canadian Journal of Botany* 40, 869–886.
Staudt, G. (1973) *Fragaria iturupensis*: a new species of strawberry from East Asia. *Willdenowia* 7, 101–104.
Staudt, G. (1999) *Systematics and Geographic Distribution of the American Strawberry Species: Taxonomic Studies in the Genus Fragaria (Rosaceae: Potentilleae)*. University of California Press, Berkeley, California.
Stewart, P.J. and Folta, K.M. (2010) A review of photoperiodic flowering research in strawberry (*Fragaria* spp.). *Critical Reviews in Plant Sciences* 29, 1–13.
UPOV (2013) International Union for the Protection of New Varieties of Plants. Available at: http://www.upov.int/pluto/en/ (accessed 11 May 2015).
USDA (2015) Summary statistics of holdings as of 10 May 2015. Composition by site for *Fragaria*. Available at: https://npgsweb.ars-grin.gov (accessed 13 April 2016).
Wang, S.Y. and Lewers, K.S. (2007) Antioxidant capacity and flavonoid content in wild strawberries. *Journal of the American Society for Horticultural Science* 132, 629–637.
Weebadde, C.K., Wang, D., Finn, C.E., Lewers, K.S., Luby, J.J., Bushakra, J., Sjulin, T.M. and Hancock, J.F. (2008) Using a linkage mapping approach to identify QTL for day-neutrality in the octoploid strawberry. *Plant Breeding* 127, 94–101.
Wilhelm, S. and Sagen, J.E. (1974) *A History of the Strawberry from Ancient Gardens to Modern Markets*. University of California Press, Berkeley, California.
Zafra-Stone, S., Yasmin, T., Bagchi, M., Chatterjee, A., Vinson, J.A., and Bagchi, D. (2007) Berry anthocyanins as novel antioxidants in human health and disease prevention. *Molecular Nutrition and Food Research* 51, 675–683.

3 Strawberry Breeding

Walther Faedi and Gianluca Baruzzi*
Consiglio per la Ricerca in Agricoltura e L'Analisi dell'economia agraria, Unità di Ricerca per la Frutticoltura, Forlì, Italy

3.1. Introduction

The strawberry (*Fragaria* × *ananassa*) is an octoploid species derived from the hybridization of two wild octoploid species, *Fragaria chiloensis* and *Fragaria virginiana*, in the middle of the 1700s in a botanical garden of Versailles in France. The few plants born from this interspecific cross showed a large-sized fruit, larger than either of the parental fruits. Antoine Nicolas Duchesne, working at the Court of France at that time, was the first botanist to describe the hybrid *F.* × *ananassa* in his *Histoire Naturelle des Frasiers* in 1766. These first plants were the 'ancestors' of all current cultivars now cultivated in many areas of the world, from the coldest to the warmest regions. The high level of ploidy and important interactions with the environment still allow a large phenotypic variability. Since the early 19th century, a strong breeding programme aimed at creating new cultivars with innovative and improving characteristics of plants, and particularly of fruits, has been carried out.

3.2. Historical Breeding

The first breeding activities were carried out in the UK in the 19th century by Michael Keens and Andrew Knight. Keens had a more practical approach and, in 1821, released the

* gianluca.baruzzi@crea.gov.it

strawberry cultivar 'Keens' Seedling', which was then grown successfully for about a century. This cultivar also became a common parent in the pedigrees of the cultivars that followed.

In the early 19th century, Knight, a founding member of the Royal Horticultural Society, with a more scientific and academic approach, bred two important strawberry cultivars ('Downton' and 'Elton') in the UK, both of which were cultivated largely by the English growers of that time. 'Laxton Noble' (1887), and especially 'Royal Sovereign' (1898), the latest cultivars originating from a 'Laxton Nobl' × 'King of Earliest' cross, can be considered the first important cultivars at a European level. 'Royal Sovereign' has been widely cultivated starting from the early 20th century in France, Germany, the UK and Holland.

In France, Duchesne's studies did not improve on the breeding activity. For several years, selections of the original *F.* × *ananassa* hybrid were cultivated; later, the French growers cultivated 'Keens' Seedling'. The first successful French cultivar was probably 'Vicomtesse Héricart de Thury' obtained by J.L. Jamin in 1849 from a random pollination of 'Keens' Seedling'. Other important cultivars were 'Duc de Malakoff' (1854), 'Docteur Morère' (1867) and 'Madame Moutot' (1906), the latter bringing about an important improvement in fruit size. 'Madame Moutot', obtained from a 'Royal Sovereign' × 'Docteur Morère' cross, had wide success across Europe. Until the 1960s, it was considered the first strawberry cultivar to be cultivated at an 'industrial' level.

Some successful cultivars were obtained in Germany, starting from the early 20th century: 'Deutsch Evern' (1902), 'Königen Luise' (1905), 'Osterfee' (1917), 'Oberschlesien' (1919) and 'Frau Mieze Schindler' (1933). These cultivars bred in different countries, together with 'Lucida Perfecta', 'Johannes Müller', 'Markee', 'Sieger', 'Deutsch Evern', 'Tardiva di Leopoldo', 'Surprise des Halles', 'Surprise de Campentras', 'Ville de Paris', 'Macherauch Spaternte D.P. Wallbaum', 'Regina', 'Cambridge Favourite', 'Senga Precosana', 'Souvenir de Charles Machiroux', 'Senga Sengana' and 'Belrubi', now represent an important heritage of the old European cultivar germplasm.

In the USA, breeding activity was started by C.M. Hovey who bred cross combinations among clones of *F. virginiana* and *F. chiloensis* and the English cultivar 'Keens' Seedling'. With the first US cultivar, obtained in 1834 and named 'Hovey', the first US strawberry cultures started in the Atlantic coast (Boston area). In the mid-1850s, James Wilson obtained the cultivar 'Wilson', characterized by a larger fruit than 'Hovey', which contributed to extend strawberry culture beyond the Atlantic coast (in 1850–1860 about 1000 ha was cultivated). The strawberry cultivar 'Dunlap' (1890) was later released and cultivated for more than 70 years.

At the end of the 19th century, Arthur Howard started another breeding programme leading to the release of 'Howard 17' (1907), a parent now present in the pedigree of many modern cultivars. At the beginning of the 20th century, Albert Etter selected new cultivars using some clones of *F. chiloensis* in his breeding programme. He obtained 'Ettesburgh 80', known as 'Huxley' in Europe and widely cultivated in the UK. In the USA, the success of 'Marshall' was also remarkable: this cultivar, characterized by high yield and high adaptability to many different cultural areas, was obtained in 1890 by F. Marshall and cultivated until 1930–1940.

3.3. Modern Breeding

In the 20th century, strawberry breeding activity expanded further around the world (Rosati, 1993). Initially, programmes were carried out at public research institutes, universities in particular, but many private programmes were also then started as a result of the sale of plant propagation and commercialization rights. The programmes have become so numerous that a recent survey registered that more than 900 new cultivars were obtained from 1982 to 2008 worldwide (Della Strada and Fideghelli, 2011). The USA topped the list, with more than 190 cultivars, followed by Italy (74), France (70), Japan (65), the UK (56), Canada (51) and other countries (33).

The number of cultivars has increased gradually, especially starting from the early 1990s, and has stabilized from 2000 (40–50 new cultivars per year). The number of cultivars obtained by private breeding programmes has increased progressively, reaching, in 2008, more than 50% of the number of cultivars released by public programmes. The cultivars introduced up until 1999 were developed by 79 public research institutes and by 32 private companies. This production of new cultivars derived from the development of breeding programmes currently present in more than 40 countries (Faedi *et al.*, 2002).

Most of the cultivars that have been introduced belong to the octoploid species *F.* × *ananassa*, but there also exist cultivars with diploid (*Fragaria vesca*), hexaploid (*Fragaria moschata*) and decaploid (*Fragaria* × *vescana*) chromosomes. Some *Fragaria* × *Potentilla* intergeneric hybrids have also been obtained. All breeding programmes generally aim to increase the plant yield, as well as an increase in fruit size, fruit organoleptic quality and resistance/tolerance to pests and diseases, and extension of the ripening calendar. Some studies have quantified the results obtained in terms of yield, fruit size and flesh firmness (Faedi *et al.*, 1997; Shaw and Larson, 2008). In California, during 50 years of breeding activity at the University of California, large increases in plant yield, average fruit weight and flesh firmness were obtained (Table 3.1). Similar results were obtained during more than 30 years of activity by the Italian breeding programme carried out in the Po Valley, Italy.

3.3.1. Yield

The high yield of the plant continues to be a characteristic of key importance in strawberry breeding. Yield depends on the combination of a series of characteristics including the number of flowers, and consequently of fruits, their size, the plant crown number, hardiness and resistance to disease (Hancock, 1999). The results obtained by some programmes questioned whether it would be possible to increase plant yield further, especially as it has a negative effect on fruit sugar content. However, in some areas (especially those with temperate climate and fresh summers, such as southern Canada and central and northern Europe), the day-neutrality characteristic was used to increase the plant yield. This trait allows an extended harvest calendar (3–4 months or more) and allows the production of a larger number of inflorescences per plant (Dale, 2005).

3.3.2. Harvesting time and fruiting habit

Most breeding programmes aim to extend the harvest calendar of the June-bearing (short-day) cultivars to both early and late seasons. This has becoming increasingly important to achieve a culture deseasonalization and better management of farm labour. Earliness is a very important characteristic in breeding programmes carried out in warm winter areas (e.g. California, Florida, Spain, Israel, southern Italy), searching for genotypes with low winter chilling requirements; in areas that use bare-root

Table 3.1. Results obtained in the breeding programmes carried out in California, USA, and Po Valley, Italy. (From Shaw and Larson, 2008.)

Location		Yield (g per plant)	Fruit size (g)	Firmness (N)
California[a]	1945–1966	595	14.9	0.245
	1993–2004	1.429	24.9	0.456
Italy[b]	1970–1980	768	16.5	0.389
	2000–2010	1.390	26.1	0.572

[a]Average data from the two cultural techniques of winter planting and summer planting of the reference varieties.
[b]Average data from the five best advanced selections (summer planting).

plants (winter planting system), the harvest time is further extended compared with cold-stored plants (summer planting). In areas with particularly warm winters (subtropical areas), low winter chilling requirement cultivars are able to fruit in winter time.

Despite the early ripening time, a late harvest time is particularly important in countries with colder temperatures, especially in winter. New genotypes have been obtained recently that have a very late ripening season, being able to postpone the blooming and harvesting time of traditional cultivars for more than a month, due to the characteristic of extended dormancy present in the European cultivars 'Malwina' and 'Judibell' (Simpson *et al.*, 2009). Many programmes aim to extend harvest time using the 'ever-bearing' characteristic. One of the best-known and common criteria of cultivar classification is fruiting habit (June-bearing and ever-bearing), which depends on the reaction of the plant to photoperiod. The June-bearing cultivars bloom only once in a year following differentiation and require a daily photoperiod of 14 h and temperatures higher than 15°C for flower induction. The ever-bearing cultivars are divided into two branches depending on their reaction to photoperiod: long-day and day-neutral plants. However, classification of the ever-bearing genotypes into one of the two categories is not always easy (Nicoll and Galletta, 1987). Various authors have described different possible origins of the ever-bearing characteristic, which have been summarized by Hancock (1999):

1. The ever-bearing long-day characteristic present in the spontaneous diploid *F. vesca* in Europe could have been transferred in the first octoploid accessions introduced into Europe in the 19th century; an example is 'Climax', one of the first ever-bearing cultivars grown in Europe, which can rebloom although with difficulty when summer temperatures are too high – typical behaviour of the long-day ever-bearing type.

2. The ever-bearing characteristic present in the cultivar 'PanAmerican' (which can be considered the first ever-bearing cultivar having an American origin) could have originated from a mutation of 'Bismark' discovered at the end of the 19th century.

3. The ever-bearing, day-neutral characteristic is present in *F. virginiana* subsp. *glauca* growing spontaneously in the Rocky Mountains; some clones have been used in crosses with the cultivar 'PanAmerican,' obtaining ever-bearing genotypes but with intermediate characteristics between the long-day and day-neutral types; a single clone of *F. virginiana* subsp. *glauca*, found in 1955 by R. Bringhurst in the mountains of Utah near Salt Lake City, had remarkable success in strawberry breeding. Through backcrosses with the best June-bearing accessions of *F.* × *ananassa*, breeders at the University of California introduced this characteristic, which is now present in almost all of the main ever-bearing cultivars cultivated worldwide.

The first day-neutral cultivars, which were introduced commercially in 1979 ('Aptos', 'Brighton' and 'Hecker'), were derived from the third generation of backcrosses where a single clone of *F. virginiana* subsp. *glauca* was used as the first pollinating parent, i.e. the donor of the day-neutral characteristic. Figure 3.1 shows the pedigree of the Californian ever-bearing day-neutral cultivar, 'Brighton'.

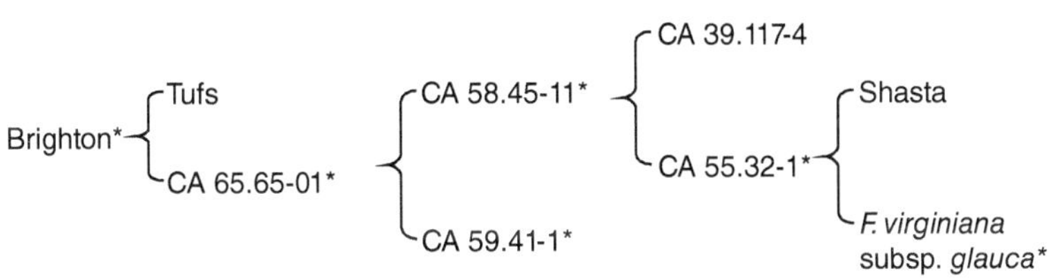

Fig. 3.1. Pedigree of the Californian ever-bearing day-neutral cultivar 'Brighton'. *, Day-neutral genotype.

Following the release of 'Brighton', the University of California introduced the cultivars 'Selva' (1893), 'Seascape' (1991), 'Diamante' (1997), 'Albio' (2004) and more recently 'Monterey', 'Portola' and 'San Andreas' (2008).

High summer temperatures negatively affect flower bud formation in most day-neutral types (Durner, 1984); however, this aspect has improved remarkably in the cultivars obtained recently from breeding programmes carried out mainly in Europe; the problem of the limited production of runners by day-neutral types (Simpson and Sharp, 1988) has also partly been resolved, with the most recent cultivars generally not showing this problem.

The genetics underlying the ever-bearing trait are still not clear. Initially, it was thought that this characteristic was controlled by a single dominant gene (Bringhurst *et al.*, 1989). However, recent studies have demonstrated polygenic control (Shaw and Famula, 2005). There exists great variability in the expression of this characteristic depending on both genotype and environmental factors, especially temperature (Faedi *et al.*, 1994; Maltoni *et al.*, 1996).

Most breeding programmes aim to create both 'June-bearing short-day' and 'ever-bearing day-neutral' cultivars. Currently, the Californian strawberry industry is dominated by day-neutral cultivars, and in Europe the new ever-bearing cultivars will play an increasingly important role, especially in the strawberry industry of northern Europe where the climate is temperate (fairly cool summer) and the harvest could last four or five summer/autumn months using cold-stored plants planted from the middle of March to early April.

Some June-bearing cultivars, in specific environmental conditions, can have a second flowering after the main bloom caused by a second period of differentiation, taking place in spring when there are the right temperature and photoperiod conditions (before the end of March in the northern hemisphere). This trait is common in southern areas, but happens occasionally in northern areas, most frequently only in protected culture. In the Po Valley in Italy, some crossed combinations between very productive short-day genotypes and day-neutral ones have produced weak day-neutral selections that perform like short-day plants (Faedi *et al.*, 2002).

3.3.3. Average fruit weight

In the past, many programmes had the specific objective of increasing fruit size in order to improve plant production and reduce harvest costs. In Italy in particular, the constant increase in labour costs accelerated the research, aiming to improve this characteristic in order to limit the costs of production. The increase in average fruit weight has certainly improved the harvested yield per hour, leading to remarkable savings in labour. In Italy, in 1985, a study carried out by researchers at the University of Bologna showed that a 1 g increase in average fruit weight allowed savings of 1 million Lire (about 500 Euros) per hectare to get the best harvested yield per hour, increasing from 11–13 kg/h with the cultivar 'Gorella' to 18–19 kg/h with the 'Addie' (Malagoli and Pirazzoli, 1986). A similar study carried out in 1998 with 'Onda' fruits showed a further increase in the harvest output of about 4 kg/h – from 19.6 to 23.4 kg/h – obtained by increasing the average fruit weight (Baruzzi, 1998). This trend of increasing fruit size can, however, lead to some problems in commercializing fruits, especially in packaging.

In order to reduce picking costs, long fruit stems (pedicels) is a sought-after characteristic, particularly in southern areas where fresh plants (winter planting system) are used, as they have limited plant development during the coldest months.

3.3.4. Organoleptic quality of fruit

Combining high yield and high fruit quality has always been a main objective in breeding programmes but is difficult to achieve. In some programmes, the fruit quality has been considered for a long time as a 'secondary objective', with the preference being to obtain large fruits to improve the yield

(Faedi *et al.*, 2002). A remarkable genetic variability has been recorded for all of the main fruit quality traits, all of which are quantitatively controlled: flesh firmness and skin resistance (correlated positively) (Ourecky and Bourne, 1968), fruit sweetness, skin and flesh colour (slightly correlated) (Shaw, 1991), acidity and vitamin C content, and flavour.

The negative correlation between yield and sugar level (measured on the Brix scale) (Wenzel, 1980) makes it difficult to select new genetic material with both high plant yield and very sweet fruits. The levels of aromatic substances and of flesh firmness are also negatively correlated with increased size, so the trend to improve flesh firmness results in a progressive loss in fruit aroma (Ulrich *et al.*, 2014). However, an increase in both flesh firmness and skin resistance allows a lower rot susceptibility to be obtained, resulting in a longer shelf-life.

The nutritional parameters (antioxidant compounds and phenolic content) seem to be the most innovative characteristics on which research is focusing its attention due to their benefits on human health. The variability between the existing germplasms is remarkable, and these can be used to improve these parameters in specific breeding programmes (Wang and Lewers, 2007; Capocasa *et al.*, 2008b; Battino *et al.*, 2009; Olbricht *et al.*, 2011; Giampieri *et al.*, 2012). Some specific breeding programmes have already been started, aiming to improve these characteristics using clones of wild species (in particular of *F. virginiana* subsp. *glauca*) as parents (Capocasa *et al.*, 2008a).

3.3.5. Pest and disease resistance

Disease resistance is a primary objective for all the main breeding programmes and will be even more important in the future with the increasing quest for sustainable agriculture. The pathogens causing problems to strawberries have different levels of importance in different cultural areas. For this reason, breeding programmes focus on different pathogens depending on their importance in the area where the programme is carried out.

One of the first goals pursued was resistance to *Phytophthora fragariae* (red core or red stele), which has caused serious damage in strawberry fields of northern Europe and the USA since the mid-1950s. In the middle of the 20th century, two programmes were specifically finalized to combat this pathogen; one of these programmes was carried out in Scotland, UK, and the other one at USDA, Maryland, USA. 'Auchincruive Climax' (released in 1947); 'Stelemaster' (1954) and 'Surecrop' (1956) were the first cultivars resistant to *Phytophthora fragariae* released in Europe and northern America. Later, programmes aimed at resistance against other important pathogens: *Colletotrichum acutatum* (anthracnose), *Phytophthora cactorum* (crown rot), *Verticillium dahliae* (wilt), *Fusarium oxysporum* (wilt), *Sphaerotheca macularis* (mildew), *Mycosphaerella fragariae* (leaf spot), *Diplocarpon earliana* (scorch), *Botrytis cinerea* (grey mould) and *Xanthomonas fragariae* (angular leaf spot). However, many successful cultivars do not show complete or specific resistance to a particular disease. Breeding has focused more on obtaining cultivars that are tolerant, or at least less susceptible, to more pathogens at the same time, rather than to a particular pathogen.

The phase-out of methyl bromide in soil fumigation has added incentive to the search to identify genotypes able to give good results in non-fumigated, old and organic soils. From the mid-1990s, many institutions started the selection of genetic material in non-fumigated soils. The screening of Californian genetic material carried out in the mid-1990s on non-fumigated soils showed a high susceptibility of all tested genotypes to soilborne pathogens (Larson and Shaw, 1995). However, in other breeding programmes, similar studies showed genotypes being able to adapt to non-fumigated soils, registering a high performance both in fumigated and non-fumigated soils (Baruzzi *et al.*, 2009).

3.4. Future Breeding Programmes

In future, breeding programmes are likely to be carried out more frequently by private organizations only or in collaboration with public organizations. The private organizations, especially growers' associations, will be increasingly involved in the final evaluation of new genetic material. The primary goals are likely to remain yield, extension of the harvest period (especially with day-neutral genotypes), fruit quality (nutraceutical aspects) and, in particular, research on plant resistance to reduce the use of pesticides, especially soil fumigants, which will undergo further restrictions.

An increasingly important aim will be to reduce the labour necessary for the harvest through the selection of new genotypes adapted to a faster harvest (easy-to-pick fruits, compact plant habits and long-stem fruits). Breeding programmes will need to consider the climate changes in progress, especially the progressive increases and the sudden change in temperatures. The genotypes with a low winter chilling requirement traditionally grown in temperate areas could be interesting in the future in northern areas, especially for the earliest productions. However, the resistance to spring cold temperatures (causing frost damage during blooming) resulting in serious damage to flowers also needs to be considered. It will be important to select genetic material characterized by a small interval between blooming and harvest in order to have a late blooming and a medium to early ripening. Moreover, the increase in temperatures in summer causes a loss of flesh consistency in some genotypes (e.g. in European 60-day cultures of the cultivar 'Elsanta') with serious consequences at a commercial level. Some genotypes have been shown to be tolerant to high summer temperatures than others, maintaining high flesh firmness and skin resistance.

It will be necessary to take into account the continuous evolution of cultural techniques. For example, the increasingly important programmed cultures (60-day cultures) in central and northern Europe have led to the development of a breeding programme aiming to select genotypes fully adapted to this type of cultural technique (Whitehouse *et al.*, 2009).

The classic strawberry breeding method, which has been well summarized by Chandler *et al.* (2012), based on crossing, selecting the obtained seedlings and then evaluating the selected seedlings, will continue to be used, but it will be increasingly combined with the use of genetic maps and molecular markers, which will speed up the selection times. Currently, strawberry molecular markers are used mainly for analysis of genetic diversity, in particular for cultivar identification (fingerprinting), which is important for the nursery industry and patent protection, but it is reasonable to think that in the future it will be possible to use the marker-assisted selection for the most important agronomic and quality traits. It will be possible to make a first selection in the laboratory by the identification of genotypes presenting one or more molecular markers (quantitative trait loci) associated with the characteristic of interest (e.g. fruit quality and plant resistance). In this way, during the selection phase in the field, it will be possible to focus only on material having the specific traits required by the breeder. Continual refinement of these markers will improve their use in genome mapping. Some molecular markers associated with important traits such as anthracnose resistance and photoperiodic flowering have already been identified and even used in breeding (Whitaker, 2011). However, their application in selection work in breeding programmes is still limited to few research centres.

The recent completion of the genome sequence of the diploid wild strawberry will be important to determining the identification of new markers and also new genes of interest (Shulaev *et al.*, 2011). Some linkage maps of *F.* × *ananassa* are already available (Lerceteau-Köhler *et al.*, 2003; Sargent *et al.*, 2009, 2012; Spigler *et al.*, 2010; Isobe *et al.*, 2013).

This new knowledge will improve the further application of the gene transfer (resulting in genetically modified organisms (GMOs)) and the validation of genes controlling important characters (e.g. resistance, quality and nutritional characteristics). Some genes have already been identified and transferred to strawberry (e.g. for the synthesis of anthocyanins), showing the possibilities available to improve the nutritional quality of fruits without modifying the other cultivar characteristics. It is already known that the limit of GMO technology in strawberry, as for other crops, remains the lack of acceptance by the consumers of many countries.

3.5. The Main Breeding Programmes and Released Cultivars Worldwide

There are currently more than 40 active breeding programmes worldwide, with most being located in North America and Europe (Chandler *et al.*, 2012), although the number of programmes being carried out in other areas, especially China, has increased recently. Many breeding programmes have had much success and their activities have led to the release of cultivars adapted to various cultural areas.

One of the most successful programmes has been that carried out by the University of California, which released in the past some successful low winter chilling requirement cultivars such as 'Tioga' (1963), 'Aliso' (1967), 'Sequoia' (1968), 'Toro' (1975), 'Pajaro' (1979), 'Douglas' (1979), 'Chandler' (1983), 'Camarosa' (1992), 'Ventana' (1997), 'Albion' (2004), Monterey, Portola and San Andreas (2008). The last four cultivars are day-neutral plants, with 'Albion' and 'San Andreas' currently being predominant in California (Table 3.2). The latest released cultivars are 'Benicia' and 'Mojave', both June-bearing, representing the current main objectives of the Californian programme: early season harvest, easy cultivation, high-quality fruit and limited plant development to reduce harvest costs.

There are some important Californian private breeding programmes. Proprietary cultivars developed by these private programmes represent now 42% of the estimated 16,500 total acreage of plantings (Table 3.2). The most active private programmes are carried out by Driscoll Strawberry Associates and Plant Sciences (and its affiliated company, Berry Genetics). Driscoll's headquarters are located in Watsonville, California, but it selects cultivars adapted to specific growing regions throughout the world. Proprietary cultivars are grown only in associated farms (club system) in some of the world's growing regions in northern and southern America, Europe, Africa and Australia. To date, several cultivars, both June-bearing and ever-bearing, have been released and the main cultivars are 'Alafia', 'Amado', 'Del Ray', 'El Dorado', 'Pasadena', 'Pilgrim' and 'San Juan'.

Table 3.2. Evolution of the Californian strawberry cultivar standard from 2008 to 2013. (From California Strawberry Commission: http://www.calstrawberry.com.)

	Dominant cultivars within California (%)					
Cultivar	2008	2009	2010	2011	2012	2013
'Albion'	34.7	39.5	34.6	33.8	30.7	21.8
'Benicia'	–	–	–	0.2	2.7	3.4
'Camarosa'	6.8	3.3	1.6	0.9	0.1	0.3
'Camino Real'	3.5	4.9	4.0	2.1	0.4	0.2
'Chandler'	0.4	0.4	0.2	0.3	0.5	0.2
'Monterey'	–	0.3	0.9	2.4	3.0	6.9
'Palomar'	0.1	1.1	1.6	1.8	1.3	0.4
'Portola'	0.8	0.5	0.7	1.7	3.2	5.1
Proprietary cultivars	39.6	36.0	39.4	40.1	40.9	42.9
'Radiance'	–	–	–	0.1	0.5	1.1
'San Andreas'	0.1	3.3	8.1	11.6	13.8	15.0
'Ventana'	12.4	10.6	8.5	4.8	2.7	2.4
Others	1.6	0.1	0.4	0.2	0.2	0.3
Total state acreage (ha)	14,779	15,634	15,220	15,109	15,529	16,518

Since the mid-1980s, Plant Science Inc. of Watsonville has been carrying out a programme to select new early cultivars characterized by low chilling requirements and adapted to mild climate conditions. Currently, the most important cultivars are 'Promise', 'Splendor', 'Valor' and 'Virtue'. 'Splendor' has also rapidly spread in Spain (Huelva area) where in 2013 it represented the main cultivar (25%; Table 3.3).

In Florida, there is an active breeding programme aiming to select very low chilling requirement cultivars that are able to produce fruits in the winter time in the central to western areas of the country. Its main goal is to obtain cultivars producing very regular-shaped fruits in the difficult winter weather conditions (low temperatures and frequent rainfall). In 2000, this programme released the cultivar 'Florida Festival', which is gaining in importance in many strawberry areas throughout the world. 'Florida Radiance' (marketed as 'Florida Fortuna' in Europe), released in 2009, has become popular in Spain and in other Mediterranean areas (e.g. Sicily) due to its very early ripening season.

In Australia, the University of Queensland and the Horticulture and Forestry Science are carrying out a breeding programme focused on subtropical areas. The most recent cultivar is 'Kalinda' (day-neutral). Another Australian breeding programme is being carried out at the Department of Primary Industries in Victoria. It has been breeding strawberries that are better adapted to conditions in southern Australia.

In Spain, a public-private breeding programme is being carried out by the Institute of Agricultural Research and Training (IFAPA). It is supported by private partners representing most of the growers in Huelva and on Spanish nursery farms, and aims to select improved early cultivars adapted to the winter planting culture of the Huelva area. The cultivar 'Amiga', released in 2006, represents now a small share of the cultivar standard of the important Spanish strawberry industry. The most recent cultivars are 'Santaclara' and 'Fontanilla' (Domínguez *et al.*, 2012). The Spanish private company Fresas Nuevos Materiales (FNM) – as well as co-financing the Spanish public project IFAPA – has been carrying out its own private breeding programme since 1999. 'Antilla', 'Primoris' and 'Niebla' are the most recent commercially diffused cultivars (Refoyo and Arenas, 2009). 'Antilla' is a medium to late cultivar similar to the cultivar 'Sabrosa', and is becoming widespread in the Huelva area (Table 3.3). Another private programme in Spain is being conducted by Planasa, which is successful mostly for its cultivar 'Sabrosa', currently dominating in some Mediterranean areas (e.g. southern Italy), although in Spain its production is now

Table 3.3. Dominant cultivars in the Huelva area (%). (Data from Institute of Agricultural Research and Training (IFAPA), https://www.researchgate.net/institution/Institute_of_Agricultural_Research_and_Training_IFAPA.)

Cultivar	2009	2010	2011	2012	2013
'Amiga'				1	1
'Antilla'				2	3
'Benicia'				3	4
'Camarosa'	45	35	22	16	4
'Festival'	10	4	3	1	1
'Fortuna'		9	5	13	17
'Primoris'			4	4	5
'Sabrina'				7	23
'Sabrosa'	35	35	40	26	10
'San Andreas'				2	3
'Splendor'	3	15	22	22	25
'Ventana'	2	1	2	2	2
'Virtue'					1
Others	5	1	2	1	1

strongly reduced. 'Sabrina', which fruits earlier than 'Sabrosa', has already fully established itself in Spain where it represents, together with 'Splendor', half of the strawberry plantings. The latest released cultivars are 'Sahara' and 'Safari', both June fruit bearers, and the ever-bearing day-neutral 'Cristal'.

The private programme of Darbonne (France) has recently joined the programme of Planasa (Spain) to select improved genotypes with a medium to high chilling requirement. 'Darselect' released in 1996 is largely grown in central to northern Europe. 'Dream', 'Deluxe' and 'Starlette' (June-bearing) and 'Amandine' (day-neutral) are the latest diffused cultivars. Another successful private programme in France has been carried out by Marionnet. The most interesting result was the release of the ever-bearing cultivar 'Mara des Bois' characterized by an excellent organoleptic quality of the fruit. Recently, two new cultivars were released: 'Mariguette' (ever-bearing) and 'Magnum' (June-bearing), both having a similar aroma to 'Mara des Bois'. The French programme of Creation de Variétés Fraises et Fruits Rouges (CIREF) aims to select cultivars, both short-day and ever-bearing (day-neutral), with excellent fruit characteristics, similar to the old cultivars 'Gariguette' (Roudeillac and Markocic, 1997). Recently, the project has released the cultivar 'Rubis Des Jardin'.

In the UK, the historical programme based at East Malling Research Station aimed to create both June-bearing and ever-bearing cultivars. Since 2008, the breeding programme has been supported financially by the East Malling Strawberry Breeding Club and the Department for Environment, Food and Rural Affairs (DEFRA). During this period, three new cultivars have been released; two short-day types, 'Serenity' and 'Malling Centenary', and the ever-bearing 'Buddy'. 'Serenity' (2012) is a somaclonal variant of the late season cultivar 'Florence'. 'Malling Centenary' (2013) is an early to mid-season cultivar with outstanding fruit quality, particularly in terms of its flavour and appearance. 'Buddy' (2012) produces fruit during the summer but concentrates its harvest earlier than 'Everest' – the reference cultivar for UK strawberry areas.

The UK private programme of Edward Vinson Plants aims mainly to obtain ever-bearing cultivars. The most successful cultivars are 'Everest' and 'Evie 2', which are now grown in many European areas. The most recently released cultivars are 'Sweet Eve', 'Velvet' and 'Verity'. The June-bearing cultivar 'Viva Patricia', adapted to southern areas, has been recently released.

The Dutch cultivar 'Elsanta', released in 1981 in the framework of the historical public breeding programme from the University of Wageningen, The Netherlands, is still largely cultivated in northern European areas. In the past, the same programme released the cultivar 'Gorella' (1960), which met success in several European strawberry areas. Currently, the cultivar 'Sonata' (2002) is partly replacing 'Elsanta'. This programme has been renamed Fresh Forward, and involves the participation of both public and private groups. The programme recently released the cultivars 'Vivaldi', 'Musica' and 'Jive'. In the Netherlands, at least two other private programmes are now in progress. One, carried out by Vissers, recently released the cultivar 'Elianny', while the other, carried out by Goossens Flevoplants, has most recently released the cultivars 'Felicita', 'Florin Florentina', 'Favori', 'Pink Extra', 'Sussette' and 'Fleurette'.

In Poland, a breeding programme is being carried out at the Research Institute of Horticulture of Skierniewice. A new medium to early season cultivar named 'Grandarosa' has been released recently (Masny *et al.*, 2015).

In Finland, since 1991, the breeding programme of the Maa-ja elintarviketalouden tutkimuskeskus (MTT) Agrifood Research Finland in Piikkio has as its main purpose the creation of cultivars adapted to polyannual cultures, resistant to low winter temperatures and to powdery mildew. The most recent cultivars are 'Suvetar', 'Valotar' and 'Lumotar' (Hietaranta and Karhu, 2014).

In Norway, the public programme was recently fully privatized and named Graminor Breeding Ltd. It aims to obtain cultivars adapted to extreme weather conditions, such as very cold and dry winters, often without snow. Two medium to late season cultivars, resistant to powdery mildew, have recently been released: 'Gudleif' and 'Blink', which are adapted to the industrial transformation (Alsheikh and Sween, 2012).

In Italy, the private organization Consorzio Italiano Vivaisti (CIV) in Ferrara is carrying out breeding programmes that have released some very popular cultivars in both Italy and Europe ('Marmolada' and 'Clery'). The programme aims to select new improved cultivars adapted to central and northern European areas and to southern Mediterranean areas. CIV has recently released many new cultivars, both June-bearing and day-neutral (day-neutral: 'Murano', 'Spargi', 'Lipari', 'Vivara', 'Gavi', 'Ischia', 'Linosa' and 'Capri', all adapted to northern areas; June-bearing: 'Dely' and 'Joly' for northern areas, 'Nabila', 'Kami' and 'Rania' for the Mediterranean areas, and 'Rubinociv' and 'DipRed' for processing use. In the Po Valley (Italy), the private company NewFruits in Cesena has released in the past very successful June-bearing cultivars such as 'Alba' and 'Roxan', which have spread widely in the Po Valley and in central and northern Europe. Recently, NewFruits released the June-bearing cultivar 'Alina' and the day-neutral cultivars 'Thelma' and 'Louise'. The Department of Agricultural, Food and Environmental Sciences of the Università Politecnica delle Marche is carrying out its own breeding programme in Agugliano, Ancona, which has led to the release of two new June-bearing cultivars, 'Romina' and 'Cristina'. The CRA Unità di Ricerca per la Frutticoltura in Forlì coordinates many public/private breeding programmes carried out in a number of Italian cultural areas. In the past, some very successful cultivars have been released in Italy ('Eva' is the main cultivar of the Verona and Po Valley areas). Two new cultivars have recently been released for southern areas ('Pircinque' and 'Jonica') and two for northern areas ('Garda' and 'Brilla') (Faedi *et al.*, 2014).

In China, interest in strawberries has constantly increased, confirmed by the high number of public institutes carrying out breeding programmes aiming to obtain cultivars adapted to cropping areas often characterized by very different weather conditions. Some breeding programmes are currently in progress in various areas of China. One has been conducted since 2003 in Zhejiang area, where the harvest takes place from November to May in protected culture, and this programme has recently released two new cultivars, 'Yuezhu' and 'Yueli', obtained from crosses among Japanese cultivars. They are both early-season cultivars, with light-coloured fruits that are very sweet with little acid taste, typical of the cultivars of Japanese origin. Another breeding programme is being carried out in the province of Hebei by the Baoding Strawberry Research Institute and focuses on the protected cultures of northern China (Duan *et al.*, 2012). 'Baotong' is the most recent cultivar obtained from a Japanese cultivar ('Benihomalei'). In the province of Shanxi, a programme is being carried out by the Pomology Institute. The selection CM05-1-2 has recently been obtained and released. At the Jiangsu Academy of Agricultural Science, another breeding programme focusing on the warmest areas is being carried out (Wang *et al.*, 2012). 'Ningyu' is the most recent released cultivar obtained, also from Japanese cultivars ('Sachinoka' × 'Akihime'). 'Shimei-7' is a new cultivar released in the framework of a programme conducted by the Shijiazhuang Pomology Institute (Yang *et al.*, 2014). Obtained from a cross with the Japanese cultivar 'Tochiotome', which has a low winter chilling requirement, it is well adapted to forced cultures. The breeding programme conducted at the University of Shenyang works on ornamental cultivars characterized by violet-coloured flowers (Xue *et al.*, 2014). 'Pink Beauty' and 'Pretty Beauty' are two recently released cultivars obtained through crosses made with 'Pink Panda', a pink-flowered cultivar released in the UK in 1971. Another programme being carried out in Chengdu in the province of Sichuan at the Academy of Agricultural Science is aiming to obtain new genetic material by carrying out crosses between commercial cultivars and native clones of the species *Fragaria nilgerrensis*, which is particularly interesting for its fruit aroma (Li *et al.*, 2012).

A strong breeding programme is being carried out in Japan. In northern Japan (Hokkaido and Tohoku districts), two breeding programmes for late June-bearers (in cold regions: Hokkaido, Akita and Yamagata Prefectures) and early June-bearers (in warm regions: the Pacific coast of Aomori, Miyagi and Fukushima Prefectures) have been conducted (Takahashi *et al.*, 2009; Morishita, 2012). Many cultivars were released in the framework of these programmes. The programme aim to release a decaploid interspecific hybrid is particularly interesting. ‘Karume’, obtained from a cross between an octoploid cultivar and a diploid wild species (*F. nilgerrensis*) (Noguchi *et al.*, 2009), presents a strong peach-like sweet fragrance. The cultivar ‘Tokun’ was released recently (Noguchi *et al.*, 2011).

Many breeding programmes are conducted in North America, although compared with the past they have been considerably reduced (Hancock, 1999; Hancock *et al.*, 2008). Most programmes to produce new cultivars are carried out through collaborations among many research institutes. The programme carried out at the US Department of Agriculture/Agricultural Research Service (USDA-ARS) in Beltsville (Maryland) and Corvallis (Oregon) is certainly the longest running worldwide, and stands out for having released many cultivars characterized by resistance to soilborne pathogens. Some recently released cultivars are ‘Valley Red’, ‘Charm’ and ‘Sweet Bliss’, all short-day cultivars adapted to the northwestern Pacific area (Finn *et al.*, 2013). Other programmes are now in progress at Washington State University (latest cultivar released: ‘Puget Crimson’) and North Carolina State University (latest cultivar released: ‘Galletta’).

In Canada, many breeding programmes are now in progress, often working in collaboration with US institutions. In British Columbia, a breeding programme is being carried out at the Pacific Agri-Food Research Centre, Agassiz, part of Agriculture and Agri-Food Canada. ‘Nisgaa’ and ‘Stolo’ are the two most recently released June-bearing cultivars, and ‘Puget Crimson’ (short-day) was commercially released in collaboration with Washington State University and USDA-ARS, Corvallis, confirming the large collaboration among different breeding programmes in northern America. In Kentville, Nova Scotia, the Atlantic Food and Horticulture Research Centre of Agriculture and Agri-Food Canada recently released ‘Valley Sunset’ (2009) and ‘ACC Lila’ (2013). In Ontario, at the Ottawa Eastern Cereal and Oilseed Research Centre of Agriculture and Agri-Food Canada, some selections (LL0210-60, LL0311-43 and LL0312-23) were released. In Quebec, a breeding programme carried out at the Horticultural Research and Development Center is now in progress. As well as some cultivars (e.g. ‘Stolo’) released in collaboration with other research centres, this programme releases also ornamental cultivars such as ‘Roseberry’ (day-neutral) with attractive pink blooms and aromatic fruits (Kempler *et al.*, 2011).

References

Alsheikh, M. and Sween, R. (2012) Advances in the Norwegian Strawberry Breeding Program. In: *VII International Strawberry Symposium*, 18–22 February, Beijing, China, Poster 8.

Baruzzi, G. (1998) Aspetti del miglioramento genetico della fragola in Italia. *Frutticoltura* 12, 55–62.

Baruzzi, G., Faedi, W., Lucchi, P., Migani, M., Sbrighi, P. and Turci, P. (2009) Performance of Italian strawberry genotypes on fumigated, non-fumigated soils and organic culture. *Acta Horticulturae* 842, 553–556.

Battino, M., Beekwilder, J., Denoyes-Rothan, B., Laimer, M., McDougall, G.J. and Mezzetti, B. (2009) Bioactive compounds in berries relevant to human health. *Nutrition Reviews* 67, S145–S150.

Bringhurst, R.S., Ahmadi, H. and Voth, V. (1989) Inheritance of the day-neutral trait in strawberries. *Acta Horticulturae* 265, 35–41.

Capocasa, F., Diamanti, J., Mezzetti, B., Tulipani, S. and Battino, M. (2008a) Breeding strawberry (*Fragaria* × *ananassa* Duch) to increase fruit nutritional quality. *Biofactors* 34, 67–72.

Capocasa, F., Scalzo, J., Mezzetti, B. and Battino, M. (2008b) Combining quality and antioxidant attributes in the strawberry: the role of genotype. *Food Chemistry* 111, 872–878.

Chandler, C., Folta, K., Dale, A., Whitaker, V. and Herrington, M. (2012) Strawberry. In: Badenes, M.L. and Byrne, D.H. (eds) *Fruit Breeding*. Springer-Verlag, New York, pp. 305–325.

Dale, A. (2005) Future trends in strawberry breeding in North America. In: Khanizadeh, S. and DeEll, J. (eds) *Our strawberries – Les Fraisiers de chez nous*. Agriculture and Agri-Food Canada, St. Jean-sur-Richelieu, Quebec, Canada, pp. 70–83.

Della Strada, G. and Fideghelli, C. (2011) *The Fruit Varieties Released in the World from 1980 through 2008*. Centro di Ricerca per la Frutticoltura and Ministero delle Politiche Agricole, Alimentari e Forestali, Rome, Italy.

Domínguez, P., Medina, J.J., López-Aranda, J.M., Ariza, M.T., Miranda, L., Amaya, I., Sánchez-Sevilla, J.F., Villalba, R. and Soria, C. (2012) 'Santaclara' strawberry. *HortScience* 47, 948–951.

Duan, L., Tian, F., Zhang, X., Liu, Y., Wang, C. and Xu, X. (2012) 'Baotong' – a new, early maturing and high quality strawberry cultivar. *Acta Horticulturae* 1049, 253–254.

Durner, E.F. (1984) Photoperiod and temperature effects on flower and runner development in day-neutral, June-bearing, and ever-bearing strawberries. *Journal of the American Society for Horticultural Science* 109, 396–400.

Faedi, W., Simpson, D.W., Dradi, R., Maltoni, M.L. and Persano, S. (1994) Influenza ambientale sul carattere 'rifiorente' in fragola. In: *II Giornate Scientifiche SOI*, 22–24 June, San Benedetto del Tronto, Italy, pp. 43–44.

Faedi, W., Baruzzi, G., Dradi, R., Rosati, P. and Lucchi, P. (1997) Strawberry breeding in Italy. *Acta Horticulturae* 439, 121–128.

Faedi, W., Mourgues, F. and Rosati, C. (2002) Strawberry breeding and varieties: situation and perspectives. *Acta Horticulturae* 567, 51–59.

Faedi, W., Baruzzi, G., Lucchi, P., Magnani, S., Maltoni, M.L., Migani, M., Sbrighi, P., Ballini, L., Baroni, G., Cossio, F., Capriolo, G. and Carullo, A. (2014) Updates on Italian strawberry breeding programs coordinated by CRA-FRF. *Acta Horticulturae* 1049, 877–882.

Finn, C.E., Moore, P.P., Yorgey, B.M., Lee, J., Strik, B.C., Kempler, C. and Martin, R.R. (2013) 'Charm' strawberry. *HortScience* 48, 1184–1188.

Giampieri, F., Tulipani, S., Alvarez-Suarez, J.M., Quiles, J.L., Mezzetti, B. and Battino, M. (2012) The strawberry: composition, nutritional quality, and impact on human health. *Nutrition* 28, 9–19.

Hancock, J.F. (1999) *Strawberries*. Crop Production Science in Horticulture Series, CABI Publishing, Wallingford, UK.

Hancock, J.F., Sjulin, T.M. and Lobos, G.A. (2008) Strawberries. In: Hancock, J.F. (ed.) *Temperate Fruit Crop Breeding*. Springer, The Netherlands, pp. 393–437.

Hietaranta, T.P. and Karhu, S.T. (2014) Enhancing strawberry production at high latitudes. *Acta Horticulturae* 1049, 73–76.

Isobe, S.N., Hirakawa, H., Sato, S., Maeda, F., Ishikawa, M., Mori, T., Yamamoto, Y., Shirasawa, K., Kimura, M., Fukami, M., Hashizume, F., Tsuji, T., Sasamoto, S., Kato, M., Nanri, K., Tsuruoka, H., Minami, C., Takahashi, C., Wada, T., Ono, A., Kawashima, K., Nakazaki, N., Kishida, Y., Kohara, M., Nakayama, S., Yamada, M., Fujishiro, T., Watanabe, A. and Tabata, S. (2013) Construction of an integrated high density simple sequence repeat linkage map in cultivated strawberry (*F.* × *ananassa*) and its applicability. *DNA Research* 20, 79–92.

Kempler, C., Daubeny, H.A., Harding, B., Frey, L., Baumann, T.E., Finn, C.E., Khanizadeh, S., Jamieson, A.R., MacKenzie, K., Moore, P.P. and Sweeney, M. (2011) 'Stolo' strawberry. *HortScience* 46, 1046–1048.

Larson, K.D. and Shaw, D.V. (1995) Relative performance of strawberry genotypes on fumigated and nonfumigated soils. *Journal of the American Society for Horticultural Science* 120, 274–277.

Lerceteau-Köhler, E., Guérin, G., Laigret, F. and Denoyes-Rothan, B. (2003) Characterization of mixed disomic and polysomic inheritance in the octoploid strawberry (*F.* × *ananassa*) using AFLP mapping. *Theoretical and Applied Genetics* 107, 619–628.

Li, H.W., Liu, J.J., Chen, K.L., He, J., Guan, B. and Wang, J.H. (2012) Investigation on wild resources of *Fragaria* in Sichuan and neighboring regions. *Journal of Plant Genetic Resources* 13, 946–951.

Malagoli, C. and Pirazzoli, C. (1986) Ipotesi di recupero della competitivita della coltura della fragola in Emilia-Romagna. *Informatore Agrario* 14, 8–11.

Maltoni, M.L., Faedi, W., Simpson, D.W., Dradi, R. and Persano, S. (1996) Influenza dell'ambiente sulla espressione del carattere rifiorenza in Fragola. *Italus Hortus* 3, 30–36.

Masny, A., Żurawicz, E. and Markowski, J. (2015) 'Grandarosa' strawberry. *HortScience* 50, 1401–1404.

Morishita, M. (2012) The status of strawberry breeding and cultivation in Japan. *Acta Horticulturae* 1049, 125–131.
Nicoll, M.F. and Galletta, G.J. (1987) Variation in growth and flowering habits of June-bearing and ever-bearing strawberries. *Journal of the American Society for Horticultural Science* 112, 872–880.
Noguchi, Y., Muro, T. and Morishita, M. (2009) The possibility of using decaploid interspecific hybrids (*Fragaria* × *ananassa* × *F. nilgerrensis*) as a parent for a new strawberry. *Acta Horticulturae* 842, 447–450.
Noguchi, Y., Morishita, M., Muro, T., Kojima, A., Sakata, Y., Yamada, T. and Sugiyama, K. (2011) 'Tokun': a new aromatic decaploid interspecific hybrid strawberry. *Bulletin of the National Institute of Vegetable and Tea Science* 10, 59–67.
Olbricht, K., Ulrich, D., Weiss, K. and Grafe, C. (2011) Variation in the amounts of selected volatiles in a model population of *F.* × *ananassa* Duch. as influenced by harvest year. *Journal of Agricultural and Food Chemistry* 59, 944–952.
Ourecky, D.K. and Bourne, M.C. (1968) Measurement of strawberry texture with an Instron machine. *Proceedings of the American Society for Horticultural Science* 93, 317–325.
Refoyo, A. and Arenas, J.M. (2009) Cultivars developed in the strawberry breeding program of Fresas Nuevos Materiales SA. *Acta Horticulturae* 842, 439–442.
Rosati, P. (1993) Recent trends in strawberry production and research: an overview. *Acta Horticulturae* 23–44.
Roudeillac, P. and Markocic, M. (1997) Improved flavour and soil disease tolerance: two major aims of the breeding programme at CIREF for new advanced selections combining good commercial potential. *Acta Horticulturae* 439, 139–144.
Sargent, D.J., Fernandéz-Fernandéz, F., Ruiz-Roja, J.J., Sutherland, B.G., Passey, A., Whitehouse, A.B. and Simpson, D.W. (2009) A genetic linkage map of the cultivated strawberry (*F.* × *ananassa*) and its comparison to the diploid *Fragaria* reference map. *Molecular Breeding* 24, 293–303.
Sargent, D.J., Passey, T., Šurbanovski, N., Lopez Girona, E., Kuchta, P., Davik, J., Harrison, R., Passey, A., Whitehouse, A.B. and Simpson, D.W. (2012) A microsatellite linkage map for the cultivated strawberry (*F.* × *ananassa*) suggests extensive regions of homozygosity in the genome that may have resulted from breeding and selection. *Theoretical and Applied Genetics* 124, 1229–1240.
Shaw, D.V. (1991) Genetic variation for objective and subjective measures of fresh fruit color in strawberries. *Journal of the American Society for Horticultural Science* 116, 894–898.
Shaw, D.V. and Famula, T.R. (2005) Complex segregation analysis of day-neutrality in domestic strawberry (*Fragaria* × *ananassa* Duch.). *Euphytica* 145, 331–338.
Shaw, D.V. and Larson, K.D. (2008) Performance of early-generation and modern strawberry cultivars from the University of California breeding programme in growing systems simulating traditional and modern horticulture. *Journal of Horticultural Science and Biotechnology* 83, 648–652.
Shulaev, V., Sargent, D.J., Crowhurst, R.N., Mockler, T.C., Folkerts, O., Delcher, A.L., Jaiswal, P., Mockaitis, K., Liston, A., Mane, S.P., Burns, P., Davis, T.M., Slovin, J.P., Bassil, N., Hellens, R.P., Evans, C., Harkins, T., Kodira, C., Desany, B., Crasta, O.R., Jensen, R.V., Allan, A.C., Michael, T.P., Setubal, J.C., Celton, J.-M., Rees, D.J.G., Williams, K.P., Holt, S.H., Rojas, J.J.R., Chatterjee, M., Liu, B., Silva, H., Meisel, L., Adato, A., Filichkin, S.A., Troggio, M., Viola, R., Ashman, T.-L., Wang, H., Dharmawardhana, P., Elser, J., Raja, R., Priest, H.D., Bryant, D.W., Fox, S.E., Givan, S.A., Wilhelm, L.J., Naithani, S., Christoffels, A., Salama, D.Y., Carter, J., Girona, E.L., Zdepski, A., Wang, W., Kerstetter, R.A., Schwab, W., Korban, S.S., Davik, J., Monfort, A., Denoyes-Rothan, B., Arus, P., Mittler, R., Flinn, B., Aharoni, A., Bennetzen, J.L., Salzberg, S.L., Dickerman, A.W., Velasco, R., Borodovsky, M., Veilleux, R.E. and Folta, K.M. (2011) The genome of woodland strawberry (*Fragaria vesca*). *Nature Genetics* 43, 109–116.
Simpson, D.W. and Sharp, D.S. (1988) The inheritance of fruit yield and stolon production in everbearing strawberries. *Euphytica* 38, 65–74.
Simpson, D.W., Hammond, K.J., Johnson, A.W., Passey, A.J. and Whitehouse, A.B. (2009) Six new strawberry cultivars from East Malling Research. *Acta Horticulturae* 842, 525–528.
Spigler, R.B., Lewers, K.S., Johnson, A.L. and Ashman, T.-L. (2010) Comparative mapping reveals autosomal origin of sex chromosome in octoploid *Fragaria virginiana*. *Journal of Heredity* 101, S107–S117.
Takahashi, H., Yamasaki, A., Shoji, K., Kawagishi, K., Taguchi, T., Yoshida, Y. and Morishita, M. (2009) Present status and prospects of strawberry breeding and cropping type in northern Japan. *Acta Horticulturae* 842, 475–478.
Ulrich, D., Olbricht, K. and Thomas, E. (2014) Diversity and dynamic of volatile patterns in *Fragaria*. *Acta Horticulturae* 1049, 897–901.
Wang, J., Zhao, M.Z., Wang, Z.W., Yu, H.M. and Huang, W.Y. (2012) 'Ningyu': a Chinese new cultivar with high level reducing sugar, ascorbic acid and antioxidant capacity. *Acta Horticulturae* 1049, 231–235.

Wang, S.Y. and Lewers, K.S. (2007) Antioxidant capacity and flavonoid content in wild strawberries. *Journal of the American Society for Horticultural Science* 132, 629–637.
Wenzel, W.G. (1980) Correlation and selection index components. *Canadian Journal of Genetics and Cytology* 13, 42–50.
Whitaker, V.M. (2011) Applications of molecular markers in strawberry. *Journal of Berry Research* 1, 115–127.
Whitehouse, A.B., Passey, A.J. and Simpson, D.W. (2009) Developing a breeding strategy for improved performance in programmed cropping systems. *Acta Horticulturae* 842, 503–506.
Xue, L., Lei, J.J., Dai, H.P. and Deng, M.Q. (2014) Two new red-flowered strawberry cultivars 'Pink Beauty' and 'Pretty Beauty'. *Acta Horticulturae* 1049, 219–224.
Yang, L., Li, L., Yang, L., Zhang, J., Li, H. and Ha, B. (2014) Strawberry cultivar 'Shimei-7' for sightseeing and fruit-picking tourism. *Acta Horticulturae* 1049, 247–248.

4 Chilean White Strawberry: Phenolic Profiling of its Different Parts

Guillermo Schmeda-Hirschmann[1]*, Ezequiel Viveros-Valdez[1,2], Rudy Montenegro[3], Cristina Quispe[1,4] and Mario Simirgiotis[5]

[1]*Instituto de Química de Recursos Naturales, Universidad de Talca, Talca, Chile;* [2]*Universidad Autónoma de Nuevo León, Nuevo León, México;* [3]*Universidad Austral de Chile, Valdivia, Chile; Facultad de Ciencias de la Salud, Instituto de Etnofarmacologia (IDE), Universidad Arturo Prat, Iquique, Chile;* [5]*Universidad de Antofagasta, Antofagasta, Chile*

4.1. Introduction

The white Chilean strawberry, *Fragaria chiloensis* subsp. *chiloensis* (L.) Mill. f. *chiloensis* (Rosaceae), is becoming an interesting crop in the coastal range of Chile. The plant was selected and kept by small farmers and peasants due to its colour, aroma and excellent taste, and is now cultivated in a small area in central Chile, between the latitudes 35 and 39°S. (Fig. 4.1) (Retamales *et al.*, 2005). The chemistry of this native berry has shown differences compared with the red-coloured commercial strawberry (*Fragaria* × *ananassa*), as well as with other *Fragaria* spp. (Cheel *et al.*, 2005, 2007; Simirgiotis *et al.*, 2009b). Recent studies have aimed to disclose the molecular basis of the *F. chiloensis* response under different stress conditions (González *et al.*, 2009, 2013), as well as the genes involved in the fruit colour (Saud *et al.*, 2009; Salvatierra *et al.*, 2010, 2013) and softening (Opazo *et al.*, 2010). However, less is known about the chemical response of the plant under biotic and abiotic stress. Research done by Saud *et al.* (2009) and Salvatierra *et al.* (2010, 2013) has demonstrated the potential of combining information on plant genomics with that on some of the phenolic plant constituents responsible for the fruit colour. Therefore, both genomic and metabolomic approaches to the Chilean white strawberry

* schmeda@utalca.cl

Fig. 4.1. (a) White strawberry culture in Contulmo, Región del Bio-Bio, Chile. (b) Infected plant. (c) Commercialization of white and red strawberry at the Putu market, Región del Maule, Chile. (a, b) From Rudi Montenegro (Universidad Austral de Chile, Chile); (c) from Cristina Theoduloz (Universidad de Talca, Chile).

are relevant for its development as an industrial crop.

Metabolomics is the large-scale chemical analysis of plants, cells or tissues, and includes metabolite target analysis, metabolic profiling and metabolic fingerprinting. The compounds that constitute the metabolome are the end products of gene expression and are related to the chemical phenotype of the living organisms (Sumner *et al.*, 2003). Several responses in living organisms that involve altered gene expression, such as those arising from environmental stimuli in plants, result in qualitative changes in the secondary metabolite pools. Metabolomics requires selected chemical profiling technologies that allow fast and replicable separation of complex mixtures and a fast, reliable characterization of the individual constituents using databases. The chemical profiles are compared to disclose differences in gene expression that are associated with changes in the secondary metabolite pool. Due to the chemical complexity of the metabolome, the inherent biological variability in organisms of the same species and the limitations of the instrumental approaches, a comprehensive profiling of the complete metabolome is not possible. At present, there is no single analytical technique able to profile all of the metabolome.

From an application point of view, metabolomics focuses on high-throughput characterization of selected groups of small-sized metabolites (<1500 Da) in biological matrices. Following this definition, Wishart (2008) referred to it as 'the collection of all small molecule metabolites or chemicals that can be found in a cell, organ or organisms'. These selected compounds can, for instance, be phenolics, terpenes or alkaloids. Metabolomics is also known as metabolic profiling, and requires systems for fast separation of compounds, mixtures and instruments for the spectrometric or spectroscopic characterization of the compounds in the sample. If required, selected compounds can be isolated from the mixture for further identification using standard methods.

Polyphenols are one of the principal compounds related to the benefits of fruits, vegetables, wine and herbal teas consumed in the diet because of their antioxidant properties and other biological activities. Several analytical methods have been proposed for analysis of phenolic compounds in agricultural products, differing in the extraction, separation and quantification techniques. A rapid dereplication of constituents in complex mixtures is desirable to establish similarities and differences that can be used to assess metabolomic changes in the target organism. High-performance liquid chromatography with diode array detection coupled with tandem mass spectrometry (HPLC-DAD-MS/MS) is the method of choice for fast dereplication of complex mixtures, using low amounts of samples. The technique has been applied successfully for fast characterization of phenolics in berries (Määttä *et al.*, 2003; Määttä-Riihinen *et al.*, 2004), including strawberries (Seeram *et al.*, 2006; Simirgiotis and Schmeda-Hirschmann, 2010a; Schmeda-Hirschmann *et al.*, 2011), edible fruits (Simirgiotis *et al.*, 2009a,b, 2013a,b,c), herbal teas (Simirgiotis and Schmeda-Hirschmann, 2010b; Quispe *et al.*, 2012; Simirgiotis *et al.*, 2012), propolis (Agüero *et al.*, 2010) and spices (Viveros-Valdez *et al.*, 2008).

This chapter reviews all relevant information that will be relevant to set up a database of phenolic constituents of the white Chilean strawberry for metabolomic studies.

4.2. Phenolic Profile of Cultivated Chilean White Strawberry

The plant samples were taken from commercial plantations at Contulmo, Región del Bio-Bio, Chile, and can be considered representative of the white strawberry plants regarded as healthy by producers. No visible damage was observed for the botanical material employed as the baseline for comparison purposes. The plants were taken to the laboratory, separated into fruits, leaves, flowers and rhizomes, washed with tap water and immediately extracted three times with methanol:formic acid (99:1) in a 1 g sample:20 ml solvent ratio. The extracts were filtered and taken to dryness for HPLC and HPLC-DAD-MS/MS analysis. Plants presenting leaf damage associated with a fungal pathogen were collected and processed as described above for comparison purposes (Fig. 4.1b).

Different organs from the cultivated Chilean white strawberry were analysed separately for phenolic compounds using HPLC-DAD-MS/MS. The HPLC system was Merck-Hitachi equipment (LaChrom, Tokyo, Japan) consisting of an L-7100 pump, an L-7455 UV diode array detector and a D-7000 chromatointegrator. A 250 mm × 4.60 mm internal diameter, 5 μm Kromasil 100-5C18 column (Eka Chemicals, Brewster, NY) maintained at 25°C was used. The HPLC analysis was performed using a linear gradient solvent system consisting of 1% formic acid (A) and acetonitrile (B) as follows: 90 to 75% A over 30 min, followed by 75 to 40% A from 30 to 45 min at a flow rate of 1 ml min^{-1}. The extracts were dissolved in methanol:H_2O (1:1, v/v) (final concentration ~5 mg ml^{-1}), filtered through a 0.45 μm PTFE filter (Waters, Milford, MA) and submitted to HPLC-DAD and HPLC-electrospray ionization (ESI)-MS analysis. The volume injected was 20 μl. The compounds were monitored at 250 nm, and UV spectra were recorded at 200–600 nm for peak characterization.

HPLC coupled to MS was performed using a Squire 4000 plus ion trap spectrometer fitted with ESI (Bruker Daltonics, Billerica, MA). The capillary voltage was 4000 V. Nitrogen was used as nebulizer gas at 350°C at a flow of 8.0 l min^{-1}, and the nebulizer pressure was 27.5 psi. Spectra were recorded between *m*/*z* 150 and 2000 in positive-ion mode. Collision-induced dissociation spectra were obtained with a fragmentation amplitude of 1.00 V (MS/MS) using helium as the collision gas.

Analysis of the constituents of the different organs was carried out under the same experimental conditions to allow comparison of the 'baseline' plants (i.e. plants considered healthy) with those presenting phytopathogen-induced damage. While the first approach renders information on the main constituents and fingerprinting of different organs, comparison with diseased plants might confirm whether the method is suitable for the detection of a chemical response, measured as changes in the qualitative and/or quantitative composition of the plant phenolics.

Comparison of phenolics in leaves was preferred for obvious reasons, as correlation with damage and the pathogen can both be carried out using chemical markers or molecular biology tools. The HPLC chromatograms of the different plant parts are shown in Fig. 4.2, while the distribution of phenolic compounds in the different plant parts is summarized in Table 4.1.

4.2.1. HPLC analysis of white strawberry leaves

The profiles of the leaves showed mainly tannins and quercetin 3-*O*-glucuronide (Q-3-Gluc), with kaempferol coumaroyl hexosides as minor constituents (Table 4.2). The main compounds were ellagic acid, quercetin 3-*O*-glucuronide and ellagitannins. The ellagitannins with an $[M+1]^+$ ion at *m*/*z* 785 and 769 were tentatively assigned to platycaryanin D or pedunculagin, and as a balanophotannin (Jiang *et al.*, 2005), respectively. Ellagic acid sulphate conjugate (Souleman *et al.*, 1998) was also present. Quercetin and kaempferol coumaroyl hexoside were described in the leaves of the wild-growing red-coloured form, *F. chiloensis* var. *chiloensis* (Simirgiotis *et al.*, 2009b).

4.2.2. HPLC analysis of white strawberry flowers

The phenolic constituents of the white strawberry flowers were described for the first time, showing ellagic acid derivatives and flavonoids as the main compounds. The flavonoids included quercetin and kaempferol glycosides, with glucuronic acid as the sugar moieties in both flavonoids.

4.2.3. HPLC analysis of white strawberry rhizomes

The rhizomes contained mainly condensed and hydrolysable tannins (Table 4.2) with dimers and trimers based on catechin/epicatechin, in agreement with data in the literature (Yoshida *et al.*, 1987; Nonaka *et al.*, 1989; Terashima *et al.*, 1990; de Souza *et al.*,

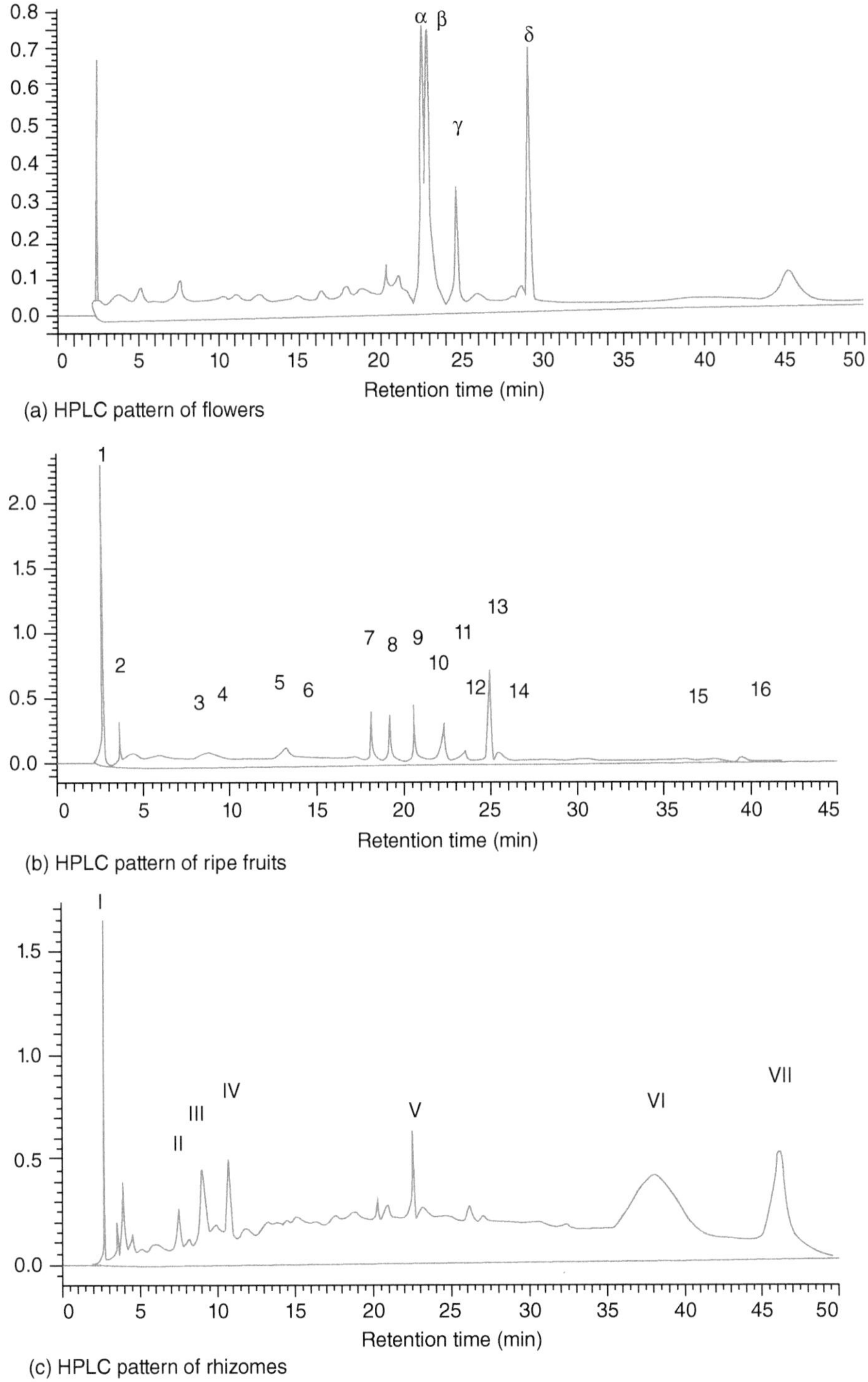

Fig. 4.2. High-performance liquid chromatography (HPLC) with diode array detection chromatograms at 250 nm of flowers (a), ripe fruits (b), rhizomes (c), healthy leaves (d) and infected leaves (e) of *Fragaria chiloensis* subsp. *chiloensis* f. *chiloensis*. See Table 4.2 for a description of the peak labels.

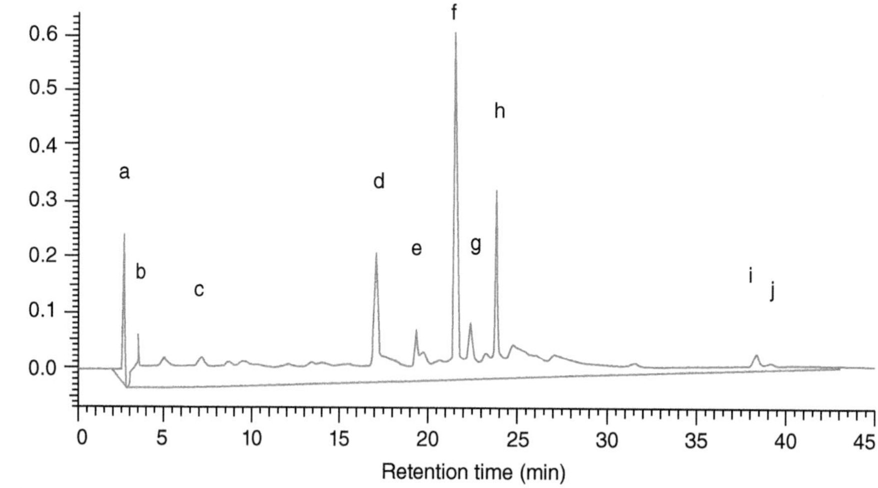

(d) HPLC pattern of healthy leaves

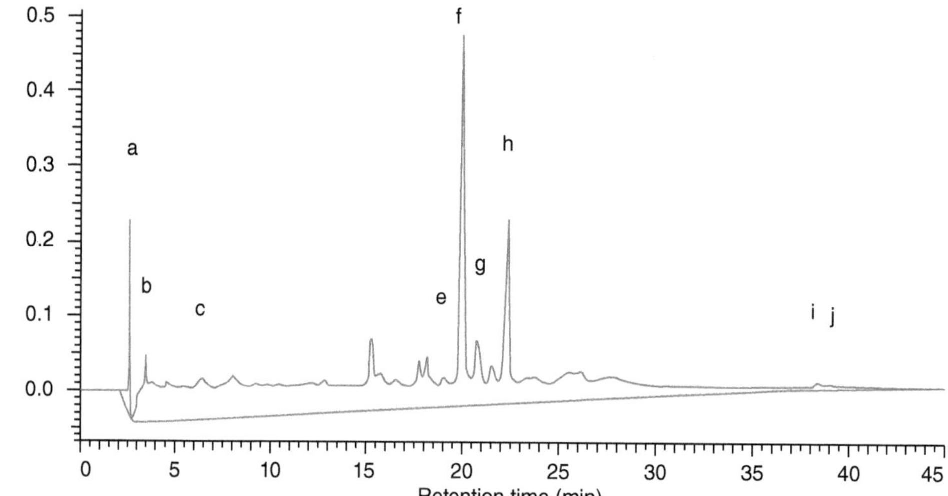

(e) HPLC pattern of infected leaves

Fig. 4.2. Continued.

2008; Appeldoorn *et al.*, 2009). Three trimers and two dimers were tentatively identified, while two additional related compounds remained unidentified. The same trend was observed for the rhizomes of *F. chiloensis* f. *patagonica*, where tannins were the predominant compounds.

4.2.4. HPLC analysis of white strawberry fruits

The HPLC pattern of the ripe white strawberry fruits showed caffeoylquinic acid and several procyanidin trimers as minor compounds. The main constituents were quercetin derivatives, including quercetin-3-*O*-glucuronide and ellagic acid pentoside (Table 4.1).

4.3. The Plant Response under Fungal Infection

The main phenolics detected in healthy leaves of the cultivated white strawberry were a hydrolysable tannin of molecular

Table 4.1. Phenolic compound distribution in leaves, flowers, rhizomes and fruit of cultivated Chilean white strawberry.

		Plant part			
Molecular weight	Tentative compounds identification	L	Fl	Rh	F
354	Caffeoylquinic acid/chlorogenic acid				F
354	Caffeoylquinic acid/chlorogenic acid				F
224	Sinapinic acid				F
620	Hydroxybenzoic acid derivative				F
578	Procyanidin dimer isomer 1			Rh	
578	Procyanidin dimer isomer 2			Rh	
864	Procyanidin trimer isomer			Rh	
866	Procyanidin trimer isomer 1			Rh	
866	Procyanidin trimer isomer 1				F
866	Procyanidin trimer isomer 2				F
866	Procyanidin trimer isomer 3				F
866	Procyanidin trimer isomer 4				F
866	Procyanidin trimer isomer			Rh	
302	Ellagic acid[a]	L			
396	Methyl ellagic acid sulphate	L			
434	Ellagic acid pentoside				F
478	Digalloylshikimic acid	L			
628	Ellagic acid hexoside		Fl		
656	Ellagic acid derivative		Fl		
766	Ellagitannin	L			
768	Balanophotannin	L			
784	Platycaryanin D or pedunculagin	L			
566	K derivative				F
594	K 3-*O*-coumaroyl hexoside isomer 1	L			
594	K 3-O-coumaroyl hexoside isomer 2	L			
624	K 3-*O*-glucuronide-hexoside		Fl		
478	Q 3-*O*-glucuronide[a]	L	Fl		F
576	Q derivative				F
578	Q hexose + unknown moiety				F
866	Q derivative				F
926	Taxifolin derivative				F
380	Unknown	L			
684	Unknown			Rh	
704	Unknown			Rh	

L, leaves; Fl, flowers; Rh, rhizomes; F, fruit; Q, quercetin; K, kaempferol.
[a]Identified by spiking experiment with standard phenolic compound.

weight 784 Da (platycaryanin D or pedunculagin, peak 4, Rt 17.01 min), free ellagic acid (as the main constituent, with Rt of 21.52 min) and quercetin 3-*O*-glucuronide (Rt 23.84 min). After infection, the main changes in the chromatogram were a decrease in the peak area of the tannin and an increase in the ellagic acid content of the sample. The biological significance of the findings should be investigated further to determine whether the modifications in the phenolic pattern are specific to the phytopathogenic organism (fungus, bacterium or virus) or whether this is a general chemical response of the plant under microbial challenge. The role of ellagitannins in human health has been reviewed recently by Landete (2011). Ellagic acid is released from tannins after hydrolysis and plays several relevant biological roles as an antioxidant as well as in modulating the cell response.

Table 4.2. High-performance liquid chromatography with diode array detection coupled with tandem mass spectrometry data of extracts from flower, fruit, rhizome and leaf of white Chilean strawberry plants.

Peak	Part	Rt (min)	UV max	$[M^+H]^+$ and MS^n ions	M	Tentative identification
α	Flower	22.64	364, 295 sh, 252	657, 397, 303	656	EA derivative
β	Flower	22.93	260 sh, 238	629, 465, 447, 303	628	EA hexose
γ	Flower	24.69	352, 295 sh, 265 sh, 255	479, 303	478	Q 3-*O*-glucuronide[a]
δ	Flower	29.17	344, 295 sh, 265	625, 463, 287	624	K 3-*O*-glucuronide-hexoside
1	Fruit	2.59	276, 240	355, 183, 163, 85	354	CQ/chlorogenic acid isomer 1
2	Fruit	3.57	273, 235	355, 171, 153, 136	354	CQ/chlorogenic acid isomer 2
3	Fruit	9.23	ND	867, 579, 427, 409, 289	866	P-trimer isomer 1
4	Fruit	9.95	ND	867, 579, 427, 409, 289	866	P-trimer isomer 2
5	Fruit	13.44	ND	867, 579, 409, 287, 271	866	P-trimer isomer 3
6	Fruit	14.16	ND	867, 579, 437, 287, 271	866	P-trimer isomer 4
7	Fruit	18.19	350, 290 sh, 251	579, 303	578	Q hexose+unknown moiety
8	Fruit	19.25	358, 290 sh, 251	867, 579, 303	866	Q derivative
9	Fruit	20.64	364, 290 sh, 251	577, 303, 289, 98	576	Q derivative
10	Fruit	22.05	ND	379, 333, 131, 103	378	
11	Fruit	22.32	352, 285 sh, 254	479, 303	478	Q 3-*O*-glucuronide[a]
12	Fruit	23.47	ND	927, 895, 459, 437, 305, 259, 231	926	Taxifolin derivative
13	Fruit	24.88	ND	621, 379, 333, 311, 149, 131	620	Hydroxybenzoic acid derivative
14	Fruit	26.16	280, 228	435, 303, 131, 103	434	EA pentoside
15	Fruit	36.48	ND	567, 287, 131	566	K derivative
16	Fruit	40.91	ND	225, 143, 100, 83	224	Sinapinic acid ($C_{11}H_{12}O_5$)

I	Rhizome	2.64	278, 242	685, 523, 505, 381, 343, 325, 163	684	Unknown
II	Rhizome	7.47	278, 237	865, 632, 579, 409, 291	864	P-trimer
III	Rhizome	9.04	278, 240	867, 631, 579, 427, 409, 301, 289	866	P-trimer isomer 1
IV	Rhizome	10.69	278, 239	579, 409, 344, 291, 207, 139	578	P-dimer isomer 1
V	Rhizome	22.59	278, 239	867, 579, 289	866	P-trimer isomer 2
VI	Rhizome	38.13	278, 239	579, 409, 303	578	P-dimer isomer 2
VII	Rhizome	46.19	278, 239	*703*, 423, 303, 240	704	Unknown
a	Leaf	2.62	235	381, 325, 163, 145, 127	380	Unknown
b	Leaf	3.57	236	479, 381, 187, 146	478	Digalloylshikimic acid
c	Leaf	7.12	276, 216	767, 465, 187, 146	766	Ellagitannin
d	Leaf	17.01	237, 260 sh	785, 617, 447, 303, 187	784	Platycaryanin D or Pedunculagin
e	Leaf	19.33	276	769, 599, 465, 303, 187, 146	768	Balanophotannin
f	Leaf	21.52	238, 260 sh	303, 187, 146	302	Ellagic acid[a]
g	Leaf	22.37	251, 363	397, 303, 187, 146	396	Methyl EA sulphate
h	Leaf	23.84	255, 305 sh, 352	479, 303	478	Q 3-*O*-glucuronide[a]
i	Leaf	38.43	238, 265, 313, 350	595, 309, 287, 187, 147	594	K 3-*O*- coumaroyl hexoside
j	Leaf	39.17	265, 305, 350	595, 287, 187, 146	594	K 3-*O*- coumaroyl hexoside

Rt, retention time; M, molecular mass (Da); sh, shoulder; EA, ellagic acid; Q, quercetin; K, kaempferol; CQ, caffeoyl quinic acid; P, procyanidin; ND, not detected by UV–visible photodiode array detector, or compound showing a weak UV–visible spectrum.

[a]Identified by spiking experiment with standard phenolic compound.

4.4. Future Perspectives

This chapter has presented the first insights into the phenolic compound composition of different plant organs from the cultivated Chilean white strawberry. The study was undertaken with plants showing no signs of disease, and the fingerprints obtained were compared with plants that were naturally infected with fungi. The chemical profile of the leaves showed clear changes according to the plant health status. However, additional work should be carried out with virus-free plants as well as with individuals selectively inoculated with different fungal and bacterial infections to ascertain whether the changes observed in the phenolic profiles can be related to the infection source (virus, fungus or bacterium) or whether the response is non-specific. Quantitative studies measuring the individual phenolic constituents content is required to obtain a more robust basis for a metabolomics study, giving support to a better association between the plant genome, biotic and abiotic responses, and the plant chemistry.

Acknowledgements

C. Quispe and E. Viveros-Valdez thank the PBCT Program PSD-50 for a postdoctoral grant. Financial support from the FONDECYT Project 1120096 is gratefully acknowledged.

References

Agüero, M.B., Gonzalez, M., Lima, B., Svetaz, L., Sánchez, M., *et al.* (2010) Argentinean propolis from *Zuccagnia punctata* Cav. (Caesalpinieae) exudates: phytochemical characterization and antifungal activity. *Journal of Agricultural and Food Chemistry* 58, 194–201.

Appeldoorn, M.M., Vincken, J.-P., Sanders, M., Hollman, P.C.H. and Gruppen, H. (2009) Combined normal-phase and reversed-phase liquid chromatography/ESI-MS as a tool to determine the molecular diversity of A-type procyanidins in peanut skins. *Journal of Agricultural and Food Chemistry* 57, 6007–6013.

Cheel, J., Theoduloz, C., Rodríguez, J., Saud, G., Caligari, P.D.S. and Schmeda-Hirschmann, G. (2005) E-cinnamic acid derivatives and phenolics from Chilean strawberry fruits, *Fragaria chiloensis* ssp. *chiloensis*. *Journal of Agricultural and Food Chemistry* 53, 8512–8518.

Cheel, J., Theoduloz, C., Rodríguez, J.I., Caligari, P.D.S. and, Schmeda-Hirschmann, G. (2007) Free radical scavenging activity and phenolic content in achenes and thalamus from *Fragaria chiloensis* ssp. *chiloensis*, *F. vesca* and *F.* × *ananassa* cv. Chandler. *Food Chemistry* 102, 36–44.

de Souza, L.M., Cipriani, T.R., Iacomini, M., Gorin, P.A.J. and Sassaki, G.L. (2008) HPLC/ESI-MS and NMR analysis of flavonoids and tannins in bioactive extract from leaves of *Maytenus ilicifolia*. *Journal of Pharmaceutical and Biomedical Analysis* 47, 59–67.

González, G., Moya, M., Sandoval, C. and Herrera, R. (2009) Genetic diversity in Chilean strawberry (*Fragaria chiloensis*): differential response to *Botrytis cinerea* infection. *Spanish Journal of Agricultural Research* 7, 886–895.

González, G., Fuentes, L., Moya-León, M.A., Sandoval, C. and Herrera, R. (2013) Characterization of two PR genes from *Fragaria chiloensis* in response to *Botrytis cinerea* infection: a comparison with *Fragaria* × *ananassa*. *Physiological and Molecular Plant Pathology* 82, 73–80.

Jiang, Z.-H., Tanaka, T., Iwata, H., Sakamoto, S., Hirose, Y. and Kouno, I. (2005) Ellagitannins and lignan glycosides from *Balanophora japonica* (Balanophoraceae). *Chemical and Pharmaceutical Bulletin* 53, 339–341.

Landete, J.M. (2011) Ellagitannins, ellagic acid and their derived metabolites: a review about source, metabolism, functions and health. *Food Research International* 44, 1150–1160.

Määttä-Riihinen, K.R., Kamal-Eldin, A. and Törrönen, A.R. (2003) High-performance liquid chromatography (HPLC) analysis of phenolic compounds in berries with diode array and electrospray ionization mass spectrometric (MS) detection: *Ribes* species. *Journal of Agricultural and Food Chemistry* 51, 6736–6744.

Määttä-Riihinen, K.R., Kamal-Eldin, A. and Törrönen, A.R. (2004) Identification and quantification of phenolic compounds in berries of *Fragaria* and *Rubus* species (family Rosaceae). *Journal of Agricultural and Food Chemistry* 52, 6178–6187.

Nonaka, G.-I., Ishimatsu, M., Ageta, M. and Nishioka, I. (1989) Tannins and related compounds. LXXVI. Isolation and characterization of cercidinins A and B and cuspinin, unusual 2, 3-(R)-hexahydroxydiphenoyl glucoses from *Cercidiphyllum japonicum* and *Castanopsis cuspidata* var. *sieboldii*. *Chemical and Pharmaceutical Bulletin* 37, 50–53.

Opazo, M.C., Figueroa, C.R., Henríquez, J., Herrera, R., Bruno, C., Valenzuela, P.D.T. and Moya-León, M.A. (2010) Characterization of two divergent cDNAs encoding xyloglucan endotransglycosylase/hydrolase (XTH) expressed in *Fragaria chiloensis* fruit. *Plant Science* 179, 479–488.

Quispe, C., Viveros-Valdez, E. and Schmeda-Hirschmann, G. (2012) Phenolic constituents of the Chilean herbal tea *Fabiana imbricata* R. et P. *Plant Foods for Human Nutrition* 67, 242–246.

Retamales, J.B., Caligari, P.D.S., Carrasco, B. and Saud, G. (2005) Current status of the Chilean native strawberry and the research needs to convert the species into a commercial crop. *HortScience* 40, 1633–1634.

Salvatierra, A., Pimentel, P., Moya-Leon, M.A., Caligari, P.D.S. and Herrera, R. (2010) Comparison of transcriptional profiles of flavonoid genes and anthocyanin contents during fruit development of two botanical forms of *Fragaria chiloensis* ssp. *chiloensis*. *Phytochemistry* 71, 1839–1847.

Salvatierra, A., Pimentel, P., Moya-León, M.A. and Herrera, R. (2013) Increased accumulation of anthocyanins in *Fragaria chiloensis* fruits by transient suppression of *FcMYB1* gene. *Phytochemistry* 90, 25–36.

Saud, G., Carbone, F., Perrotta, G., Figueroa, C.R., Moya, M., Herrera, R, Retamales, J.B., Carrasco, B., Cheel, J., Schmeda-Hirschmann, G. and Caligari, P.D.S. (2009) Transcript profiling suggests transcriptional repression of the flavonoid pathway in the white-fruited Chilean strawberry, *Fragaria chiloensis* (L.) Mill. *Genetic Resources and Crop Evolution* 56, 895–903.

Schmeda-Hirschmann, G., Simirgiotis, M. and Cheel, J. (2011) Chemistry of the Chilean Strawberry (*Fragaria chiloensis* spp. *chiloensis*). *Genes, Genomes and Genomics* 5, 85–90.

Seeram, N.P., Lee, R., Scheuller, H.S. and Heber, D. (2006) Identification of phenolic compounds in strawberries by liquid chromatography electrospray ionization mass spectroscopy. *Food Chemistry* 97, 1–11.

Simirgiotis, M.J. and Schmeda-Hirschmann, G. (2010a) Determination of phenolic composition and antioxidant activity in fruits, rhizomes and leaves of the white strawberry (*Fragaria chiloensis* spp. *chiloensis* form *chiloensis*) using HPLC-DAD-ESI-MS and free radical quenching techniques. *Journal of Food Composition and Analysis* 23, 545–553.

Simirgiotis, M.J. and Schmeda-Hirschmann, G. (2010b) Direct identification of phenolic constituents in Boldo Folium (*Peumus boldus* Mol.) infusions by high-performance liquid chromatography with diode array detection (HPLC-DAD) and electrospray ionization tandem mass spectrometry (HPLC-MSn). *Journal of Chromatography A* 1217, 443–449.

Simirgiotis, M.J., Caligari, P.D.S. and Schmeda-Hirschmann, G. (2009a) Identification of phenolic compounds from the fruits of the mountain papaya *Vasconcellea pubescens* A. DC. grown in Chile by liquid chromatography–UV detection–mass spectrometry. *Food Chemistry* 115, 775–784.

Simirgiotis, M.J., Theoduloz, C., Caligari, P.D.S. and Schmeda-Hirschmann, G. (2009b) Comparison of phenolic composition and antioxidant properties of two native Chilean and one domestic strawberry genotypes. *Food Chemistry* 113, 377–385.

Simirgiotis, M.J., Silva, M., Becerra, J. and Schmeda-Hirschmann, G. (2012) Direct characterisation of phenolic antioxidants in infusions from four Mapuche medicinal plants by liquid chromatography with diode array detection (HPLC-DAD) and electrospray ionisation tandem mass spectrometry (HPLC-ESI–MS). *Food Chemistry* 131, 318–327.

Simirgiotis, M.J., Bórquez, J. and Schmeda-Hirschmann, G. (2013a) Antioxidant capacity, polyphenolic content and tandem HPLC–DAD–ESI/MS profiling of phenolic compounds from the South American berries *Luma apiculata* and *L. chequén*. *Food Chemistry* 139, 289–299.

Simirgiotis, M.J., Ramirez, J.E., Schmeda Hirschmann, G. and Kennelly, E.J. (2013b) Bioactive coumarins and HPLC-PDA-ESI-ToF-MS metabolic profiling of edible queule fruits (*Gomortega keule*), an endangered endemic Chilean species. *Food Research International* 54, 532–543.

Simirgiotis, M.J., Schmeda-Hirschmann, G., Bórquez, J. and Kennelly, E.J. (2013c) The *Passiflora tripartita* (banana passion) fruit: a source of bioactive flavonoid C-glycosides isolated by HSCCC and characterized by HPLC–DAD–ESI/MS/MS. *Molecules* 18, 1672–1692.

Souleman, A.A., Barakat, H.H., Hussein, S.A., El-Mousallamy, A.M. and Nawwar, M.A. (1998) Unique phenolic sulphate conjugates from the flowers of *Tamarix amplexicaulis*. *Natural Product Sciences* 4, 245–252.

Sumner, L.W., Mendes, P. and Dixon, R.A. (2003) Plant metabolomics: large-scale phytochemistry in the functional genomics era. *Phytochemistry* 62, 817–836.

Terashima, S., Shimizu, M., Nakayama, H., Ishikura, M., Ueda, Y., Imai, K., Suzui, A. and Morita, N. (1990) Studies on aldose reductase inhibitors from medicinal plant of "sinfito," *Potentilla candicans*, and further synthesis of their related compounds. *Chemical and Pharmaceutical Bulletin* 38, 2733–2736.

Viveros-Valdez, E., Rivas-Morales, C., Carranza-Rosales, P., Mendoza, S. and Schmeda-Hirschmann, G. (2008) Free radical scavengers from the Mexican herbal tea "Poleo" (*Hedeoma drummondii*). *Zeitschrift für Naturforschung C* 63, 341–346.

Wishart, D.S. (2008) Metabolomics: applications to food science and nutrition research. *Trends in Food Science and Technology* 19, 482–493.

Yoshida, T., Chen, X.-M., Hatano, T., Fukushima, M. and Okuda, T. (1987) Tannins and related polyphenols of Rosaceous medicinal plants. IV. Roxbins A and B from *Rosa roxburghii* fruits. *Chemical and Pharmaceutical Bulletin* 35, 1817–1822.

5 Role of Fungal Avirulent Pathogens in the Defence Response of Strawberry

Juan C. Díaz Ricci*, Carlos F. Grellet-Bournonville, Nadia R. Chalfoun, Ursula Tonello, Gustavo G. Martos, Verónica Hael-Conrad, Marisa Perato and Gustavo Martínez Zamora

Instituto Superior de Investigaciones Biológicas (INSIBIO), CONICET-UNT, and Instituto de Química Biológica 'Dr Bernabé Bloj', Facultad de Bioquímica, Química y Farmacia, Universidad Nacional de Tucumán, Tucumán, Argentina

5.1. Introduction

Plants are constantly being challenged by potential pathogens; however, disease is a relatively rare event because they have evolved sophisticated defence mechanisms, which include preformed physical and chemical barriers, as well as inducible defence mechanisms (Hammond-Kosack and Parker, 2003; Jones and Dangl, 2006). The induction of plant defence/plant immune system occurs at two levels in plant cells: membrane and cytosol. First, microbial or pathogen-associated molecular patterns (MAMPs or PAMPs) are recognized by transmembrane pattern-recognition receptors (PRRs) (Chisholm *et al.*, 2006; He *et al.*, 2007; Jones and Dangl, 2006). PAMPs are highly conserved motifs of structures that play an essential role in microbial lifestyle. Several PAMP-like proteins have been identified from phytopathogens, including flagellin and 'harpins' from bacteria, xylanase from fungi, invertase from yeast, and Pep13 and

* juan@fbqf.unt.edu.ar

elicitins from oomycetes, as well as necrosis and ethylene-inducing protein (NEP)-like proteins (Nürnberger *et al.*, 2004). Plants can also recognize products generated from damaged tissues, usually produced after interaction with hydrolytic enzymes secreted by phytopathogens, which are known as damage-associated molecular patterns (DAMPs). Cell wall fragments released by microbial xylanases, pectate lyases and endopolygalacturonases are classic examples of DAMPs (Darvill and Albersheim, 1984; Lotze *et al.*, 2007; Boller and Felix, 2009). Chemically pure oligogalacturonides and cutin monomers can also act as endogenous elicitors (Galletti *et al.*, 2008, 2011; Ferrari *et al.*, 2013). If recognition of PAMPs or/and DAMPs occurs on the plant cell surface, PAMP-triggered immunity (PTI) is activated. Usually, PTI is sufficient to prevent the invasion of the plant. However, virulent strains of phytopathogens have evolved mechanisms to suppress PTI by interfering with the recognition at the plasma membrane or by secreting effector proteins into the plant cell cytosol, altering the defence signalling and leading to effector-triggered susceptibility (ETS) and therefore to disease (Jones and Dangl, 2006). None the less, plants can still resist by activating the second branch of the innate immunity, which occurs at the cytosol, known as effector-triggered immunity (ETI). ETI is a yet more sophisticated mechanism that recognizes the effectors or its function by plant resistance (R) proteins, which in turn activate an R-protein-mediated resistance, suppressing the microbial growth (Jones and Dangl, 2006; Chisholm *et al.*, 2006). When a virulent effector is recognized by a plant R-protein, the former is known as an avirulence (Avr) factor.

The term elicitor is commonly applied to agents that can stimulate any type of defence response in plants, resulting in an enhanced resistance towards the invading pathogen (Nürnberger and Brunner, 2002). Elicitors of diverse chemical nature have been characterized including (poly)peptides, glycoproteins, lipids, glycolipids and oligosaccharides, which can all be derived from different phytopathogenic microbes (Jones and Dangl, 2006; Chalfoun *et al.*, 2011; Henry *et al.*, 2012; Wiesel *et al.*, 2014) or host plants (Bonas and Lahaye, 2002). The broad structural and functional diversity exhibited by elicitors validates the view that these molecules and the genes involved in their synthesis are the target of the evolutionary forces that drive the antagonistic interplay between pathogen and host (Kamoun, 2007).

Following PRR or R-protein-mediated recognition, a complex network of defence mechanisms is activated either locally or systemically, accompanied by many metabolic changes. MAMPs are able to stimulate a rapid and transient defence response characterized by ion fluxes (e.g. K^+, Ca^{2+}), the production of reactive oxygen species (ROS) (mainly H_2O_2 and $O_2^{\bullet-}$), mitogen-activated protein kinase signalling and the accumulation of antimicrobial compounds (Jones and Dangl, 2006). After these early events, the response continues bringing about an increase in the expression of defence-associated genes (i.e. *NPR1*) (Pieterse and van Loon, 2004), pathogenesis-related (PR) genes (Tsuda *et al.*, 2008; Vlot *et al.*, 2009) and deposition of callose and lignin to reinforce the cell wall at sites of infection, all of which contribute to hinder microbial growth (Nürnberger *et al.*, 2004). On the other hand, if Avr proteins are involved, they usually trigger a stronger defence reaction characterized primarily by rapid apoptotic cell death and local necrosis, known as a hypersensitive response (HR). When the plant recognizes an Avr protein and the reaction is enough to halt the pathogens intent to invade the cell, the interaction is called incompatible, and the pathogen behaves as avirulent, and the plant cultivar behaves as resistant. In contrast, if the plant does not recognize the Avr protein, no defence reaction takes place and the plant acquires the disease. In this case, the interaction is called compatible, and the pathogen behaves as virulent and the plant as a susceptible host (Chaturvedi and Shah, 2007).

The defence initial stimulus can subsequently be amplified by pathogen- or elicitor-induced secondary signal molecules synthesized by the plant, such as salicylic acid (SA), ethylene, and jasmonic acid (JA) and its derivatives. SA is a signal molecule in plant defence responses effective against biotrophic and hemibiotrophic pathogens (Glazebrook, 2005; Pieterse and Dicke, 2007; Koornneef and Pieterse, 2008; Vlot *et al.*, 2009). It is required for both local defence at the site of pathogen attack and for triggering a systemic response, known as systemic acquired resistance (SAR) (Malamy *et al.*, 1990; Métraux *et al.*, 1990; Delaney *et al.*, 1994; Mauch-Mani and Métraux, 1998). SAR is a secondary defence response occurring at distal plant tissues but is activated by local defence responses. SAR is long lasting and effective against a broad spectrum of pathogens, including fungi, bacteria and viruses (Zhang *et al.*, 2010). Finally, the plant can also activate different signalling pathways that include the synthesis of proteins such as chitinase, glucanase and lipid-transfer proteins, and secondary metabolites such as phytoalexins, with antimicrobial activity (Pieterse and Dicke, 2007). Two phytoanticipins present in strawberry leaves have been reported and were shown to exhibit not only antimicrobial properties but also defence activation activity (Filippone *et al.*, 1999; Mamaní *et al.*, 2012).

Anthracnose disease, caused by fungal species of the genus *Colletotrichum*, represents one of the major fungal diseases of strawberry (*Fragaria*×*ananassa* Duch.), affecting all tissues of the plant: fruits, flowers, leaves, runners, roots and crowns (Howard *et al.*, 1992; Freeman and Katan, 1997). There are three principal *Colletotrichum* spp. involved in the disease: *Colletotrichum acutatum* J.H. Simmonds, *Colletotrichum fragariae* A.N. Brooks and *Colletotrichum gloeosporioides* (Penz.) Penz. & Sacc. (Smith and Black, 1990; Howard *et al.*, 1992).

Chemical control of the disease relies on foliar application of fungicides. However, the use of fungicides often yields poor results when the weather conditions favour the disease development, or the treatment must be limited due to restrictions imposed on pesticide usage (Freeman *et al.*, 1997). Therefore, elicitor-induced plant resistance represents one sustainable strategy to develop new methods to control anthracnose. However, to achieve reliable results, it is necessary to understand the mechanisms activated in plants during the interaction with pathogens.

Using double infection experiments, it was reported that an avirulent isolate of *C. fragarie* (F7) provided strawberry plants of the cultivar 'Pájaro' with efficient protection against a virulent isolate of *C. acutatum* (M11); the protection was due to activation of a plant defence response (Salazar *et al.*, 2007). Later, it was demonstrated that the resistance acquired by strawberry plants treated with this avirulent isolate was of the SAR type and that it could be passed from mother to daughter plants through the runners (Salazar *et al.*, 2013). Recently, it was reported that this type of resistance could also be induced by other avirulent pathogens, namely: isolate M23 of *C. fragariae* (Chalfoun *et al.*, 2011) and isolate SS71 of *Acremonium strictum* (Chalfoun *et al.*, 2013), suggesting that avirulent fungal pathogens may use a common mechanism of activation of the defence response.

To analyse further the mechanisms involved in the establishment of a defence response against fungal pathogens in strawberries, the efforts of our group have been directed towards studying the molecular and biochemical bases of the defence signalling and towards identifying putative elicitor molecules.

5.2. Temporal Accumulation of Salicylic Acid Mediated by the Avirulent Strain M23 of *C. fragariae* in Strawberry

As described above, it is well known that strawberry plants can induce defence mechanisms against pathogens; however, little is known about phytohormone signalling pathways. Some reports have demonstrated that application of benzothiadiazole *S*-methyl ester (BTH), an analogue of SA (Eikemo *et al.*, 2003; Hukkanen *et al.*, 2007), or chitosan (Bhaskara Reddy *et al.*, 2000) to strawberries could induce resistance against pathogens.

It was reported that the content of SA in strawberry plants of cultivar 'Pájaro' inoculated with the avirulent isolate of *C. fragariae* (M23) increased by almost twofold with respect to control non-treated plants at 48 h post-treatment (p.t.) and decreased afterwards (Fig. 5.1). These plants also showed an early accumulation of ROS. In contrast, plants inoculated with a virulent isolate of *C. acutatum* (M11) did not exhibit an early accumulation of ROS after infection, and their SA content did not change over the time interval studied (Grellet-Bournonville *et al.*, 2012).

As no disease symptoms were observed in plants treated with the avirulent strain, it was suggested that an SA-mediated defence mechanism was taking place. This agrees with the results of studies in various other species, such as *Arabidopsis*, tobacco, cucumber and barley (Malamy *et al.*, 1990; Métraux *et al.*, 1990; Rasmussen *et al.*, 1991; Summermatter *et al.*, 1995; Delaney, 1997; Shah *et al.*, 1997; Vallelian-Bindschedler *et al.*, 1998).

The association between SA accumulation and activation of a defence response was confirmed indirectly with plants treated with the virulent isolate M11. In this case, plants that did not accumulate ROS and SA exhibited disease symptoms after 72 h p.t. (Grellet-Bournonville *et al.*, 2012).

5.2.1. Expression analysis of the *PR-1* gene

PR proteins have been defined as host proteins that are induced either local and systemically in response to an avirulent or virulent pathogen (van Loon and van Strien, 1999). SA accumulation often parallels or precedes the increase in expression of PR genes (An and Mou, 2011). In particular, expression of the gene *PR-1* is closely associated with SA accumulation in several plant species (Kawano *et al.*, 2004; An and Mou, 2011).

In strawberry plants, clear upregulation of the gene *FaPR1* was observed during the first 48 h p.t. with the avirulent isolate M23 but was abruptly downregulated after this time point. In contrast, the expression of *FaPR1* in M11 (virulent strain)-treated plants showed an increasing downregulation over the period studied (Grellet-Bournonville *et al.*, 2012) (Fig. 5.2).

These results provide clear evidence that accumulation of SA in strawberry triggers *PR-1* expression, and both events are probably required for the full establishment of the defence response and for plant protection. Furthermore, these outcomes strongly suggest that M23 is responsible for the activation of an SA-dependent defence mechanism in plants of the cultivar 'Pájaro'.

The fact that *FaPR1* expression was also downregulated after 48 h p.t. with M23 suggests that a time-dependent regulatory process is taking place. It is well known that SA exerts an antagonistic effect on jasmonate and

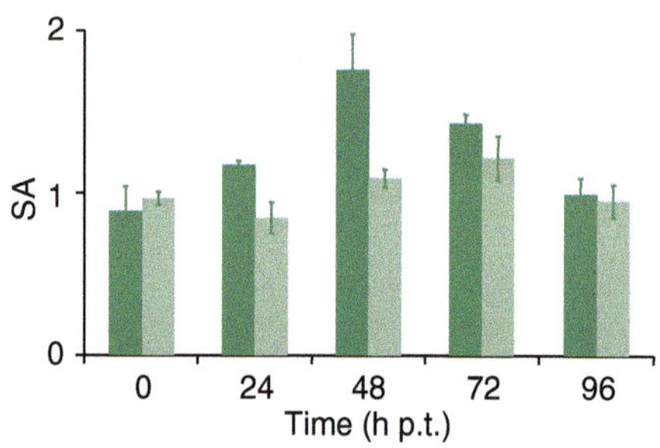

Fig. 5.1. Changes in salicylic acid (SA) content in strawberry plants of the cultivar 'Pájaro' inoculated with the fungal pathogen *C. fragariae* M23 (dark bars) or *C. acutatum* M11 (light bars) at different times post-treatment (h p.t.), relative to water-treated plants (control). Values represent the means ± standard deviation of three independent experimental replicates (n = 6).

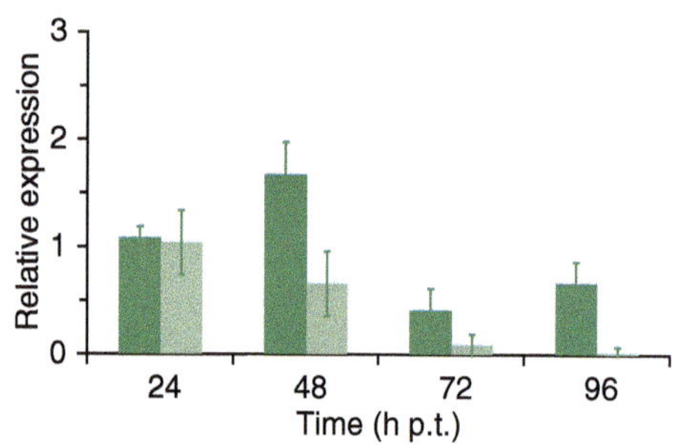

Fig. 5.2. Expression of the *FaPR1* gene at different times after treatment of plants with *C. fragariae* M23 (avirulent; dark bars) or *C. acutatum* M11 (virulent; light bars). Results are expressed as the ratio of the level of expression of the *FaPR1* gene in inoculated plants compared with control (water-treated) plants. Values represent the means ± standard deviation of four replicates and two independent experiments (n = 6).

ethylene signalling pathways (Pieterse *et al.*, 2009; Verhage *et al.*, 2010). Therefore, we may hypothesize that the activation of these signalling pathways could take place at 48 h p.t.

5.2.2. Expression analysis of the *PAL-3* gene

SA in plants can be generated via two distinct enzymatic pathways that require chorismate: the isochorismate synthase and the phenylalanine ammonia-lyase (PAL) pathways. Whereas the former is the main source of SA synthesis during SAR in plants, the latter is proposed to be responsible for the rapid production of SA associated with local cell death (Wildermuth *et al.*, 2001). However, there is evidence that supports an important role of the PAL pathway in pathogen-induced SA formation in many plant species (Elkind *et al.*, 1990; Meuwly *et al.*, 1995; Mauch-Mani and Slusarenko, 1996; Pallas *et al.*, 1996; Coquoz *et al.*, 1998; Huang *et al.*, 2010). Therefore, both SA biosynthetic pathways appear to participate in basal and pathogen-induced SA production, and the importance of PAL and isochorismate synthase for SA biosynthesis relies on the intermediates generated in both pathways (An and Mou, 2011).

Grellet-Bournonville *et al.* (2012) reported that the strawberry gene *FaPAL3* in M23-treated plants shows a particular expression profile, exhibiting a discrete upregulation during the first 72 h p.t. but becoming more widespread at 96 h p.t. (Fig. 5.3). They suggested that, although the first induction of *FaPAL3* could contribute to synthesis of SA at a very early stage of the induction, it is later downregulated by SA (Fig. 5.1); none the less, as soon as the content of SA decreases, the control over the PAL pathway is released.

A further explanation of this behaviour was proposed based on the fact that the M23 isolate of *C. fragariae* is a hemibiotrophic pathogen. The authors speculated that M23 is making a switch from the biotrophic to the necrotrophic phase after 48 h p.t.; this event is perceived by the plant and, with the aim of protecting itself, it induces the phenylpropanoid pathway to strengthen cell wall components (e.g. callose, lignin). In contrast, plants challenged with the virulent isolate M11 do not exhibit significant changes in *FaPAL3* expression over the period studied (Fig. 5.3). This hypothesis is also supported by the fact that *FaPAL3* induction at 96 h p.t. in M23-treated plants was almost fourfold that of plants infected with M11.

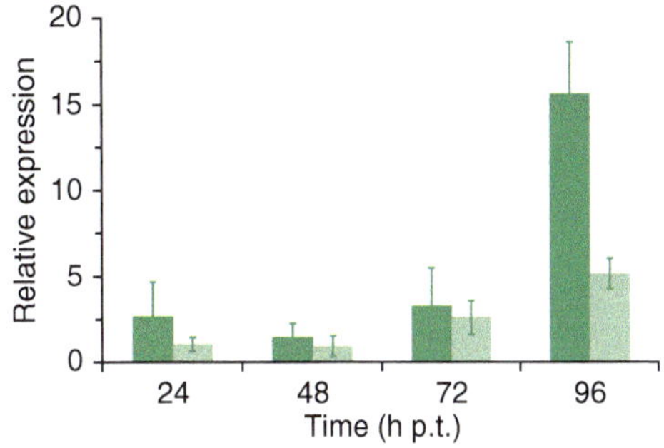

Fig. 5.3. Expression of the *FaPAL3* gene at different times after treatment with M23 (dark bars) or M11 (light bars). Values are expressed as the ratio of the level of expression of the *FaPAL3* gene in inoculated plants compared with control (water-treated) plants. Values represent the means ± standard deviation of four replicates and two independent experiments ($n = 6$).

5.3. Salicylic Acid-induced Protection

Previous reports have shown that application of exogenous SA or BTH activates expression of PR genes and induces disease resistance against viral, bacterial, oomycete and fungal pathogens in a variety of plant species (An and Mou, 2011).

In agreement with results obtained in infection experiments, Grellet-Bournonville *et al.* (2012) observed a clear induction of the *FaPR1* gene at 72 h p.t. when SA was applied exogenously (Fig. 5.4). In contrast, the *FaPAL3* gene was not upregulated at any time point evaluated (Fig. 5.4), suggesting that it is regulated through an SA-independent signalling pathway or that SA can actually downregulate the PAL-dependent SA

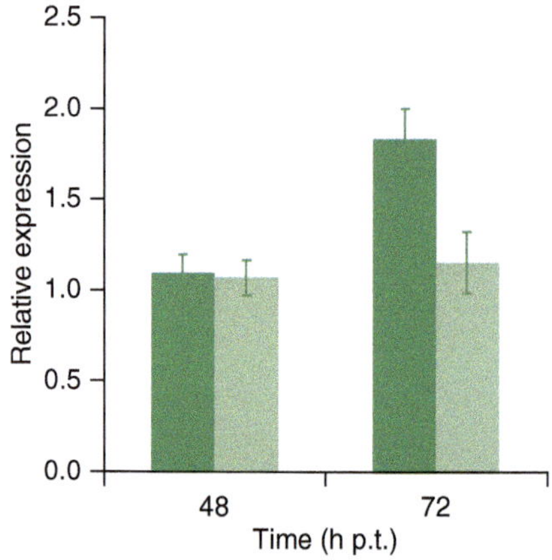

Fig. 5.4. Relative expression of *FaPR1* (dark bars) and *FaPAL3* (light bars) genes at different times after treating with salicylic acid. Values are expressed as the ratio of the level of expression of the genes *FaPR1* and *FaPAL3* in treated compared with non-treated plants (controls). Results correspond to means ± standard deviation of four replicates and two independent experiments (n = 6).

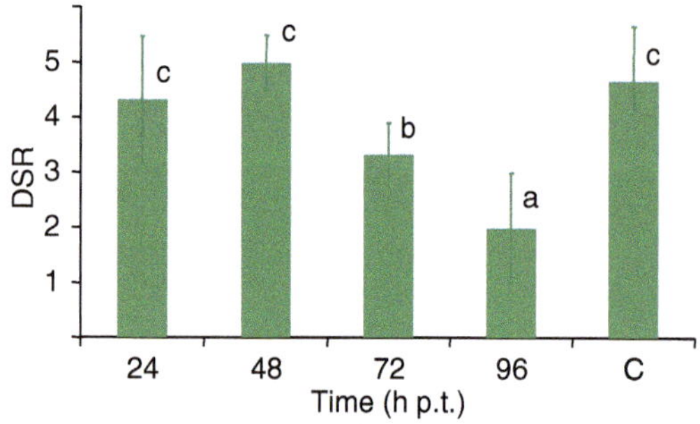

Fig. 5.5. Protection mediated by the ectopic application of SA to strawberry plants (cultivar 'Pájaro'). Susceptibility was measured as disease severity rating (DSR) of plants treated or not (C, control) with SA (5 mM) at different times before inoculation with M11. Control plants were treated with 10% methanol. DSR was evaluated over 30 days after inoculation with M11. Bars represent the means ± standard deviation of three independent experimental replicates (n = 6). Different lower-case letters above bars represent statistically different values (Fisher's test, $P < 0.05$).

pathway. However, the latter does not agree with previous reports showing that exogenous SA application could restore disease resistance in plant species in which the *PAL* gene was inhibited or mutated (Elkind *et al.*, 1990; Meuwly *et al.*, 1995; Mauch-Mani and Slusarenko, 1996; Pallas *et al.*, 1996; Coquoz *et al.*, 1998; Nawrath and Métraux, 1999; Wildermuth *et al.*, 2001; Nawrath *et al.*, 2002).

The effect of the exogenous application of SA on plant resistance/protection was also tested (Grellet-Bournonville *et al.*, 2012). The results showed that SA exerted an induced protection on strawberry plants towards the virulent isolate M11 and that the level of protection increased with time (Fig. 5.5). Similar results were reported by Eikemo *et al.* (2003) when inducing strawberry resistance to the pathogens *Phytophthora cactorum* and *Phytophthora fragariae* by BTH application; however, a gene expression study was not reported. An effect of time elapsed between plant treatment and challenge with the virulent pathogen was also observed by Salazar *et al.* (2007) when using an avirulent strain of *C. fragariae* before infecting the plants with a virulent strain of *C. acutatum*.

5.4. The Systemic Acquired Resistance Response

As expression of the *FaPR1* gene in strawberry plants treated with the avirulent isolate M23 was accompanied by accumulation of SA, Chalfoun *et al.* (2011) investigated the occurrence of SAR in strawberry plants of the cultivar 'Pájaro'; plants were treated in one isolated leaf with M23, and 96 h later the whole plant was treated with the virulent M11. The results showed that plants infected with M23 activated a systemic resistance, providing total protection against M11. The SAR was confirmed by analysing ROS, callose accumulation and the expression of *FaPR1* in distal tissues. Similar results were also obtained when the avirulent isolate SS71 of the opportunist strawberry fungal pathogen *A. strictum* (Racedo *et al.*, 2013) was used as the inducer of the defence and M11 as the challenging pathogen. Interestingly, these authors reported that conidial extracts of both avirulent isolates were also able to induce a systemic defence response and to grant protection against the virulent isolate M11 of *C. acutatum* (Chalfoun *et al.*, 2011, 2013).

The results presented in this section provide strong evidence that strawberry plants have an SA-signalling-associated defence mechanism that can be activated by avirulent fungal pathogens, and that the defence response is effective against the *C. acutatum* anthracnose causal agent.

Further investigations will be required to elucidate the events that precede and follow the accumulation of SA to explain and understand the behaviour of SA observed in plants challenged with avirulent pathogens.

5.5. Transcriptional Changes in Strawberry Leaves in Response to *A. strictum* Interaction

To investigate the genes involved in the plant defence response reported above, a differential display PCR technique was used to compare the expression profiles of strawberry leaves infected with the avirulent isolate SS71 of *A. strictum* and the virulent isolate M11 of *C. acutatum* at 48 h p.t. (Agüero *et al.*, 2003). This condition was chosen because infection assays had shown that, at this time point, plants treated previously with this isolate of *A. strictum* did not develop anthracnose disease symptoms, even when they were reinfected with a conidial suspension of the M11 isolate. By using a combination of arbitrary and anchoring primers, many expressed sequence tags (ESTs) were isolated. Among these, only 13 showed clear and consistent differential expression in the incompatible interaction with SS71 compared with the compatible interaction with M11. Based on their homology to sequences included in the National Center for Biotechnology Information protein database (http://www.ncbi.nlm.nih.gov/protein), putative identification and function prediction was possible. The 13 ESTs sequences encoded: a class tau glutathione *S*-transferase (GST), with roles in cellular protection against stress (Dixon *et al.*, 2002); two calmodulin-like (CML) proteins involved in sensing the cytoplasmic Ca^{2+} concentration (McCormack and Braam, 2003); a protein kinase HT1 that regulates stomatal movements in response to CO_2 (Hashimoto *et al.*, 2006); a THI1 enzyme involved in thiazole biosynthesis (Ribeiro *et al.*, 1996); an H^+-ATPase from the plasma membrane (Ohno *et al.*, 2004); an asparagine-rich protein expressed in endoplasmic reticulum stress (Ludwig and Tenhaken, 2001); a hybrid proline-rich protein from the cell walls (Dvořáková *et al.*, 2007); a chloroplast DGEP5 protease involved in the degradation of damaged proteins (Sun *et al.*, 2007); an enoyl-CoA hydratase-isomerase involved in defence against pathogens (Reumann *et al.*, 2007); a PR protein activated by BTH treatment (Bovie *et al.*, 2004); a chlorophyll *a*/*b*-binding protein (LHCII) involved in stress defence (Kerchev *et al.*, 2011); and a uridine/cytidine monophosphate kinase (Zhou *et al.*, 1998). Among these sequences, full-length cDNAs were obtained for the GST (FaGSTU1) and the two CML proteins (FaCML1 and FaCML2) were selected for further analysis.

GSTs are multifunctional proteins encoded by a large gene family involved in protecting cells from biotic and abiotic stress such as pathogen attack, heavy metal toxins, oxidative damage and ultraviolet radiation, and detoxification of xenobiotics (Frova, 2003). In plants, GSTs are divided into seven classes: theta, zeta, tau, phi, lambda, dehydroascorbate reductase and tetrachlorohydroquinone dehalogenase (Lan *et al.*, 2009). The proteins of the tau and phi classes are dimeric and catalyse the conjugation of endogenous compounds and herbicides (Edwards and Dixon, 2005). Each subunit consists of a conserved N-terminal glutathione-binding domain (G-site) and a variable C-terminal α-helical domain that contains the binding site for the hydrophobic substrate (H-site). The predicted FaGSTU1 subunit has 223 aa and shares 50% amino acid identity with LeGSTU3 from *Lycopersicon esculentum*. This protein was reported to be involved in the control of cell death induced by oxidative stress (Kilili *et al.*, 2004). The transcript level of *FaGSTU1* showed clear upregulation in the interaction with SS71, and was repressed during infection with the *C. acutatum* M11 isolate (Tonello *et al.*, 2013). In contrast, *FaCML1* and *FaCML2* exhibited similar expression patterns in response to both fungal isolates; the level of transcripts increased in leaves treated with the avirulent SS71 isolate, but was slightly downregulated when the infection was performed with the virulent M11 isolate (Tonello *et al.* 2013) (Fig. 5.6).

Ca^{2+} has a fundamental role as a secondary messenger during pathogen- and elicitor-triggered immunity in plants (Lecourieux *et al.*, 2006; Kudla *et al.*, 2010; Reddy *et al.*, 2011; Stael *et al.*, 2012). It has been reported that an increase in the cytosolic Ca^{2+} level is an essential signalling event during plant–pathogen interactions, and variations in its concentration are sensed by Ca^{2+}-sensor proteins such as CMLs (Hashimoto and Kudla, 2011). In plants, CML proteins have a variable number of EF-hand motifs (Grabarek, 2006). The binding of Ca^{2+} induces a conformational change in the protein that promotes its interaction with downstream effectors (Perochon *et al.*, 2011). The deduced sequences of the FaCML1 and FaCML2 proteins contained 125 and 157 aa and three and four EF-hand calcium-binding motifs, respectively (Tonello and Díaz Ricci, 2015). The protein *FaCML1* shares 79.6% amino acid identity with *FaCML2*, and the absence of the fourth EF-hand motif in *FaCML1* was

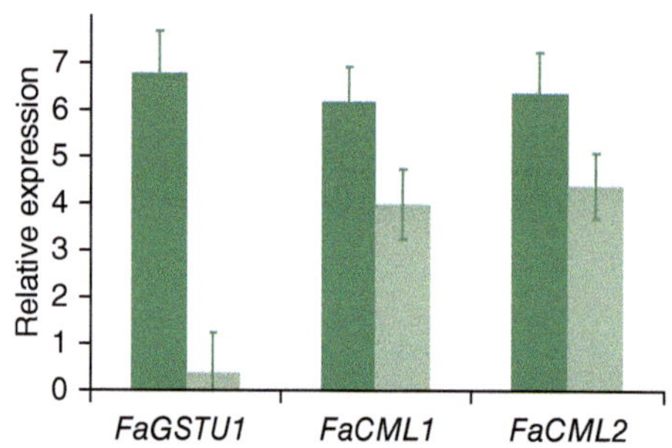

Fig. 5.6. Relative expression levels for the *FaGST1*, *FaCML1* and *FaCML2* genes in strawberry leaves of the cultivar 'Pájaro' at 48 h after treatment with the avirulent SS71 isolate of *A. strictum* (dark bars) and the virulent M11 isolate of *C. acutatum* (light bars) relative to control (water-treated) plants. Values represent the means ± standard deviation of four replicates and two independent experiments (n = 6).

due to a deletion of 96 nt in the transcript (Tonello and Díaz Ricci, 2015). Additionally, *FaCML2* shares 65% amino acid identity with AtCML24 from *Arabidopsis thaliana*, with a role in the pathogen perception signalling network leading to innate immunity (Ma *et al.*, 2008). The two transcripts encoding members of the CML protein family, *FaCML1* and *FaCML2*, were differentially expressed during the incompatible interaction of strawberry with *A. strictum* (Tonello and Díaz Ricci, 2015). Recent research has revealed the role of Ca^{2+}, CML proteins and another signalling molecule as nitric oxide (NO) in the development of HR in plants (Ma *et al.*, 2008). A loss-of-function mutant of *A. thaliana*, *Atcml24-4*, inhibited pathogen-induced NO generation and HR, suggesting that Ca^{2+} influx to the cytosol activates *AtCML24*, which in turn induces NO synthesis, leading to innate immunity in plants (Ma *et al.*, 2008).

These results revealed the participation of the *FaGSTU1*, *FaCML1* and *FaCML2* genes in the strawberry defence response during the interaction with the SS71 isolate of *A. strictum*. The expression analysis confirmed that these genes are upregulated at lower levels in plants infected with the virulent M11 isolate of *C. acutatum* and at higher levels in plants infected with the avirulent SS71 isolate. The latter is in concordance with the peak accumulation of SA and *FaPR1* reported previously and shown above (Grellet-Bournonville *et al.*, 2012). Taken together, these results strongly suggest an active role for these genes in the resistance to *Colletrotrichum* in strawberry. Plant GSTs are induced by diverse biotic and abiotic stimuli, and are important for protecting plants against oxidative damage. The upregulation of orthologous genes has been reported in several plant–pathogen interactions. A tomato GST, Bax-inhibitor (BI)-GST, has been identified as an inhibitor of Bax lethality in yeast, and also enhanced the resistance to ROS (Kilili *et al.*, 2004).

A proteomic analysis in strawberry seedlings at different times after infection with a virulent strain of *C. fragariae* showed noticeable changes in plant metabolism and redox balance, suggesting that, although it was a compatible interaction, activation of a defence response took place (Fang *et al.*, 2012). Although many proteins were analysed, the authors did not observe expression of GSTs or CML proteins. A possible explanation of these observations is that, as the interaction was compatible, the expression level of GSTs and CML proteins may have been very low (and undetectable with the approach used), as observed in strawberry plants infected with the virulent M11 isolate of *C. acutatum* (Fig. 5.6). Alternatively, it might be explained because the authors analysed protein levels rather than RNA synthesis. Nevertheless, a transcriptomic analysis of the genes differentially expressed in two strawberry cultivars infected with a virulent strain of *C. acutatum* carried out by Casado-Díaz *et al.* (2006) did not report detection of either GSTs or CML proteins. These authors showed that, although many genes were upregulated (i.e. *FaWRKY-1*, *FaLRRP-1*, *FaPR5-1* and *FaLPR10-1*) or downregulated (i.e. *FaβGLN-1*, *FaPROX-1*, *FagTHIO-1* and *FaCHIT-1*) in a complex manner, only one corresponded to a calcium-dependent protein kinase (GenBank accession no. AJ871791) and none corresponded to a GST (Casado-Díaz *et al.*, 2006). However, as these experiments were performed with a virulent strain of *C. acutatum*, perhaps the level of transcription of GSTs and CML proteins did not display any difference with respect to non-infected plants (controls).

As expression of GSTs was not observed in strawberry plants infected with virulent strains of *Colletotrichum* (Casado-Díaz *et al.*, 2006; Fang *et al.*, 2012), we may conclude that GSTs participate more actively in an incompatible interaction by controlling the oxidative stress and inducing plant defence mechanisms. The latter would include a plant cell death programme, usually regarded as an HR, that would contribute to enhance the resistance against virulent biotrophic or hemibiotrophic pathogens, as is the case with the fungus *Colletotrichum*.

5.6. Defence Elicitor Produced by *A. strictum*

The induction of a defence response mediated by an avirulent pathogen results from recognition of elicitors, many of which have been identified, although only a few have been characterized biochemically (Nürnberger *et al.*, 2004; Kamoun, 2007). Recently a new elicitor protein, designated AsES, produced by the avirulent SS71 isolate of *A. strictum*, was identified and characterized (Chalfoun *et al.*, 2013). The protein corresponded to a subtilisin of 34 kDa and pI 8.8. Chalfoun *et al.* (2013) reported that the AsES protein not only induced a local defence response in treated leaf but also induced systemic protection against the virulent M11 isolate of *C. acutatum*, the causal agent of anthracnose (Fig. 5.7).

The complete cDNA sequence for the AsES protein was obtained, showing a gene of 1167 nt. The deduced protein sequence demonstrated a high amino acid similarity with other extracellular serine proteases produced by several fungal species, and belonged to the subfamily of proteinase K-like subtilisins (S8A) (Chalfoun *et al.*, 2013). Subtilisin-like proteinases are considered to be important virulence factors in the infection process of entomopathogenic, nematophagous and mycoparasitic fungi (Wang *et al.*, 2009; Yang *et al.*, 2011), but little is known about those produced by phytopathogenic fungi. It has been reported that proteins of the family of subtilases exhibit the ability to activate the innate immune system in *Arabidopsis* (Rautengarten *et al.*, 2005; Ramírez *et al.*, 2013).

Although the AsES protein exhibited high similarity with subtilisins of many microorganisms, it nevertheless showed a relatively low similarity (<45%) to two serine proteases previously reported for *Acremonium* spp. (Liu *et al.*, 2007). These proteases, AS-E1 and AS-E2, which are 34.4 and 32 kDa, respectively, were biochemically characterized, and it was shown that they were able to proteolytically activate prothrombin to meizothrombin (desF1)-like molecules and inhibit plasma

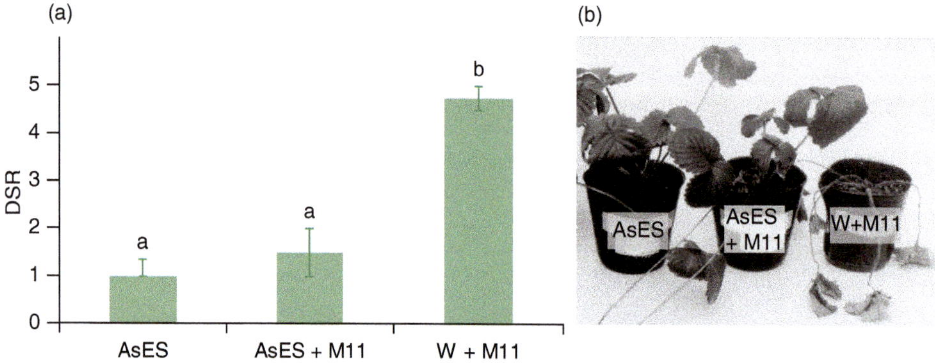

Fig. 5.7. Plant protection elicited by the protein AsES against anthracnose in strawberry plants of the cultivar 'Pájaro'. (a) Susceptibility was evaluated as disease severity rating (DSR) at 21 days p.t. on plants sprayed on one leaf with: AsES only (AsES), AsES and with the virulent isolate M11 of *C. acutatum* (AsES+M11), and with water followed by infection with M11 7 days later (W+M11). Bars represent the means ± standard deviation of three independent experimental replicates (n = 6). Different lower-case letters above bars represent statistically different values (Fisher's test, $P < 0.05$). (b) Examples of strawberry plants treated as described in (a) at 21 days p.t.

clotting (Liu *et al.*, 2007). However, there was no mention of their ability to activate plant defences, and only their N-terminal regions were sequenced.

Bioinformatic analysis has suggested that AsES is synthesized as a 388 aa protein precursor, comprising a signal peptide of 15 aa and two domains that are highly conserved in precursors of subtilisin-like serine proteases: a peptidase inhibitor I9 domain, and a peptidase S8 domain (Fig. 5.8). Maturation would involve two proteolytic steps: first, cleavage of the secretory signal peptide, which directs the protein to the extracellular medium, and secondly, removal of the peptidase inhibitor I9 domain. In this way, the mature 283 aa active protein would be obtained, which, in addition to its plant defence-eliciting activity, retains its protease activity *in vitro*.

It has been reported that AsES can trigger a strong defence reaction in strawberry plants, which is characterized by an early transient oxidative burst (see below), followed by the accumulation of SA and the upregulation of defence-related genes such as *FaPR1* and *FaChi2-1* (an SA-dependent class II chitinase) (Chalfoun *et al.*, 2013). The systemic characteristic of the resistance induced by AsES was tested by inoculating a single isolated leaf of strawberry plants (cultivar 'Pájaro') and then challenging the whole plant with the virulent hemibiotrophic isolate M11 of *C. acutatum* or treating the same leaf with the virulent necrotrophic isolate B1 of *Botrytis cinerea*. The results showed that plants acquired a systemic resistance against the former isolate (Fig. 5.7). Additional studies revealed that AsES can also confer protection to different strawberry cultivars against different virulent isolates of *C. acutatum* (Chalfoun *et al.*, 2013). The latter was further confirmed with plants of *Arabidopsis* treated with AsES, where accumulation of ROS (e.g. H_2O_2 and $O_2^{\bullet-}$) and callose was also observed, further suggesting that the protein may act as a PAMP elicitor.

Experiments with AsES-treated plants, when the elicitor was previously irreversibly inhibited with PMSF (phenylmethanesulphonyl fluoride), suggested that the protease activity of AsES would be required for the defence elicitation (Chalfoun *et al.*, 2013). Thus, the question about whether other proteases of the family would also exhibit a defence-eliciting character was addressed in experiments with plants treated with a protease of the same family as the proteinase K. The results showed that the eliciting activity was exhibited only by AsES. These results disagree with previous experiments carried out with an endoxylanase

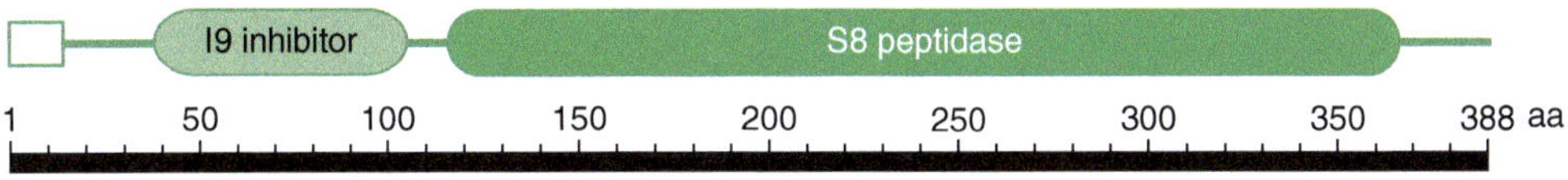

Fig. 5.8. Scheme showing the structure of the protein AsES. The secretory signal peptide is indicated by an open square.

from *Trichoderma viride*, which stimulated HR, ethylene and phytoalexin production in tobacco and tomato (Sharon *et al.*, 1993). The eliciting activity of the endoxylanase was independent of enzyme activity. In a few cases, the defence-eliciting activity was found to be determined by small fragments of the intact elicitor molecule of PAMP, suggesting the recognition of 'epitope'-like structures by receptors at the plant cell surface (de Wit, 1997). Examples of the latter are: an 8 aa glycopeptide fragment derived from yeast invertase (gp 8c) (Basse *et al.*, 1993), and an internal peptide of 13 aa (Pep-13) derived from the 42 kDa cell wall glycoprotein of *Phytophthora sojae*, which exhibits Ca^{2+}-dependent transglutaminase activity (Sacks *et al.*, 1995; Brunner *et al.*, 2002). In fungi, defence-inducing activity was detected in a single group of proteases (Chisholm *et al.*, 2006). The cultivar-specific *Avr-Pita* gene from *Magnaporthe oryzae* encodes a secreted pre-protein with homology to fungal zinc-dependent neutral metalloproteases (Valent and Chumley, 1994), and point mutations in the putative protease catalytic residues resulted in gain of virulence in rice cultivars carrying the cognate *R* gene *Pi-ta*, although direct biochemical evidence for the protease activity of Avr-Pita is still required (de Wit, 1997; Orbach *et al.*, 2000).

This scenario led Chalfoun *et al.* (2013) to propose that AsES may function by releasing an active elicitor from a plant precursor molecule, rather than being an elicitor itself. The latter has been described for the gene *avrD* from *Pseudomonas syringae* pv. *tomato*, whose protein product is responsible for the generation of syringolide elicitors (Keen *et al.*, 1990) and for some peptides such as systemin, hydroxyproline-containing glycopeptides and rapid alkalinization inducing factor, which appear to come from protein precursors constitutively present in plant cell walls or the cytoplasm and are activated by proteases following cell injury, acting as elicitors of the DAMP type (Yamaguchi and Huffaker, 2011). As elicitor activity was found only in AsES from the *A. strictum* SS71 isolate and was not found in a homologous subtilisin, Chalfoun *et al.* (2013) concluded that the proteolytic activity is necessary but is not sufficient to induce defence, and suggested that AsES might induce defence by means of proteolysis of one or multiple host proteins that are specific targets of this protease. Further experiments using recombinant AsES protein, site-directed mutants or synthetic peptides derived from AsES are necessary to elucidate its defence activation mechanism.

AsES can be considered, therefore, a new member of the fungal protein elicitors, and its discovery is relevant, as it may contribute to the development of possible strategies for controlling diseases in the field.

5.7. Early Reactive Oxygen Species Response of Strawberry Cells to the AsES Elicitor Protein

In plants, ROS are continuously produced as by-products of various metabolic pathways localized in different cellular compartments (Foyer *et al.*, 1994), predominantly in chloroplasts, mitochondria and peroxisomes (Apel and Hirt, 2004). One of the earliest events in the defence against pathogen attack is oxidative burst, characterized by the production of ROS, mainly $O_2^{\bullet-}$ and H_2O_2 (Apostol *et al.*, 1989; Mittler *et al.*, 2011). Whereas H_2O_2 can translocate through the cell membrane through aquaporins (Dynowski *et al.*, 2008; Jang *et al.*, 2012), the superoxide anion $O_2^{\bullet-}$ cannot because it is reactive in a hydrophobic environment such as the interior of the cell membrane. Although the effect of the oxidative burst in plants can be dependent on tissue localization, timing and the level of ROS attained (Mittler, 2002; De Gara *et al.*, 2003), after the induction of a defence response, $O_2^{\bullet-}$ and H_2O_2 are found inside and outside of the cells. ROS can exert two opposite but not mutually exclusive effects: they can activate pathways aimed at protecting the cell from oxidative damage (Pastore *et al.*, 1999; Pastore *et al.*, 2000; Di Cagno *et al.*, 2001; Mittler, 2002; Vranová *et al.*, 2002), and/or they can impair the cellular redox balance inducing programmed cell death (van Breusegem and Dat, 2006). Therefore, as ROS accumulation is required for the onset of the defence responses, a tight control of ROS production and scavengers is required to avoid unwanted effects on cell metabolism (Foyer *et al.*, 1994).

As described above, Chalfoun *et al.* (2013) reported that strawberry plants treated with the elicitor AsES exhibited an early and transient ROS accumulation that was observed at 4 h p.t. measured as H_2O_2 using two different dyes: diaminobenzidine and 2′,7′-dichlorodihydrofluorescein diacetate (H_2DFFDA). The latter is interesting because, whereas diaminobenzidine detects intra- and extracellular H_2O_2, the fluorescent probe H_2DFFDA only detects intracellular H_2O_2 (Fig. 5.9). This result agrees with previous reports indicating that strawberry plants treated with the avirulent *Colletotrichum* isolates F7 and M23, and with SS71 of *A. strictum*, also exhibited a notorious accumulation of H_2O_2 and $O_2^{\bullet-}$ between 6 and 8 h p.t. (Salazar *et al.*, 2007; Grellet-Bournonville *et al.*, 2012), but disagrees with others who proposed that a biphasic production of apoplastic ROS is a hallmark of successful recognition of plant pathogens (Lamb and Dixon, 1997; Grant *et al.*, 2000; Nürnberger *et al.*, 2004).

To address this issue, and because in previous experiments the detection of ROS was performed histochemically in plant leaves (Fig. 5.9), analyses were conducted on disaggregated leaf tissue to evaluate with more precision the evolution of H_2O_2 in strawberry (cultivar 'Pájaro') cells treated with AsES (Martos *et al.*, 2012). By using the fluorescent probe H_2DFFDA, Martos *et al.* (2012) observed that when the cell suspension was treated with AsES, the accumulation of H_2O_2 was biphasic, with a first peak at 2 h p.t., and another at 7 h p.t. (Fig. 5.10). This outcome thus supports the proposed hallmark of a successful plant–pathogen interaction described above (Lamb and Dixon, 1997; Grant *et al.*, 2000; Nürnberger *et al.*, 2004), although in plant leaves a single event of oxidative burst was detected, as reported also by Ponce de León and Montesano (2013), and Iakimova *et al.* (2013) in *Physcomitrella patens* and *Malus domestica* leaves, respectively.

5.8. Conclusions and Future Perspective

The development of alternative strategies for the biocontrol of diseases requires an understanding of the mechanisms induced in plants during their interaction with

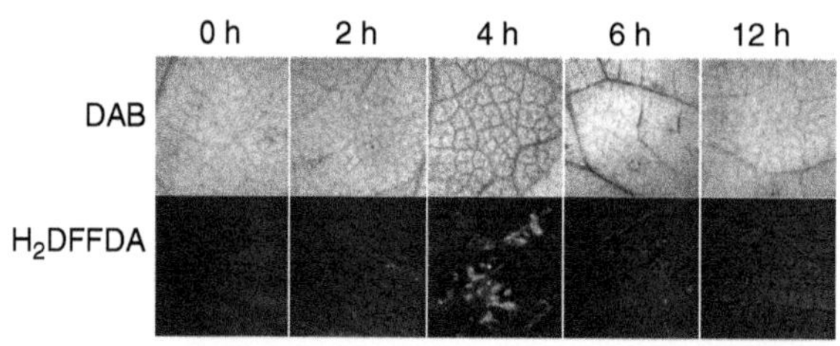

Fig. 5.9. H_2O_2 detected with diaminobenzidine (DAB) and the fluorescent probe 2′,7′-dichlorodihydrofluorescein diacetate (H_2DFFDA) (excitation wavelength 485 nm) in strawberry leaves of cultivar 'Pájaro' treated with AsES at different times post-treament.

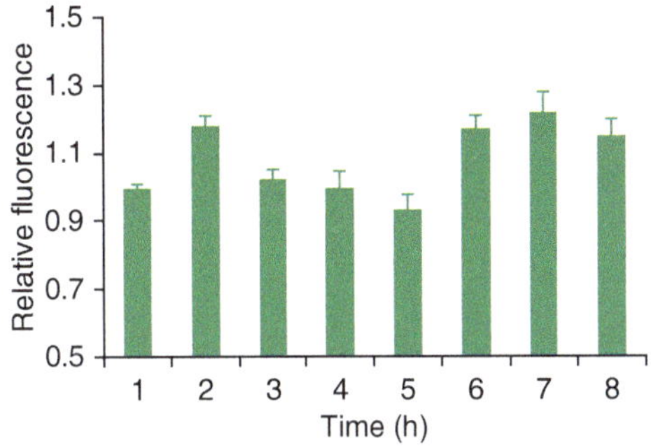

Fig. 5.10. Evolution of intracellular H_2O_2 detected in strawberry cell suspensions (cultivar 'Pájaro') treated with the protein AsES, using the H_2DFFDA fluorescent probe (λ_{ex} = 485 nm, λ_{em} = 525 nm). Values are expressed as the ratio of the fluorescence intensity of cells treated with AsES with respect to the fluorescence produced by untreated cells, at different times. Mean values ± standard deviation of two independent experiments (*n* = 6) are presented.

pathogens. In this chapter, we have reviewed the advances obtained by our group regarding the induction of the defence response and SAR in strawberry plants mediated by two avirulent fungal isolates of *C. fragariae* and *A. strictum*. Evaluation of plant susceptibility showed that plants treated with both isolates acquired a strong resistance against the virulent M11 isolate of *C. acutatum*, the causal agent of anthracnose disease. Plants treated with the avirulent strains exhibited an early oxidative burst, detected as the accumulation of H_2O_2 and $O_2^{\bullet -}$, which was followed by an accumulation of SA, callose and lignin. The upregulation of two genes related to SAR: *FaPR1*, an SA-regulated PR protein, and *FaPAL3*, an enzyme involved in the phenylpropanoid pathway and the synthesis of many defence-related secondary metabolites including SA, was observed. Upregulation of the *FaPR1* gene during the first 48 h p.t. was correlated with the increase in SA in the phloem sap of infected plants. At 48 h p.t. upregulation of the transcripts of two CML genes (*FaCML1* and *FaCML2*) and a GST gene (*FaGST*), which are critical regulators of defence responses, were also observed. Finally, characterization of the fungal elicitor AsES obtained from cultures of *A. strictum* is reported. Plants treated with this elicitor exhibited a similar behaviour to those treated with the avirulent isolates.

The results presented in this chapter confirm that strawberry plants exposed to avirulent pathogens induce an efficient defence mechanism that endows plants not only with a suitable artillery to cope with the stress provoked by the attack of a virulent pathogen but also, and most importantly, with an effective protection against it. The time course analysis reported revealed that the defence response activated by the avirulent pathogens takes place in at least two stages: a first stage in which the SA level is high, and a second stage in which the SA level is low; and that the switch takes place precisely when the SA level decreases. This scenario provides a new mode to interpret how the defence response proceeds when strawberry plants are challenged with an avirulent fungal pathogen. The latter suggests therefore that the signalling pathway activating the defence response can change with time according to the level of SA attained.

Although these outcomes contribute to our understanding of how strawberry plants defend themselves against pathogens, they represent only a small step forward in the more extensive issue of plant defence responses. Hence, more work is needed to articulate the valuable but disperse biochemical and molecular data available. We also expect that the availability of a plant defence elicitor such as the AsES protein, obtained from an avirulent fungus, will help us to understand the early events involved in plant–pathogen interactions and the activation of the defence response.

References

Agüero, T., Tonello, U., Salazar, S., Castagnaro, A. and Díaz Ricci, J.C. (2003) Identification of strawberry genes envolved in the defensive response against *Colletotrichum* sp. *Bio Cell* 27, 1.

An, C. and Mou, Z. (2011) Salicylic acid and its function in plant immunity. *Journal of Integrative Plant Biology* 53, 412–428.

Apel, K. and Hirt, H. (2004) Reactive oxygen species: metabolism, oxidative stress, and signal transduction. *Annual Review of Plant Biology* 55, 373–399.

Apostol, I., Heinstein, P.F. and Low, P.S. (1989) Rapid stimulation of an oxidative burst during elicitation of cultured plant cells: role in defense and signal transduction. *Plant Physiology* 90, 109–116.

Basse, C.W., Fath, A. and Boller, T. (1993) High affinity binding of a glycopeptide elicitor to tomato cells and microsomal membranes and displacement by specific glycan suppressors. *Journal of Biological Chemistry* 268, 14724–14731.

Bhaskara Reddy, M.V., Belkacemi, K., Corcuff, R., Castaigne, F. and Arul, J. (2000) Effect of pre-harvest chitosan sprays on post-harvest infection by *Botrytis cinerea* and quality of strawberry fruit. *Postharvest Biology and Technology* 20, 39–51.

Boller, T. and Felix, G. (2009) A renaissance of elicitors: perception of microbe-associated molecular patterns and danger signals by pattern-recognition receptors. *Annual Review of Plant Biology* 60, 379–406.

Bonas, U. and Lahaye, T. (2002) Plant disease resistance triggered by pathogen-derived molecules: refined models of specific recognition. *Current Opinion in Microbiology* 5, 44–50.

Bovie, C., Ongena, M., Thonart, P. and Dommes, J. (2004) Cloning and expression analysis of cDNAs corresponding to genes activated in cucumber showing systemic acquired resistance after BTH treatment. *BMC Plant Biology* 4, 15.

Brunner, F., Rosahl, S., Lee, J., Rudd, J.J., Geiler, C., Kauppinen, S., Rasmussen, G., Scheel, D. and Nürnberger, T. (2002) Pep-13, a plant defense-inducing pathogen-associated pattern from *Phytophthora transglutaminases*. *EMBO Journal* 21, 6681–6688.

Casado-Díaz, A., Encinas-Villarejo, S., Santos, B.d.l., Schilirò, E., Yubero-Serrano, E.-M., Amil-Ruíz, F., Pocovi, M.I., Pliego-Alfaro, F., Dorado, G., Rey, M., Romero, F., Muñoz-Blanco, J. and Caballero, J.-L. (2006) Analysis of strawberry genes differentially expressed in response to *Colletotrichum* infection. *Physiologia Plantarum* 128, 633–650.

Chalfoun, N.R., Castagnaro, A.P. and Díaz Ricci, J.C. (2011) Induced resistance activated by a culture filtrate derived from an avirulent pathogen as a mechanism of biological control of anthracnose in strawberry. *Biological Control* 58, 319–329.

Chalfoun, N.R., Grellet-Bournonville, C.F., Martínez-Zamora, M.G., Díaz-Perales, A., Castagnaro, A.P. and Díaz-Ricci, J.C. (2013) Purification and characterization of AsES protein: a subtilisin secreted by *Acremonium strictum* is a novel plant defence elicitor. *Journal of Biological Chemistry* 288, 14098–14113.

Chaturvedi, R. and Shah, J. (2007) Salicylic acid in plant disease resistance. In: Hayat, S. and Ahmad, A. (eds) *Salicylic Acid: a Plant Hormone*. Springer, Amsterdam, the Netherlands, pp. 335–370.

Chisholm, S.T., Coaker, G., Day, B. and Staskawicz, B.J. (2006) Host–microbe interactions: shaping the evolution of the plant immune response. *Cell* 124, 803–814.

Coquoz, J.-L., Buchala, A. and Métraux, J.-P. (1998) The biosynthesis of salicylic acid in potato plants. *Plant Physiology* 117, 1095–1101.

Darvill, A.G. and Albersheim, P. (1984) Phytoalexins and their elicitors – a defense against microbial infection in plants. *Annual Review of Plant Physiology* 35, 243–275.

De Gara, L., de Pinto, M.C. and Tommasi, F. (2003) The antioxidant systems vis-à-vis reactive oxygen species during plant–pathogen interaction. *Plant Physiology and Biochemistry* 41, 863–870.

de Wit, P.J.G.M. (1997) Pathogen avirulence and plant resistance: a key role for recognition. *Trends in Plant Science* 2, 452–458.

Delaney, T.P. (1997) Genetic dissection of acquired resistance to disease. *Plant Physiology* 113, 5–12.

Delaney, T.P., Uknes, S., Vernooij, B., Friedrich, L., Weymann, K., Negrotto, D., Gaffney, T., Gut-Rella, M., Kessmann, H., Ward, E. and Ryals, J. (1994) A central role of salicylic acid in plant disease resistance. *Science* 266, 1247–1250.

Di Cagno, R., Guidi, L., De Gara, L. and Soldatini, G.F. (2001) Combined cadmium and ozone treatments affect photosynthesis and ascorbate-dependent defences in sunflower. *New Phytologist* 151, 627–636.

Dixon, D.P., Lapthorn, A. and Edwards, R. (2002) Plant glutathione transferases. *Genome Biology* 3, 3004.1–3004.10.

Dvořáková, L., Cvrčková, F. and Fischer, L. (2007) Analysis of the hybrid proline-rich protein families from seven plant species suggests rapid diversification of their sequences and expression patterns. *BMC Genomics* 8, 412.

Dynowski, M., Schaaf, G., Loque, D., Moran, O. and Ludewig, U. (2008) Plant plasma membrane water channels conduct the signalling molecule H_2O_2. *Biochemical Journal* 414, 53–61.

Edwards, R. and Dixon, D.P. (2005) Plant glutathione transferases. In: Helmut, S. and Lester, P. (eds) *Methods in Enzymology*. Academic Press, New York, pp. 169–186.

Eikemo, H., Stensvand, A. and Tronsmo, A.M. (2003) Induced resistance as a possible means to control diseases of strawberry caused by *Phytophthora* spp. *Plant Disease* 87, 345–350.

Elkind, Y., Edwards, R., Mavandad, M., Hedrick, S.A., Ribak, O., Dixon, R.A. and Lamb, C.J. (1990) Abnormal plant development and down-regulation of phenylpropanoid biosynthesis in transgenic tobacco containing a heterologous phenylalanine ammonia-lyase gene. *Proceedings of the National Academy of Sciences USA* 87, 9057–9061.

Fang, X., Chen, W., Xin, Y., Zhang, H., Yan, C., Yu, H., Liu, H., Xiao, W., Wang, S., Zheng, G., Liu, H., Jin, L., Ma, H. and Ruan, S. (2012) Proteomic analysis of strawberry leaves infected with *Colletotrichum fragariae*. *Journal of Proteomics* 75, 4074–4090.

Ferrari, S., Savatin, D.V., Sicilia, F., Gramegna, G., Cervone, F. and De Lorenzo, G., (2013) Oligogalacturonides: plant damage-associated molecular patterns and regulators of growth and development. *Frontiers in Plant Science* 4, 49.

Filippone, M.P., Díaz Ricci, J., Mamaní de Marchese, A., Farías, R.N. and Castagnaro, A. (1999) Isolation and purification of a 316 Da preformed compound from strawberry (*Fragaria ananassa*) leaves active against plant pathogens. *FEBS Letters* 459, 115–118.

Foyer, C.H., Descourvières, P. and Kunert, K.J. (1994) Protection against oxygen radicals: an important defence mechanism studied in transgenic plants. *Plant, Cell & Environment* 17, 507–523.

Freeman, S. and Katan, T. (1997) Identification of *Colletotrichum* species responsible for anthracnose and root necrosis of strawberry in Israel. *Phytopathology* 87, 516–521.

Freeman, S., Nizani, Y., Dotan, S., Even, S. and Sando, T. (1997) Control of *Colletotrichum acutatum* in strawberry under laboratory, greenhouse, and field conditions. *Plant Disease* 81, 749–752.

Frova, C. (2003) The plant glutathione transferase gene family: genomic structure, functions, expression and evolution. *Physiologia Plantarum* 119, 469–479.

Galletti, R., Denoux, C., Gambetta, S., Dewdney, J., Ausubel, F.M., de Lorenzo, G. and Ferrari, S. (2008) The AtrbohD-mediated oxidative burst elicited by oligogalacturonides in *Arabidopsis thaliana* is dispensable for the activation of defense responses effective against *Botrytis cinerea*. *Plant Physiology* 148, 1695–1706.

Galletti, R., Ferrari, S. and De Lorenzo, G. (2011) Arabidopsis MPK3 and MPK6 play different roles in basal and oligogalacturonide- or flagellin-induced resistance against *Botrytis cinerea*. *Plant Physiology* 157, 804–814.

Glazebrook, J. (2005) Contrasting mechanisms of defense against biotrophic and necrotrophic pathogens. *Annual Review of Phytopathology* 43, 205–227.

Grabarek, Z. (2006) Structural basis for diversity of the EF-hand calcium binding proteins. *Journal of Molecular Biology* 359, 509–525.

Grant, M., Brown, I., Adams, S., Knight, M., Ainslie, A. and Mansfield, J. (2000) The *RPM1* plant disease resistance gene facilitates a rapid and sustained increase in cytosolic calcium that is necessary for the oxidative burst and hypersensitive cell death. *The Plant Journal* 23, 441–450.

Grellet-Bournonville, C.F., Martínez-Zamora, M.G., Castagnaro, A.P. and Díaz-Ricci, J.C. (2012) Temporal accumulation of salicylic acid activates the defense response against *Colletotrichum* in strawberry. *Plant Physiology and Biochemistry* 54, 10–16.

Hammond-Kosack, K.E. and Parker, J.E. (2003) Deciphering plant–pathogen communication: fresh perspectives for molecular resistance breeding. *Current Opinion in Biotechnology* 14, 177–193.

Hashimoto, K. and Kudla, J. (2011) Calcium decoding mechanisms in plants. *Biochimie* 93, 2054–2059.

Hashimoto, M., Negi, J., Young, J., Israelsson, M., Schroeder, J.I. and Iba, K. (2006) *Arabidopsis* HT1 kinase controls stomatal movements in response to CO_2. *Nature Cell Biology* 8, 391–397.

He, P., Shan, L. and Sheen, J. (2007) Elicitation and suppression of microbe-associated molecular pattern-triggered immunity in plant–microbe interactions. *Cellular Microbiology* 9, 1385–1396.

Henry, G., Thonart, P. and Ongena, M. (2012) PAMPs, MAMPs, DAMPs and others: an update on the diversity of plant immunity elicitors. *Biotechnology, Agronomy, Society and Environment* 16, 257–268.

Howard, C.M., Maas, J.L., Chandler, C.L. and Albregts, E.A. (1992) Anthracnose of strawberry caused by the *Colletotrichum* complex in Florida. *Plant Disease* 76 976–981.

Huang, J., Gu, M., Lai, Z., Fan, B., Shi, K., Zhou, Y.-H., Yu, J.-Q. and Chen, Z. (2010) Functional analysis of the *Arabidopsis PAL* gene family in plant growth, development, and response to environmental stress. *Plant Physiology* 153, 1526–1538.

Hukkanen, A.T., Kokko, H.I., Buchala, A.J., McDougall, G.J., Stewart, D., Kärenlampi, S.O. and Karjalainen, R.O. (2007) Benzothiadiazole induces the accumulation of phenolics and improves resistance to powdery mildew in strawberries. *Journal of Agricultural and Food Chemistry* 55, 1862–1870.

Iakimova, E.T., Sobiczewski, P., Michalczuk, L., Węgrzynowicz-Lesiak, E., Mikiciński, A. and Woltering, E.J. (2013) Morphological and biochemical characterization of *Erwinia amylovora*-induced hypersensitive cell death in apple leaves. *Plant Physiology and Biochemistry* 63, 292–305.

Jang, J.Y., Rhee, J.Y., Chung, G.C. and Kang, H. (2012) Aquaporin as a membrane transporter of hydrogen peroxide in plant response to stresses. *Plant Signaling and Behavior* 7, 1180–1181.

Jones, J.D.G. and Dangl, J.L. (2006) The plant immune system. *Nature* 444, 323–329.

Kamoun, S. (2007) Groovy times: filamentous pathogen effectors revealed. *Current Opinion in Plant Biology* 10, 358–365.

Kawano, T., Furuichi, T. and Muto, S. (2004) Controlled salicylic acid levels and corresponding signaling mechanisms in plants. *Plant Biotechnology* 21, 319–335.

Keen, N., Tamaki, S., Kobayashi, D., Gerhold, D., Stayton, M., Shen, H., Gold, S., Lorang, J., Thordal-Christensen, H. and Dahlbeck, D. (1990) Bacteria expressing avirulence gene D produce a specific elicitor of the soybean hypersensitive reaction. *Molecular Plant–Microbe Interactions* 3, 112–121.

Kerchev, P.I., Pellny, T.K., Vivancos, P.D., Kiddle, G., Hedden, P., Driscoll, S., Vanacker, H., Verrier, P., Hancock, R.D. and Foyer, C.H. (2011) The transcription factor ABI4 is required for the ascorbic acid-dependent regulation of growth and regulation of jasmonate-dependent defense signaling pathways in *Arabidopsis*. *The Plant Cell* 23, 3319–3334.

Kilili, K.G., Atanassova, N., Vardanyan, A., Clatot, N., Al-Sabarna, K., Kanellopoulos, P.N., Makris, A.M. and Kampranis, S.C. (2004) Differential roles of tau class glutathione *S*-transferases in oxidative stress. *Journal of Biological Chemistry* 279, 24540–24551.

Koornneef, A. and Pieterse, C.M.J. (2008) Crosstalk in defense signaling. *Plant Physiology* 146, 839–844.

Kudla, J., Batistic, O. and Hashimoto, K. (2010) Calcium signals: The LEAD currency of plant information processing. *The Plant Cell* 22, 541–563.

Lamb, C. and Dixon, R.A. (1997) The oxidative burst in plant disease resistance. *Annual Review of Plant Physiology and Plant Molecular Biology* 48, 251–275.

Lan, T., Yang, Z.-L., Yang, X., Liu, Y.-J., Wang, X.-R. and Zeng, Q. Y. (2009) Extensive functional diversification of the *Populus* glutathione *S*-transferase supergene family. *The Plant Cell* 21, 3749–3766.

Lecourieux, D., Ranjeva, R. and Pugin, A. (2006) Calcium in plant defence-signalling pathways. *New Phytologist* 171, 249–269.

Liu, C., Matsushita, Y., Shimizu, K., Makimura, K. and Hasumi, K. (2007) Activation of prothrombin by two subtilisin-like serine proteases from *Acremonium* sp. *Biochemical and Biophysical Research Communications* 358, 356–362.

Lotze, M.T., Zeh, H.J., Rubartelli, A., Sparvero, L.J., Amoscato, A.A., Washburn, N.R., DeVera, M.E., Liang, X., Tör, M. and Billiar, T. (2007) The grateful dead: damage-associated molecular pattern molecules and reduction/oxidation regulate immunity. *Immunological Reviews* 220, 60–81.

Ludwig, A. and Tenhaken, R. (2001) A new cell wall located N-rich protein is strongly induced during the hypersensitive response in *Glycine max* L. *European Journal of Plant Pathology* 107, 323–336.

Ma, W., Smigel, A., Tsai, Y.-C., Braam, J. and Berkowitz, G.A. (2008) Innate immunity signaling: cytosolic Ca^{2+} elevation is linked to downstream nitric oxide generation through the action of calmodulin or a calmodulin-like protein. *Plant Physiology* 148, 818–828.

Malamy, J., Carr, J.P., Klessig, D.F. and Raskin, I. (1990) Salicylic acid: a likely endogenous signal in the resistance response of tobacco to viral infection. *Science* 250, 1002–1004.

Mamaní, A., Filippone, M.P., Grellet, C., Welin, B., Castagnaro, A.P. and Ricci, J.C.D. (2012) Pathogen-induced accumulation of an ellagitannin elicits plant defense response. *Molecular Plant–Microbe Interactions* 25, 1430–1439.

Martos, G.G., Chalfoun, N.R. and Díaz Ricci, J.C. (2012) The defence response mediated by the elicitor of *A. strictum* does not require the apoplast. *Biocell* 36, 82.

Mauch-Mani, B. and Métraux, J.P. (1998) Salicylic acid and systemic acquired resistance to pathogen attack. *Annals of Botany* 82, 535–540.

Mauch-Mani, B. and Slusarenko, A.J. (1996) Production of salicylic acid precursors is a major function of phenylalanine ammonia-lyase in the resistance of *Arabidopsis* to *Peronospora parasitica*. *The Plant Cell* 8, 203–212.

McCormack, E. and Braam, J. (2003) Calmodulins and related potential calcium sensors of *Arabidopsis*. *New Phytologist* 159, 585–598.

Métraux, J.P., Signer, H., Ryals, J., Ward, E. and Wyss-Benz, M. (1990) Increase in salicylic acid at the onset of systemic acquired resistance in cucumber. *Science* 250, 1004–1006.

Meuwly, P., Molders, W., Buchala, A. and Métraux, J.P. (1995) Local and systemic biosynthesis of salicylic acid in infected cucumber plants. *Plant Physiology* 109, 1107–1114.

Mittler, R. (2002) Oxidative stress, antioxidants and stress tolerance. *Trends in Plant Science* 7, 405–410.

Mittler, R., Vanderauwera, S., Suzuki, N., Miller, G., Tognetti, V.B., Vandepoele, K., Gollery, M., Shulaev, V. and van Breusegem, F. (2011) ROS signaling: the new wave? *Trends in Plant Science* 16, 300–309.

Nawrath, C. and Métraux, J.-P. (1999) Salicylic acid induction-deficient mutants of *Arabidopsis* express *PR-2* and *PR-5* and accumulate high levels of camalexin after pathogen inoculation. *The Plant Cell* 11, 1393–1404.
Nawrath, C., Heck, S., Parinthawong, N. and Métraux, J.-P. (2002) EDS5, an essential component of salicylic acid–dependent signaling for disease resistance in *Arabidopsis*, is a member of the MATE transporter family. *The Plant Cell* 14, 275–286.
Nürnberger, T. and Brunner, F. (2002) Innate immunity in plants and animals: emerging parallels between the recognition of general elicitors and pathogen-associated molecular patterns. *Current Opinion in Plant Biology* 5, 318–324.
Nürnberger, T., Brunner, F., Kemmerling, B. and Piater, L. (2004) Innate immunity in plants and animals: striking similarities and obvious differences. *Immunological Reviews* 198, 249–266.
Ohno, T., Nakahira, S., Suzuki, Y., Kani, T., Hara, T. and Koyama, H. (2004) Molecular characterization of plasma membrane H^+-ATPase in a carrot mutant cell line with enhanced citrate excretion. *Physiologia Plantarum* 122, 265–274.
Orbach, M.J., Farrall, L., Sweigard, J.A., Chumley, F.G. and Valent, B. (2000) A telomeric avirulence gene determines efficacy for the rice blast resistance gene *Pi-ta*. *The Plant Cell* 12, 2019–2032.
Pallas, J.A., Paiva, N.L., Lamb, C. and Dixon, R.A. (1996) Tobacco plants epigenetically suppressed in phenylalanine ammonia-lyase expression do not develop systemic acquired resistance in response to infection by tobacco mosaic virus. *The Plant Journal* 10, 281–293.
Pastore, D., Stoppelli, M.C., Di Fonzo, N. and Passarella, S. (1999) The existence of the K^+ channel in plant mitochondria. *Journal of Biological Chemistry* 274, 26683–26690.
Pastore, D., Fratianni, A., Di Pede, S. and Passarella, S. (2000) Effects of fatty acids, nucleotides and reactive oxygen species on durum wheat mitochondria. *FEBS Letters* 470, 88–92.
Perochon, A., Aldon, D., Galaud, J.-P. and Ranty, B. (2011) Calmodulin and calmodulin-like proteins in plant calcium signaling. *Biochimie* 93, 2048–2053.
Pieterse, C.M.J. and Dicke, M. (2007) Plant interactions with microbes and insects: from molecular mechanisms to ecology. *Trends in Plant Science* 12, 564–569.
Pieterse, C.M.J. and van Loon, L.C. (2004) NPR1: the spider in the web of induced resistance signaling pathways. *Current Opinion in Plant Biology* 7, 456–464.
Pieterse, C.M.J., Leon-Reyes, A., van der Ent, S. and van Wees, S.C.M. (2009) Networking by small-molecule hormones in plant immunity. *Nature Chemical Biology* 5, 308–316.
Ponce de León, I. and Montesano, M. (2013) Activation of defense mechanisms against pathogens in mosses and flowering plants. *International Journal of Molecular Sciences* 14, 3178.
Racedo, J., Salazar, S.M., Castagnaro, A.P. and Díaz-Ricci, J.C. (2013) A strawberry disease caused by *Acremonium strictum*. *European Journal of Plant Pathology* 137, 649–654.
Ramírez, V., López, A., Mauch-Mani, B., Gil, M.J. and Vera, P. (2013) An extracellular subtilase switch for immune priming in *Arabidopsis*. *PLoS Pathogens* 9, e1003445.
Rasmussen, J.B., Hammerschmidt, R. and Zook, M.N. (1991) Systemic induction of salicylic acid accumulation in cucumber after inoculation with *Pseudomonas syringae* pv *syringae*. *Plant Physiology* 97, 1342–1347.
Rautengarten, C., Steinhauser, D., Büssis, D., Stintzi, A., Schaller, A., Kopka, J. and Altmann, T. (2005) Inferring hypotheses on functional relationships of genes: analysis of the *Arabidopsis thaliana* subtilase gene family. *PLoS Computational Biology* 1, e40.
Reddy, A.S.N., Alim G.S., Celesnik, H. and Day, I.S. (2011) Roles of calcium- and calcium/calmodulin-regulated gene expression. *The Plant Cell* 23, 2010–2032.
Reumann, S., Babujee, L., Ma, C., Wienkoop, S., Siemsen, T., Antonicelli, G.E., Rasche, N., Lüder, F., Weckwerth, W. and Jahn, O. (2007) Proteome analysis of *Arabidopsis* leaf peroxisomes reveals novel targeting peptides, metabolic pathways, and defense mechanisms. *The Plant Cell* 19, 3170–3193.
Ribeiro, A., Praekelt, U., Akkermans, A.D.L., Meacock, P.A., van Kammen, A., Bisseling, T. and Pawlowski, K. (1996) Identification of *agthi1*, whose product is involved in biosynthesis of the thiamine precursor thiazole, in actinorhizal nodules of *Alnus glutinosa*. *The Plant Journal* 10, 361–368.
Sacks, W., Nürnberger, T., Hahlbrock, K. and Scheel, D. (1995) Molecular characterization of nucleotide sequences encoding the extracellular glycoprotein elicitor from *Phytophthora megasperma*. *Molecular and General Genetics* 246, 45–55.
Salazar, S.M., Castagnaro, A.P., Arias, M.E., Chalfoun, N., Tonello, U. and Díaz Ricci, J.C. (2007) Induction of a defense response in strawberry mediated by an avirulent strain of *Colletotrichum*. *European Journal of Plant Pathology* 117, 109–122.
Salazar, S.M., Grellet, C.F., Chalfoun, N.R., Castagnaro, A.P. and Díaz Ricci, J.C. (2013) Avirulent strain of *Colletotrichum* induces a systemic resistance in strawberry. *European Journal of Plant Pathology* 135, 877–888.

Shah, J., Tsui, F. and Klessig, D.F. (1997) Characterization of a salicylic acid-insensitive mutant (*sai1*) of *Arabidopsis thaliana*, identified in a selective screen utilizing the SA-inducible expression of the *tms2* gene. *Molecular Plant–Microbe Interactions* 10, 69–78.
Sharon, A., Fuchs, Y. and Anderson, J.D. (1993) The elicitation of ethylene biosynthesis by a *Trichoderma* xylanase is not related to the cell wall degradation activity of the enzyme. *Plant Physiology* 102, 1325–1329.
Smith, B. and Black, L. (1990) Morphological, cultural, and pathogenic variation among *Colletotrichum* species isolated from strawberry. *Plant Disease* 74, 69–76.
Stael, S., Wurzinger, B., Mair, A., Mehlmer, N., Vothknecht, U.C. and Teige, M. (2012) Plant organellar calcium signalling: an emerging field. *Journal of Experimental Botany* 63, 1525–1542.
Summermatter, K., Sticher, L. and Métraux, J.P. (1995) Systemic responses in *Arabidopsis thaliana* infected and challenged with *Pseudomonas syringae* pv *syringae*. *Plant Physiology* 108, 1379–1385.
Sun, X., Peng, L., Guo, J., Chi, W., Ma, J., Lu, C. and Zhang, L. (2007) Formation of DEG5 and DEG8 complexes and their involvement in the degradation of photodamaged photosystem II reaction center D1 protein in *Arabidopsis*. *The Plant Cell* 19, 1347–1361.
Tonello, U. and Díaz Ricci, J.C. (2015) Proteínas tipo calmodulina revelan la participación del calcio en la respuesta al estrés en frutilla. In: *3° Reunión Conjunta de Sociedades de Biología de Argentina*, 9 September, Tucumán, Argentina, p. 216.
Tonello, U., Castagnaro, A.P. and Díaz Ricci, J.C. (2013) Two class tau glutathione transferases from strawberry leaves are differentially expressed in response to biotic and abiotic stresses. *Biocell* 37, 75.
Tsuda, K., Sato, M., Glazebrook, J., Cohen, J.D. and Katagiri, F. (2008) Interplay between MAMP-triggered and SA-mediated defense responses. *The Plant Journal* 53, 763–775.
Valent, B. and Chumley, F.G. (1994) Avirulence genes and mechanisms of genetic instability in the rice blast fungus. In: Zeigler, R.S., Leong, S.A. and Teng, P.S. (eds) *Rice Blast Disease*. CAB International, Wallingford, UK, pp. 111–134.
Vallelian-Bindschedler, L., Mösinger, E., Métraux, J.-P. and Schweizer, P. (1998) Structure, expression and localization of a germin-like protein in barley (*Hordeum vulgare* L.) that is insolubilized in stressed leaves. *Plant Molecular Biology* 37, 297–308.
van Breusegem, F. and Dat, J.F. (2006) Reactive oxygen species in plant cell death. *Plant Physiology* 141, 384–390.
van Loon, L.C. and van Strien, E.A. (1999) The families of pathogenesis-related proteins, their activities, and comparative analysis of PR-1 type proteins. *Physiological and Molecular Plant Pathology* 55, 85–97.
Verhage, A., van Wees, S.C.M. and Pieterse, C.M.J. (2010) Plant immunity: it's the hormones talking, but what do they say? *Plant Physiology* 154, 536–540.
Vlot, A.C., Dempsey, D.M.A. and Klessig, D.F. (2009) Salicylic acid, a multifaceted hormone to combat disease. *Annual Review of Phytopathology* 47, 177–206.
Vranová, E., Inzé, D. and van Breusegem, F. (2002) Signal transduction during oxidative stress. *Journal of Experimental Botany* 53, 1227–1236.
Wang, B., Liu, X., Wu, W., Liu, X. and Li, S. (2009) Purification, characterization, and gene cloning of an alkaline serine protease from a highly virulent strain of the nematode-endoparasitic fungus *Hirsutella rhossiliensis*. *Microbiological Research* 164, 665–673.
Wiesel, I., Newton, A.C., Elliott, I., Booty, D., Gilroy, E.M., Birch, P.R.J. and Hein, I. (2014) Molecular effects of resistance elicitors from biological origin and their potential for crop protection. *Frontiers in Plant Science* 5, 1–13.
Wildermuth, M.C., Dewdney, J., Wu, G. and Ausubel, F.M. (2001) Isochorismate synthase is required to synthesize salicylic acid for plant defence. *Nature* 414, 562–565.
Yamaguchi, Y. and Huffaker, A. (2011) Endogenous peptide elicitors in higher plants. *Current Opinion in Plant Biology* 14, 351–357.
Yang, J., Zhao, X., Liang, L., Xia, Z., Lei, L., Niu, X., Zou, C. and Zhang, K.-Q. (2011) Overexpression of a cuticle-degrading protease Ver112 increases the nematicidal activity of *Paecilomyces lilacinus*. *Applied Microbiology and Biotechnology* 89, 1895–1903.
Zhang, Y., Xu, S., Ding, P., Wang, D., Cheng, Y.T., He, J., Gao, M., Xu, F., Li, Y. and Zhu, Z. (2010) Control of salicylic acid synthesis and systemic acquired resistance by two members of a plant-specific family of transcription factors. *Proceedings of the National Academy of Sciences USA* 107, 18220–18225.
Zhou, L., Lacroute, F. and Thornburg, R. (1998) Cloning, expression in *Escherichia coli*, and characterization of *Arabidopsis thaliana* UMP/CMP kinase. *Plant Physiology* 117, 245–254.

6 *Agrobacterium*-mediated Genetic Transformation of Strawberry

Amjad M. Husaini[1]* and Y. Wen Xu[2]

[1]*Centre for Plant Biotechnology, Sher-e-Kashmir University of Agricultural Sciences and Technology of Kashmir, Shalimar, Jammu and Kashmir, India;* [2]*Pharmaceutical Botany, Sichuan Agricultural University, Ya'an, PR China*

6.1. Introduction

Cultivated strawberry (*Fragaria* × *ananassa* Duch.) is an economically important berry crop with immense demand for its fresh fruit, as well as in the fruit-processing industry, while its wild relative, the woodland strawberry (*Fragaria vesca* L.), is of scientific importance due to its small genome size and sequencing of its genome (Shulaev *et al.*, 2008). *F. vesca* is diploid (2n = 2x = 14) and has a small genome (240 Mb) (Shulaev *et al.*, 2011), while *F.* × *ananassa* has a complicated octoploid (2n = 8x = 56) genome and, due to genetic limitations associated with high heterozygosity and polyploidy, it has limited potential for improvement using traditional breeding methods. The application of plant tissue culture and genetic engineering has therefore been of special significance for strawberry improvement (Husaini and Srivastava, 2006a). Octoploid strawberry accessions are extremely variable from genotype to genotype, and hence the variation in transformation and regeneration abilities is as wide as the agromorphological characters (Folta and Dhingra, 2006). The response to factors affecting genotype-specific regeneration makes standardization of an efficient regeneration system for each strawberry genotype an indispensable prerequisite for the successful development of transgenic plants using *Agrobacterium*-mediated transformation (Li and Tang, 2006; Husaini *et al.*, 2008).

Transgenic technology allows plants with specific qualities to be developed in a much shorter period of time than when using conventional plant breeding, and also makes possible the introduction of characteristics that cannot be achieved through plant breeding

* amjadhusaini@skuastkashmir.ac.in

alone. With its extensive host range, *Agrobacterium* is the favoured tool for plant genetic engineering and is used extensively for transformation of several crops. Over more than three decades of *Agrobacterium* biotechnology research, numerous genetic transformation protocols have been established for various plant species, and a wide range of *Agrobacterium*-related patents have been claimed. *Agrobacterium*-mediated transformation is a single-cell transformation system, avoiding the production of mosaic plants. Replacing its bacterial T-DNA genes with genes of interest does not affect the transformation process and is the molecular basis for almost all *Agrobacterium*-mediated genetic transformation protocols. As the *Agrobacterium* transformation process involves many stages, different pre- and post-agroinfection strategies have been employed by workers to achieve maximum transformation efficiency. In this chapter, we will discuss these strategies in the context of strawberry transformation.

6.2. Factors Affecting Transformation and Regeneration of Transformants

6.2.1. Robust *in vitro* regeneration protocol

Establishment of a regeneration system for efficient recovery of transformed cells following agroinfection is of foremost importance in *Agrobacterium*-mediated transformation of strawberry (Husaini and Srivastava, 2006b; Debnath and Teixeira da Silva, 2007; Husaini *et al.*, 2008). Use of a high-efficiency regeneration system greatly enhances the induction of shoot organogenesis from the transformed cells. The better the regeneration system, the greater are the chances of successful recovery of transgenic plants (Husaini, 2010).

In vitro regeneration is a complex process and is influenced by a number of genetic and environmental factors. As every species seems to have its own specific requirements, there are a number of reports about the substances and conditions that help cells to differentiate (Guha-Mukherjeem, 2002; Liu *et al.*, 2011). Regeneration via shoot organogenesis has been described in different cultivars of strawberry, and many workers have investigated various factors influencing organogenesis. These studies have demonstrated the importance of various factors including the balance of plant growth regulators, incubation culture conditions, genotype and explant type on successful plant regeneration (Liu and Sanford, 1988; Nehra *et al.*, 1989, 1990c; Sorvari *et al.*, 1993; Flores *et al.*, 1998; Schaart *et al.*, 2002; Passey *et al.*, 2003; Zhao *et al.*, 2004; Qin *et al.*, 2005a; Husaini and Abdin, 2007; Husaini *et al.*, 2008; Zhang *et al.*, 2010; Liu *et al.*, 2011). Likewise, the importance of plant growth regulators and growth media in achieving regeneration via somatic embryogenesis is well established (Wang *et al.*, 1984; Lis, 1987; Donnoli *et al.*, 2001; Husaini and Abdin, 2007; Husaini *et al.*, 2008). Wang *et al.* (1984) found that the most effective medium for inducing somatic embryos contained 2,4-dichlorophenoxyacetic acid (2,4-D; 22.62 μM), benzyladenine (BA; 2.22 μM) and casein hydrolysate (500 mg l^{-1}), while Lis (1987) reported the formation of adventitious buds and somatic embryos using the medium of Lee and de Fossard (1977). Husaini and Abdin (2007) for the first time achieved shoot regeneration in strawberry plants simultaneously through both somatic embryogenesis and shoot bud formation. Leaf explants cultured on relatively higher concentration of thidiazuron (TDZ) (18.16 μM) achieve better results for somatic embryogenesis, while at half concentration (9.08 μM) they show more bud formation. The protocol was later improved further, fine-tuning the effect of temperature on the induction and maintenance of somatic embryogenesis (Husaini *et al.*, 2008).

Among a vast number of protocols developed for *in vitro* regeneration of strawberry plants (Table 6.1), some protocols enable strawberry regeneration in a single step where shoot multiplication and rooting take place in the same culture medium (Debnath, 2006; Husaini *et al.*, 2008). In cultivar ‘Chandler’, after induction of somatic embryos on a medium containing TDZ, embryos successfully germinated and developed small shoots and roots on a medium containing kinetin (Husaini *et al.*, 2008).

Table 6.1. Studies on standardization of regeneration protocols in strawberry cultivars/genotypes.

Species	Cultivar	Explant used	Reference
Fragaria × ananassa	–	Flower buds	Lis (1987)
Fragaria × ananassa	'Allstar' and 'Honeoye'	Leaf, runner	Liu and Sanford (1988)
Fragaria × ananassa	'Sengana' and 'Canoga'	Leaf mesophyll protoplasts	Nyman and Wallin (1988)
Fragaria × ananassa	–	–	Sarwar (1989)
Fragaria × ananassa	'Redcoat'	Leaf	Nehra *et al.* (1989)
Fragaria × ananassa	'Redcoat'	Immature leaf	Nehra *et al.* (1990b)
Fragaria × ananassa	'Chandler' and 'Totem'	Achenes (cotyledon explants)	Miller and Chandler (1990)
Fragaria × ananassa	'Senga Sengana'	Meristem tip from stolon buds	Petrovic and Jacimovic (1990)
Fragaria vesca		Leaf, petiole	Greene *et al.* (1990)
Fragaria × ananassa	–	Protoplasts isolated from shoots	Nyman and Wallin (1991, 1992)
Fragaria vesca		Callus	Bois (1992)
Fragaria vesca	'Alpine'	Protoplast	Infante *et al.* (1993)
Fragaria × ananassa	'Surprise des Halles', 'Redgountlet' and 'Sanga Sengana'	Stolons, peduncles	Lis (1993)
Fragaria × ananassa	'Saladin'	Anthers	Foley and Hennerty (1993)
Fragaria × ananassa	'Hiku' and 'Jonsok'	Leaf discs	Sorvari *et al.* (1993)
Fragaria × ananassa	'Aiko', 'Dana', 'Gorella', 'Premial'	Leaf, petiole	Isac *et al.* (1994)
Fragaria × ananassa	'Redcoat' and 'Veestar'	Meristem, callus, leaf discs	Nehra *et al.* (1994)
Fragaria × ananassa	'Nyohou'	Shoot apex	Saito and Sakamori (1995)
Fragaria × ananassa	'Elsanta', 'Kama' and 'Senga Sengana'	Apical tip	Lis and Chlebowska (2000)
Fragaria × ananassa	'Chandler', 'Honeoye' and 'Redchief'	Anther	Owen and Miller (1996)
Fragaria × ananassa	'Samahberi'	Meristem tip	Kim *et al.* (1998)
Fragaria × ananassa	'Selva'	Runner tip	Okasha *et al.* (1996)
Fragaria × ananassa	-	Anthers	Zhou *et al.* (1996)
Fragaria × ananassa	'Camarosa', 'Parker' and 'Pajaro'	Apical and lateral buds	Sutter *et al.* (1997)
Fragaria × ananassa	'Teodora' and 'Clea'	Whole leaves, blade fragments, petioles, stipules	Damiano *et al.* (1997)
Fragaria × ananassa, *Fragaria vesca*	–	Protoplasts	Wallin (1997)
Fragaria × ananassa	'Gorella'	Petiole, leaf	Popescu *et al.* (1997)
Fragaria × ananassa	'Gorella'	Axillary and adventitious stipular shoots	Jemmali *et al.* (1997)
Fragaria moschata		Leaf disc, petiole	Infante *et al.* (1998)
Fragaria × ananassa	'Honeoye'	Axillary buds	Hammoudeh *et al.* (1998)
Fragaria × ananassa	'Elsanta', 'Senga Sengana', 'Kaster' and clones K-1349 and K1356	Leaf lamina, whole leaf with petiole and stipule	Wawrzyńczak *et al.* (1998)
Fragaria × ananassa	'Konvoy-Cascata' and 'Chandler'	Leaf discs	Flores *et al.* (1998)
Fragaria × ananassa	'Fengxiang'	Dormant bud	Zhao *et al.* (2009)

Continued

Table 6.1. Continued.

Species	Cultivar	Explant used	Reference
Fragaria indica		Nodal explants	Bhatt and Dhar (2000)
Fragaria × ananassa	'Redcoat', 'Pajaro', 'OsaGrande', 'Grace' and 'Selva'	Leaf explant, runner tip, crown node	Dhir *et al.* (2001)
Fragaria × ananassa	'Calypso', 'Bolero', 'Pegasus', 'Elsanta' and 'Tango'	Leaf disc, petiole, root, stipule	Passey *et al.* (2003)
Fragaria × ananassa	'Sweet Charlie' and 'Pajaro'	Leaf disc	Singh and Pandey (2004)
Fragaria × ananassa	'Senga Sengana'	Leaf disc, petiole	Litwinczuk and Zubel (2005)
Fragaria × ananassa	'Hecker' and 'La Sans Rivale'	Leaf disc, petiole	Zhao *et al.* (2004)
Fragaria × ananassa	'Bounty'	Leaf disc, petiole, sepal	Debnath (2005)
Fragaria × ananassa	'Toyonoka'	Leaf disc	Qin *et al.* (2005a,b)
Fragaria × ananassa	'Bounty', 'Jonsok', 'Polka', 'Korona' and 'Zephyr'	Leaf disc	Hanhineva *et al.* (2009)
Fragaria × ananassa	'Chandler'	Leaf disc, petiole	Husaini and Srivastava (2006b)
Fragaria × ananassa	'LF9'	Leaf disc, petiole, stolon	Folta *et al.* (2006)
Fragaria × ananassa	'Chandler'	Leaf disc	Husaini and Abdin (2007); Husaini and Abdin (2008a,b)
Fragaria × ananassa	'Darserlect', 'Toyonoka' and 'Allstar'	Shoot tip	Zhang *et al.* (2010)
Fragaria × ananassa	'Benihoppe'	Stolon tip, leaf, petiole, stem	Feng (2011)
Fragaria vesca	'Hawaii 4'	Leaf	Feng (2011)
Fragaria × ananassa	'Benihoppe'	Stem tip	Chen *et al.* (2011)
Fragaria × ananassa	'Tochiotome', 'Benihoppe', 'Lijinv' and 'Wandde'	Shoot tip, creeping stem	Liu *et al.* (2011)
Fragaria nilgerrensis		Shoot tip	Wang *et al.* (2012c)
Fragaria × ananassa	'Miaozi'	Stalk section, shoot tip	Liu *et al.* (2012)
Fragaria × ananassa	'Selva'	Stem tip	Wang *et al.* (2012d)
Fragaria × ananassa	'Allstar'	Stem tip	Zhang (2013)
Fragaria × ananassa	–	Anther	Li *et al.* (2013)
Fragaria × ananassa	'Totem'	Stem tip	He *et al.* (2013)
Fragaria × ananassa	'Benihoppe'	Shoot tip	Dong (2013)
Fragaria × ananassa	'Benihoppe'	Stem tip	Liu (2013)
Fragaria × ananassa	'Ningyu'	Stem tip	Xia *et al.* (2014)
Fragaria × ananassa	'Toyonoka', 'Tochiotome', 'Sweet Charlie', 'Benihoppe' and 'Camarosa'	Buds	Xue *et al.* (2014)
Fragaria × ananassa	–	Stem tip, anther	Yuan *et al.* (2015)
Fragaria × ananassa	'Benihoppe' and 'Sweet Charlie'	Stolon tips	Zhai *et al.* (2015)

Some of the factors affecting transformation efficiency are discussed below.

Explant type

There are a large number of reports on shoot organogenesis from leaf discs (Jones *et al.*, 1988; Liu and Sanford, 1988; Nehra *et al.*, 1989, 1990c; Sorvari *et al.*, 1993; Flores *et al.*, 1998; Passey *et al.*, 2003; Debnath, 2005, 2006; Qin *et al.*, 2005a,b; Husaini and Srivastava, 2006b; Husaini and Abdin, 2007; Husaini *et al.*, 2008), stems (Graham *et al.*, 1995; Chen *et al.*, 2011; Wang J. *et al.*, 2012a; Wang Y. *et al.*, 2012; Liu, 2013; Zhang, 2013; Xia *et al.*, 2014; Yuan *et al.*, 2015), petioles (Foucault and Letouze, 1987; Rugini, 1992; Isac *et al.*, 1994; Damiano *et al.*, 1997; Popescu *et al.*, 1997; Infante *et al.*, 1998; Passey *et al.*, 2003; Debnath, 2005, 2006; Feng, 2011), peduncles (Foucault and Letouze, 1987; Lis, 1993), stipules (Rugini, 1992; Passey *et al.*, 2003), stolons (Lis, 1993; Feng, 2011; Zhai *et al.*, 2015), anthers (Owen and Miller, 1996; Li *et al.*, 2013; Yuan *et al.*, 2015), embryos (Wang *et al.*, 1984), runners (Liu and Sanford, 1988), roots (Rugini, 1992; Passey *et al.*, 2003), buds (Xue *et al.*, 2014) and protoplasts (Nyman and Wallin, 1988) of strawberries.

Most of the work in strawberry regeneration has been achieved using leaf discs and petioles as explants. Of the 62 studies listed in Table 6.1, 26 (42%) used leaf as the explant material. Leaf tissue has the greatest regeneration capacity of strawberry plant tissue (Jones *et al.*, 1988; Liu and Sanford, 1988; Nehra *et al.*, 1989, 1990c; Jelenkovic *et al.*, 1991; Popescu *et al.*, 1997; Passey *et al.*, 2003). Passey *et al.* (2003) compared regeneration in seven commercial cultivars of strawberry using leaf discs, petioles, roots and stipules as explant material. Leaf discs had the highest regeneration rates for all cultivars, with greater than 90% of explants producing shoots. Husaini and Abdin (2007) also achieved 85% regeneration using leaf discs as explants. Callus production was also more prolific from leaf tissue than from petioles (Husaini and Srivastava, 2006b).

Physical factors

DARK PERIOD. The problem of darkening of culture medium of *in vitro*-cultured strawberry explants is well known and is attributed to phenolic compounds exuding from these tissues. This process is initiated by browning of the surface of plant tissues due to the oxidation of phenolic compounds, resulting in the formation of quinines, which are highly reactive and toxic to plant tissue (Taji and Williams, 1996; Xu *et al.*, 2011a). Incubation of leaf explants in the dark decreases browning of the culture medium caused by exudation of phenolics by explants (Nehra *et al.*, 1989; Rugini, 1992; Blando *et al.*, 1993; George, 1993; Popescu *et al.*, 1997; Barceló *et al.*, 1998; Titov *et al.*, 2006; Husaini and Abdin, 2007; Xu *et al.*, 2011b). In addition, this seems to enhance organogenesis in strawberry (cultivar 'Chandler') leaf explants (Liu and Sanford, 1988; Barceló *et al.*, 1998; Husaini and Abdin, 2007). It also enhances somatic embryogenesis in leaf explants of the cultivars 'Clea' (Donnoli *et al.*, 2001) and 'Chandler' (Husaini and Abdin, 2007). Explant cultivation in the dark during the first experimental step could be a key factor influencing somatic embryogenesis (Fiore *et al.*, 1997), especially when the negative effect of light on somatic embryo induction in strawberry has been reported in 'Clea' (Donnoli *et al.*, 2001). In 'Chandler', dark treatment increased the number of somatic embryos significantly in leaf explants cultured on 18.16 µM TDZ and later incubated under a 16 h photoperiod (Husaini and Abdin, 2007). Surprisingly, when the leaf explants of 'Chandler' were cultured on 9.08 µM TDZ (half concentration) and kept under a 16 h photoperiod, dark treatment increased shoot organogenesis considerably. However, when darkness was accompanied by chilling, a decrease in the number of shoots was observed (Husaini and Abdin, 2007).

LIGHT AND PHOTOPERIOD. Light affects somatic embryogenesis through its effect on induction (Verhagen and Wann, 1989) and influences the morphological characteristics of differentiated somatic embryos (Halperin,

1966; Ammirato and Steward, 1971). Despite these powerful effects of light, little attention has been devoted to its role in *in vitro* culture (Torné *et al.*, 2001) and particularly in somatic embryogenesis. Photoperiod has been implicated in the regulation of cytokinin levels (Forsline and Langille, 1975), as well as in photoconversion of phytochromes (Torné *et al.*, 1996). Incubation with a 16 h photoperiod is commonly used for shoot organogenesis (Nehra *et al.*, 1989; Nehra *et al.*, 1990a,b,c; Barceló *et al.*, 1998), as well as for somatic embryogenesis (Husaini and Abdin, 2007; Husaini *et al.*, 2008). A comparison of photoperiods (24, 16 and 12 h) used for incubation of strawberry leaf discs has revealed that a 16 h photoperiod is the optimum for shoot organogenesis (Husaini and Abdin, 2007). However, it should be noted that when explants were cultured on 18.16 µM TDZ and given chilling treatment, a 12 h photoperiod proved better than many other treatments of 16 h incubation (Husaini and Abdin, 2007). This has led to the conclusion that, when chilling and other treatments are applied to strawberry explants, the photoperiod requirement may also alter due to unknown reasons. More studies are needed to understand the interplay between photoperiod and other factors.

In contrast, variation in light intensities for incubation of cultures is not so uncommon. For example in cultivar 'Chandler' after a dark pre-treatment period (1–4 weeks), Barceló *et al.* (1998) transferred the leaf discs into a light intensity of 40 µmol m^{-2} s^{-1}, while Husaini and Abdin (2007) transferred the leaf discs into a light intensity of 48 µmol m^{-2} s^{-1}. A study by Nehra *et al.* (1990c) in which identical sets of cultures of cultivar 'Redcoat' were incubated at 12.5 and 65.5 µmol m^{-2} s^{-1} revealed that calli from *in vitro* leaves did not form shoots under high light intensity on any of the culture media, but at low light intensity some calli developed into shoots.

CHILLING. Chilling is a type of stress treatment that may increase the embryogenic potential of strawberry (Husaini and Abdin, 2007). Chilling may have some effect on the microtubule network of the cytoskeleton in strawberry, as has been reported in chicory (Decout *et al.*, 1994). While Husaini *et al.* (2008) demonstrated that TDZ concentration is the primary factor responsible for induction of somatic embryogenesis in strawberry, nevertheless incubation at a temperature regime of 10 ± 1°C has a complementary effect on increasing the number of somatic embryos per explant. Incubation at an optimum temperature causes the redirection of the developmental programme in cells and acquisition of competence in the embryogenic process, most likely due to temperature-regulated proteins (Giuliano *et al.*, 1984; Lo Schiavo *et al.*, 1990; De Jong *et al.*, 1992; Baldan *et al.*, 1997).

Chemical factors

PLANT GROWTH REGULATORS. There are various plant hormones used for regeneration of strawberry. Each cultivar has specific requirements that are vital for regeneration. BA and indole-3-acetic acid (IAA) were used for shoot regeneration in the cultivars 'Redcoat' (Nehra *et al.*, 1989), 'Sweet Charlae' and 'Pajaro' (Singh and Pandey, 2004), while BA and indole-3-butyric acid (IBA) gave good results in 'Jonsok', 'Hiku' (Sorvari *et al.*, 1993), 'Chandler' (Barceló *et al.*, 1998; Husaini and Srivastava, 2006b) and 'Totem' (He *et al.*, 2013). Finstad and Martin (1995) obtained success with 2,4-D and BA in cultivars 'Hood' and 'Totem', while Qin *et al.* (2005a,b) succeeded with IBA and TDZ in 'Toyonoka'. BA and 1-naphthaleneacetic acid (NAA) were best for proliferation of cultivars 'Benihoppe', 'Sweet Charlie', 'Selva' and 'Ningyu' and *Fragaria nilgerrensis* (Wang Y. *et al.*, 2012; Wang Z.L. *et al.*, 2012; Dong, 2013; Liu, 2013; Zhang, 2013; Xia *et al.*, 2014; Zhai *et al.*, 2015), while IBA was best for rooting of cultivars 'Benihoppe', 'Sweet Charlie', 'Selva' and 'Allstar' and *F. nilgerrensis* (Wang *et al.*, 2012c,d; Dong, 2013; Liu, 2013; Zhang, 2013; Zhai *et al.*, 2015). NAA was best for rooting of 'Ningyu' (Xia *et al.*, 2014). Chen *et al.* (2011) demonstrated that the effect on rooting of 'Benihoppe' was NAA > IBA > IAA.

There has been increased interest in the use of TDZ for strawberry regeneration.

TDZ greatly enhances the shoot regeneration percentage in strawberry leaf discs (Nyman and Wallin, 1992; Sutter *et al.*, 1997; Flores *et al.*, 1998; Hammoudeh *et al.*, 1998; Schaart *et al.*, 2002; Passey *et al.*, 2003; Zhao *et al.*, 2004; Qin *et al.*, 2005a,b; Husaini and Abdin, 2007; Nasri and Bahramnejad, 2013; Haddadi *et al.* 2015). The differences among these reports are in the optimum concentration of TDZ for shoot regeneration in different strawberry cultivars. Interestingly, TDZ concentration in the range of 18 ± 2 μM (which approximately equals 4 mg l^{-1}) has been found to be most successful for shoot regeneration from leaf explants of strawberry (Husaini and Abdin, 2007; Husaini *et al.* 2008; Nasri and Bahramnejad, 2013; Haddadi *et al.* 2015).

VITAMINS AND SALTS. Murashige and Skoog vitamins (Murashige and Skoog, 1968) or Gamborg B_5 vitamins (Gamborg *et al.*, 1968) are generally combined with Murashige and Skoog micro/macrosalts to generate basal media. These are then supplemented with organic substances, agar, auxins and cytokinins to yield complete medium for growth of plant tissue culture. Murashige and Skoog salts are most commonly used in strawberry regeneration studies, although there are exceptions. For example, Lis (1987) used the medium of Lee and de Fossard (1977) for regeneration of adventitious buds and somatic embryos in strawberry, while Barceló *et al.* (1998) used Lopez-Aranda *et al.* (1994) for regeneration of the strawberry cultivar 'Chandler'. Sucrose is generally used as a carbon source, but glucose and table sugar have also been used occasionally.

6.2.2. Robust transformation protocol

Over the last three decades, a number of transformation studies with the objective of standardizing *Agrobacterium*-mediated transformation protocols for different strawberry cultivars have been undertaken (Table 6.2) (reviewed by Husaini and Srivastava, 2006a; Husaini *et al.*, 2011). A general protocol for efficient *Agrobacterium*-mediated transformation of strawberry, which may be tailored by individual workers according to their need, is described elsewhere (Husaini *et al.*, 2011; Schaart, 2014). Table 6.3 shows the scope of alterations and adjustments that can be made in the different medium components to suit the different stages of regeneration of strawberry transformants.

It is well known that a rigorous transformation process can reduce the regeneration capacity of a strawberry tissue (leaf) drastically, and may slash it from approximately 95% to 1–6% (Passey *et al.*, 2003). In *Agrobacterium*-mediated transformation, a sufficient quantity of bacteria during co-cultivation, a long enough co-cultivation period, use of *vir* gene inducers such as acetosyringone and stringent selection pressure are important to obtain stable transformations (Sriskandarajah *et al.*, 2004). Using the 'leaf disc transformation technique', as defined by Horsch *et al.* (1985), Husaini (2010) examined a number of variables that affect transformation of strawberry, such as choice and concentration of antibiotics in the selection medium, effect of pre-culture and/or pre-selection on the regeneration of kanamycin (Kan)-resistant shoots, effect of the plant phenolic acetosyringone and length of the co-cultivation period. This aided in the improvement of the transformation protocol developed by Barceló *et al.* (1998) (Table 6.4).

Efficient Agrobacterium strain

The efficiency of shoot production varies with the type of *Agrobacterium tumefaciens* strain and vector, even on the same selection medium (Vergauwe *et al.*, 1998). Most binary vectors used to transform strawberry are derived from pBIN19 (Bevan *et al.*, 1989) and contain the neomycin phosphotransferase II (*nptII*) gene for Kan selection of transgenic shoots (Mercado *et al.*, 2007). As a combination, *Agrobacterium* strain LBA4404 and gene construct pBI121 have been used extensively in strawberry transformation of the cultivars 'Rapella' (James *et al.*, 1990), 'Melody', 'Rhapsody', 'Symphony' (Graham *et al.*, 1995), 'Chandler' (Barceló *et al.*,

Table 6.2. Studies on *Agrobacterium*-mediated reporter/marker gene transfer in strawberry cultivars/genotypes.

Species and cultivar/genotype	Explant	Reporter gene(s)	Reference
Fragaria × *ananassa* 'Rapella'	Leaf, petiole	*nos/nptII*	James *et al.* (1990)
Fragaria × *ananassa* 'Redcoat'	Leaf	*nptII/gus*	Nehra *et al.* (1990a,b)
Fragaria × *ananassa* 'Tristar'	Leaf	*nptII/gus*	Mathews *et al.* (1995)
Fragaria vesca	Leaf	*nptII/gus*	El Mansouri *et al.* (1996)
Fragaria × *ananassa* 'Marmolada Onebor'	Leaf, stipule, petioles	*nptII*	Martinelli *et al.* (1997)
Fragaria × *ananassa* 'Skelta'	Leaf	*nptII*	Du Plessis *et al.* (1997)
Fragaria × *ananassa* 'Chandler'	Leaf	*nptII/gus*	Barceló *et al.* (1998)
Fragaria × *ananassa* 'Chandler'	Leaf	*nptII/gus*	Harpster *et al.* (1998)
Fragaria × *ananassa* 'Totem'	Leaf	*nptII/gus*	Mathews *et al.* (1998)
F. vesca 'Alpine' accession FRA 197	Leaf, petiole	*nptII/gus*	Haymes and Davis (1998)
Fragaria × *ananassa* 'Gariguette', 'Polka', 'Elsanta' and CPRO-DLO breeding line 88312	Leaf	*nptII/gus*	Puite and Schaart (1998)
Fragaria × *ananassa* 'Tudla'	Leaf discs	*gus*	Zhang and Wu (1998)
Fragaria vesca	Leaf	*nptII/gus*	de Mesa *et al.* (2000)
Fragaria × *ananassa* 'Elsanta', 'Kaster', 'K-1349', 'K1476' and 'Senga Sengana'	Leaf	*gus*	Wawrzyńczak *et al.* (2000)
Fragaria × *ananassa* 'M14'	Leaf discs	*gus/npt II*	Deng and Hu (2000)
Fragaria × *ananassa* strains BHN FL90031-30 and BHN 92664-501	Leaf	*nptII/gus*	Dhir *et al.* (2001)
Fragaria × *ananassa* 'Clea', 'Irvine', 'Paros', 'Alpine', 'Ilaria' and 'Regina'	Leaf	*nptII*	Donnoli *et al.* (2001)
Fragaria × *ananassa* 'Redcoat', 'Pajaro', 'Osa Grande', 'Grace' and 'Selva'	Leaf, Runner tip	*nptII/gus/hpt*	Dhir *et al.* (2001)
Fragaria × *ananassa* 'Gariguette', 'Polka' and Breeding line no. 88312	Leaf	*nptII/gus*	Schaart *et al.* (2002)
Fragaria × *ananassa* 'Chandler'	Leaf	*nptII/gus*	Jiménez-Bermúdez *et al.* (2002)
Fragaria × *ananassa* 'Teodara' and 'Egla'	Stipule	*nptII/gus*	Monticelli *et al.* (2002)
Fragaria vesca	Leaf, petiole	*nptII/gus*	Alsheikh *et al.* (2002)
Fragaria × *ananassa* 'Pajaro'	Leaf	*nptII/gus*	Ricardo *et al.* (2003)
Fragaria × *ananassa* 'Elista', 'Wega', ' Senga Precosa', 'Kama', 'Induka', 'Maria', 'Redgauntlet', 'Zao Hang Guang', 'Dukat' and 'Favette'	Leaf	*nptII/gus*	Gruchała *et al.* (2004b)
Fragaria × *ananassa* 'Induka' and 'Elista'	Leaf	*nptII/gus*	Gruchała *et al.* (2004a)
Fragaria × *ananassa* 'Hecker' and 'La Sans Rivale'; *Fragaria vesca* 'Alpine'	Leaf, petiole	*nptII/gus*	Zhao *et al.* (2004)
Fragaria vesca accessions FRA197 and 198	Leaf, petiole	*nptII/gus*	Zhao *et al.* (2004)

Continued

Table 6.2. Continued.

Species and cultivar/genotype	Explant	Reporter gene(s)	Reference
Fragaria × *ananassa* 'Tiogar'	Leaf	*nos/nptII*	Khammuang *et al.* (2005)
Fragaria × *ananassa* 'Allstar'	Leaf	*nptII/gus*	Zhang and Wang (2005)
Fragaria × *ananassa* 'LF9'	Petiole	*nptII*	Folta and Dhingra (2006)
Fragaria × *ananassa* 'Chandler'	Leaf	*nptII*	Husaini and Srivastava (2006)
Fragaria × *ananassa* 'Anther'	Leaf	*nptII*	Park *et al.* (2006)
Fragaria × *ananassa* 'Pajaro'	Leaf	*nptII*	Vellicce *et al.* (2006)
Fragaria vesca accession P1	Leaf	*hpt/gfp*	Oosumi *et al.* (2006)
Fragaria × *ananassa* 'Zaohong'	Leaf	*gus*	Wang *et al.* (2009)
Fragaria × *ananassa* 'Chandler'	Leaf	*nptII*	Husaini (2010)
Fragaria × *ananassa* 'Kurdistan', 'Camarosa' and 'Paros'	Leaf discs	*uidA, nptII*	Nasri and Bahramnejad (2013)
Fragaria × *ananassa* 'Festival', 'Sweet Charlie' and 'Florida'	Leaf discs	*uidA, bar*	Zakaria *et al.* (2014)
Fragaria × *ananassa* 'Camarosa'	Leaf discs	*luc, nptII*	Haddadi *et al.* (2015)
Fragaria vesca 'Hawaii 4'	Leaf	*gfp*	Zhou *et al.* (2015)

Table 6.3. Example illustrating the scope of using different combinations and concentrations of media components for creating specialized media suitable at different stages of transformation process. (From Husaini *et al.*, 2011, with kind permission from Global Science Books).

Medium	Components
MS liquid medium (MSL)	MS salts and vitamins + 3% sucrose
Regeneration medium	MS salts + B_5 vitamins + 2% glucose + 4 mg l^{-1} TDZ
Shoot elongation medium	MS salts + B_5 vitamins + 2% sucrose + 1% glucose + 0.1 mg l^{-1} BA + 0.05 mg l^{-1} kinetin + 2 mg l^{-1} GA_3
Pre-selection regeneration media	MS salts + B_5 vitamins + 2% glucose + 4 mg l^{-1} TDZ + 500 µg ml^{-1} cefotaxime
	MS salts + B_5 vitamins + 2% glucose + 4 mg l^{-1} TDZ + 500 µg ml^{-1} timentin
	MS salts + B_5 vitamins + 2% glucose + 4 mg l^{-1} TDZ + 250 µg ml^{-1} cefotaxime + 250 µg ml^{-1} timentin
Selective regeneration media	
S_LM_{IA}	MS salts + B_5 vitamins + 2% glucose + 4 mg l^{-1} TDZ + 500 µg ml^{-1} cefotaxime + 50 µg ml^{-1} Kan
S_LM_{IB}	MS salts + B_5 vitamins + 2% glucose + 4 mg l^{-1} TDZ + 500 µg ml^{-1} timentin + 50 µg ml^{-1} Kan
S_LM_{IC}	MS salts + B_5 vitamins + 2% glucose + 4 mg l^{-1} TDZ + 250 µg ml^{-1} cefotaxime + 250 µg ml^{-1} timentin + 50 µg ml^{-1} Kan
Selective shoot elongation media	MS salts + B_5 vitamins + 2% sucrose + 1% glucose + 0.1 mg l^{-1} BA + 0.05 mg l^{-1} kinetin + 2 mg l^{-1} GA_3 + 25 µg ml^{-1} Kan
	MS salts + B_5 vitamins + 2% sucrose + 1% glucose + 1.0 mg l^{-1} kinetin + 25 µg ml^{-1} Kan
Root induction medium	MS salts + B_5 vitamins + 2% sucrose + 1% glucose + 1.0 mg l^{-1} kinetin

MS, Murashige and Skoog; TDZ, thidiazuron; BA, benzyladenine; GA_3, gibberellic acid; Kan, kanamycin.

Table 6.4. Differences in transformation protocols using the same explant material (leaf) and genotype (cultivar 'Chandler'), and the resultant cumulative interactive effect on transformation percentage in the studies by Barceló *et al.* (1998) and Husaini (2010). (From Husaini *et al.*, 2011, with kind permission from Global Science Books.)

Parameter	Barceló *et al.* (1998)	Husaini (2010)
Source of explant	Variable	20-Day-old plantlets maintained on MS salts + B_5 vitamins + glucose (2%) + agar (0.9%) + kinetin (1 mg l^{-1})
Shoot regeneration medium	Lopez-Aranda *et al.* (1994)	Murashige and Skoog (1962) + B_5 vitamins + 2% glucose
Efficiency of regeneration system (%)	66.7	100
Agrobacterium tumefaciens strain	LBA 4404	GV 2260
Binary vector	pBI121	pBinAR
Acetosyringone (μM)	0	100
Co-cultivation duration (h)	72	72
Kanamycin in selection medium (mg l^{-1})	25	50 and 25
Agrobactericidal antibiotics (mg l^{-1})	Carbenicillin 500	Cefotaxime 250 + Timentin 250
Osmoprotectant (μM)	0	Validamycin A 100
Pre-culture/pre-incubation duration (days)	3 and 10	7
Pre-selection (days)	0	5
Transformation % based on number of explants regenerating shoots on Kan	4.2	10

1998; Husaini and Srivastava, 2006b), 'Lijinv'(Qian, 2010; Lin *et al.*, 2012), 'Virginia' (Bai *et al.*, 2012) and 'Allstar' (Song, 2009) and in *F. vesca* (Alsheikh *et al.*, 2002). In addition, *Agrobacterium* strain GV2260 has also been used successfully for genetic transformation of the strawberry cultivars 'Selekta' (Du Plessis *et al.*, 1997), 'Mermolada onebor' (Martinelli *et al.*, 1997), *F.* × *ananassa* breeding selection 'AN 93.231.53' (Mezzetti *et al.*, 2004) and cultivar 'Chandler' (Husaini and Abdin, 2008a b). *Agrobacterium* strain GV3101 has also been used successfully for genetic transformation of the strawberry cultivars *F. vesca* 'Hawaii 4' (Zhou *et al.*, 2015), 'Xingdu 2' and 'Allstar' (Sun *et al.*, 2009). Finally, the supervirulent *Agrobacterium* strain Agl0 (Lazo *et al.*, 1991) and derivatives EHA101 and EHA105 (Hood *et al.*, 1986) have been used successfully in transformation of cultivars 'Totem', 'Tristan' (Mathews *et al.*, 1995), 'Elsanta' (Puite and Schaart, 1998), 'Gariguette', 'Polka', breeding line 88312 (Schaart *et al.*, 2002), 'M14' (Deng and Hu, 2000; Na *et al.*, 2006), 'Toyonoka' (Zhu, 2008; Chen, 2014), 'Tudla' (Zhang and Wu, 1998; Qin *et al.*, 2005b; Zheng, 2008; Wang J.H. *et al.*, 2012b), 'Akihime' (Wang, 2012) and 'Calypso' (Schaart *et al.*, 2010).

Physical factors

PRE-CULTURE (PRE-INCUBATION). Before agroinfection, explants are sometimes incubated on an *in vitro* regeneration medium. The period of incubation generally varies from 1 to 10 days, and allows these explants to adjust to the regeneration medium. This practice is beneficial in most cases (Sorvari *et al.*, 1993; El Mansouri *et al.*, 1996; Asao *et al.*, 1997; Barceló *et al.*, 1998; de Mesa *et al.*, 2000; Alsheikh *et al.*, 2002; Husaini, 2010). Pre-culture improves the transformation percentage, probably by increasing the number of plant cells competent for regeneration and transgene integration (Table 6.5). The percentage of explants regenerating putative transgenic shoots as well as the percentage of transgenic shoots per se increase significantly by pre-culturing.

CO-CULTIVATION. One of the most crucial steps in gene transfer is the co-cultivation of explant material with genetically engineered *Agrobacterium*. A critical balance in the *Agrobacterium* population is needed, as an excessive number of bacteria imposes stress on the plant cells, negatively affecting their regeneration potential, while lower numbers reduce the frequency of T-DNA transfer. Increased co-cultivation duration can increase the percentage of transfection events but may cause stress and tissue necrosis. Agroinfection time generally varies between 15 min (Nehra *et al.*, 1990a) and 2 h (Mathews *et al.*, 1998), while the period for co-cultivation is between 24 and 72 h in the dark (Zhang and Wang, 2005; Folta and Dhingra, 2006; Husaini, 2010).

PRE-SELECTION. Selective agents such as Kan interfere with the regeneration of transformants (van Wordragen and Dons, 1992; Husaini, 2010). Thus, the introduction of a brief intervening period on a non-selective medium is sometimes done to allow recovery of the transformed cells from the infection/transformation process, and to allow enough time for stable integration and expression of the selectable marker gene (Alsheikh *et al.*, 2002; Zhao *et al.*, 2004). The leaf discs inoculated with *Agrobacterium* regenerate shoots at a low frequency when subjected to selection pressure immediately after co-cultivation, but with the introduction of a pre-selection phase, the percentage of leaf discs regenerating shoots increases significantly (Nehra *et al.*, 1990a,b; Alsheikh *et al.*, 2002; Husaini, 2010). The percentage of leaf discs regenerating shoots increased by almost sixfold with a 5-day pre-selection phase in cultivar 'Chandler' (Table 6.5). The average percentage of leaf discs regenerating shoots on selection medium increased from 0.5 (no pre-selection) to 3.1% (with pre-selection), while the average percentages of putative transgenic shoots per regenerating leaf disc increased from 0.5 (no pre-selection) to 6.1% (with pre-selection), respectively (Husaini, 2010).

Chemical factors

ANTIBIOTICS. After agroinfection, explants are exposed to two classes of antibiotics, one for selection of transformed cells and the other to eliminate the *Agrobacterium*. The proportions, combination and duration of exposure are always optimized to achieve a compromise between producing transgenics and screening out escapes. Kan is the most widely used antibiotic for selection of transformants in strawberry, but hygromycin (Nyman and Wallin, 1992; Mathews *et al.*, 1995; Oosumi *et al.*, 2006), geneticin (Mathews *et al.*, 1995) and phosphinothricin (Wang *et al.*, 2004; Folta *et al.*, 2006) have also been used. Kan concentration in the selective regeneration medium is of vital importance for optimum transformation efficiencies. Shoot regeneration from leaf discs is impaired at Kan concentrations as low as 10 mg l^{-1} (El Mansouri *et al.*, 1996; Barceló *et al.*, 1998; Gruchała *et al.*, 2004a), and a higher concentration of Kan in the selective medium significantly reduces shoot regeneration (Alsheikh *et al.*, 2002; Husaini, 2010). The presence of Kan at 10 mg l^{-1} drastically decreases the organogenic capacity of leaf discs (Barceló *et al.*, 1998) of cultivar 'Chandler', which remain 100% viable in the absence of Kan, but are only 21% viable in its presence. The concentration of Kan sensitivity varies among cultivars, explants and the procedure employed. For example, in cultivar 'Redcoat', leaf explants were subjected to Kan selection at 25–50 mg l^{-1} (Nehra *et al.*, 1990a, b), 'Rapella' petioles at 25 mg l^{-1} (James *et al.*, 1990), 'Melody', 'Rhapsody' and 'Symphony' stems at 20 mg l^{-1} (Graham *et al.*, 1995), 'Chandler' leaves at 25 mg l^{-1} (Barceló *et al.*, 1998; de Mesa *et al.*, 2000; Jiménez-Bermúdez *et al.*, 2002), 'Teodora' and 'Egla' stipules at 50 mg l^{-1} (Monticelli *et al.*, 2002), *F. vesca* leaves and petioles at 10–25 mg l^{-1} (Alsheikh *et al.*, 2002) and 'Chandler' leaf at 25–50 mg l^{-1} (Husaini and Abdin, 2008b; Husaini,

Table 6.5. Effect of different factors on genetic transformation of explants cultured on selective regeneration media (S_LM_{IA}, S_LM_{IB}, and S_LM_{IC}). (Adapted and modified from Husaini, 2010.)

	Percentage of explants regenerating putative transgenic shoots on three selective regeneration media (8 weeks) (% E_{SPT})				Percentage of putative transgenic shoots on three selective regeneration media (8 weeks) (% S_{PT})			
Treatment	S_LM_{IA}[a]	S_LM_{IB}[a]	S_LM_{IC}[a]	CD at 5%	S_LM_{IA}[a]	S_LM_{IB}[a]	S_LM_{IC}[a]	CD at 5%
Pre-incubation period (days)								
0	1.20 ± 0.07	1.64 ± 0.09	4.04 ± 0.10	0.886	1.68 ± 0.21	2.46 ± 0.14	11.90 ± 0.17	3.561
7	3.48 ± 0.10	5.43 ± 0.17	6.90 ± 0.12		5.58 ± 0.17	6.91 ± 0.18	23.33 ± 0.23	
Co-cultivation duration (h)								
48	1.70 ± 0.05	3.03 ± 0.19	5.41 ± 0.17	1.436	2.43 ± 0.24	4.05 ± 0.12	16.23 ± 0.26	2.895
72	3.05 ± 0.07	4.26 ± 0.17	5.54 ± 0.10		4.94 ± 0.19	5.60 ± 0.20	19.03 ± 0.22	
Agroinfection bacterial suspension								
MSL	1.92 ± 0.19	4.0 ± 0.28	5.68 ± 0.11	1.287	2.65 ± 0.10	5.33 ± 0.16	17.95 ± 0.17	1.958
MSL + AS (100 μM)	2.85 ± 0.14	3.29 ± 0.24	5.25 ± 0.21		4.76 ± 0.18	4.30 ± 0.18	24.00 ± 0.23	
Pre-selection (days) + validamycin A (μM)								
0 + 0	0.33 ± 0.07	0.42 ± 0.08	0.83 ± 0.10	0.953	0.33 ± 0.07	0.42 ± 0.08	0.83 ± 0.10	3.310
5 + 0	1.20 ± 0.02	3.21 ± 0.07	4.82 ± 0.07		1.60 ± 0.20	3.92 ± 0.18	12.75 ± 0.31	
5 + 100	5.61 ± 0.09	7.05 ± 0.11	9.67 ± 0.10		9.12 ± 0.26	9.80 ± 0.25	35.16 ± 0.35	

Values are shown as mean ± standard error.

CD, confidence distribution; MSL, Murashige and Skoog liquid medium; AS, acetosyringone.

[a]See Table 6.3 for details of media.

2010). Husaini (2010) demonstrated that, with the addition of Kan (50 mg l^{-1}) to a regeneration medium containing cefotaxime (500 mg l^{-1}) or timentin (500 mg l^{-1}), the percentages of leaf discs regenerating dropped drastically from 75.0 to 3.5%, and the average number of shoots per regenerating leaf disc dropped from 8 to 1.5. Interestingly, however, in the cultivar 'Calypso', a Kan concentration of 150 mg l^{-1} was used for selection of transgenic plants (Schaart *et al.*, 2010), indicating that some octoploid genetic lines may be highly resistant to this compound.

Use of Kan is related to the risk of formation of chimeras (shoots containing transgenic and non-transgenic sections) (Mathews *et al.*, 1998; Schestibratov and Dolgov, 2005), especially when stipules are used as explants (Monticelli *et al.*, 2002; Chalavi *et al.*, 2003). This could be due to a high antibiotic tolerance of the particular cultivar, as non-transformed shoots (control) were also able to grow and proliferate at the Kan concentration used for selection (Mercado *et al.*, 2007).

Three methods are employed to induce transgenic shoots on selection medium: first, where the concentrations of Kan are kept constant (non-iterative method); secondly, where Kan levels are increased gradually during subculture (iterative method); and thirdly, where a higher Kan concentration is used in the start of selection, while later reducing it to half (reverse-iterative method). Both iterative (Mathews *et al.*, 1998; Houde *et al.*, 2004) and reverse-iterative (Husaini 2010) methods successfully recover non-chimeric transgenic strawberry plants.

The next important step is the elimination of *Agrobacterium* from the regeneration media. Various antibiotics are used to kill the *Agrobacterium*, but these are phytotoxic, especially at higher concentrations. For effective transformation, the antibiotic regime should control bacterial overgrowth without inhibiting the regeneration of the plant cells (Graham *et al.*, 1995; Alsheikh *et al.*, 2002; Wang *et al.*, 2009; Bai *et al.*, 2012; Chen, 2014; Zhou *et al.*, 2015). Use of carbenicillin to control *Agrobacterium* after transformation of strawberry leaf explants (cultivar 'Totem') resulted in stunted top and root growth of plantlets, while with timentin (a mixture of ticarcillin (96%) and clavulanic acid (4%)) the regenerated plantlets showed vigorous, healthy top and root growth (Finstad and Martin, 1995). In contrast, in the octoploid strawberry genotype LF9, timentin, although found to be effective in curbing *Agrobacterium* growth, slowed its growth and differentiation slightly (Folta *et al.*, 2006). Alsheikh *et al.* (2002) studied the regeneration of *F. vesca* and *F. vesca semperflorens* in the presence of different concentrations of cefotaxime, carbenicillin and cefoxitin and found that carbenicillin was the least phytotoxic of the three. In contrast, a study comparing the effect of timentin, cefotaxime, carbenicillin and ampicillin on leaf explants of cultivar 'Camarosa' showed the best shoot regeneration on timentin (300 mg l^{-1}) and cefotaxime (150 mg l^{-1}) (Haddadi *et al.* 2015). Combination of agrocidal antibiotics with synergistic effects has proven to be less phytotoxic and better in eliminating *Agrobacterium* than when used in isolation at identical concentrations (Tanprasert and Reed, 1998; Husaini, 2010) (Table 6.6). Husaini (2010) showed that a combination of timentin and cefotaxime at half concentrations each (250 mg l^{-1} each) was less phytotoxic to leaf discs of cultivar 'Chandler' than use of either of these antibiotics at higher concentrations (500 mg l^{-1}). This provides evidence of the complementary effect of these two antibiotics in eliminating *Agrobacterium* without being overly phytotoxic to leaf discs of strawberry (cultivar 'Chandler'). It can therefore be concluded that the interaction of antibiotics with plant species is genotype dependent, and that variations may occur because of *Agrobacterium* strain and explant type.

ACETOSYRINGONE. In most strawberry cultivars, the addition of acetosyringone during pre-culture and co-cultivation has shown a synergistic effect on *Agrobacterium*-mediated transformation and is reported to increase

Table 6.6. A typical *Agrobacterium*-mediated genetic transformation protocol illustrating the various factors involved in optimization of transformation efficiency, and different ways of calculating transformation success. (Adapted and modified from Husaini, 2010.)

					Pre-selection (days)		Explants on selection medium (S_LM_{IC}) (Cef_{250} + Tim_{250} + Kan_{50})			Transformation percentage based on:				$T\eta$
Pre-incubation in the dark (days)	No. of agroinfected explants (NE_{Ai})	Agroinfection medium	Co-cultivation in the dark (h)	Washing solution composition	Cef_{250} + Tim_{250}	Cef_{250} + Tim_{250} + Val_{100}	Initial no. (NE_{SM})	No. of explants regenerating shoots (8 wks) (NE_{SPT})	No. of putative transgenic shoots regenerated (8 wks) (NS_{PT})	$(NS_{PT}/NE_{Ai}) \times 100$	$(NE_{SPT}/NE_{Ai}) \times 100$	$(NS_{PT}/NE_{SM}) \times 100$	$(NS_{PCR}/NE_{Ai}) \times 100$	$(NS_{PT}/NE_{Ai}) \times (NE_{nsm}/NS_{nsm}) \times 100$[a]
0	75	MSL	48	H_2O + Cef_{250} + Tim_{250}	0	0	69	0	0	0	0	0	0	0 (6.9)
0	55	MSL	72	H_2O + Cef_{250} + Tim_{250}	0	0	49	0	0	0	0	0	0	0 (8.2)
7	65	MSL	48	H_2O + Cef_{250} + Tim_{250}	5	0	60	5	15	23.07	7.69	25.00	13.84	3.16 (7.3)
7	90	MSL	72	H_2O + Cef_{250} + Tim_{250}	5	0	84	6	18	20.00	6.66	21.42	11.11	2.73 (7.3)
0	75	MSL	48	H_2O + Cef_{250} + Tim_{250}	0	5	70	7	21	28.00	9.33	30.00	18.66	3.29 (8.5)
0	80	MSL	72	H_2O + Cef_{250} + Tim_{250}	0	5	74	7	25	31.25	8.75	33.78	18.75	3.43 (9.1)
7	60	MSL + AS	48	H_2O + Cef_{250} + Tim_{250}	0	0	55	1	1	1.66	1.66	1.81	1.66	0.21 (7.6)
7	50	MSL + AS	72	H_2O + Cef_{250} + Tim_{250}	0	0	46	1	1	2.00	2.00	2.17	2.00	0.25 (7.7)
0	75	MSL + AS	48	H_2O + Cef_{250} + Tim_{250}	5	0	70	2	2	2.66	2.66	2.85	2.66	0.36 (7.2)
0	60	MSL + AS	72	H_2O + Cef_{250} + Tim_{250}	5	0	53	1	2	3.33	1.66	3.77	1.66	0.46 (7.2)
7	75	MSL + AS	48	H_2O + Cef_{250} + Tim_{250}	0	5	70	8	30	40.00	10.66	42.85	29.33	4.34 (9.2)
7	80	MSL + AS	72	H_2O + Cef_{250} + Tim_{250}	0	5	74	8	33	41.25	10.00	44.59	31.25	4.43 (9.3)

NE_{Ai}, number of explants agroinfected; Cef_{250}, 250 mM cefotaxime; Tim_{250}, 250 mM timentin; Val_{100}, 100 μM validamycin A; NE_{SM}, number of explants cultured on selection medium; NE_{SPT}, number of explants regenerating putative transgenic shoots on selection medium; NS_{PT}, number of putative transgenic shoots regenerated on selection medium; NS_{PCR}, number of PCR-positive shoots; NE_{nsm}, number of explants cultured on non-selective (kanamycin-free) medium; NS_{nsm}, number of shoots regenerated on non-selective medium; $T\eta$, transformation efficiency; MSL, Murashige and Skoog liquid medium; AS, acetosyringone.

[a]Numbers in parentheses indicate the number of shoots or shoot clusters per explant on non-selective medium.

the number of transformed cells in target tissues (James *et al.*, 1993; Alsheikh *et al.*, 2002; Gruchała *et al.*, 2004a; Husaini, 2010) (Table 6.5). There was a 15% increase in transformation percentage in *F. vesca semperflorens* (Alsheikh *et al.*, 2002), an 8–11% increase in *Fragaria* × *ananassa* 'Chandler' (Husaini, 2010), 2–3% in *F.* × *ananassa* 'Induka' and 'Elista' (Gruchała *et al.*, 2004a), while James *et al.* (1990) did not detect any effect of acetosyringone on the percentage of strawberry leaf explants that produced Kan-resistant callus after 100 days of culture. This shows that there is no consistency in the degree of response, which could be due to extreme genotype dependence, variability in regeneration/transformation rates for different cultivars (Alsheikh *et al.*, 2002; Quesada *et al.*, 2007) and suppression of virulence in some strain–plant species interactions (Godwin *et al.*, 1991).

TREHALOSE AND VALIDAMYCIN A. The transformation process causes considerable stress in cells, which could hamper their post-transformation regenerative potential. One effective mechanism to reduce damage from stress is the accumulation of high intracellular levels of trehalose (Crowe *et al.*, 1984; Drennan *et al.*, 1993; Goddijn and van Dun, 1999; Argüelles, 2000). Although it is not known exactly how trehalose interacts with stress pathways, its addition to culture media results in the accumulation of trehalose in transgenic plants (Goddijn *et al.*, 1997). In an interesting study, a specific trehalase inhibitor, validamycin A, was added to the pre-selection medium with the aim of reducing the stress in transformed cells (imposed by the transformation process) and to facilitate the recovery of Kan-resistant putative transformants (Husaini, 2010). There was an unprecedented influence following its addition to the pre-selection medium (Table 6.5). The percentage of putative transgenic shoots regenerating per leaf disc on selection medium increased from 6.1 to 54.1% (ninefold), while the average percentage of leaf discs regenerating shoots on selection medium increased from 3.1 to 7.4% (2.5-fold) (Husaini, 2010). As the plant trehalose biosynthesis pathway is tightly regulated by multiple stress signals such as dehydration (Drennan *et al.*, 1993), osmotic stress tolerance (Styrvold and Strøm, 1991; Hounsa *et al.*, 1998), heat and desiccation tolerance (Hottiger *et al.*, 1987) and chilling stress (Pramanik and Imai, 2005), it probably reduces the 'transformation stress' on cells and results in their increased survival percentage.

6.3. How to Estimate Transformation Success – Efficiency and Percentage are not Synonymous

Examination of the literature indicates a wide difference in regeneration and transformation efficiency. The number of independent shoots generated per explant is usually referred to as the 'regeneration efficiency', while the percentage of explants that produce a transgenic shoot is referred as the 'transformation efficiency' (Folta and Dhingra, 2006). However, this definition leads to many different formulae (Table 6.7). The first four formulae in Table 6.7 give more weight to the regeneration system used, and hence do not reflect the actual transformation efficiency. These formulae aim to calculate the 'number of transformation events' that 'successfully regenerate shoots/plantlets' after application of an appropriate selection pressure, and assume that 'every single shoot' represents a 'unique transformation event'. There is a difference between transformation efficiency and transformation percentage; the two are not synonymous. The former describes the number of transgenic shoots that arise on selection media compared with the number of regeneration events that occur in the absence of selection (Husaini, 2010). However, reporting 'transformation efficiency' as transformants per explant distorts the representation, as Oosumi *et al.* (2006) and Folta *et al.* (2006) reported transformation efficiencies greater than 100%. This metric simply means that each explant produced at least one transgenic shoot. Actually, the transformation efficiency described by Folta *et al.* (2006) is quite low (1–3%). In

Table 6.7. Explanation of the formulae used in Table 6.6. (Adapted from Husaini *et al*., 2011, with kind permission from Global Science Books.)

Formula	How to calculate	Remarks	Reference(s)
$NS_{PT}/NE_{Ai} \times 100$	Number of putative transgenic shoots regenerated on selection medium (usually after 8 weeks)/number of explants agroinfected	This formula can be used at early stages of regeneration but may also include escapes	Husaini (2010); Husaini *et al*. (2011)
$NE_{SPT}/NE_{Ai} \times 100$	Number of explants regenerating putative transgenic shoots on selection medium (usually after 8 weeks)/number of explants agroinfected	This formula ignores multiple transformation events occurring on the same explant but at separate loci	Zhao *et al*. (2004)
$NS_{PT}/NE_{SM} \times 100$	Number of putative transgenic shoots regenerated on selection medium (usually after 8 weeks)/number of explants added to selection medium	This formula ignores the positive/negative effect of the pre-selection strategy (pre-culture, pre-selection, agroinfection) on transformation	Nehra *et al*. (1990a,b)
$NS_{PCR}/NE_{Ai} \times 100$	Number of PCR-confirmed transgenic shoots regenerated on selection medium (usually after 8 weeks)/number of explants agroinfected	This is the most accurate formula, and can be used at a later stage when sufficient tissue material becomes available for DNA isolation/PCR. However, as some transformants die in various stages of development, such transformation events are not taken into account	Husaini (2010)
$(NS_{PT}/NE_{Ai}) \times (NE_{nsm}/NS_{nsm}) \times 100$	(Number of putative transgenic shoots regenerated on selection medium (usually after 8 weeks)/number of explants agroinfected) × (number of explants cultured on non-selective (kanamycin-free) medium/number of shoots regenerated on non-selective medium)	Technically, this is the most appropriate formula for describing transformation 'efficiency', as it compares the relative regeneration capacities of agroinfected and normal (non-agroinfected) explants	Gruchała *et al*. (2004a,b)

our opinion, reporting a transformation 'efficiency' that is greater than 100% (Folta *et al*., 2006; Oosumi *et al*., 2006) is mathematically incorrect, as efficiency cannot be greater than 1, i.e. 100%. However, a transformation 'percentage' of greater than 100% is quite possible, because each leaf explant can regenerate multiple shoots/shoot clusters in strawberry. When calculating transformation percentage, we actually aim to calculate the 'number of transformation events' that 'successfully regenerate shoots/plantlets' when exposed to appropriate selection pressure. Some explants regenerate multiple shoots but 'one shoot per initiation site', and hence each site represents a 'single' transformation event.

Each worker uses his own metric to describe the success of transformation. Nehra *et al*. (1990b) inoculated leaf discs of the cultivar 'Redcoat', which formed light green meristematic regions on selection medium with 50 mg ml^{-1} Kan, later developing into transformed shoots at a frequency of 6.5% (on selection medium with 25 μg ml^{-1} Kan). In another study in the same cultivar, they obtained transformed calluses at 3% on selection medium (with 50 μg ml^{-1} Kan), and 20% of these 'transformed calli' regenerated shoots (Nehra *et al.*, 1990a). Similarly, Jelenkovic *et al.* (1991) showed 40, 30 and 20% transformed callus formation from runner segments, hypocotyl segments and leaf discs, respectively. It should be noted that, when working at the callus level, it is most appropriate to calculate the transformation percentage as 'calli-regenerating explants' on selection medium, but when the same metric is used for the percentage of shoot-regenerating explants, it can be misleading.

Some reports of transformation describe efficiency as the number of transgenic shoots that arise compared with the number of regeneration events that occur in the absence of selection. In other words, it is possible to have efficient regeneration and no transformation, as the latter is dependent on the ability to separate transgenic antibiotic-resistant tissue from non-transformed antibiotic-sensitive tissue. Gruchała *et al.* (2004b) analysed 25 strawberry cultivars to select genotypes most suitable for transformation. The number of regenerated shoots varied from 0.7 to 11 shoots per 100 transformed explants in comparison with 3.5–259.3 shoots from control tissues grown on medium without antibiotics. Transformation/regeneration efficiency, expressed as transformant number per 100 explants, was high for cultivars such as 'Elista' (9.5), 'Wega' (7.3), 'Senga Precosa' (6.5), 'Kama' (6.2), 'Induka' (4.5) and 'Maria' (3.0), which, when expressed as a transformation percentage, were 0.09, 0.07, 0.06, 0.06, 0.04 and 0.03%, respectively. In another study, Gruchała *et al.* (2004a) compared regeneration and transformation conditions for two strawberry cultivars, reporting 8.3 shoots per 100 explants (0.08%) in 'Elista' and 4.2 shoots per 100 explants (0.04%) in 'Induka'.

Zhao *et al.* (2004) used the term 'transformation rate' to calculate transformation success, and calculated it as the percentage of explants that regenerated shoots on the selection medium in 8 weeks. A closer examination revealed that the transformation success was actually calculated as the percentage of 'putative transgenic shoots' on the selection medium. Forty-seven and 36 independent putative transgenic lines, capable of growth on 50 mg l^{-1} Kan, were generated from transformed cells derived from 73 and 53 explants of diploid strawberry accessions FRA197 and FRA198, respectively, representing a transformation frequency of 64% for FRA197 and 68% for FRA198. Under the same conditions, only five and four such putative transgenic lines were obtained from 48 and 54 leaf explants of the octoploid strawberry cultivars 'Hecker' and 'La Sans Rivale', respectively, representing a transformation frequency of 10.4% for 'Hecker' and 7.4% for 'La Sans Rivale'.

Transformation efficiency reports the 'relative' regeneration capacities of agroinfected and control explants, while transformation percentage measures the 'success' in recovering transgenic shoots only. Hence, the parameters taken for calculation of transformation percentage are extremely important when reporting the extent of transformation proficiency. From our previous results, shown in Table 6.6, it is evident that, based on the method of calculation, different values for transformation percentages can be derived. The most widely used formula for transformation percentage is based on the 'number of shoots regenerated per 100 explants on selection medium' and in our opinion is a fairly accurate one. The most accurate is that based on 'number of PCR-positive shoots' recovered. The only thing to be kept in mind in both cases is to take the 'initial number' of explants into account and not the number of explants added to the selection medium, because in the latter case the effect of a pre-selection strategy is ignored.

6.4. Conclusions and Future Perspectives

Despite several non-*Agrobacterium*-based techniques and the use of other bacterial strains, *Agrobacterium*-based techniques still seem to be the most popular among strawberry researchers aiming to introduce novel characteristics into existing cultivars (Table 6.8). Standardization of an efficient regeneration system for each strawberry genotype is an indispensable prerequisite for the successful development of transgenic plants. Factors such as type of explant, darkness treatment, photoperiod, growth regulators and chilling treatment are vital for efficient regeneration of complete plantlets under *in vitro* conditions. As in nature, *Agrobacterium*-mediated genetic transformation is normally a single-cell event. The reasonably good transformation percentage in a strawberry cultivar is generally achieved due to the favourable medium modifications, which foster successful organogenesis from these transformed cells. Sufficient numbers of *Agrobacterium* during co-cultivation, an adequate co-cultivation period, the use of *vir* gene inducers such as acetosyringone, introduction of a pre-selection phase between co-cultivation and selection, and optimum selection pressure are all important factors to obtain stable transformants. An excellent post-agroinfection strategy is the incubation of explants for some days on pre-selection medium, possibly containing an osmoprotectant such as validamycin A. This makes it possible for transformed cells to overcome the stress imposed by agroinfection and also allows a sufficient time gap until the selectable marker gene is able to be expressed effectively in the transformed cells. One more point that needs emphasis is that the metrics employed by different workers for measuring the success of transformation need to be uniform, and use of the terms transformation 'efficiency' and 'percentage' interchangably should be avoided.

Some consumers as well as regulatory authorities express concern about the environmental safety of genetically modified organisms and demand that commercial transgenic plants be free of marker genes (antibiotic resistance) or vector backbone sequences. Using different models, diverse techniques of gene transfer have been reported recently, which may be used in strawberry systems with appropriate modifications, in order to provide an alternative to the complex patent landscape of *Agrobacterium* technology and create an 'open source' platform. Broothaerts *et al.* (2005) took the first step in harnessing non-*Agrobacterium* species for plant genetic engineering. They designed *Rhizobium* sp. NGR234, *Sinorhizobium meliloti* and *Mesorhizobium loti* with the machinery needed for the transformation process: a set of virulence genes encoded by an *Agrobacterium* Ti plasmid and the T-DNA segment residing on a separate, small binary plasmid. The T-DNA segment carries genes encoding a hygromycin selection marker and a *uidA* reporter. Broothaerts *et al.* (2005) then used these bacterial species to transiently and stably transform various plant species, targeting different tissues (e.g. leaf discs, calli, ovule tissue). The results demonstrated that these non-*Agrobacterium* species can genetically transform dicots as well as monocots. Although, to the best of our knowledge, there are no reports as yet that have determined the viability of these '*Agrobacterium*-like' species for the transformation of commercially important strawberry plant species, these findings may open a new way forward with technological and legal advantages over *Agrobacterium* technology. Recently, a novel method of genome editing called CRISPR (clustered regulatory interspaced short palindromic repeats) and a CRISPR-associated (Cas) system has been successfully employed to introduce useful genome modifications, mostly in model plants. As strawberry is octoploid and its complete genome sequence is not yet available, the use of this technology may take a while for successful use in strawberry genome editing.

Acknowledgements

A.M. Husaini is grateful to Dr Jaime A. Teixeira da Silva for permission to reproduce the content from Global Science Books resources. Financial support from the Council of Scientific and Industrial Research, New Delhi, India, is also duly acknowledged.

Table 6.8. Agronomically and economically beneficial traits transferred into strawberry using *Agrobacterium*-mediated gene transfer technology.

Species and cultivar/genotype	Explant	Transgene(s)	Trait studied	Reference
Fragaria × *ananassa* 'Rapella'	–	Cowpea protease inhibitor (*CpTi*)	Insect resistance	James *et al.* (1992, 1993)
Fragaria × *ananassa*	–	Coat protein	Mild yellow edge virus	Finstad and Martin (1995)
Fragaria × *ananassa*' Symphony'	–	Cowpea protease inhibitor (*CpTi*)	Insect resistance	Graham *et al.* (1995)
Fragaria × *ananassa* 'Toyonoka'	–	Chitinase from rice (*rcc2*)	Fungal resistance	Asao *et al.* (1997)
Fragaria × *ananassa* 'Honeoye'	–	*CBF1*	Cold tolerance	Owens *et al.* (2002, 2003)
Fragaria × *ananassa* 'Joliette'	Stipules	Chitinase (*pcht28*)	Fungal resistance	Chalavi *et al.* (2003)
Fragaria × *ananassa* 'Chambly'	Shoot	nptII/acidic dehydrin (*wcor410a*)	Drought tolerance	Houde *et al.* (2004)
Fragaria × *ananassa* 'Toyonaka'	–	Late embryogenesis abundance (*LEA3*)	Drought tolerance	Wang *et al.* (2004)
Fragaria × *ananassa* 'Firework'	Leaf	Thaumatin II (*thauII*)	Resistance against grey mould	Schestibratov and Dolgov (2005)
Fragaria × *ananassa* 'Camarosa'	–	Chitinase and glucanase	Fungal resistance	Mercado *et al.* (2005)
Fragaria × *ananassa* 'Tioga'	–	Antifreeze protein (*AFP*)	Cold tolerance	Khammuang *et al.* (2005)
Fragaria × *ananassa* 'Toyonoka'	Leaf discs	Pathogeny inducing promer (*Prp1-1*)/glucose oxidase (*GO*)	Resistance against grey mould	Jin *et al.* (2005)
Fragaria × *ananassa* 'Pajaro'	–	Chitinase (*ch5B*), glucanase (*gln2*) and thaumatin-like protein (*ap24*)		Vellicce *et al.* (2006)
Fragaria × *ananassa* 'M14'	Leaf discs	*annfaf*	Keeping quality of fruits	Na *et al.* (2006)
Fragaria × *ananassa* 'Toyonaka'	Leaf	*nptII*/Antibacterial peptide-D (*APD*)	Bacteria resistance	Qin and Zhang (2007)
Fragaria × *ananassa* 'Honeoye'	Leaf	*CBF1*	Cold tolerance	Jin *et al.* (2007); Dong *et al.* (2009)
Fragaria × *ananassa* 'Virginia'	Leaf discs	*cNHX1*	Drought tolerance	Wei *et al.* (2007)
Fragaria × *ananassa* 'Kinuama'	Leaf discs	*FaEtr1*, *FaEtr2*, *FaErs1*	Keeping quality of fruits	Zhu (2008)
Fragaria × *ananassa* 'Benihope'	Leaf discs	*PcFT*	Regulation of flowering time	Zheng (2008)
Fragaria × *ananassa* 'Chandler'	Leaf	Osmotin	Salt and drought tolerance	Husaini and Abdin (2008b; Husaini *et al.* (2012)
Fragaria × *ananassa* 'Chandler'	Leaf discs	*FaEtr2*	Keeping quality of fruits	Song (2009)
Fragaria × *ananassa* 'Xingdu 2', 'Allstar'	Leaf discs, petioles	Pre-, pro- and mature antifreeze protein (*AFP*)	Freezing tolerance	Sun *et al.* (2009)
Fragaria × *ananassa* 'Honeoye'	Leaf discs	Resveratrol synthase (*RS*)	Fungal resistance and nutritional quality	Qian (2010)
Fragaria × *ananassa* 'Benihope'	Leaf discs	*AtCYP2*	Nutritional quality	Feng (2011)
Fragaria × *ananassa* 'Akihime'	Leaf	Brazzein, *PpydFN1*	Improved sweetness	Wang (2012)
Fragaria × *ananassa* 'Toyonoka'	Leaf discs	*dhn4*	Drought and cold tolerance	Wang *et al.* (2012b)

Continued

Table 6.8. Continued.

Species and cultivar/genotype	Explant	Transgene(s)	Trait studied	Reference
Fragaria × *ananassa* 'Benihope'	Leaf discs	*RdreB1BI*	Cold tolerance	Gu *et al.* (2012)
Fragaria × *ananassa* 'Lijinv'	Leaf discs	Resveratrol synthase (*RS*)	Fungal resistance and nutritional quality	Lin *et al.* (2012)
Fragaria × *ananassa* 'Sengana'	Leaf discs	*DHAR*	Nutritional quality	Zhang *et al.* (2012)
Fragaria × *ananassa* 'Virginia'	Leaf discs	Tissue-type plasminogen activator (*t-PA*)	Nutritional quality	Bai *et al.* (2012)
Fragaria × *ananassa* 'Fumei 1' and 'Hong shimei'	Leaf discs	*psy*, *pds*	Nutritional quality	Chen (2014)
Fragaria × *ananassa*' Camarosa'	–	*Bgn13.1*	Fungal resistance	Mercado *et al.* (2015)
Fragaria × *ananassa*' Camarosa' and 'Kurdistan'	–	*P5CS*	Drought tolerance	Bahramnejad *et al.* (2015)
Fragaria vesca	–	*AtNPR1*	Fungal resistance	Silva *et al.* (2015)

References

Alsheikh, M.K., Suso, H.P., Robson, M., Battey, N.H. and Wetten, A. (2002) Appropriate choice of antibiotic and *Agrobacterium* strain improves transformation of antibiotic-sensitive *Fragaria vesca* and *F. v. semperflorens*. *Plant Cell Reports* 20, 1173–1180.

Ammirato, P.V. and Steward, F.C. (1971) Some effects of environment on the development of embryos from cultured free cells. *Botanical Gazette* 132, 149–158.

Argüelles, J.C. (2000) Physiological roles of trehalose in bacteria and yeasts: a comparative analysis. *Archives of Microbiology* 174, 217–224.

Asao, H., Nishizawa, Y., Arai, S., Sato, T., Hirai, M., Yoshida, K., Shinmyo, A. and Hibi, T. (1997) Enhanced resistance against a fungal pathogen *Sphaerotheca humuli* in transgenic strawberry expressing a rice chitinase gene. *Plant Biotechnology* 14, 145–150.

Bahramnejad, B., Nasri, S., Mozafari, A. and Siosemardeh, A. (2015) *Agrobacterium tumefaciens*-mediated transformation of two economically important strawberry cultivars with *P5CS* gene. *Journal of Plant Physiology and Breeding* 5, 55–66.

Bai, X., Di, S., Zhang, X. and Zhu, Y. (2012) Establishment of strawberry regeneration and transformation system through leaf discs. *Journal of Northeast Agricultural University* 43, 141–150.

Baldan, B., Guzzo, F., Filippini, F., Gasparian, M., LoSchiavo, F., Vitale, A., de Vries, S.C., Mariani, P. and Terzi, M. (1997) The secretory nature of the lesion of carrot cell variant *ts11*, rescuable by endochitinase. *Planta* 203, 381–389.

Barceló, M., El-Mansouri, I., Mercado, J.A., Quesada, M.A. and Pliego Alfaro, F. (1998) Regeneration and transformation via *Agrobacterium tumefaciens* of the strawberry cultivar Chandler. *Plant Cell, Tissue and Organ Culture* 54, 29–36.

Bevan, M., Shufflebottom, D., Edwards, K., Jefferson, R. and Schuch, W. (1989) Tissue-and cell-specific activity of a phenylalanine ammonia-lyase promoter in transgenic plants. *EMBO Journal* 8, 1899.

Bhatt, I.D. and Dhar, U. (2000) Micropropagation of Indian wild strawberry. *Plant Cell, Tissue and Organ Culture* 60, 83–88.

Blando, F., Niglio, A., Frattarelli, A., Speranza, S. and Damiano, C. (1993) Cell suspension cultures in strawberry: growth characterization and variability. *Acta Horticulturae* 336, 257–262.

Bois, F. (1992) The influence of some natural cell-wall derived precursors on organogenesis and differentiation of wild strawberry (*Fragaria vesca* L.) callus cultures. *Plant Cell, Tissue and Organ Culture* 28, 91–96.

Broothaerts, W., Mitchell, H.J., Weir, B., Kaines, S., Smith, L.M., Yang, W., Mayer, J.E., Roa-Rodríguez, C. and Jefferson, R.A. (2005) Gene transfer to plants by diverse species of bacteria. *Nature* 433, 629–633.

Chalavi, V., Tabaeizadeh, Z. and Thibodeau, P. (2003) Enhanced resistance to *Verticillium dahliae* in transgenic strawberry plants expressing a *Lycopersicon chilense* chitinase gene. *Journal of the American Society for Horticultural Science* 128, 747–753.

Chen, H., Gao, L., Lu, H., Chen, M. and Zhang, L. (2011) Studies of stem tip culture of strawberry virus-free tissue culture technology. *Beijing Agriculture* 3, 18–20.

Chen, M. (2014) Studies on transformation of carotenoids *psy*, *pds* genes in strawberry. Masters thesis, Fujian Agriculture and Forestry University, Fuzhou, Fujian, China.

Crowe, J.H., Crowe, L.M. and Chapman, D. (1984) Preservation of membranes in anhydrobiotic organisms: the role of trehalose. *Science* 223, 701–703.

Damiano, C., Monticelli, S., Frattarelli, A., Nicolini, S. and Corazza, L. (1997) Somaclonal variability and *in vitro* regeneration of strawberry. *Acta Horticulturae* 447, 87–94.

De Jong, A.J., Cordewener, J., Lo Schiavo, F., Terzi, M., Vandekerckhove, J., Van Kammen, A. and De Vries, S.C. (1992) A carrot somatic embryo mutant is rescued by chitinase. *The Plant Cell* 4, 425–433.

de Mesa, M.C., Jiménez-Bermúdez, S., Pliego-Alfaro, F., Quesada, M.A. and Mercado, J.A. (2000) *Agrobacterium* cells as microprojectile coating: a novel approach to enhance stable transformation rates in strawberry. *Functional Plant Biology* 27, 1093–1100.

Debnath, S. (2005) Strawberry sepal: another explant for thidiazuron-induced adventitious shoot regeneration. *In Vitro Cellular and Developmental Biology – Plant* 41, 671–676.

Debnath, S.C. (2006) Zeatin overcomes thidiazuron-induced inhibition of shoot elongation and promotes rooting in strawberry culture in vitro. *Journal of Horticultural Science and Biotechnology* 81, 349–354.

Debnath, S.C. and Teixeira da Silva, J.A. (2007) Strawberry culture *in vitro*: applications in genetic transformation and biotechnology. *Fruit, Vegetable and Cereal Science and Biotechnology* 1, 1–12.

Decout, E., Dubois, T., Guedira, M., Dubois, J., Audran, J.-C. and Vasseur, J. (1994) Role of temperature as a triggering signal for organogenesis or somatic embryogenesis in wounded leaves of chicory cultured *in vitro*. *Journal of Experimental Botany* 45, 1859–1865.
Deng, X. and Hu, W.Y. (2000) Establishment of regeneration and genetic transformation system of strawberry leaf. *Chinese Bulletin of Botany* 17, 174–178.
Dhir, S., Hinchee, M.A.W., Layton, J.G. and Oakes, J.V. (2001) Methods for strawberry transformation using *Agrobacterium tumefaciens*. USA Patent 6274791.
Dong, J., Zhang, Y., Pan, Q., Li, S. and Jin, W. (2009) Genetic analysis of foreign transferring gene *CBF1* in strawberry. *Molecular Plant Breeding* 7, 63–66.
Dong, J.C. (2013) Tissue culture of stem tip and rapid propagation of *Fragaria* × *ananassa* Duch. 'Benihoppe'. *Northern Horticulture* 24, 106–108.
Donnoli, R., Sunseri, F., Martelli, G. and Greco, I. (2001) Somatic embryogenesis, plant regeneration and genetic transformation in *Fragaria* spp. *Acta Horticulturae* 560, 235–240.
Drennan, P.M., Smith, M.T., Goldsworthy, D. and van Staden, J. (1993) The occurrence of trehalose in the leaves of the desiccation-tolerant angiosperm *Myrothamnus flabellifolius* Welw. *Journal of Plant Physiology* 142, 493–496.
Du Plessis, H.J., Brand, R.J., Glyn-Woods, C. and Goedhart, M.A. (1997) Efficient genetic transformation of strawberry (*Fragaria* × *ananassa* Duch.) cultivar Selekta. *Acta Horticulturae* 447, 289–294.
El Mansouri, I., Mercado, J.A., Valpuesta, V., López-Aranda, J.M., Pliego-Alfaro, F. and Quesada, M.A. (1996) Shoot regeneration and *Agrobacterium*-mediated transformation of *Fragaria vesca* L. *Plant Cell Reports* 15, 642–646.
Feng, L. (2011) Studies on in vitro regeneration and transformation of strawberry (*Fragaria ananassa* Duch.) cultivar 'Benihoppe' and its application. Masters thesis, Zhejiang University, Hangzhou, Zhejiang, China.
Finstad, K. and Martin, R.R. (1995) Transformation of strawberry for virus resistance. *Acta Horticulturae* 385, 86–90.
Fiore, M.C., Trabace, T. and Sunseri, F. (1997) High frequency of plant regeneration in sunflower from cotyledons via somatic embryogenesis. *Plant Cell Reports* 16, 295–298.
Flores, R., Oliveira, M.d.F., Camargo, J.T., Andrade, L., Peters, J.A. and Fortes, G.R.D.L. (1998) *In vitro* callogenesis and organogenesis in two strawberry cultivars. *Agropecuaria-Clima-Temperado* 1, 155–161.
Foley, I. and Hennerty, M.J. (1993) Field performance of androgenic strawberry plants. *Acta Horticulturae* 348, 188–195.
Folta, K. and Dhingra, A. (2006) Transformation of strawberry: the basis for translational genomics in Rosaceae. *In Vitro Cellular and Developmental Biology – Plant* 42, 482–490.
Folta, K.M., Dhingra, A., Howard, L., Stewart, P.J. and Chandler, C.K. (2006) Characterization of LF9, an octoploid strawberry genotype selected for rapid regeneration and transformation. *Planta* 224, 1058–1067.
Forsline, P.L. and Langille, A.R. (1975) Endogenous cytokinins in *Solanum tuberosum* as influenced by photoperiod and temperature. *Physiologia Plantarum* 34, 75–77.
Foucault, C. and Letouze, R. (1987) *In vitro*: régénération de plants de fraisier à partir de fragments de pétioles et de bourgeons floraux. *Biologia Plantarum* 29, 409–414.
Gamborg, O.L., Miller R.A., Ojima, K. (1968) Nutrient requirements of suspension cultures of soybean root cells. *Experimental Cell Research* 50, 151–158.
George, E.F. (1993) *Plant Propagation by Tissue Culture. Part 1: The Technology*. Exegetics Ltd, Edington, UK.
Giuliano, G., Lo Schiavo, F. and Terzi, M. (1984) Isolation and developmental characterization of temperature-sensitive carrot cell variants. *Theoretical and Applied Genetics* 67, 179–183.
Goddijn, O.J.M. and van Dun, K. (1999) Trehalose metabolism in plants. *Trends in Plant Science* 4, 315–319.
Goddijn, O.J.M., Verwoerd, T.C., Voogd, E., Krutwagen, R.W.H.H., de Graff, P.T.H.M., Poels, J., van Dun, K., Ponstein, A.S., Damm, B. and Pen, J. (1997) Inhibition of trehalase activity enhances trehalose accumulation in transgenic plants. *Plant Physiology* 113, 181–190.
Godwin, I., Todd, G., Ford-Lloyd, B. and Newbury, H.J. (1991) The effects of acetosyringone and pH on *Agrobacterium*-mediated transformation vary according to plant species. *Plant Cell Reports* 9, 671–675.
Graham, J., McNicol, R.J. and Greig, K. (1995) Towards genetic based insect resistance in strawberry using the cowpea trypsin inhibitor gene. *Annals of Applied Biology* 127, 163–173.
Greene, G.M., Gaus, A.G. and Lehman, L.J. (1990) World's largest, small chamber, recirculating, controlled atmosphere (CA), storage research facility. *HortScience* 25, 1092.
Gruchała, A., Korbin, M. and Żurawicz, E. (2004a) Conditions of transformation and regeneration of 'Induka' and 'Elista' strawberry plants. *Plant Cell, Tissue and Organ Culture* 79, 153–160.
Gruchała, A., Korbin, M. and Żurawicz, E. (2004b) Suitability of selected strawberry cultivars for genome modification by *Agrobacterium tumefaciens*. *Acta Horticulturae* 663, 491–494.

Gu, X.B., Wang, F., Gao, Z.H., Qiao, Y.S., Zhang, Z., Lin, Z.L., Mi, L. and Li, J.F. (2012) Establishment of *Agrobacterium*-mediated transformation system for 'Benihope' strawberry. *Acta Agriculturae Jiangxi* 24, 11–14.

Guha-Mukherjeem, S. (2002) Level of endogenous free amino acids during various stages of culture of *Vigna mungo* (L.) Hepper – somatic embryogenesis, organogenesis and plant regeneration. *Current Science* 82, 429–433.

Haddadi, F., Abd Aziz, M., Abdullah, S.N.A., Tan, S.G. and Kamaladini, H. (2015) An efficient *Agrobacterium*-mediated transformation of strawberry cv. Camarosa by a dual plasmid system. *Molecules* 20, 3647–3666.

Halperin, W. (1966) Alternative morphogenetic events in cell suspensions. *American Journal of Botany* 53, 443–453.

Hammoudeh, H.Y., Suwwan, M.A., Abu Quoud, H.A. and Shibli, R.A. (1998) Micropropagation and regeneration of Honeoye strawberry. *Dirasat: Agricultural Sciences* 25, 170–178.

Hanhineva, K., Kokko, H. and Kärenlampi, S. (2009) Shoot regeneration from leaf explants of five strawberry (*Fragaria* × *ananassa*) cultivars in temporary immersion bioreactor system. *In Vitro Cellular and Developmental Biology – Plant* 41, 826–831.

Harpster, M.H., Brummell, D.A. and Dunsmuir, P. (1998) Expression analysis of a ripening-specific, auxin-repressed endo-1,4-β-glucanase gene in strawberry. *Plant Physiology* 1307–1316.

Haymes, K.M. and Davis, T.M. (1998) *Agrobacterium*-mediated transformation of 'Alpine' *Fragaria vesca*, and transmission of transgenes to R1 progeny. *Plant Cell Reports* 17, 279–283.

He, X.Y., Bi, H.L., Xue, R.G., Tao, B. and He, Q.J. (2013) Study on tissue culture and rapid propagation technique of *Fragaria ananassa* Totem. 26, 701–704.

Hood, E.E., Helmer, G.L., Fraley, R.T. and Chilton, M.D. (1986) The hypervirulence of *Agrobacterium tumefaciens* A281 is encoded in a region of pTiBo542 outside of T-DNA. *Journal of Bacteriology* 168, 1291–1301.

Horsch, R.B., Fry, J.E., Hoffmann, N.L., Eichholtz, D., Rogers, S.G. and Fraley, R.T. (1985) A simple and general method for transferring genes into plants. *Science* 227, 1229–1231.

Hottiger, T., Boller, T. and Wiemken, A. (1987) Rapid changes of heat and desiccation tolerance correlated with changes of trehalose content in *Saccharomyces cerevisiae* cells subjected to temperature shifts. *FEBS Letters* 220, 113–115.

Houde, M., Dallaire, S., N'Dong, D. and Sarhan, F. (2004) Overexpression of the acidic dehydrin WCOR410 improves freezing tolerance in transgenic strawberry leaves. *Plant Biotechnology Journal* 2, 381–387.

Hounsa, C.-G., Brandt, E.V., Thevelein, J., Hohmann, S. and Prior, B.A. (1998) Role of trehalose in survival of *Saccharomyces cerevisiae* under osmotic stress. *Microbiology* 144, 671–680.

Husaini, A.M. (2010) Pre- and post-agroinfection strategies for efficient leaf disk transformation and regeneration of transgenic strawberry plants. *Plant Cell Reports* 29, 97–110.

Husaini, A.M. and Abdin, M.Z. (2007) Interactive effect of light, temperature and TDZ on the regeneration potential of leaf discs of *Fragaria* × *ananassa* Duch. *In Vitro Cellular and Developmental Biology – Plant* 43, 576–584.

Husaini, A.M. and Abdin, M.Z. (2008a) Overexpression of tobacco osmotin gene leads to salt stress tolerance in strawberry (*Fragaria* × *ananassa* Duch.) plants. *Indian Journal of Biotechnology* 7, 465–472.

Husaini, A.M. and Abdin, M.Z. (2008b) Development of transgenic strawberry (*Fragaria* × *ananassa* Duch.) plants tolerant to salt stress. *Plant Science* 174, 446–455.

Husaini, A.M. and Srivastava, D.K. (2006a) Genetic transformation in strawberry-A review. *Asian Journal of Microbiology, Biotechnology and Environmental Sciences* 8, 75–81.

Husaini, A.M. and Srivastava, D.K. (2006b) Plant regeneration and *Agrobacterium*-mediated gene transfer studies in strawberry tissues (*Fragaria* × *ananassa*). *Asian Journal of Microbiology Biotechnology and Environmental Sciences* 8, 671–678.

Husaini, A.M., Aquil, S., Bhat, M., Qadri, T., Kamaluddin, T.A. and Abdin, M.Z. (2008) A high-efficiency direct somatic embryogenesis system for strawberry (*Fragaria* × *ananassa* Duch.) cultivar Chandler. *Journal of Crop Science and Biotechnology* 11, 107–110.

Husaini, A.M., Mercado, J.A., Texeira da Silva, J.A. and Schaart, J.G. (2011) Review of factors affecting organogenesis, somatic embryogenesis and *Agrobacterium tumefaciens*-mediated transformation of strawberry. Global Science Books, UK, pp. 1–11.

Husaini, A.M., Abdin, M.Z., Khan, S., Xu, Y.W., Aquil, S. and Anis, M. (2012) Modifying strawberry for better adaptability to adverse impact of climate change. *Current Science* 102, 1660–1673.

Infante, R., Iasi, L., Berardi, G. and Rosati, P. (1993) Isolation and culture of blackberry protoplasts (*Rubus* sp.) var. 'Hull Thornless'. *Acta Horticulturae* 352, 345–352.

Infante, R., Mazzara, M. and Rosati, P. (1998) Growth estimation of *in vitro* cultured callus and plant regeneration from leaf disk and petiole callus of musk strawberry (*Fragaria moschata* Duch.) subcultured for 18 months. *Journal of the Japanese Society for Horticultural Science* 67, 39–43.

Isac, V., Popescu, A.N. and Coman, M. (1994) Studies on plant regeneration from tissue derived callus in *Fragaria × ananassa* Duch. In: Schmidt, H. and Kellerhals, M. (eds) *Progress in Temperate Fruit Breeding. Springer*, The Netherlands, pp. 395–398.

James, D.J., Passey, A.J. and Barbara, D.J. (1990) *Agrobacterium*-mediated transformation of apple and strawberry using disarmed Ti-binary vectors. *Acta Horticulturae* 280, 495–502.

James, D.J., Passey, A.J., Eastebrook, M.A., Solomon, M.G. and Barbara, D.J. (1992) Transgenes for pest and disease resistance: progress in the introduction of transgenes for pest and disease resistance into strawberries. *Phytoparasitica* 20, S83–S87.

James, D.J., Uratsu, S., Cheng, J., Negri, P., Viss, P. and Dandekar, A.M. (1993) Acetosyringone and osmoprotectants like betaine or proline synergistically enhance *Agrobacterium*-mediated transformation of apple. *Plant Cell Reports* 12, 559–563.

Jelenkovic, G., Chin, C., Billings, S., Eberhardt, J., Dale, A. and Luby, J.J. (1991) Transformation studies in cultivated strawberry, *Fragaria × ananassa* Duch. In: *Proceedings of the Third North American Strawberry Conference*, 14–16 February 1990, Houston, Texas, pp. 91–97.

Jemmali, A., Robbe, A. and Boxus, P. (1997) Histochemical events and other fundamental differences between axillary and adventitious stipular shoots in micropropagated strawberry. *Acta Horticulturae* 439, 341–346.

Jiménez-Bermúdez, S., Redondo-Nevado, J., Muñoz-Blanco, J., Caballero, J.L., López-Aranda, J.M., Valpuesta, V., Pliego-Alfaro, F., Quesada, M.A. and Mercado, J.A. (2002) Manipulation of strawberry fruit softening by antisense expression of a pectate lyase gene. *Plant Physiology* 128, 751–759.

Jin, W., Yin, S., Lu, R., Yuan, W., Dong, J. and Wang, G. (2005) *GO* gene transformation of strawberry and its resistance to gray mold. *Molecular Plant Breeding* 6, 797–800.

Jin, W.M., Dong, J., Yin, S.P., Yan, A.L. and Chen, M.X. (2007) *CBF1* gene transgenic strawberry and increase freezing tolerance. *Acta Botanica Boreali-Occidentalia Sinica* 27, 223–227.

Jones, O.P., Waller, B. and Beech, M.G. (1988) The production of strawberry plants from callus cultures. *Plant Cell, Tissue and Organ Culture* 12, 235–241.

Khammuang, S., Dheeranupattana, S., Hanmuangjai, P. and Wongroung, S. (2005) *Agrobacterium*-mediated transformation of modified antifreeze protein gene in strawberry. *Songklanakarin Journal of Science and Technology* 27, 693–703.

Kim, S.S., Kim, S.Y., Kim, D.W., Shin, S.G. and Chang, K.S. (1998) Moisture sorption characteristics of composite foods filled with strawberry jam. *Food Science and Technology* 31, 397–401.

Lazo, G.R., Stein, P.A. and Ludwig, R.A. (1991) A DNA transformation-competent Arabidopsis genomic library in *Agrobacterium. Nature Biotechnology* 9, 963–967.

Lee, E.C.M. and de Fossard, R.A. (1977) Some factors affecting multiple bud formation of strawberry (*Fragaria ananassa* Duchesne) *in vitro. Acta Horticulturae* 78, 187–196.

Li, X. and Tang, H. (2006) Progress on genetic transformation of strawberry mediated by *Agrobacterium tumefaciens. Biotechnology Bulletin* 2006, 23–27.

Li, X.J., Zheng, Y.C. and Chi, L.H. (2013) Screening of the propagation medium of *Fragaria × ananassa* seedlings *in vitro. Journal of Jilin Agricultural Sciences* 38, 63–65.

Lin, B., Qian, K., Lian, X., Lin, Z. and Zhang, Y. (2012) Transformation of resveratrol synthase gene in *Fragaria ananassa. Chinese Journal of Tropical Crops* 33, 848–853.

Lis, E.K. (1987) Somatic embryogenesis and plantlets formation from flower buds cultivated *in vitro. Acta Horticulturae* 212, 731.

Lis, E.K. (1993) Strawberry plant regeneration by organogenesis from peduncle and stolon segements. *Acta Horticulturae* 348, 435–438.

Lis, E. and Chlebowska, D. (2000) Growth characteristics of micropropagated strawberry plants in field cultivation. *Zeszyty Naukowe Instytutu Sadownictwa i Kwiaciarstwa w Skierniewicach* 8, 263–267.

Litwinczuk, W. and Zubel, A. (2005) Growth in vitro cultures of strawberry (*Fragaria × ananassa* Duch.) depending on different photoperiods. *Folia Horticulturae* 17, 81–87.

Liu, C.X., Cui, M., Chen, X.S. and Peng, F.T. (2012) Study on tissue culture and rapid propagation techniques for new strawberry line Miaozi. *Shandong Agricultural Sciences* 44, 17–19.

Liu, R. (2013) Standardization of tissue culturing and temporary immersion bioreactor and strong seedling with aperture disk in 'Hong Jia' strawberry. Masters thesis, Zhejiang Sci-Tech University, Hangzhou, Zhejiang, China.

Liu, Y.T., Liu, C.Q., Feng, X.J., Xi, G.C., Sun, X., Pu, B.Z., Liu, F.S. and Wang, Q.L. (2011) Study on influence factors of rapid propagation of strawberry. *Journal of Hebei Agricultural Sciences* 15, 41–43, 45.

Liu, Z.R. and Sanford, J.C. (1988) Plant regeneration by organogenesis from strawberry leaf and runner tissue. *HortScience* 23, 1057–1059.

Lo Schiavo, F., Giuliano, G., de Vries, S.C., Genga, A., Bollini, R., Pitto, L., Cozzani, F., Nuti-Ronchi, V. and Terzil, M. (1990) A carrot cell variant temperature sensitive for somatic embryogenesis reveals a defect in the glycosylation of extracellular proteins. *Molecular and General Genetics* 223, 385–393.
Lopez-Aranda, J.M., Pliego-Alfaro, F., Lopez-Navidad, I. and Barcelo-Munoz, M. (1994) Micropropagation of strawberry (*Fragaria* × *ananassa* Duch.). Effect of mineral salts, benzyladenine levels and number of subcultures on the in vitro and field behaviour of the obtained microplants and the fruiting capacity of their progeny. *Journal of Horticultural Science* 69, 625–637.
Martinelli, A., Gaiani, A. and Cella, R. (1997) *Agrobacterium*-mediated transformation of strawberry cultivar Marmolada® onebar*. *Acta Horticulturae* 439, 169–174.
Mathews, H., Wagoner, W., Kellogg, J. and Bestwick, R. (1995) Genetic transformation of strawberry: stable integration of a gene to control biosynthesis of ethylene. *In Vitro Cellular and Developmental Biology – Plant* 31, 36–43.
Mathews, H., Dewey, V., Wagoner, W. and Bestwick, R.K. (1998) Molecular and cellular evidence of chimaeric tissues in primary transgenics and elimination of chimaerism through improved selection protocols. *Transgenic Research* 7, 123–129.
Mercado, J.A., Martín-Pizarro, C., Pascual, L., Quesada, M.A., Pliego-Alfaro, F., Santos, B.D.L., Romero, F., Galvez, J., Rey, M., Viña, G.D.L., Llobell, A., Yubero-Serrano, E.-M., Muñoz-Blanco, J. and Caballero, J.L. (2005) Evaluation of tolerance to *Colletotrichum acutatum* in strawberry plants transformed with *Trichoderma*-derived genes. *Acta Horticulturae* 738, 383–388.
Mercado, J.A., Pliego-Alfaro, F. and Quesada, M.A. (2007) Strawberry. In: Pua, E.C. and Davey, M.R. (eds) *Transgenic Crops V. Biotechnology in Agriculture and Forestry*. Springer, Berlin, pp. 309–328.
Mercado, J.A., Barceló, M., Pliego, C., Rey, M., Caballero, J., Muñoz-Blanco, J., Ruano-Rosa, D., López-Herrera, C., de los Santos, B., Romero-Muñoz, F. and Pliego-Alfaro, F. (2015) Expression of the β-1,3-glucanase gene *bgn13.1* from *Trichoderma harzianum* in strawberry increases tolerance to crown rot diseases but interferes with plant growth. *Transgenic Research* 24, 979–989.
Mezzetti, B., Costantini, E., Chionchetti, F., Landi, L., Pandolfini, T. and Spena, A. (2004) Genetic transformation in strawberry and raspberry for improving plant productivity and fruit quality. *Acta Horticulturae* 649, 107–110.
Miller, A.R. and Chandler, C.K. (1990) Plant regeneration from excised cotyledons of mature strawberry schenes. *HortScience* 25, 569–571.
Monticelli, S., Gentile, A. and Damiano, C. (2002) Regeneration and *Agrobacterium*-mediated transformation in stipules of strawberry. *Acta Horticulturae* 567, 105–107.
Murashige, T. and Skoog, F. (1962). A revised medium for rapid growth and bio assays with tobacco tissue cultures. *Physiologia Plantarum* 15, 473–497.
Na, J., Wang, G.L., Xia, R. and Yang, H.Y. (2006) Construction of anti-sense gene of annfaf and genetic transformation of strawberry. *Scientia Agricultura Sinica* 39, 582–586.
Nasri, S. and Bahramnejad, B. (2013) Regeneration and *Agrobacterium*-mediated transformation of three economically important strawberry cultivars Kurdistan, Camarosa and Paros. *Progress in Biological Sciences* 3, 60–66.
Nehra, N.S., Stushnoff, C. and Kartha, K. (1989) Direct shoot regeneration from strawberry leaf disks. *Journal of the American Society for Horticultural Science* 114, 1014–1018.
Nehra, N.S., Kartha, K.K., Stushnoff, C. and Giles, K.L. (1994) Effect of in vitro propagation methods on field performance of two strawberry cultivars. *Euphytica* 76, 107–115.
Nehra, N.S., Chibbar, R.N., Kartha, K.K., Datla, R.S., Crosby, W.L. and Stushnoff, C. (1990a) *Agrobacterium*-mediated transformation of strawberry calli and recovery of transgenic plants. *Plant Cell Reports* 9, 10–13.
Nehra, N.S., Chibbar, R.N., Kartha, K.K., Datla, R.S.S., Crosby, W.L. and Stushnoff, C. (1990b) Genetic transformation of strawberry by *Agrobacterium tumefaciens* using a leaf disk regeneration system. *Plant Cell Reports* 9, 293–298.
Nehra, N.S., Stushnoff, C. and Kartha, K.K. (1990c) Regeneration of plants from immature leaf-derived callus of strawberry (*Fragaria* × *ananassa*). *Plant Science* 66, 119–126.
Nyman, M. and Wallin, A. (1988) Plant regeneration from strawberry (*Fragaria* × *ananassa*) mesophyll protoplasts. *Journal of Plant Physiology* 133, 375–377.
Nyman, M. and Wallin, A. (1991) Protoplast technology in strawberry (*Fragaria*× *ananassa*). *Physiologia Plantarum* 82, A34.
Nyman, M. and Wallin, A. (1992) Improved culture technique for strawberry (*Fragaria* × *ananassa* Duch.) protoplasts and the determination of DNA content in protoplast derived plants. *Plant Cell, Tissue and Organ Culture* 30, 127–133.

Okasha, K.A., Ragab, M.E., Mekail, A.F. and El-Bahr, M.K. (1996) In vitro propagation of strawberry plants from runners. *Annals of Agricultural Science Special Issue*, 163–171.
Oosumi, T., Gruszewski, H.A., Blischak, L.A., Baxter, A.J., Wadl, P.A., Shuman, J.L., Veilleux, R.E. and Shulaev, V. (2006) High-efficiency transformation of the diploid strawberry (*Fragaria vesca*) for functional genomics. *Planta* 223, 1219–1230.
Owen, H.R. and Miller, A.R. (1996) Haploid plant regeneration from anther cultures of three North American cultivars of strawberry (*Fragaria* × *ananassa* Duch.). *Plant Cell Reports* 15, 905–909.
Owens, C.L., Thomashow, M.F., Hancock, J.F. and Iezzoni, A.F. (2002) *CBF1* orthologs in sour cherry and strawberry and the heterologous expression of *CBF1* in strawberry. *Journal of the American Society for Horticultural Science* 127, 489–494.
Owens, C.L., Iezzoni, A.F. and Hancock, J.F. (2003) Enhancement of freezing tolerance of strawberry by heterologous expression of *CBF1*. *Acta Horticulturae* 626, 93–100.
Park, J.-I., Lee, Y.-K., Chung, W.-I., Lee, I.-H., Choi, J.-H., Lee, W.-M., Ezura, H., Lee, S.-P. and Kim, I.-J. (2006) Modification of sugar composition in strawberry fruit by antisense suppression of an ADP-glucose pyrophosphorylase. *Molecular Breeding* 17, 269–279.
Passey, A., Barrett, K. and James, D. (2003) Adventitious shoot regeneration from seven commercial strawberry cultivars (*Fragaria* × *ananassa* Duch.) using a range of explant types. *Plant Cell Reports* 21, 397–401.
Petrovic, D. and Jacimovic, P.M. (1990) Propagation of the strawberry cultivar Senga Sengana by in vitro meristem culture. *Practice* 20, 11–18.
Popescu, A.N., Isac, V.S., Coman, M.S. and Radulescu, M.S. (1997) Somaclonal variation in plants regenerated by organogenesis from callus culture of strawberry (*Fragaria* × *ananassa*). *Acta Horticulturae* 439, 89–96.
Pramanik, M.H.R. and Imai, R. (2005) Functional identification of a trehalose 6-phosphate phosphatase gene that is involved in transient induction of trehalose biosynthesis during chilling stress in rice. *Plant Molecular Biology* 58, 751–762.
Puite, K. and Schaart, J. (1998) *Agrobacterium*-mediated transformation of the apple cultivars 'Gala', 'Golden Delicious' and 'Elstar', and the strawberry cultivars 'Gariguette', 'Polka' and 'Elsanta'. *Acta Horticulturae* 484, 547–556.
Qian, K. (2010) Transformation of resveratrol synthase gene in strawberry. Masters thesis, Fujian Agriculture and Forestry University, Fuzhou, Fujian, China.
Qin, Y.H. and Zhang, S.L. (2007) Factors influencing the efficincy of *Agrobacterium*-mediated transformation in strawberry cultivar Toyonoka. *Journal of Nuclear Agricultural Sciences* 21, 461–465.
Qin, Y., Zhang, S., Asghar, S., Zhang, L., Qin, Q., Chen, K. and Xu, C. (2005a) Regeneration mechanism of Toyonoka strawberry under different color plastic films. *Plant Science* 168, 1425–1431.
Qin, Y., Zhang, S., Zhang, L.X., Zhu, D. and Asghar, S. (2005b) Response of strawberry cv. Toyonoka *in vitro* to silver nitrate ($AgNO_3$). *HortScience* 40, 747–751.
Quesada, M.A., Martín-Pizarro, C., García-Gago, J.A., Posé, S., Santiago, N., Sesmero, R., Pliego-Alfaro, F. and Mercado, J.A. (2007) Transgenic strawberry: current status and future perspectives. *Transgenic Plant Journal* 1, 280–288.
Ricardo, V.G., Coll, Y., Castagnaro, A. and Ricci, J.C.D. (2003) Transformation of a strawberry cultivar using a modified regeneration medium. *HortScience* 38, 277–280.
Rugini, E. (1992) High efficiency shoot regeneration from calluses of strawberry (*Fragaria* × *ananassa* Duch.) stipules of *in vitro* shoot cultures. *Journal of Horticultural Science* 67, 577–582.
Saito, Y. and Sakamori, M. (1995) A survey on the mutation of plantlets regenerated from multiple shoots in strawberry [*Fragaria chiloensis*]. *Research Bulletin of the Aichi-ken Agricultural Research Center* No. 27, 181–186.
Sarwar, M. (1989) The role of medium composition and light intensity on in vitro root formation of strawberry. *Pakistan Journal of Botany* 21, 24–30.
Schaart, J.G. (2014) *Agrobacterium*-mediated transformation of strawberry. *Bio-protocol* 4, e1022.
Schaart, J., Salentijn, E. and Krens, F. (2002) Tissue-specific expression of the β-glucuronidase reporter gene in transgenic strawberry (*Fragaria* × *ananassa*) plants. *Plant Cell Reports* 21, 313–319.
Schaart, J.G., Salentijn, E.M.J., Pelgrom, K.T.B., Aharoni, A. and Krens, F.A. (2010) Isolation and characterisation of a strawberry fruit-specific promoter. In: Husaini, A.M. and Mercado, J.A. (eds) *Genomics, Transgenics, Molecular Breeding and Biotechnology of Strawberry*. Global Science Books, Japan/UK, pp. 108–114.
Schestibratov, K.A. and Dolgov, S.V. (2005) Transgenic strawberry plants expressing a thaumatin II gene demonstrate enhanced resistance to *Botrytis cinerea*. *Scientia Horticulturae* 106, 177–189.
Shulaev, V., Korban, S.S., Sosinski, B., Abbott, A.G., Aldwinckle, H.S., Folta, K.M., Iezzoni, A., Main, D., Arús, P., Dandekar, A.M., Lewers, K., Brown, S.K., Davis, T.M., Gardiner, S.E., Potter, D. and Veilleux, R.E. (2008) Multiple models for Rosaceae genomics. *Plant Physiology* 147, 985–1003.

Shulaev, V., Sargent, D.J., Crowhurst, R.N., Mockler, T.C., Folkerts, O., Delcher, A.L., Jaiswal, P., Mockaitis, K., Liston, A., Mane, S.P., Burns, P., Davis, T.M., Slovin, J.P., Bassil, N., Hellens, R.P., Evans, C., Harkins, T., Kodira, C., Desany, B., Crasta, O.R., Jensen, R.V., Allan, A.C., Michael, T.P., Setubal, J.C., Celton, J.-M., Rees, D.J.G., Williams, K.P., Holt, S.H., Rojas, J.J.R., Chatterjee, M., Liu, B., Silva, H., Meisel, L., Adato, A., Filichkin, S.A., Troggio, M., Viola, R., Ashman, T.-L., Wang, H., Dharmawardhana, P., Elser, J., Raja, R., Priest, H.D., Bryant, D.W., Fox, S.E., Givan, S.A., Wilhelm, L.J., Naithani, S., Christoffels, A., Salama, D.Y., Carter, J., Girona, E.L., Zdepski, A., Wang, W., Kerstetter, R.A., Schwab, W., Korban, S.S., Davik, J., Monfort, A., Denoyes-Rothan, B., Arus, P., Mittler, R., Flinn, B., Aharoni, A., Bennetzen, J.L., Salzberg, S.L., Dickerman, A.W., Velasco, R., Borodovsky, M., Veilleux, R.E. and Folta, K.M. (2011) The genome of woodland strawberry (*Fragaria vesca*). *Nature Genetics* 43, 109–116.

Silva, K.J.P., Brunings, A., Peres, N.A., Mou, Z. and Folta, K.M. (2015) The *Arabidopsis NPR1* gene confers broad-spectrum disease resistance in strawberry. *Transgenic Research* 24, 693–704.

Singh, A.K. and Pandey, S.N. (2004) Genotypic variation among strawberry cultivars for shoot organogenesis. *Acta Horticulturae* 662, 277–280.

Song, C.L. (2009) Study of strawberry ethylene receptor *FaEtr2* gene RNAi plant expression vector construction and genectic transformation. PhD thesis, Agricultural University of Hebei, Baoding, Hebei, China.

Sorvari, S., Ulvinen, S., Hietaranta, T. and Hiirsalmi, H. (1993) Preculture medium promotes direct shoot regeneration from micropropagated strawberry leaf disks. *HortScience* 28, 55–57.

Sriskandarajah, S., Frello, S., Jørgensen, K. and Serek, M. (2004) *Agrobacterium tumefaciens*-mediated transformation of *Campanula carpatica*: factors affecting transformation and regeneration of transgenic shoots. *Plant Cell Reports* 23, 59–63.

Styrvold, O.B. and Strøm, A.R. (1991) Synthesis, accumulation, and excretion of trehalose in osmotically stressed *Escherichia coli* K-12 strains: influence of amber suppressors and function of the periplasmic trehalase. *Journal of Bacteriology* 173, 1187–1192.

Sun, R.F., Li, T.R., An, Y.L., LI, K., Shi, H.Q., Niu, S.Q. and Zhao, J.H. (2009) The genetic transformation of strawberry with winter flounder antifreeze protein gene. *Journal of Nuclear Agricultural Sciences* 23, 423–428.

Sutter, E.G. Ahmadi, H. and Labavitch, J.M. (1997) Direct regeneration of strawberry (*Fragaria* × *ananassa* Duch.) from leaf disks. *Acta Horticulturae* 447, 243–246.

Taji, A.M. and Williams, R.R. (1996) Overview of plant tissue culture. In: Taji, A.M. and Williams, R.R. (eds) *Tissue Culture of Australian Plants: Past, Present and Future*. University of New England Press, Armidale, Australia, pp. 1–15.

Tanprasert, P. and Reed, B. (1998) Detection and identification of bacterial contaminants of strawberry runner explants. *Plant Cell, Tissue and Organ Culture* 52, 53–55.

Titov, S., Bhowmik, S.K., Mandal, A., Alam, M.S. and Uddin, S.N. (2006) Control of phenolic compound secretion and effect of growth regulators for organ formation from *Musa* spp. cv. Kanthali floral bud explants. *American Journal of Biochemistry and Biotechnology* 2, 97–104.

Torné, J.M., Moysset, L., Claparols, I. and Simón, E. (1996) Photocontrol of somatic embryogenesis and polyamine content in *Araujia sericifera* petals. *Physiologia Plantarum* 98, 413–418.

Torné, J.M., Moysset, L., Santos, M. and Simón, E. (2001) Effects of light quality on somatic embryogenesis in *Araujia sericifera*. *Physiologia Plantarum* 111, 405–411.

van Wordragen, M.F. and Dons, H.J.M. (1992) *Agrobacterium tumefaciens*-mediated transformation of recalcitrant crops. *Plant Molecular Biology Reporter* 10, 12–36.

Vellicce, G., Ricci, J.D., Hernández, L. and Castagnaro, A. (2006) Enhanced resistance to *Botrytis cinerea* mediated by the transgenic expression of the chitinase gene *ch5B* in strawberry. *Transgenic Research* 15, 57–68.

Vergauwe, A., van Geldre, E., Inzé, D., van Montagu, M. and van den Eeckhout, E. (1998) Factors influencing *Agrobacterium tumefaciens*-mediated transformation of *Artemisia annua* L. *Plant Cell Reports* 18, 105–110.

Verhagen, S.A. and Wann, S.R. (1989) Norway spruce somatic embryogenesis: high-frequency initiation from light-cultured mature embryos. *Plant Cell, Tissue and Organ Culture* 16, 103–111.

Wallin, A. (1997) Somatic hybridization in *Fragaria*. *Acta Horticulturae* 439, 63–66.

Wang, D., Wergin, W.P. and Zimmerman, R.H. (1984) Somatic embryogenesis and plant regeneration from immature embryos of strawberry. *HortScience* 19, 71–72.

Wang, J., Zhao, M.Z., Wang, Z.W., Yu, H.M. and Huang, W.Y. (2012a) 'Ningyu': a Chinese new cultivar with high level reducing sugar, ascorbic acid and antioxidant capacity. *Acta Horticulturae* 1049, 231–235.

Wang, J.H., Liu, J.J., Chen, K.L., Li, H.W., He, J. and Guan, B. (2012b) Studies on strawberry transformation mediated by *Agrobacterium*. *Southwest China Journal of Agricultural Sciences* 25, 1011–1013.

Wang, J.L., Ge, H.B., Peng, S.Q., Zhang, H.M., Chen, P.L. and Xu, J.R. (2004) Transformation of strawberry (*Fragaria ananassa* Duch.) with late embryogenesis abundant protein gene. *Journal of Horticultural Science and Biotechnology* 79, 735–738.
Wang, Y. (2012) *Agrobacterium*-mediated transformation of sweet protein *Brazzein* gene in *Fragaria ananassa* Duch. 'Akihime'. Masters thesis, Nanjing Agricultural University, Nanjing, Jiangsu, China.
Wang, Y., Hao, J. and Jia, J. (2009) Establishment of recipient system of high frequency for genetic transformation of 'Zaohong' strawberry. *Genomics and Applied Biology* 28, 990–997.
Wang, Y., Chen, B.Y., Zhang, Z., Gao, Z.H. and Qiao, Y.S. (2012c) Study on tissue culture and rapid propagation technique of *Fragaria nilgerrensis*. *Southwest China Journal of Agricultural Sciences* 25, 252–256.
Wang, Z.L., Yan, F.F., Wang, J. and Lin, M.J. (2012d) Research on the rapid propagation technology of tissue culture seedling of strawberry. *Northern Horticulture* 11, 130–132.
Wawrzynczak, D., Sowik, I. and Michalczuk, L. (1998) Shoot regeneration from in vitro leaf explants of five strawberry genotypes (*Fragaria* × *ananassa* Duch.). *Journal of Fruit and Ornamental Plant Research* 6, 63–71.
Wawrzyńczak, D., Sowik, I. and Michalczuk, L. (2000) *Agrobacterium*-mediated transformation of five strawberry genotypes. *Journal of Fruit and Ornamental Plant Research* 8, 1–8.
Wei, Z.L., Dong, L., Jiao, C.Z. and Chen, Y.J. (2007) cNHX1 gene expression level and copy number analyze of different transgenic strawberry lines by using real time PCR. *Northern Horticulture* 5, 37–39.
Xia, J., Zhao, M.Z. and Meng, X.F. (2014) Study on stem-tip tissue culture and rapid propagation techniques of strawberry 'Ningyu'. *Acta Agriculturae Jiangxi* 26, 16–17.
Xu, Y.W., Zeng, J.W., Zou, Y.T., Husaini, A.M., Yao, R.Y., Wu, D.G. and Wu, W. (2011a) Combined effect of dark and wounding on regeneration potential of *Houttuynia cordata* Thunb. leaves. *Indian Journal of Experimental Biology* 49, 540–546.
Xu, Y.W., Zou, Y.T., Husaini, A.M., Zeng, J.W., Guan, L.L., Liu, Q. and Wu, W. (2011b) Optimization of potassium for proper growth and physiological response of *Houttuynia cordata* Thunb. *Environmental and Experimental Botany* 71, 292–297.
Xue, Q.Q., Li, M.Q., LV, J.F., Pei, H.L., Wang, X.C. and Yang, T.H. (2014) Study on tissue culture and rapid propagation technique of different genotypes strawberry. *Northern Horticulture* 21, 110–113.
Yuan, Q., Shen, S.H., Liu, Z.L., Peng, C.F., Zhou, Q.P., Yuan, Z.H. and Zheng, H.C. (2015) Study on tissue culture of *Fragaria ananassa* Duch. *Modern Agricultural Science and Technology* 2015, 90–91.
Zakaria, H., Hussein, G.M., Abdel-Hadi, A.H. and Abdallah, N.A. (2014) Improved regeneration and transformation protocols for three strawberry cultivars. *GM Crops and Food* 5, 27–35.
Zhai, T., Liu, C., Yuan, Y., Qu, H., Shen, J. and Liu, Q. (2015) Rapid propagation system of meristem culture for strawberry. *Journal of Anhui Agricultural University* 42, 545–548.
Zhang, B.X., Li, Z.G., Gao, Y., Liu, D. and Yu, G.Q. (2012) Relationship between DHAR and AsA accumulation in strawberry and its DHAR RNAi genetic transformation. *Journal of Southern Agriculture* 43, 1626–1632.
Zhang, H. (2013) Study on simplify the process of tissue culture and rapid propagation of strawberry virus-free seedling and reduce production cost. *Journal of Hebei Agricultural Sciences* 17, 68–71.
Zhang, H.M. and Wang, J.L. (2005) Establishment of genetic transformation system of 'Allstar' strawberry leaf. *Biotechnology* 15, 68–70.
Zhang, L.Y., Li, H.N., Zhang, X., Xie, X.M., Wang, P.Y. and Zhang, H.W. (2010) Experimental study on simplifying the technologies and reducing the cost on strawberry tissue culture. *Northern Horticulture* 11, 147–148.
Zhang, Z. and Wu, L. (1998) Development of genetic transformation system of the commercial strawberry cultivar 'Tudla'. *Journal of Agricultural Biotechnology* 6, 200–204.
Zhao, M.Z., Wang, Z.W., Qian, Y.M., Yuan, J. and Wu, W.M. (2009) Experiment of forced culture of 4 strawberry cultivars under low temperature and short daylight conditions. *China Fruits* 1, 19–22.
Zhao, Y., Liu, Q.Z. and Davis, R.E. (2004) Transgene expression in strawberries driven by a heterologous phloem-specific promoter. *Plant Cell Reports* 23, 224–230.
Zheng, J.J. (2008) Establishment of strawberry regeneration system and transgene of early flower gene *PcFT*. Masters thesis, Huazhong Agricultural University, Wuhan, China.
Zhou, H., Zhang, W., Zhang, Q., Cao, Q., Shen, Y., Qin, L. and Xing, Y. (2015) Establishment of high-efficiency transformation of the woodland strawberry (*Fragaria vesca*, Hawaii 4). *Journal of Beijing University of Agriculture* 30, 10–14.
Zhou, Y., Zhu, B., Guo, M., Bofan and Liang (1996) Formation and morphogenesis of anther callus of strawberry. *Acta Horticulturae Sinica* 2, 119–122.
Zhu, H.S. (2008) Studies on transformation of antisense ethylene receptor gene in strawberry (*Fragaria* × *ananassa* Duch.). PhD thesis, Fujian Agriculture and Forestry University, Fuzhou, China.

7 Plant Architecture in Different Cultivation Systems

Francesca Massetani[1] and Davide Neri[1,2]*
[1]*Dipartimento di Scienze Agrarie, Alimentari e Ambientali, Università Politecnica delle Marche, Ancona, Italy;* [2]*Centro di Ricerca per la Frutticoltura – Consiglio per la Ricerca in Agricoltura e l'Analisi dell'Economia Agraria, Roma, Italy*

7.1. Introduction

Plant architecture represents the morphological expression of a genotype in a certain period (Hallé *et al.*, 1978). It can be defined according to different criteria (growth type, branching model, position of the reproductive structure) detected through the observation of the development of the shoots and other elementary parts along space–time scales (Godin, 2000), and gives information about the rules of growth and flowering of the plant (Hallé and Oldeman, 1970). The architectural model is a dynamic description of the growth pattern of a species from the birth of a plant, in the absence of limiting factors (Hallé *et al.*, 1978).

Strawberry plant architecture shows some constant features related to its determined growth pattern that always stops with

* davide.neri@crea.gov.it or d.neri@univpm.it

a terminal inflorescence (Guttridge, 1955) and enables further growth only from lateral branches (sympodial growth) showing the same fractal organization, whereas the vegetative shoots develop as stolons. Variability of plant architecture is related to the distribution and position of the vegetative and reproductive structures along its short axis (rosette plant). These features of the growth habit change according to genotype, plant age, growing environment and cultivation technique, due to the plasticity of the species.

The response of the plants can be studied dynamically, providing elements to predict and manipulate their behaviour (Savini and Neri, 2004). In fact, the reproductive and vegetative behaviour of the strawberry plants is sensitive to environmental (Serçe and Hancock, 2005; Sønsteby and Heide, 2006; Verheul *et al.*, 2006) and agronomic and nutritional factors (Savini and Neri, 2004; Savini *et al.*, 2005, 2006b; van Delm *et al.*, 2009; Bosc *et al.*, 2012).

Plants can be part of a programmed production system in the nursery to produce in predicted periods under different cultivation systems and, using specific propagation techniques, they can bear a different number and distribution of shoots, inflorescences or flowers and stolons.

7.2. Flower Induction and Environmental Control

Flower induction in strawberry is sensitive to thermophotoperiod according to plant genotypes, which are classified into remontant or recurrent (ever-bearing or long day and day-neutral plants) and non-remontant (seasonal flowering (Heide *et al.*, 2013); short day or June-bearing) types. Non-remontant cultivars provide only one harvest in spring–summer, as a result of flower induction that took place in the preceding late summer–autumn, when their thermophotoperiodic requirements for flower initiation (Darrow, 1936; Darnell and Hancock, 1996) were satisfied by short days (less than 11–16 h) (van den Muijzenberg, 1942; Borthwick and Parker, 1952; Ito and Saito, 1962; Benoit, 1975; Heide, 1977; Konsin *et al.*, 2001) or low temperatures (9–21°C, optimal 15–18°C) (Hartmann, 1947; Went, 1958; Ito and Saito, 1962; Heide, 1977; Verheul *et al.*, 2007; Opstad *et al.*, 2011). A minimum number (7–14) of short-day cycles is required for flower induction (Ito and Saito, 1962; Guttridge, 1985), according to cultivar, temperature and day length. Under long-day conditions, the terminal apex of the crown remains vegetative and many runners develop from the axillary buds (Nicoll and Galletta, 1987), as vegetative and reproductive growth have contrasting responses and requirements (Ito and Saito, 1962).

Remontant cultivars produce fruits more times per year, due to their different sensitivity to day length in relation to the temperature (Nishiyama *et al.*, 2000, 2006; Nishiyama and Kanahama, 2002; Sønsteby and Heide, 2007) for flower induction. Flower initiation is possible when day length is longer than 12 h (long-day cultivars) (Darrow and Waldo, 1934) or irrespective of photoperiod (day-neutral cultivars) (Bringhurst and Voth, 1980; Durner, 1984). Thus, a rigid classification of cultivars is not adequate (Durner, 1984) to explain all the possible complex responses to day length, whereas quantitative differences in individual plants can be seen along a continuum of obligate single-cropping and ever-bearing cultivars (Darrow, 1966; Nicoll and Galletta, 1987; Sønsteby and Nes, 1998; Savini, 2003). At low temperature, most genotypes show a day-neutral behaviour (Ito and Saito, 1962; Heide, 1977) but can be distinguished according to their response to photoperiod at higher temperatures (Heide, 1977; Verheul *et al.*, 2006). Thus, temperature affects the behaviour of both short-day and day-neutral cultivars in relation to photoperiod. Furthermore, prolonged warm temperatures (above 26–30°C) (Ito and Saito, 1962; Chabot, 1978; Durner, 1984; Durner and Poling, 1988; Okimura and Igarashi, 1997) totally (Ito and Saito, 1962; Chabot, 1978) or partially (Okimura and Igarashi, 1997;

Kadir *et al.*, 2006; Verheul *et al.*, 2006) inhibit flower formation, whatever the photoperiod (Ito and Saito, 1962; Verheul *et al.*, 2006). Therefore, in warm latitudes (tropical and equatorial), the profitable cultivation of strawberry is possible only where temperatures are lower in the highlands or where there is an eventual chill season.

The thermophotoperiodic conditions inducing flowering in short-day plants are normally effective only in summer–autumn, suggesting that other factors are involved in the control of plant physiology, such as the chilling that prevents flower induction in spring (Guttridge, 1958). Furthermore, plant meristems are sensitive to flower-inducing factors according to their physiological condition: they are more receptive when the growth of the apex is reduced (Savini *et al.*, 2005). Therefore, the behaviour of the plant (reproductive or vegetative) can be determined by the modulation of many factors interacting with the plant growth, involving abiotic, agronomic, nutritional and environmental factors or the presence of stress, and we can assume that these factors determine a main indirect effect modifying the vigour of the whole plant.

The final crop load is determined after flower induction, during the differentiation process, when the formation of flowers in the floral apex is enhanced by mild temperatures (18°C) (Le Mière *et al.*, 1996). Yield and fruit quality can be improved by modifying one or more effective factors during the growing cycles, starting from plant production in the nursery, in order to anticipate or delay flower induction and to determine the number of flowers and inflorescences.

Some general differences in plant architecture can be detected among cultivars according to their thermophotoperiodic sensitivity: short-day and day-neutral plants often show a branch crown developing from an upper axillary bud near the terminal inflorescence (Kurokura *et al.*, 2005a,b), whereas ever-bearing plants develop branch crowns mostly from their growing buds (Nicoll and Galletta, 1987).

7.3. Growing Cycles – Plant Plasticity for Harvest Seasons

Nursery techniques provide many plant types allowing the application of different planning strategies for cultivation. Different plant types have different architectures, each suitable for integrating in specific growing cycles.

Bare-root frigo (cold-stored) plants are used for summer planting (July–August in the northern hemisphere) in the field in common annual production systems that provide one short harvest season (about 30 days) in the spring. These plants bear only one differentiated flower bud in the terminal position with a single poor-quality inflorescence that often is partially damaged during the long cold storage. Therefore, after planting, the terminal inflorescence and stolons must be removed to stimulate plant establishment and growth. Growth takes place at the end of the summer and during the autumn, with the development of new crowns from lateral buds (Savini *et al.*, 2005), which receive induction signals for flower formation and provide a high yield the following spring.

Frigo plants are used also for remontant varieties, starting the growing cycle in March (in the northern hemisphere), with transplantation and removal of the first inflorescences in spring (May to early June) to promote the plant set and a strong initial vegetative growth, which is positively correlated with fruit yield (Baumann *et al.*, 1993). The harvest season lasts from the beginning of July until October.

In mild climates with warm summers, planting is delayed to avoid the high temperatures. Bare-root fresh plants of low-chilling varieties are preferred over frigo plants as the latter suffer growth problems if they are stored for too long until the September–October transplanting (for harvest from December or January). In addition, there is not enough time for the flowers to differentiate in the field and to obtain early winter production, but this is possible using flower-differentiated plants from Spanish high-elevation nurseries or from central Europe (with low temperatures for partial chilling satisfaction). These plants bear at least one flower-differentiated bud and produce fruits over an extended harvest season (January–June). Bare-root freshly dug stolon plants (fresh plants) of June-bearing varieties have also been used in the past and transplanted in autumn or spring for perennial cultivation.

For out-of-season production in central Europe, tray plants and waiting-bed and A+ frigo plants are increasingly used as programmed production plants (for '60-day' cultures) but must already have undergone flower induction.

Tray plants are available for planting in different seasons after cold storage (Lieten, 1993; Durner *et al.*, 2002), during which the plants satisfy their chilling requirement. Cold-stored tray plants may bear at least two or three flower-differentiated buds, providing good crop potential and similar or higher numbers of fruits and good yield in the open field in comparison with frigo plants (Duralija *et al.*, 2007).

Waiting-bed plants form more than one differentiated crown per plant, providing a potentially high production (500–800 g of fruit per plant) after transplanting. In the northern hemisphere, they are planted in sequence from the end of April until the middle of July for late-cropping systems, with harvest from early July to September in the open field and during the autumn under protected conditions (plastic tunnels). These plants with high crop potential must be planted early to promote their growth (Chercuitte *et al.*, 1991); later planting in spring–summer conditions may result in a decreased yield because of high temperatures.

Large cold-stored plants (commercial class A+ or A++) are generally transplanted in May–June, extending the cropping season from the beginning of July until the end of August in the northern hemisphere (Faby, 1997).

7.4. Plant Manipulation in the Nursery

The first opportunity to manipulate the plant growth and architecture, and at the same time plant vigour and fruit production, is control of the growing conditions during plant propagation, obtaining plants with a different number of lateral flower buds and shoots and at a different floral developmental phases (van Delm *et al.*, 2009).

Knowledge of effective environmental and cultural factors for plant behaviour manipulation is pivotal in applying conditioning techniques that allow anticipation or delay of flower induction and to obtain higher or lower numbers of flowers. In fact, many growing techniques available for nursery plant production, including the application of controlled stress, can be effective to manipulate the plant architecture, stimulating flower induction and differentiation, because they can interact with several aspects of plant growth, such as growth rate, shoot-to-root ratio and carbon balance. Using different growing techniques, it is possible to expose the plants to flower-inducing stimuli or to increase the plant's sensitivity to environmental inducing conditions, and thus to produce plants with different numbers of shoots, stolons, inflorescences or flowers. The timing of application of a particular technique is crucial in terms of its interaction with the physiological stage of the plant and its organs.

7.4.1. Thermophotoperiod

One of the most effective means of manipulating the equilibrium between vegetative growth and reproductive development in strawberry is modification of the thermophotoperiod, an environmental factor playing a major role in flower induction. The light availability duration can be increased using artificial light, to apply long-day conditions to short-day plants and delay flower formation (Bosc *et al.*, 2012), or decreased by adding a temporary light-proof covering (Bosc and Demené, 2009) to mimic short-day conditions. Providing artificial light during the night (night break) under short-day conditions can inhibit flower initiation in non-remontant cultivars (Ito and Saito, 1962; Vince-Prue and Guttridge, 1973) or enhance flower production in remontant cultivars (van Delm *et al.*, 2012).

The natural variation in temperature between day and night is more effective than a constant temperature. In forced cultivation, natural conditions can be reproduced applying the normal oscillation of temperatures, and flower induction can be induced earlier (Reichart, 1973; Chabot, 1978; Durner, 1984; Bish *et al.*, 1997) when the oscillation is 26.7/15.6°C day/night compared with a constant 21°C temperature (Hartmann, 1947) or with warmer temperatures (Bish *et al.*, 1996a). Temperatures below 15.6°C are common during summer–autumn in cold areas but are suboptimal for flower differentiation (Darrow, 1966), delaying the formation of flowers.

7.4.2. Location

Both photoperiodic and thermal conditions can be controlled during propagation, by transplanting the plants or moving them to specific locations (changing the altitude and latitude). The choice of environment for the nursery location and the time of propagation allow control of the vegetative and generative behaviour of the plants, resulting in different numbers of flowers per plant and programming the synchronicity and duration of the cropping.

Flower induction in short-day plants is delayed with increasing latitude, as day length reduces later in the summer (Opstad *et al.*, 2011). In Europe, fresh plants can be propagated in environments where the conditions are favourable for floral induction, which takes place earlier at higher altitude in the Alps (Savini *et al.*, 2006b) and in Spanish highland nurseries (800–1200 m) or in cold northern areas due to the thermal fluctuations between day and night and to low summer temperatures. After floral induction has taken place, at the beginning of October, and early low-temperature chilling has been provided, plants can be transplanted in mild southern areas (e.g. southern Italy or Spain) or at low altitude. In fact, a mild environment is optimal for a longer and earlier flower differentiation, in order to increase the number of flowers within the inflorescences (Savini, 2003; Savini *et al.*, 2006a) without a significant effect of photoperiod. Under these mild conditions, flower differentiation persists for the whole of the warm autumn and takes place in the second- and third-order floral apexes on the extension crowns (Savini, 2003), increasing the total number of inflorescences per plant. If the planting is deferred until temperatures are cooler and the photoperiod is shorter, vegetative growth is reduced and only one extension crown per plant is able to form. During formation of the primary buds, prolonged favourable conditions in early autumn enhance the formation of flowers within the primary inflorescence in northern Italy, resulting in more flowers (13–14 per inflorescence) in comparison with plants in the south (around 10 per inflorescence) (Savini, 2003) and with secondary inflorescences becoming differentiated during the early arrival of the cold temperatures in autumn. In the south, the number of flowers is similar in both primary and secondary inflorescences.

In contrast, if propagation takes place under a long photoperiod and relatively high temperatures, flower induction is reduced and vegetative growth is enhanced (Savini *et al.*, 2006a), with increasing formation of runners (Guttridge, 1985). The cultivation environment also modifies the growth of the plant. For instance, more crowns are formed along the principal axis of fresh plants in central and northern Italy compared with the south where the winter is warmer. The northern conditions also stimulate vegetative growth in tray plants (Savini *et al.*, 2006a).

In tropical climates, temperatures can be very high at low altitude, reducing the vegetative growth of the plants, which develop fewer leaves compared with plants at higher altitudes (Riyaphan *et al.*, 2005).

7.4.3. Chilling

Chilling temperatures (between 0 and 7–10°C) are required to overcome dormancy, but they are also effective at increasing vegetative growth (Guttridge, 1958; Voth and Bringhurst, 1958; Piringer and Scott,

1964; Wahdan and Waister, 1984; Tehranifar *et al.*, 1998) and leaf and runner formation (Bringhurst *et al.*, 1960; Porlingis and Boynton, 1961; Piringer and Scott, 1964; Bailey and Rossi, 1965; Guttridge, 1969; Braun and Kender, 1985; Rice, 1990; Kahangi *et al.*, 1992; Lieten, 1997b; Tehranifar and Battey, 1997), whereas they reduce flower induction (Guttridge, 1958) and enhance floral differentiation (Durner and Poling, 1987). Therefore, in greenhouse cultivation, cold treatments can be applied to prevent a decrease in vegetative vigour, while artificial lighting can compensate for a lack of cold (van Delm *et al.*, 2013) because long days can substitute for chilling (Lieten, 1997a; Sønsteby and Heide, 2006).

For programmed production cultivation, cold storage makes the plants available all year round and also provides the required chilling. Nevertheless, sugars and starch content can decrease if the low temperature is maintained for too long (more than 200 days), also reducing the number of emerging inflorescences and flowers (Molot and Leroux, 1973; Kinet *et al.*, 1993; Lieten *et al.*, 1995; Dradi *et al.*, 1996; Sønsteby and Hytonen, 2005), which are inhibited, especially if they are located below the last expanded leaf (Bosc *et al.*, 2012). A long duration of cold storage and the consequent low carbohydrate content induce a stress condition that may also result in earlier flowering (Lieten *et al.*, 1995), although not for all types of plant (Kinet *et al.*, 1993). Chilling before the digging date in the nursery may be advantageous for early fruit production, especially in warm regions, but extra chilling after the optimum digging date may reduce flowering (Durner and Poling, 1988).

7.4.4. Light quality

Light quality (Collins, 1966; Vince-Prue and Guttridge, 1973) and intensity may be effective in regulating flower initiation, especially in short-day plants. These features of the light reaching the plants can be modulated in different ways, using covering nets or specific artificial lights. During a decreasing photoperiod, plant shading reduces light intensity and temperature, promoting flower induction if sunlight is reduced by 75–95% (Kumakura and Shishido, 1995), but with lower light intensity, crown size (Wright and Sandrang, 1995) and leaf and inflorescence number (Awang and Atherton, 1995) can decrease, whereas increasing light intensity can promote flower differentiation (Dennis Jr *et al.*, 1970; Chabot, 1978).

Flower bud initiation is also affected by the spectral composition of the irradiation, which can be selected applying photoselective nets over the plants, with red and blue nets having an inhibitory effect (Takeda, 2012). Spectral band quality also affects the responsiveness to photoperiod. In fact, light extension to produce long-day conditions delays floral initiation if obtained using far-red light or far-red combined with low-red light, whereas it is not effective if obtained using red light (Vince-Prue and Guttridge, 1973; Guttridge, 1985); however, some discordant results have been reported (Jonkers, 1965).

7.4.5. Nutrient supply

Nutrient supply, in particular the relative ratio between nitrogen and phosphorus, plays a major role in modulating growth and the vegetative equilibrium of the plant and consequently can be a means to modify the whole plant architecture during both propagation in the nursery and further plant growth (Savini, 2003; Savini and Neri, 2004).

Nitrogen availability and the timing of supply during the growing season affect the formation and growth of new organs and interact with flower initiation, amplifying the effect of inhibitive or stimulating factors.

The nutritional protocol should be managed in different ways. Both stolon and shoot formation can be enhanced by high nitrogen levels, depending on the timing of supply and the plant growth rate (Savini, 2003; Savini and Neri, 2004). Therefore, stolon formation is induced and flower induction can be delayed (Fujimoto, 1972; Furuya *et al.*, 1988; Matsumoto, 1991; Yamasaki *et al.*, 2002) or totally prevented (Yamasaki *et al.*, 2002) if excess nitrogen is applied before the flower induction period (Sønsteby *et al.*, 2009) or when there is rapid growth of the apex.

In contrast, a high nutrient supply can increase the induction effect if applied after the beginning of short-day conditions (Lieten, 2002; Sønsteby *et al.*, 2009). It can also stimulate plant vigour after the arrest of apex growth, reactivating axillary latent meristems to form new shoots (Breen and Martin, 1981) in the basal part of the crown, increasing the total number of inflorescence sites. If nitrogen is supplied later on, shoot formation can be stimulated in the upper portion of the plant, increasing the total number of inflorescences and extending the cropping time, because the flowers are less developed compared with the terminal primary inflorescence. In the nursery, late application of foliar nitrogen during flower differentiation of freshly dug plants can advance flower development and fruiting (Kirschbaum *et al.*, 2000). Differing responses are also seen if the supply of nutrients is continuous or temporary; in fact, lateral shoot formation starts from the apical part of the crown and continues downward with continuous fertilization, whereas shoots exclusively involve the apical buds if the fertilization is suspended for 1–2 weeks during the maximum growth period, also advancing flower differentiation (Savini, 2003). During flower induction of tray plants, temporary suspension of fertilization may result in the differentiation of terminal flower buds, enhancing inflorescence development and increasing flower number (Savini, 2003), while continuous high nutrition delays flower initiation (Yoshida, 1992; Lieten, 2002). A positive effect on flower formation is also obtained by applying a pulse of extra fertilization starting shortly after the onset of inducive environmental conditions (Sønsteby *et al.*, 2013). An excess of nitrogen inhibits flower formation (Guttridge, 1985).

An early nutrient supply, before or at the beginning of inducive thermophotoperiod conditions (Sønsteby *et al.*, 2009), delays flower bud initiation more than a later application (Yamasaki and Yano, 2009). However, low nitrogen availability depresses vegetative growth and increases plant sensitivity to inducive conditions (Strik, 1985; Battey *et al.*, 1998; Lieten, 2002) promoting flower induction (Guttridge, 1985) after a few days of inducing treatment. Nevertheless, a very low nutrient supply is not favourable during flower induction and differentiation processes, especially if prolonged until mid-October, because the meristem could revert from the reproductive to the vegetative state (van den Muijzenberg, 1942) and the number of flowers decreases (Anderson and Guttridge, 1982).

If the nutritional deficiency persists after flower bud initiation and induction, further development of initiated flowers is prevented (Strik, 1985; Battey *et al.*, 1998; Lieten, 2002), irrespective of the nutrient conditions during the other growth phases, as nitrogen is required for flower differentiation (Yamasaki *et al.*, 2002). A reduction in nutrients after mid-October does not affect fruit production. Early reduced nutrition at the end of summer until mid-September can reduce the crown diameter, but increases the number of flowers and fruits inside the inflorescences (Lieten, 2002). If the early reduction of nitrogen is followed by an increased supply at the end of August or in September and by later decreased nitrogen application in October, flower initiation may be advanced in comparison with a low or high constant nitrogen supply from the beginning of August (Desmet *et al.*, 2009).

Potassium supplementary fertilization may decrease the number of flowers (Desmet *et al.*, 2009) if already suppressed under salt stress (NaCl) conditions (Awang and Atherton, 1995). Also, the source of nutritional factors seems to have effects on plant growth; in particular, the addition of organic matter (cattle, poultry, sheep or manure) enhances the formation of leaves and advances the flowering date compared with conventional fertilizers (Abu-Zahra and Tahboub, 2008).

7.4.6. Growing substrate

Year-round strawberry production is helped by the use of substrate systems in protected cultures, without using contaminated soils

(Durner *et al*., 2002; Lieten, 2012). Peat mixes are widely used substrates for soil-less cultivation and for plug plant production in nurseries, but many other growing media have been tested (Anagnostou and Vasilakakis, 1995; Paranjpe *et al*., 2003; Lieten *et al*., 2004; Bartczak *et al*., 2007; Tehranifar *et al*., 2007, 2012; Ameri *et al*., 2012). Substrates differ in physical and chemical proprieties involving pH, electrical conductivity, porosity and water-holding capacity, which may affect the water and nutrient availability for the plant and the root activity. Consequently, substrates may also modulate the vegetative vigour of the plant and the response to flower induction, with effects on the crop potential. These effects can be also detected in the plant architecture, shoot topology and number along the crown (Savini, 2003).

For soil-less propagation, fine peat is a common and optimal substrate for large runners (Kehoe *et al*., 2009), but the use of coco substrate is becoming increasingly widespread. The use of an inert substrate without peat (rock wool substrate, sand or perlite), in the absence of nutritional adjustment, reduces the vegetative growth of the plant, decreases leaf number and crown formation, decreases total yield, and advances the production of flowers and the harvest, in comparison with the use of peat (Jansen, 1997; Tehranifar *et al*., 2007), although some conflicting results have been reported (Bartczak *et al*., 2007). The reduction in plant growth on a substrate with low water-holding capacity is detectable, even when some peat is added to the substrate (Lieten, 1993; Anagnostou and Vasilakakis, 1995).

Substrates without peat may also enhance the ability of flowers to differentiate, increasing the number of flowers in the second flower flush of remontant cultivars and the number of high-order inflorescences compared with organic substrate containing peat; the effect is amplified if peat is added with the compost (Savini, 2003).

7.4.7. Water supply

The water status of the plant influences many physiological processes, interacting with carbon assimilation – with photosynthesis reduced even under mild water stress (Lenz, 1979) – nutrient uptake and the growth rate of strawberry plants (Kumar and Dey, 2011; Grant *et al*., 2012). Therefore, an adequate water supply is needed to sustain an acceptable yield, preventing a decrease in mean fruit weight and fruit number (Davies and Albrigo, 1983; Gehrmann, 1985; Peñuelas *et al*., 1992; Serrano *et al*., 1992). However, water availability may have different effects on plant development depending on the physiological processes occurring before or during flower induction and differentiation. Thus, improving the water supply during flower bud initiation and differentiation in autumn may enhance flower formation and fruit production, but abundant irrigation before the onset of inducive conditions may reduce flower production (Naumann, 1964). Mild water stress may even allow flower induction under unfavourable environmental conditions (Naumann, 1961) after the start of flowering.

Water stress affects the growth of the stolons (VanDerZanden and Cameron, 1996; Tworkoski *et al*., 2001) and also the shoots, preventing their development if the water supply is strongly diminished (25% of daily water consumption) and reducing their number under mild water stress (Gehrmann, 1985), also affecting the ability of the plant to form new inflorescences.

7.4.8. Defoliation

Leaf removal in the nursery affects vegetative growth, altering the shoot-to-root ratio, reducing the assimilating surface and the endogenous nitrogen, and promoting flower differentiation (Kim *et al*., 2011). Defoliation is also suggested to stimulate compensative growth from lateral shoots and the onset of new inflorescence sites.

7.4.9. Propagation

The use of plug transplants allows earlier plant establishment after transplanting, enhances early growth and determines earlier flowering and fruit yield compared with bare-root transplants (Hochmuth *et al.*, 1998, 2006; Ruan *et al.*, 2011), irrespective of tray size (Giménez *et al.*, 2009) and crown diameter (Cocco *et al.*, 2011) during propagation. The containerized growth makes the plant less dependent on water supply (Durner *et al.*, 2002).

The size of the stolons used for the production of tray plants seems to have no effect on fruit production during cultivation (Jansen, 1997), because one leaf is enough to detect the induction signal (Hartmann, 1947), whereas the age may interact with photoperiod sensitivity (Ito and Saito, 1962). The sequence position and the distance of the daughter plant from the mother plant along the parental stolon affect the number of leaves (Bartczak *et al.*, 2007).

The planting date in the nursery may affect shoot growth and flower differentiation, whereas this has no effect on vegetative growth (Palha *et al.*, 2012), but the effect is related to the climate conditions of the seasons (Bussell *et al.*, 2007). Advancing the plugging date of runner tips stimulated shoot growth and flower differentiation, inducing June-bearing plants to flower early (Takeda and Newell, 2007). The new shoots stimulated a prolonged, although not simultaneous, fruit production, in contrast to delayed planting. Lateral shoot growth is also prevented by high plant density (short distance apart), which benefits the uppermost buds (close to the terminal apex), which can produce a secondary inflorescence and a good second production.

Pot size determines the volume of substrate available and affects the growth of the root. Small pot volumes increase plant sensitivity to inducing conditions (Fujishige, 1994) and stimulate early flower induction during plant root system formation. Root growth can be inhibited once the roots take up the whole substrate volume after too long a growing period in the tray, resulting in stress conditions, leading to a lower flower quality. Using pots of the same size and a low planting density during plant production promotes the formation of more inflorescences and flowers (Jansen, 1997), providing a higher yield compared with production at a higher plant density.

The use of phytoregulators has also been tested in nurseries, where application of gibberellic acid biosynthesis inhibitor reduced early vegetative growth, decreasing the number of stolons (Bish *et al.*, 1996b).

7.4.10. Evaluation of plant quality

The quality of nursery plants can be evaluated according to many different parameters. Genetic and phytosanitary compliance of nursery plants is a prerequisite for certificate protocols, whereas the ability to provide the desired cultivation and fruiting performance in specific growing cycles may differentiate the quality level of the plants.

Nursery material differs in terms of size, propagation material, presence of substrate, pot type, post-digging storage and the presence of differentiated inflorescences.

When produced under the same growing conditions, larger plug plants (obtained from larger stolons) may produce more flowers than smaller ones (Cantliffe *et al.*, 2003), and therefore plants are usually graded according to their crown size (diameter) and to the length of the roots. The weight of the plant, the number of crowns and the length of roots are all parameters positively related to yield potential (Bartczak *et al.*, 2010). However, a large plant does not always equate to higher crop potential (Bish *et al.*, 2002; Bussell *et al.*, 2007), therefore plant size itself is not enough to characterize the crop potential of nursery plants. Furthermore, with size grading only of the plants, no information is given about the precocity and the duration of the potential crop.

Each plant type is different and is suitable for specific growing cycles; therefore, quality evaluation is strictly related to plant type, requiring specific elements for the expected production, and it is not possible to use the same standard evaluations for all plant types or without taking into account the treatment of the plant in cultivation. The quality of the plant from the nursery also determines its physiological condition, which affects the response to cultural and environmental factors.

Nursery plants are characterized by different crop potentials, which can be evaluated by the detection of inflorescence number and position, while earliness and synchronicity of production can be estimated by observing the developmental stages of the flowers.

In order to base the plant quality evaluation on crop potential, information about the number and position of the inflorescences must be taken into account, along with their developmental phase and the number of differentiated flowers, as the number of differentiated flowers is positively correlated to the final number of fruits (Jemmali and Boxus, 1993; Savini, 2003).

The developmental phase of the inflorescences and their differentiation are also important factors when evaluating the optimal digging time of the plants, especially for programmed production plants that will be cold stored and must have completed the differentiation process.

7.5. Plant Manipulation in the Field

Different plants have specific management needs and a higher or lower ability to produce fruits in different growing contexts. The growing technique is related not only to the plant type and size, but also according to the year of plant production and to origin (nursery), which can all affect the crop load (Faby, 1997). Knowledge about the plant quality (crop potential, plant architecture) allows planning of the planting date and density, the forcing technique and fertilization, allowing manipulation of the plant to obtain satisfactory production results.

The response of the plant to environmental signals, such temperature, varies depending on the physiological phase of the different organs and their relative positions. The date of forcing may significantly modify the plant architecture. For instance, axillary meristems at a maximum level of dormancy do not develop secondary shoots after placing the plants in the greenhouse (Bosc *et al.*, 2012). As the chilling requirement is increasingly satisfied, the axillary meristems initiate new secondary shoots, but if the temperature is too cold (delayed forcing after the end of October), lateral growth can be penalized (Savini *et al.*, 2006b).

Flower removal may promote the production of leaves (Daugaard, 1999) and runners (Scott and Marth, 1953; Robertson and Wood, 1954; Moore and Scott, 1965) in some genotypes. In other cultivars, deblossoming treatment increases runner production only in combination with defoliation, or may have no effect on runner formation (Waithaka, 1993). The removal of runners promotes branch crown development (Hancock, 1999). Transplant defoliation may affect plant growth, reducing the number of crowns and inflorescences (Chandler *et al.*, 1988; Kerkhoff *et al.*, 1988; Albregts *et al.*, 1992; Mohamed, 2002) if there is not enough time to restore the leaf surface area.

Late summer planting may reduce the crop potential, but later planting can also increase the number of flowers and inflorescences in autumn production systems (Palha *et al.*, 2012). Late autumn transplanting reduces the number of crowns, but other factors influence the total yield, affecting the fruit size (Hassell *et al.*, 2007). In remontant cultivars, the number of crowns increases by advancing the planting date and prolonging the growing period, enabling a higher yield if the cultivar is not highly crowned (Ruan *et al.*, 2011), as the inflorescences derive from the apical meristems of the plant. For delayed transplanting in the field, when environmental conditions promoting flower differentiation do not persist, it is necessary to select already differentiated plants from the nursery. For a good crop, planting density can be balanced according to the number of inflorescences per plant (flowers m^{-2}) rather than the plant size, as the former is more strongly related to the crop potential.

Small-sized plants with a few flower buds in addition to the terminal one produce a short harvest season provided by a sequence of fruits ripening inside the terminal inflorescence itself over 1 month and a possible later small production of fruits from the axillary buds. This type of plant is profitable only with high-density planting, mainly in mild climates where the plants can differentiate further flowers on the top of the lateral shoots after the first harvest.

Increasing plant density (by reducing either row or plant spacing) may reduce the crown diameter and early and total yield, with no effects on the number of leaves (Paranjpe *et al.*, 2008). At a low planting density, a higher yield is possible.

Well-differentiated plants, with the terminal inflorescence at a late developmental stage and with many less-developed lateral shoots and buds, start producing from the terminal inflorescence around 40 days after transplanting and continue producing from the other inflorescences before the end of the first production.

When plants bear vampire buds (flower-differentiated buds at the same stage as the main terminal apex) at the base of the crown, these buds can compete with the terminal inflorescence. It is useful to be aware of their presence in order to plan their removal or to adjust the growing technique, allowing a good harvest. The ability of axillary buds to grow out is related to the degree of apical dominance (Sugiyama *et al.*, 2004).

7.6. Architecture Analysis Techniques

Strawberry plants are composed of a short axis (crown) with very close nodes and leaves (rosette), ending with an inflorescence. The plant architecture facilitates the representation of the topology of the vegetative and reproductive organs along the short axis to produce a schematic model of growth (Guttridge, 1955; Savini *et al.*, 2005). Such an analysis of plant architecture detects and records the fate of all the meristems and scores the developmental stages of flower organs. The organization of the plant derives from sequential morphological processes due to the activity of the meristems, and therefore the identity and position of the buds is very important to determine the architecture of the plant. The description of the plant is based on breaking it down into elementary repeating units and on their characterization. The elementary unit of the plant structure is the phytomer, a particular type of segment of the shoot (metamer) (White, 1979), composed of one node with the subtending leaf and internode and one or more buds in the axil of the leaf.

The analysis consists of dissection of the plant, removing each leaf starting from the base of the crown, and identification of all the structures inserted in the axillary position of each node. The sequence position of the nodes can be recognized because each leaf envelops the subsequent internal leaf overlapping the stipules. The characterization proceeds in three steps, starting from identification of the plant elements (meristem fate and leaf characterization).

Leaves are characterized according to the presence of developed and living blades. Bud fates can be classified into main categories, depending on the development into a branch (side crown) or runner (long vegetative shoot), on the condition of the dormant axillary bud, and on the failure or abortion of the meristem (blind node). The fate of the terminal position of the branches and of the apical meristem inside the buds must be also identified as reproductive or vegetative in order to describe the position of the reproductive structures. Buds are enveloped and protected by modified stipules that must be opened using a scalpel and forceps, through a longitudinal and a transversal cut. The internal layers of leaf primordia are counted and removed, turning them outwards in order to expose and observe the meristematic apex under a stereomicroscope at an optimal magnification of ×40. The leaf primordia are also composed of blades and stipules enveloping the smaller primordia. In the second step, for each flower-differentiated apex, the differentiation phase must be scored according to a conventional developmental scale (Fig. 7.1) (Taylor, 2002; Savini, 2003) based on the centripetal appearance of the reproductive organs (sepals, petals, stamens, receptacle) on the primary flower of the inflorescence. The first stage (stage 0) of the scale indicates the vegetative condition of the meristem, underlying the stipules of the least developing leaf layer. At stage 1, a rising dome is

Fig. 7.1. Scale of differentiation of the phases of strawberry inflorescence. (Modified from Neri *et al.*, 2010.)

visible, revealing the appearance of the primary flower of the inflorescence. The next stages identify, respectively, the sepals initiation (stage 2), petals initiation (stage 3), sepals and petals development (stage 4), stamen formation (stage 5), enclosure by the sepals (stage 6), completion of flower organs with green anthers (stage 7), yellow colouring of the anthers (stage 8) and completion of the flower cluster (stage 9; not shown in Fig. 7.1) always with reference to the primary flower. Further investigation concerns analysis of the inflorescence architecture, because a description of flower branching provides an estimation of ripening and size sequence of the fruits, along with the total number of differentiated flowers. The inflorescence branching can be organized according to many different patterns ranging from a typical dichotomy of two secondary flowers inserted on the axis of the primary flower, or two tertiary flowers inserted on the axis of each secondary flower, to the presence of many more flowers (Darrow, 1929).

All the structures and elements of the plant are represented using conventional symbols displayed according to their position, and the plant can be represented as a schematic (Fig. 7.2) that describes the vegetative behaviour and the organ distribution for the plant quality evaluation (Fig. 7.3).

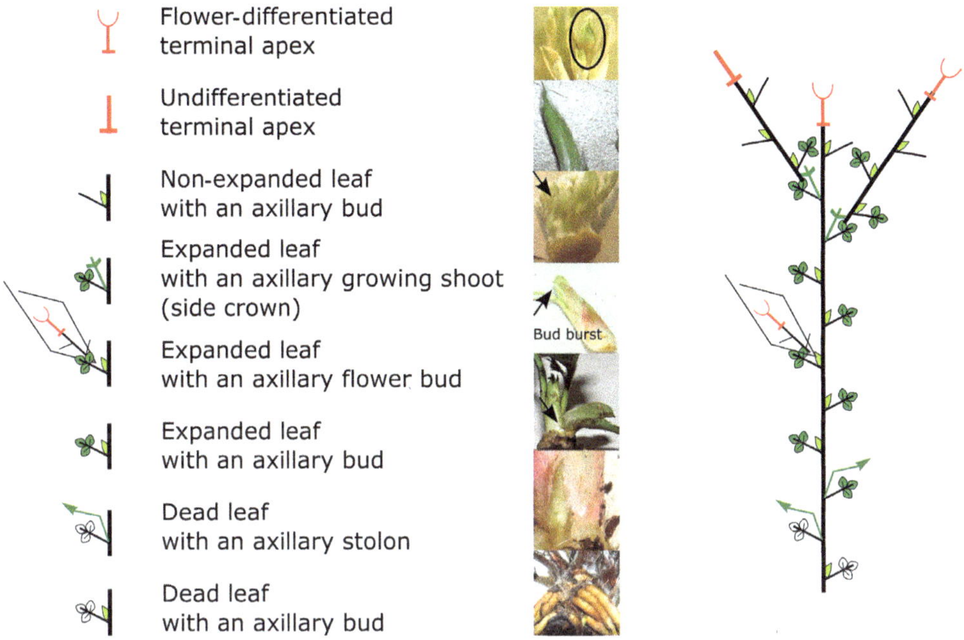

Fig. 7.2. Schematic representation of strawberry plant architecture using conventional symbols. Left: conventional symbols with their definition, illustrated in the photos; right: example of a schematic representation of a single plant.

Fig. 7.3. Example of a schematic representation of the architecture of a tray-plant of a non-remontant cultivar. The crop potential of this plant derives mainly from three well-differentiated inflorescences. The terminal one is at an advanced stage of development and is able to develop early after planting. It is composed of many flowers (21), suggesting that the late growth in the nursery took

Fig. 7.3. Continued

place in mild/warm temperatures. The other two inflorescences in the uppermost positions of the crown are on the top of extension crowns and are able to develop with a little delay and fewer flowers. The lower differentiated buds show very early stages of development and have little chances of developing and growing out, because they can remain dormant or even die, mainly after cold storage. Black numbers, flowering stage; red numbers, number of flowers.

7.7. Conclusion

The architecture of the strawberry plant reflects the high plasticity of this species due to its sensitivity to temperature and photoperiod, and also to several agronomic and nutritional factors that affect its reproductive and vegetative behaviour. Different cultivation systems can provide yields in different seasons and areas, allowing an overall year-round production, but they require the use of plants that fit into the appropriate growing cycle and have the expected productivity. The choice of appropriate plant is related primarily to genotypic sensitivity (remontant and non-remontant types) and then to the type of nursery production system, which affects the ability of the plant to establish after transplanting, the quality of the plant itself, its crop potential and timing of harvest, and finally the plant architecture.

Knowledge about the response of flowers, shoots and runners to environmental and abiotic factors allows plant manipulation during propagation in the nursery and during cultivation in the field or in protected systems. The optimal control of plant behaviour is possible only by monitoring the different vegetative and reproductive structures (plant architectural analysis) over space and time. The growth and development of the plants can be monitored throughout the season to estimate the time of optimal sensitivity to external factors, to enable transition of the plants to reproductive behaviour to obtain optimal crop potential, and to predict the best level of production of the crop (time and quantity). Furthermore, growth can be adjusted and controlled to alter the development of the plant. In the nursery, the plants can be programmed by modifying environmental conditions or modulating abiotic factors, including controlled stress conditions, and the architecture can be appropriately adapted for each farming system in different areas, modifying the time of fruit production in the field after transplanting.

References

Abu-Zahra, T.R. and Tahboub, A.A. (2008) Strawberry (*Fragaria × Ananasa* Duch) growth, flowering and yielding as affected by different organic matter sources. *International Journal of Botany* 4, 481–485.

Albregts, E.E., Howard, C.M. and Chandler, C.K. (1992) Defoliation of strawberry transplants for fruit production in Florida. *HortScience* 27, 889–891.

Ameri, A., Tehranifar, A., Davarynejad, G.H. and Shoor, M. (2012) The effects of substrate and cultivar in quality of strawberry. *Journal of Biology and Environmental Science* 6, 181–188.

Anagnostou, K. and Vasilakakis, M.D. (1995) Effect of substrate and cultivar on earliness, plant productivity, and fruit quality of strawberry. *Acta Horticulturae* 379, 267–274.

Anderson, H.M. and Guttridge, C.G. (1982) Strawberry truss morphology and the fate of high-order flower buds. *Crop Research* 22, 105–122.

Awang, Y.B. and Atherton, J.G. (1995) Growth and fruiting responses of strawberry plants grown on rockwool to shading and salinity. *Scientia Horticulturae* 62, 25–31.

Bailey, J.S. and Rossi, A.W. (1965) Effect of fall chilling, forcing temperature and day length on the growth and flowering of Catskill strawberry plants. *Proceedings of the American Society for Horticultural Science* 87, 245–252.

Bartczak, M., Pietrowska, M. and Knaflewski, M. (2007) Effects of substrate on vegetative quality of strawberry plants (*Fragaria × ananassa* Duch.) produced by soilless method. *Folia Horticulturae* 19, 39–46.

Bartczak, M., Lisiecka, J. and Knaflewski, M. (2010) Correlation between selected parameters of planting material and strawberry yield. *Folia Horticulturae* 22, 9–12.

Battey, N.H., Miere, P.l., Tehranifar, A., Cekic, C., Taylor, S., Shrives, K.J., Hadley, P., Greenland, A.J., Darby, J. and Wilkinson, M.J. (1998) Genetic and environmental control of flowering in strawberry. In: Cockshull, K.E., Gray, D., Seymour, G.B. and Thomas, B. (eds) *Genetic and Environmental Manipulation of Horticultural Crops*. CAB International, Wallingford, UK, pp. 111–131.

Baumann, T.E., Eaton, G.W. and Spaner, D. (1993) Yield components of day-neutral and short-day strawberry varieties on raised beds in British Columbia. *HortScience* 28, 891–894.

Benoit, F. (1975) Further observations on the induction of a second flowering in the strawberry cultivar Redgauntlet. *Agricultura* 23, 29–35.

Bish, E.B., Cantliffe, D.J. and Chandler, C.K. (1996a) Pretransplant temperature regime and container size alter strawberry plant morphology. *HortScience* 31, 566.

Bish, E.B., Cantliffe, D.J. and Chandler, C.K. (1996b) Strawberry plug transplants: regulation of growth and production. *Proceedings of the Florida State Horticultural Society* 109, 160–163.

Bish, E.B., Cantliffe, D.J., Hochmuth, G.J. and Chandler, C.K. (1997) Development of containerized strawberry transplants for Florida's winter production system. *Acta Horticulturae* 439, 461–468.
Bish, E.B., Cantliffe, D.J. and Chandler, C.K. (2002) Temperature conditioning and container size affect early season fruit yield of strawberry plug plants in a winter, annual hill production system. *HortScience* 37, 762–764.
Borthwick, H.A. and Parker, M.W. (1952) Light in relation to flowering and vegetative development. In: *13th International Horticultural Congress*, 8–15 September, The Royal Horticultural Society, London, pp. 801–810.
Bosc, J.P. and Demené, M.N. (2009) Floral induction duration, plant architecture and fruit production relations in strawberry cv. 'Ciflorette'. *Acta Horticulturae* 842, 667–670.
Bosc, J.P., Neri, D., Massetani, F. and Bardet, A. (2012) Relationship between plant architecture and fruit production of the short-day strawberry cultivar 'Gariguette'. *Acta Horticulturae* 926, 229–236.
Braun, J. and Kender, W. (1985) Correlative bud inhibition and growth habit of the strawberry as influenced by application of gibberellic acid, cytokinin, and chilling during short daylength. *Journal of the American Society for Horticultural Science* 110, 28–34.
Breen, P. and Martin, L. (1981) Vegetative and reproductive growth responses of three strawberry cultivars to nitrogen. *Journal American Society for Horticultural Science* 106, 266–272.
Bringhurst, R.S. and Voth, V. (1980) Six new strawberry varieties released. *California Agriculture* 34, 12–15.
Bringhurst, R.S., Voth, V. and van Hook, D. (1960) Relationship of root starch content and chilling history to performance of California strawberries. *Proceedings of American Society for Horticultural Science* 75, 373–381.
Bussell, W.T., Ennis, I.L. and Triggs, C.M. (2007) Nursery location, planting date, chilling exposure, and bed mulch treatment effects on yield performance of 'Camarosa' strawberry (*Fragaria* × *ananassa*) in Auckland, New Zealand. *New Zealand Journal of Crop and Horticultural Science* 35, 15–24.
Cantliffe, D.J., Shaw, N., Jovicich, E., Rodriguez, J.C., Secker, I. and Karchi, Z. (2003) Plantlet size affects growth and development of strawberry plug transplants. *Proceedings of the Florida State Horticultural Society* 116, 105–107.
Chabot, B.F. (1978) Environmental influences on photosynthesis and growth in *Fragaria vesca*. *New Phytologist* 80, 87–98.
Chandler, C.K., Miller, D.D. and Ferree, D.C. (1988) Influence of leaf removal, root pruning, and soil addition on the growth of greenhouse-grown strawberry plants. *Journal of the American Society for Horticultural Science* 113, 529–532.
Chercuitte, L., Sullivan, J.A., Desjardins, Y.D. and Bedard, R. (1991) Yield potential and vegetative growth of summer-planted strawberry. *Journal of the American Society for Horticultural Science* 116, 930–936.
Cocco, C., Andriolo, J.L., Cardoso, F.L., Erpen, L. and Schmitt, O.J. (2011) Crown size and transplant type on the strawberry yield. *Scientia Agricola* 68, 489–493.
Collins, W.B. (1966) Floral initiation in strawberry and some effects of red and far-red radiation as components of continuous white light. *Canadian Journal of Botany* 44, 663–668.
Darnell, R. and Hancock, J. (1996) Balancing vegetative and reproductive growth in strawberry. In: Pritts, M.V., Chandler, C.K. and Crocker, T.E. (eds) *Proceedings of the IV North American Strawberry Conference*. University of Florida, Gainesville, Florida, pp. 144–150.
Darrow, G.M. (1929) Inflorescence types of strawberry varieties. *American Journal of Botany* 16, 571–585.
Darrow, G.M. (1936) Interrelation of temperature and photoperiodism in the production of fruit-buds and runners in the strawberry. *Proceedings of the American Society for Horticultural Science* 34, 360–363.
Darrow, G.M. (1966) *The Strawberry: History, Breeding and Physiology*. Holt, Rinehart and Winston, New York.
Darrow, G.M. and Waldo, G.F. (1934) Responses of strawberry varieties and species to duration of the daily light period. *Technical Bulletin of United States Department of Agriculture* 453, 1–31.
Daugaard, H. (1999) The effect of flower removal on the yield and vegetative growth of A+ frigo plants of strawberry (*Fragaria* × *ananassa* Duch). *Scientia Horticulturae* 82, 153–157.
Davies, F.S. and Albrigo, L.G. (1983) Water relations of small fruits. In: Kozlowski, T.T. (ed.) *Water Deficits and Plant Growth*. Academic Press, New York, pp. 89–136.
Dennis, F.G. Jr, Lipecki, J. and Kiang, C.-L. (1970) Effects of photoperiod and other factors opon flowering and runner development of three strawberry cultivars. *Journal of the American Society of Horticultural Science* 95, 750–754.
Desmet, E.M., Verbraeken, L. and Baets, W. (2009) Optimisation of nitrogen fertilisation prior to and during flowering process on performance of short day strawberry 'Elsanta'. *Acta Horticulturae* 842, 675–678.
Dradi, R., Faedi, W. and Iavarone, E. (1996) Influenza della frigoconservazione sulle riserve glucidiche di piante di fragola adatte alle colture fuori suolo. *Rivista di Frutticoltura e di Ortofloricoltura* 58, 73–76.

Duralija, B., Miličević, T., Vokurka, A., Mešić, A., Jemrić, T. and Čmelik, Z. (2007) Cold-stored tray plants in open field cultivation of strawberries. *Agriculturae Conspectus Scientificus* 71, 167–170.
Durner, E.F. (1984) Photoperiod and temperature effects on flower and runner development in day-neutral, June-bearing, and everbearing strawberries. *Journal of the American Society for Horticultural Science* 109, 396–400.
Durner, E.F. and Poling, E.B. (1987) Flower bud induction, initiation, differentiation and development in the 'Earliglow' strawberry. *Scientia Horticulturae* 31, 61–69.
Durner, E.F. and Poling, E. (1988) Strawberry developmental responses to photoperiod and temperature: a review. *Advances in Strawberry Production* 7, 6–14.
Durner, E.F., Poling, E.B. and Maas, J.L. (2002) Recent advances in strawberry plug transplant technology. *HortTechnology* 12, 545–550.
Faby, R. (1997) The productivity of graded 'Elsanta' frigo plants from different origins. *Acta Horticulturae* 439, 449–456.
Fujimoto, K. (1972) Studies on physiological and ecological characteristics of strawberry 'Hoko-wase' and development of new cropping type. *Special Bulletin of the Nara Agricultural Experiment Station* 1, 1–151.
Fujishige, N. (1994) Strawberry. In: Konishi, K., Iwahori, S., Kitagawa, H. and Yakuwa, T. (eds) *Horticulture in Japan*. Asakura Publishing, Tokyo, pp. 78–81.
Furuya, S., Yamashita, M. and Yamasaki, A. (1988) Effects of nitrogen content on the flower bud initiation induced by chilling under dark condition in strawberries. *Bulletin of the National Research Institute of Vegetables, Ornamental Plants and Tea, Series D, Kurume* 1, 51–57.
Gehrmann, H. (1985) Growth, yield and fruit quality of strawberries as affected by water supply. *Acta Horticulturae* 171, 463–469.
Giménez, G., Andriolo, J.L., Janisch, D., Cocco, C. and Picio, M.D. (2009) Cell size in trays for the production of strawberry plug transplants. *Pesquisa Agropecuaria Brasileira* 44, 726–729.
Godin, C. (2000) Representing and encoding plant architecture: a review. *Annals of Forest Science* 57, 413–438.
Grant, O.M., Davies, M.J., Johnson, A.W. and Simpson, D.W. (2012) Physiological and growth responses to water deficits in cultivated strawberry (*Fragaria* × *ananassa*) and in one of its progenitors, *Fragaria chiloensis*. *Environmental and Experimental Botany* 83, 23–32.
Guttridge, C.G. (1955) Observations on the shoot growth of the cultivated strawberry plant. *Journal of Horticultural Science* 30, 1–11.
Guttridge, C.G. (1958) The effects of winter chilling on the subsequent growth and development of the cultivated strawberry plant. *Journal of Horticultural Science* 33, 127.
Guttridge, C.G. (1969) Fragaria. In: Evans, L.T. (ed.) *The Induction of Flowering*. Cornell University Press, New York, pp. 247–267.
Guttridge, C.G. (1985) *Fragaria* × *ananassa*. In: Halevy, A. (ed.) *Handbook of Flowering*. CRC Press, Boca Raton, Florida, pp. 16–33.
Hallé, F. and Oldeman, R.A.A. (1970) *Essai sur l'architecture et la dynamique de croissance des arbres tropicaux*, Monographie/Collection de Monographies de Botanique et de Biologie Vegetale, No. 6. Masson and Cie, Paris.
Hallé, F., Oldeman, R.A.A. and Tomlinson, P.B. (1978) *Tropical Trees and Forests: An Architectural Analysis*. Springer-Verlag, Berlin, Germany.
Hancock, J.F. (1999) *Strawberries, Crop Production Science in Horticulture*. CABI Publishing, Wallingford, UK.
Hartmann, H.T. (1947) Some effects of temperature and photoperiod on flower formation and runner production in the strawberry. *Plant Physiology* 22, 407–420.
Hassell, R.L., Phillips, T.L., Dufault, R.J., Hale, T.A. and Ballington, J.R. (2007) Fall transplanting date affects strawberry cultivar performance in South Carolina. *International Journal of Fruit Science* 6, 73–85.
Heide, O.M. (1977) Photoperiod and temperature interactions in growth and flowering of strawberry. *Physiologia Plantarum* 40, 21–26.
Heide, O.M., Stavang, J.A. and Sønsteby, A. (2013) Physiology and genetics of flowering in cultivated and wild strawberries – a review. *Journal of Horticultural Science and Biotechnology* 88, 1–18.
Hochmuth, G., Cantliffe, D., Chandler, C., Stanley, C., Bish, E., Waldo, E., Legard, D. and Duval, J. (2006) Containerized strawberry transplants reduce establishment-period water use and enhance early growth and flowering compared with bare-root plants. *HortTechnology* 16, 46–54.
Hochmuth, R., Leon, L.L., Crocker, T., Dinkins, D. and Hochmuth, G. (1998) *Comparison of Bare-Root and Plug Strawberry Plants in Soilless Culture in North Florida*. Institute of Food and Agricultural Sciences, University of Florida, Florida.
Ito, H. and Saito, T. (1962) Studies on the flower formation in the strawberry plants I. Effects of temperature and photoperiod on the flower formation. *Tohoku Journal of Agricultural Research* 13, 191–203.

Jansen, W.A.G.M. (1997) Growing media and plant densities for strawberry tray plants. *Acta Horticulturae* 439, 457–460.

Jemmali, A. and Boxus, P. (1993) Early estimation of strawberry floral intensity and its improvement under cold greenhouse. *Acta Horticulturae* 348, 357–360.

Jonkers, H. (1965) On the flower formation, the dormancy and the early forcing of strawberries. PhD thesis, Wageningen University, Wageningen, The Netherlands.

Kadir, S., Sidhu, G. and Al-Khatib, K. (2006) Strawberry (*Fragaria* × *ananassa* Duch.) growth and productivity as affected by temperature. *HortScience* 41, 1423–1430.

Kahangi, E.M., Fujime, Y. and Nakamura, E. (1992) Effects of chilling and growth regulators on runner production of three strawberry cultivars under tropical conditions. *Journal of Horticultural Science* 67, 381–384.

Kehoe, E., Savini, G. and Neri, D. (2009) The effects of runner grade, harvest date and peat growing media on strawberry tray plant fruit production. *Acta Horticulturae* 842, 699–702.

Kerkhoff, K.L., Williams, J.M. and Barden, J.A. (1988) Net photosynthetic rates and growth of strawberry after partial defoliation. *HortScience* 23, 1086.

Kim, D.Y., Kim, T.I., Kim, W.S., Kang, Y.I., Yun, H.K., Choi, J.M. and Yoon, M.K. (2011) Changes in growth and yield of strawberry (cv. Maehyang and Seolhyang) in response to defoliation during nursery period. *Journal of Bio-Environment Control* 20, 283–289.

Kinet, J., Parmentier, A. and Lieten, F. (1993) Changes in quality of cold-stored strawberry plants (cv. Elsanta) as a function of storage duration: the flowering response in controlled environments. *Acta Horticulturae* 348, 287–293.

Kirschbaum, D.S., Honorato, J. and Cantliffe, D.J. (2000) Strawberry waiting bed plants: a valid alternative to increase early and total yields in sub-tropical regions. *Scientia Horticulturae* 84, 83–90.

Konsin, M., Voipio, I. and Palonen, P. (2001) Influence of photoperiod and duration of short-day treatment on vegetative growth and flowering of strawberry (*Fragaria* × *ananassa* Duch.). *Journal of Horticultural Science and Biotechnology* 76, 77–82.

Kumakura, H. and Shishido, Y. (1995) Effects of temperature and light conditions on flower initiation and fruit development in strawberry. *Japan Agricultural Research Quarterly* 29, 241–250.

Kumar, S. and Dey, P. (2011) Effects of different mulches and irrigation methods on root growth, nutrient uptake, water-use efficiency and yield of strawberry. *Scientia Horticulturae* 127, 318–324.

Kurokura, T., Inaba, Y., Neri, D. and Sugiyama, N. (2005a) A morphological study of the development of the second inflorescences in strawberry (*Fragaria* × *ananassa* Duch.). *Annals of Applied Biology* 146, 511–515.

Kurokura, T., Iwama, T., Inaba, Y. and Sugiyama, N. (2005b) Effect of day-length on the developmental pattern of axillary buds in June-bearing strawberry plants. *Journal of Horticultural Science and Biotechnology* 80, 139–142.

Le Mière, P., Hadley, P., Darby, J. and Battey, N.H. (1996) Effect of temperature and photoperiod on the rate of flower initiation and the onset of dormancy in the strawberry (*Fragaria* × *ananassa* Duch.). *Journal of Horticultural Science* 71, 361–371.

Lenz, F. (1979) Wachstum und wasserverbrauch bei jungen erdbeerpflanzen ('Senga Sengana') in abhangigkeit von der wurzeltemperatur. *Erwerbsobstbau* 21, 146–148.

Lieten, F. (1993) Methods and strategies of strawberry forcing in central Europe: historical perspectives and recent developments. *Acta Horticulturae* 348, 161–170.

Lieten, F. (1997a) Relationship of digging date, chilling and root carbohydrate content to storability of strawberry plants. *Acta Horticulturae* 439, 623–626.

Lieten, F. (1997b) Effects of chilling and night-break treatment on greenhouse production of 'Elsanta'. *Acta Horticulturae* 439, 633–640.

Lieten, F., Kinet, J.-M. and Bernier, G. (1995) Effect of prolonged cold storage on the production capacity of strawberry plants. *Scientia Horticulturae* 60, 213–219.

Lieten, P. (2002) The effect of nutrition prior to and during flower differentiation on phyllody and plant performance of short day strawberry Elsanta. *Acta Horticulturae* 567, 345–348.

Lieten, P. (2012) Advances in strawberry substrate culture during the last twenty years in the Netherlands and Belgium. *International Journal of Fruit Science* 13, 84–90.

Lieten, P., Longuesserre, J. and Pivot, D. (2004) Experiences with substrates, drainage water and recirculation in strawberry culture. *Acta Horticulturae* 649, 207–212.

Matsumoto, O. (1991) Studies on the control of flower initiation and dormancy in the cultivation of strawberry, *Fragaria* × *ananassa* Duch. *Special Bulletin of the Yamaguchi Agricultural Experiment Station* 31, 1–102.

Mohamed, F.H. (2002) Effect of transplant defoliation and mulch color on the performance of three strawberry cultivars grown under high tunnel. *Acta Horticulturae* 567, 483–485.

Molot, P.M. and Leroux, J.P. (1973) Evolution des glucides dans le rhizome du fraisier en cours de conservation au frigorifique. *Comptes Rendus des Séances de l'Académie d'Agriculture de France* 59, 187–189.
Moore, J.N. and Scott, D.H. (1965) Effects of gibberellic acid and blossom removal on runner production of strawberry varieties. *Proceedings of the American Society for Horticultural Science* 87, 240–244.
Naumann, W.D. (1961) Die Wircung zeitlich begrenzter Wassergaben auf Wuchsund Ertragleistun von Erdbeeren. *Gartenbau* 26, 441–458.
Naumann, W.-D. (1964) Bewässerungstermin und Blütenknospendifferenzierung bei Erdbeeren. *Die Gartenbauwissenschaft* 29, 21–30.
Neri, D., Savini, G., Massetani, F. (2010) Architettura della pianta. In: AA.VV. *La Fragola*. Script, Bologna, pp. 142–151.
Nicoll, M.F. and Galletta, G.J. (1987) Variation in growth and flowering habits of June-bearing and everbearing strawberries. *Journal of the American Society for Horticultural Science* 112, 872–880.
Nishiyama, M. and Kanahama, K. (2002) Effects of temperature and photoperiod on flower bud initiation of day-neutral and everbearing strawberries. *Acta Horticulturae* 567, 253–255.
Nishiyama, M., Kanahama, K., Bodson, M. and Verhoyen, M.N.J. (2000) Effect of temperature and photoperiod on the development of inflorescences in everbearing strawberry (*Fragaria* × *ananassa* Duch.) plants. *Acta Horticulturae* 514, 261–267.
Nishiyama, M., Ohkawa, W., Kanayama, Y. and Kanahama, K. (2006) Critical photoperiod for flower bud initiation in everbearing strawberry 'Summerberry' plants grown at high temperatures. *Tohoku Journal of Agricultural Research* 56, 1–8.
Okimura, M. and Igarashi, I. (1997) Effects of photoperiod and temperature on flowering in everbearing strawberry seedlings. *Acta Horticulturae* 439, 605–608.
Opstad, N., Sønsteby, A., Myrheim, U. and Heide, O.M. (2011) Seasonal timing of floral initiation in strawberry: effects of cultivar and geographic location. *Scientia Horticulturae* 129, 127–134.
Palha, M.G., Campo, J.L. and Oliveira, P.B. (2012) Strawberry plant growth and dry matter partitioning as influenced by planting date and plant type in an autumn production system. *Acta Horticulturae* 926, 463–469.
Paranjpe, A., Cantliffe, D.J., Lamb, E.M., Stoffella, P.J. and Powell, C. (2003) Winter strawberry production in greenhouses using soilless substrates: an alternative to methyl bromide soil fumigation. *Proceedings of the Florida State for Horticultural Science* 116, 98–105.
Paranjpe, A.V., Cantliffe, D.J., Stoffella, P.J., Lamb, E.M. and Powell, C.A. (2008) Relationship of plant density to fruit yield of 'Sweet Charlie' strawberry grown in a pine bark soilless medium in a high-roof passively ventilated greenhouse. *Scientia Horticulturae* 115, 117–123.
Peñuelas, J., Savé, R., Marfà, O. and Serrano, L. (1992) Remotely measured canopy temperature of greenhouse strawberries as indicator of water status and yield under mild and very mild water stress conditions. *Agricultural and Forest Meteorology* 58, 63–77.
Piringer, A.A. and Scott, D.H. (1964) Interrelation of photoperiod, chilling, and flower-cluster and runner production by strawberries. *Proceedings of the American Society for Horticultural Science* 84, 295–301.
Porlingis, I.C. and Boynton, D. (1961) Growth responses of the strawberry plant, *Fragaria chiloensis* var. Ananassa, to gibberellic acid and to environmental conditions. *Proceedings of the American Society for Horticultural Science* 78, 261–269.
Reichart, G. (1973) Vegetative growth and flower and fruit development in *Fragaria* × *ananassa* cv. 'Senga Sengana' under controlled conditions. *Horticultural Abstracts* 44, 1426.
Rice Jr, R.P. (1990) Effects of cultivar and environmental interactions on runner production, fruit yield, and harvest timing of strawberry (*Fragaria* × *ananassa*) in Zimbabwe. *Acta Horticulturae* 279, 327–332.
Riyaphan, P., Pipattanawong, N. and Subhardrabandu, S. (2005) Influences of elevation on growth and yield of strawberry in Thailand. *Kasetsart Journal (Natural Science)* 39, 535–545.
Robertson, M. and Wood, C.A. (1954) Studies on the development of the strawberry. I. Flower bud initiations and development in early and late-formed runners in 1951 and 1952. *Journal of Horticultural Science* 29, 104–111.
Ruan, J., Yeoung, Y. and Larson, K. (2011) Influence of cultivar, planting date, and planting material on yield of day-neutral strawberry cultivars in highland areas of Korea. *Horticulture, Environment, and Biotechnology* 52, 567–575.
Savini, G. (2003) [Architectural model and factors implicated in the flower differentiation of strawberry plant]. Doctoral dissertation, Università Politecnica delle Marche, Marche, Italy (in Italian).
Savini, G. and Neri, D. (2004) Strawberry architectural model. *Acta Horticulturae* 649, 169–176.

Savini, G., Neri, D., Zucconi, F. and Sugiyama, N. (2005) Strawberry growth and flowering. *International Journal of Fruit Science* 5, 29–50.
Savini, G., Letouzé, A., Sabbadini, C. and Neri, D. (2006a) Evaluation of tray-plant quality in the propagation phase. *Acta Horticulturae* 708, 231–236.
Savini, G., Neri, D., Mercadante, L., Molari, G., Magnani, D. and Capriolo, G. (2006b) Meristem analysis on strawberry plants during propagation and production. *Acta Horticulturae* 708, 237–240.
Scott, D.H. and Marth, P.C. (1953) Effect of blossom removal on growth of newly set strawberry plants. *Proceedings of the American Society for Horticultural Science* 62, 255–256.
Serçe, S. and Hancock, J.F. (2005) The temperature and photoperiod regulation of flowering and runnering in the strawberries, *Fragaria chiloensis, F. virginiana*, and *F. × ananassa. Scientia Horticulturae* 103, 167–177.
Serrano, L., Carbonell, X., Savé, R., Marfà, O. and Peñuelas, J. (1992) Effects of irrigation regimes on the yield and water use of strawberry. *Irrigation Science* 13, 45–48.
Sønsteby, A. and Heide, O.M. (2006) Dormancy relations and flowering of the strawberry cultivars Korona and Elsanta as influenced by photoperiod and temperature. *Scientia Horticulturae* 110, 57–67.
Sønsteby, A. and Heide, O. (2007) Long-day control of flowering in everbearing strawberries. *Journal of Horticultural Science and Biotechnology* 82, 875–884.
Sønsteby, A. and Hytonen, T. (2005) Manipulating flower induction through temperature and photoperiod fluctuations. *International Journal of Fruit Science* 5, 17–27.
Sønsteby, A. and Nes, A. (1998) Short days and temperature effects of growth and flowering in strawberry (*Fragaria × ananassa* Duch.). *Journal of Horticultural Science and Biotechnology* 73, 730–736.
Sønsteby, A., Opstad, N., Myrheim, U. and Heide, O.M. (2009) Interaction of short day and timing of nitrogen fertilization on growth and flowering of 'Korona' strawberry (*Fragaria × ananassa* Duch.). *Scientia Horticulturae* 123, 204–209.
Sønsteby, A., Opstad, N. and Heide, O.M. (2013) Environmental manipulation for establishing high yield potential of strawberry forcing plants. *Scientia Horticulturae* 157, 65–73.
Strik, B.C. (1985) Flower bud initiation in strawberry cultivars. *Fruit Varieties Journal* 39, 5–9.
Sugiyama, N., Iwama, T., Inaba, Y., Kurokura, T. and Neri, D. (2004) Varietal differences in the formation of branch crowns in strawberry (*Fragaria ananassa*) plants. *Journal of the Japanese Society for Horticultural Science* 73, 216–220.
Takeda, F. (2012) New methods for advancing or delaying anthesis in short-day strawberry. *Acta Horticulturae* 926, 237–241.
Takeda, F. and Newell, M. (2007) Effects of runner tip size and plugging date on fall flowering in short-day strawberry (*Fragaria × ananassa* Duch.) cultivars. *International Journal of Fruit Science* 6, 103–117.
Taylor, D.R. (2002) The physiology of flowering in strawberry. *Acta Horticulturae* 567, 245–251.
Tehranifar, A. and Battey, N.H. (1997) Comparison of the effects of GA_3 and chilling on vegetative vigour and fruit set in strawberry. *Acta Horticulturae* 439, 627–632.
Tehranifar, A., Miere, P.L. and Battey, N.H. (1998) The effects of lifting date, chilling duration and forcing temperature on vegetative growth and fruit production in the June bearing strawberry cultivar Elsanta. *Journal of Horticultural Science and Biotechnology* 73, 453–460.
Tehranifar, A., Poostchi, M., Arooei, H. and Nematti, H. (2007) Effects of seven substrates on qualitative and quantitative characteristics of three strawberry cultivars under soilless culture. *Acta Horticulturae* 761, 485–488.
Tehranifar, A., Ameri, A., Shoor, M. and Davarynejad, G.H. (2012) Effect of substrate and cultivar on growth characteristic of strawberry in soilless culture system. *African Journal of Biotechnology* 11, 11960–11966.
Tworkoski, T.J., Benassi, T.E. and Takeda, F. (2001) The effect of nitrogen on stolon and ramet growth in four genotypes of *Fragaria chiloensis* L. *Scientia Horticulturae* 88, 97–106.
van Delm, T., Massetani, F., Savini, G. and Neri, D. (2009) Plant architecture of strawberry trayplants in relation to nutrient application system. In: *VII Convegno AISSA* 2–4 December, Ancona, Italy, pp. 57–58.
van Delm, T., Melis, P., Stoffels, K. and Baets, W. (2012) Pre-harvest night-interruption on everbearing cultivars in out-of-soil strawberry cultivation in Belgium. *International Journal of Fruit Science* 13, 217–226.
van Delm, T., Melis, P., Stoffels, K. and Baets, W. (2013) Pre-harvest night-interruption on everbearing cultivars in out-of-soil strawberry cultivation in Belgium. *International Journal of Fruit Science* 13, 217–226.
van den Muijzenberg, E.W.B. (1942) The influence of light and temperature on the periodic development of the strawberry and its significance in cultivation. PhD thesis, Wageningen University, Wageningen, The Netherlands.

VanDerZanden, A.M. and Cameron, J.S. (1996) Effect of water deficit stress on 11 native *Fragaria chiloensis* clones selected as ornamental groundcovers. *Scientia Horticulturae* 66, 241–253.
Verheul, M.J., Sønsteby, A. and Grimstad, S.O. (2006) Interactions of photoperiod, temperature, duration of short-day treatment and plant age on flowering of *Fragaria* × *ananassa* Duch. cv. Korona. *Scientia Horticulturae* 107, 164–170.
Verheul, M.J., Sønsteby, A. and Grimstad, S.O. (2007) Influences of day and night temperatures on flowering of *Fragaria* × *ananassa* Duch., cvs. Korona and Elsanta, at different photoperiods. *Scientia Horticulturae* 112, 200–206.
Vince-Prue, D. and Guttridge, C.G. (1973) Floral initiation in strawberry: spectral evidence for the regulation of flowering by long-day inhibition. *Planta* 110, 165–172.
Voth, V. and Bringhurst, R.S. (1958) Fruiting and vegetative response of Lassen strawberries in southern California as influenced by nursery source, time of planting and plant chilling history. *Proceedings of the American Society for Horticultural Science* 72, 186–197.
Wahdan, H.A. and Waister, P.D. (1984) Flower initiation, fruit production and vegetative development in non-induced strawberry plants exposed to outdoor conditions in Scotland. *Journal of Horticultural Science* 59, 187–196.
Waithaka, K. (1993) Growth and runner production of everbearing strawberries in Kenya. *Acta Horticulturae* 348, 357–360.
Went, F.W. (1958) The experimental control of plant growth. *Soil Science* 85, 288.
White, J. (1979) The plant as a metapopulation. *Annual Review of Ecology and Systematics* 10, 109–145.
Wright, C.J. and Sandrang, A.K. (1995) Efficiency of light utilization in the strawberry (*Fragaria* × *ananassa*) cv. Hapil. *Journal of Horticultural Science* 70, 705–711.
Yamasaki, A. and Yano, T. (2009) Effect of supplemental application of fertilizers on flower bud initiation and development of strawberry – possible role of nitrogen. *Acta Horticulturae* 842, 765–768.
Yamasaki, A., Yoneyama, T., Tanaka, F., Tanaka, K. and Nakashima (2002) Tracer studies on the allocation of carbon and nitrogen during flower induction of strawberry plants as affected by the nitrogen level. *Acta Horticulturae* 567, 349–352.
Yoshida, Y. (1992) Studies on flower and fruit development in strawberry with special reference to fruit malformation in 'Ai-Berry'. *Memoirs of Faculty of Agriculture, Kagawa University* 57, 1–94.

8 Strawberry Root Growth and Architecture in Relation to Organic Residues and Replanting Problems

Davide Neri[1,2*]

[1]Dipartimento di Scienze Agrarie, Alimentari e Ambientali – Università Politecnica delle Marche, Ancona, Italy; [2]Centro di Ricerca per la Frutticoltura – Consiglio per la Ricerca in Agricoltura e l'Analisi dell'Economia Agraria, Rome, Italy

8.1. Introduction

The interaction of the roots with the soil microbiome, soil organic matter and organic residues (debris) in ecological, sustainable and low-input cultivation systems is crucial (Lakshmanan *et al.*, 2014). Roots experience the available niches, and their own growth, through a very powerful allelopathic sensitivity (Neri, 1998, 2004). The strawberry exhibits this sensitivity and may experience 'soil sickness' and replant disease problems after a few replanting cycles resulting in the plants collapsing during very hot days and loss of fruit production (Neri *et al.*, 1998).

An understanding of the correlative interactions inside the plant, such as shoot:root ratio, carbon allocation and partitioning, root apical dominance and lateral root formation, is necessary to understand root behaviour.

* d.neri@univpm.it or davide.neri@crea.gov.it

These closely correlative interactions are the real determinants of root growth, while hormones together with nutritional substances and the additional active biocompounds of root metabolism play a subordinate role. Actively growing roots determine the synthesis and/or absorption of hormones and assimilates, not vice versa.

The biotic and abiotic external environments also determine the structure of the root systems, interacting with the plant genome (Chiatante *et al.*, 2007a,b). This stimulation may be mediated by a strict shoot–root relationship because of the self-regulation of the system, which is based on the plasticity of the stems and roots (Zucconi, 1996). Thus, in the strawberry, the number of primary roots is strictly associated with the number of leaves (Hancock, 1999), and root ramification is dependent on the rate of shoot growth, and vice versa. In addition, the plant architecture may change depending on the root behaviour (Massetani *et al.*, 2014).

8.2. Strawberry Root Growth and Architecture

Strawberry roots have a primary structure characterized by a large parenchymal layer, which has the function of storage, under the epidermis. The internal cylinder is protected by the endodermis, while the pericycle is the layer in strict contact with the vascular transport system (phloem and xylem), which is organized in two or more arches (Fig. 8.1). The largest primary roots can produce also a secondary structure (Fig. 8.1), being able to survive more than one season.

8.2.1. Apical dominance

In the shoot, the correlative inhibition of the growth of lateral (axillary) meristems is exerted by the growing apical meristem and is called apical dominance (McGraw-Hill, 2012). Partial or complete apical dominance can also occur in the roots, impeding lateral root formation along the stem of a growing root (Fig. 8.2). The correlative control of the orientation of lateral organs such as branches, leaves and stolons is called epinasty and is due to the excessive activity of the upward part of the shoot. In the roots, the same term cannot be used because the control is determined by the downward-growing root, so the term hyponasty is used.

The diverse origins of the shoot and root ramifications determine the substantial differences in the growth model between roots and shoots. In the shoot apical meristem, cell division occurs in the young newly forming leaves and along the flanks of the apical dome. As the cell number increases, the cells elongate, pushing the apical dome upwards and leaving a small fraction of the dividing cells behind the axil of each of the laterally forming leaves. This axillary meristem, called also the lateral meristem, may develop into an axillary bud protected by scales with several unfolded leaves and an axis made of short internodes. The bud may rest and grow only after satisfaction of the required chilling hours (proleptic growth). However, the axillary meristem may escape from apical dominance and elongate, while the original apex is still growing quickly, and may produce a sylleptic growing shoot (namely, a stolon in strawberry) without becoming a bud and resting.

In the roots, branching does not directly involve the root apical meristem. In fact, lateral roots originate from single cells in the pericycle layer between the endodermis and root vascular system (Fig. 8.3). Thus, lateral root formation can happen several times per growing cycle in adjacent positions starting from an undetermined number of cells and without any rest period or chilling satisfaction.

In the shoot, the intensity of apical dominance over the axillary buds is associated with the plant varieties and species. Some species, such as pea and sunflower, show strong apical dominance, which causes the growth of single branchless stems. In other plants, such as some varieties of tomato, apical dominance is particularly low, allowing the growth of many axillary buds, which determines a bushy behaviour. Several factors can break apical dominance,

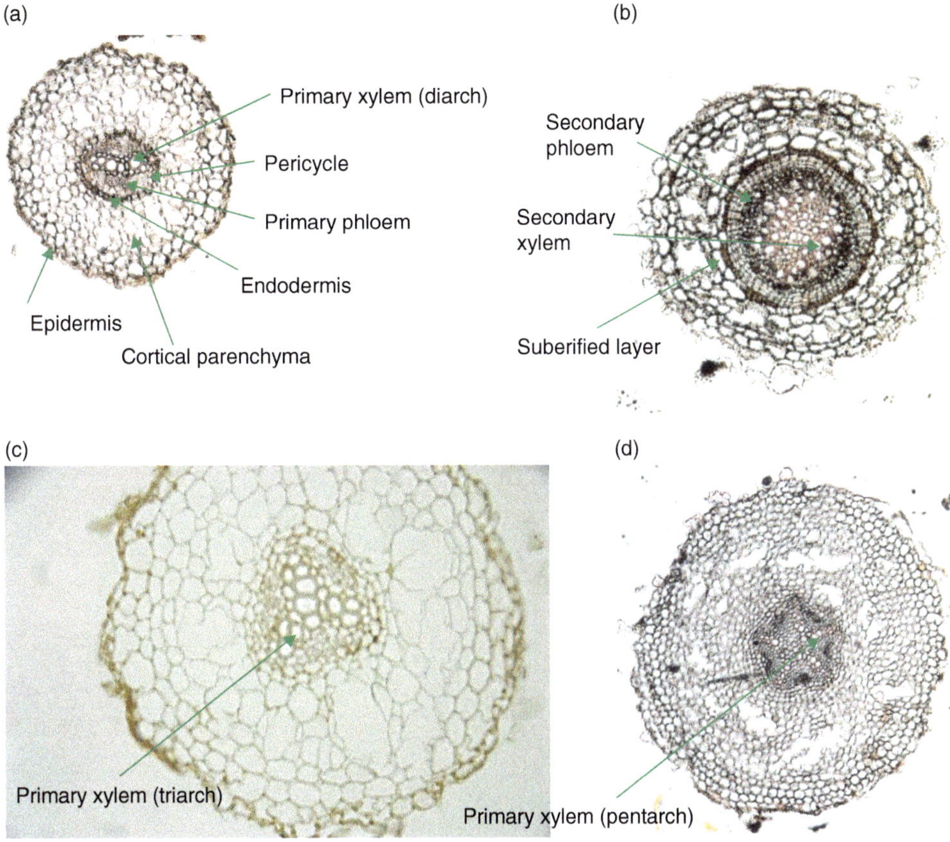

Fig. 8.1. Root anatomy. (a, c, d) Primary roots with two (a), three (c) and five (d) arches of xylem. (b) Secondary root.

including removal of the apical meristem (decapitation; Neri *et al.*, 2003), shifting the plant's main axis to a horizontal position (gravistimulation), inversion of the shoot, reduction of light intensity and/or photoperiods, and flower induction of the terminal meristem (Sugiyama *et al.*, 2004). Apical dominance may also be reduced as the plant ages.

In the root, robust dominance of the apex causes taproot growth, whereas weak dominance results in a more fibrous root system, similar to that of strawberry, but a full understanding of this mechanism is missing.

Leaves and branches that grow at a characteristic angle on the principal stem axis will grow more upright after apical bud removal. The growth of some other organs is also controlled by the apical meristem. Rhizomes (underground axillary shoots lacking chlorophyll and having rudimentary leaves) will grow upright, forming green, leafy shoots if the apical meristem and all the aboveground axillary buds are removed. Lower plants, such as mosses and ferns, as well as fungi and algae, also exhibit apical dominance.

One hypothesis for the mechanism of apical dominance is associated with the competition for nutrients between centres of growth. In agreement with this hypothesis, nutrients are preferentially transported to the apical bud, which can cause deficiencies in axillary buds. A second hypothesis for the mechanism of apical dominance proposes that one or more plant hormone, such as auxin, cytokinin or gibberellin,

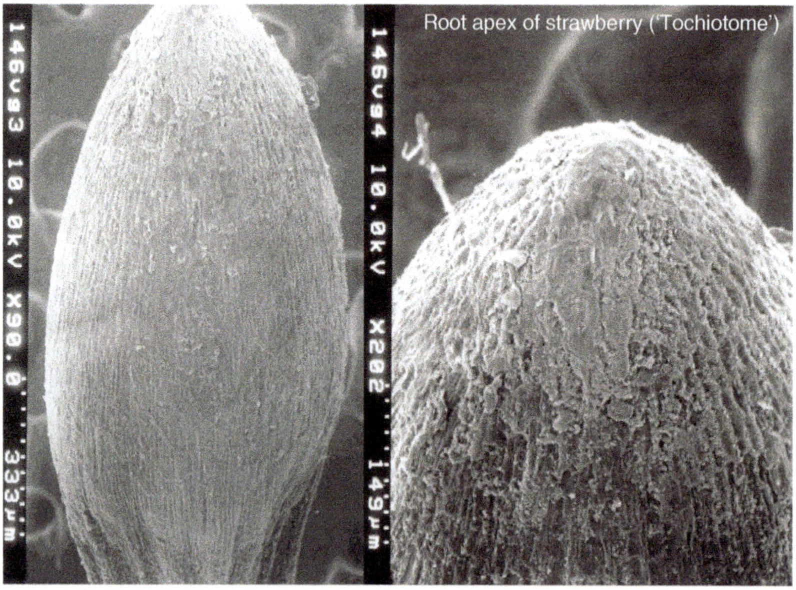

Fig. 8.2. Electron microscopic images of the root apex of strawberry. The root cap is fundamental to create the best conditions for root elongation. The bullet shape plays the same role. At this level, no laterals are formed due to strong apical dominance.

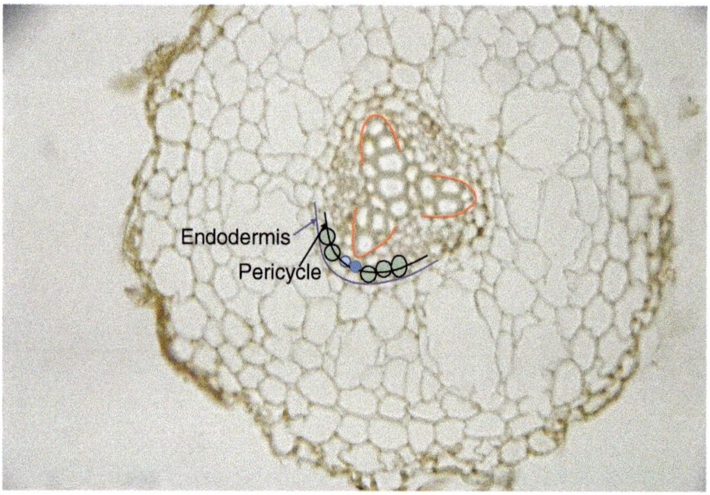

Fig. 8.3. Founder cells of lateral roots (blue circles), which are located in the pericycle layer (green circles) in front of the xylem arches (red lines). (From Neri *et al.*, 2011.)

acts as a correlative signal (Cline, 1997). We can assume that root apical dominance follows the same general behaviour but that lateral formation is fundamentally different in terms of the origin of the founder meristem.

8.2.2. Lateral root formation

Plants have the potential to originate *de novo* lateral roots in appropriate numbers and at the correct position to efficiently take up water and nutrients and to provide anchorage in a dynamic environment, the soil profile. The integrated structure of roots enables them to react in appropriate and modular ways to changing soil niches. The roots can rapidly and locally acclimate and modify their growth throughout the root system to converge their activities in the most useful niches (Hodge, 2009). This capacity can last for a long time in ageing roots and helps in responding to localized abiotic stresses, such as drought, twisting, mechanical damage and increasing slopes (Chiatante *et al.*, 2007a,b).

Lateral root formation can be described as follows: certain pericycle cells maintain their meristem properties, and once specific local conditions appear, they are able to undergo mitosis and organize a new root apical meristem (Scheres *et al.*, 2004). Therefore, the lateral roots do not originate from pre-organized 'buds', but rather arise directly from stem 'founder cells' (Laskowski *et al.*, 1995).

To study the formation of the primordia of lateral roots as early in development as possible, histone H4 expression analysis has been used, as histone synthesis occurs simultaneously along the S-phase of cell division (Neri *et al.*, 2011). Thus, changes in H4 expression indicate the localization of cell division, as H4 expression is abundant in tissues with active cell division and is rarer in older tissues. In strawberry, Neri *et al.* (2011) stimulated the growth of lateral roots using a primary root apex excision method in bench rhizotron-grown 'Camarosa' plants. This technique is very abrupt, but the response is highly predictable (in time and space) and is easily detectable. To describe the organization of the new lateral root primordia and the timing of their formation after primary root apex excision, both of which are very important for improving the survival of transplanted plants, *in situ* hybridization of histone H4 was used. The new lateral meristems were induced 24 h after excision of the root apex (1 or 8 cm from the tip), and *in situ* hybridization of histone H4 detected root primordia just below the cut (Fig. 8.4). After 48 h, new lateral roots were almost protruding along the primary roots from the specific cells in the pericycle close to the protoxylem arches (Fig. 8.5).

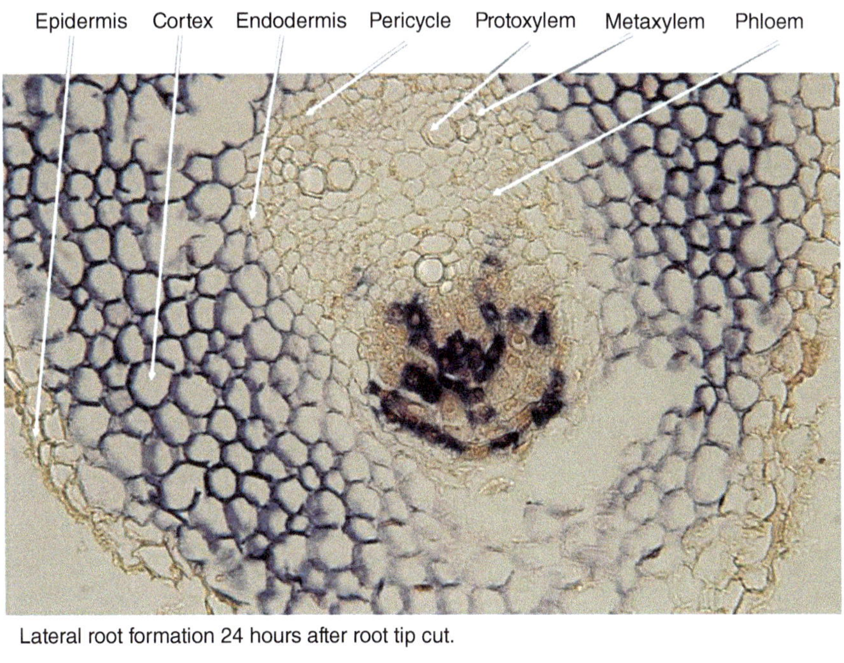

Fig. 8.4. *In situ* hybridization of H4 histone revealing organized lateral root formation 24 h after the root tip cut (decapitation). (From Neri *et al.*, 2011.)

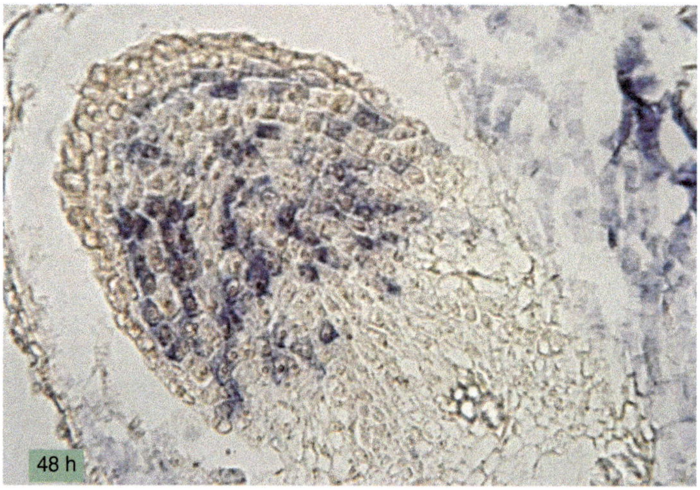

Fig. 8.5. A new lateral root, which is almost protruding at 48 h after the root tip cut. (From Neri *et al.*, 2011.)

8.2.3. Allelopathic sensitivity

The term allelopathy describes biochemical interactions that inhibit or stimulate the growth of neighbouring plants by another plant. Allelopathy is a multifaceted phenomenon related to soil conditions, root growth and biological reactions, associated chemicals and their concentrations, and phytotoxic effect. The action is governed by soil organic matter and organisms, which affect adsorption, desorption, transport and metabolism of allelochemicals (Kobayashi, 2004). It is difficult to isolate the allelopathic effects from competition among plants and/or interactions (Inderjit and Foy, 2001). All allelochemicals, directly or indirectly derived from exudates, leached substances, organic residues and decomposing debris, can be toxic to plants, and can potentially affect root growth (Zucconi, 1996, 2003; Polverigiani *et al.*, 2014; Endeshaw *et al.* 2015a,b).

To test the hypothesis that allelopathic responses can affect root growth in strawberry, Savini and Neri (2004), Neri *et al.* (2002, 2005) and Neri and Savini (2005) applied strawberry residues and eluates in substrate patches and studied the impact on root growth in bench rhizotrons (Fig. 8.6). These authors also examined the impact of residues from other species.

The strawberry residues, both solid and liquid, strongly modified root growth. They were able to disguise the natural autopathic behaviour of the roots (Fig. 8.7 and 8.8). Soon after their application, the strawberry residues induced root clustering, and the roots elongated faster, with few ramifications. When the crown roots reached a new fertile patch, lateral formation was easily recovered. Nevertheless, root migration and the effort to overcome the toxic layer resulted in a high energy and plastic cost, and carbon was allocated more towards vegetative runner production in the case of solid strawberry residues. Fruit production was higher when the roots grew in a non-hostile substrate. Thus, it can be hypothesized that, when the roots are exposed to strawberry residues, runner production fulfils the need for the plant to transmigrate into unexplored soil patches, far from the original foraging place.

8.2.4. An architectural model

Strawberry crown roots can grow very quickly and are positively geotropic. They change colour and function, becoming brown or dark brown in the second season, when they suberify. They maintain a polydermal

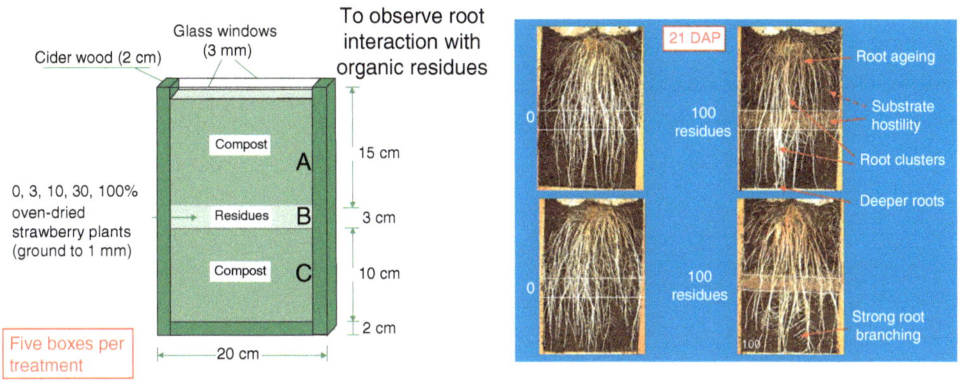

Fig. 8.6. Left: Experimental set-up to study the effect of organic residues in observation boxes. Right: Strawberry root system 21 days after planting (DAP) with or without strawberry residues in the outlined layer (O are two control plants and 100 residues are two plants with a layer of strawberry residues that had been oven dried for 2 days at 100°C and then finely ground) (From Neri *et al.*, 2005).

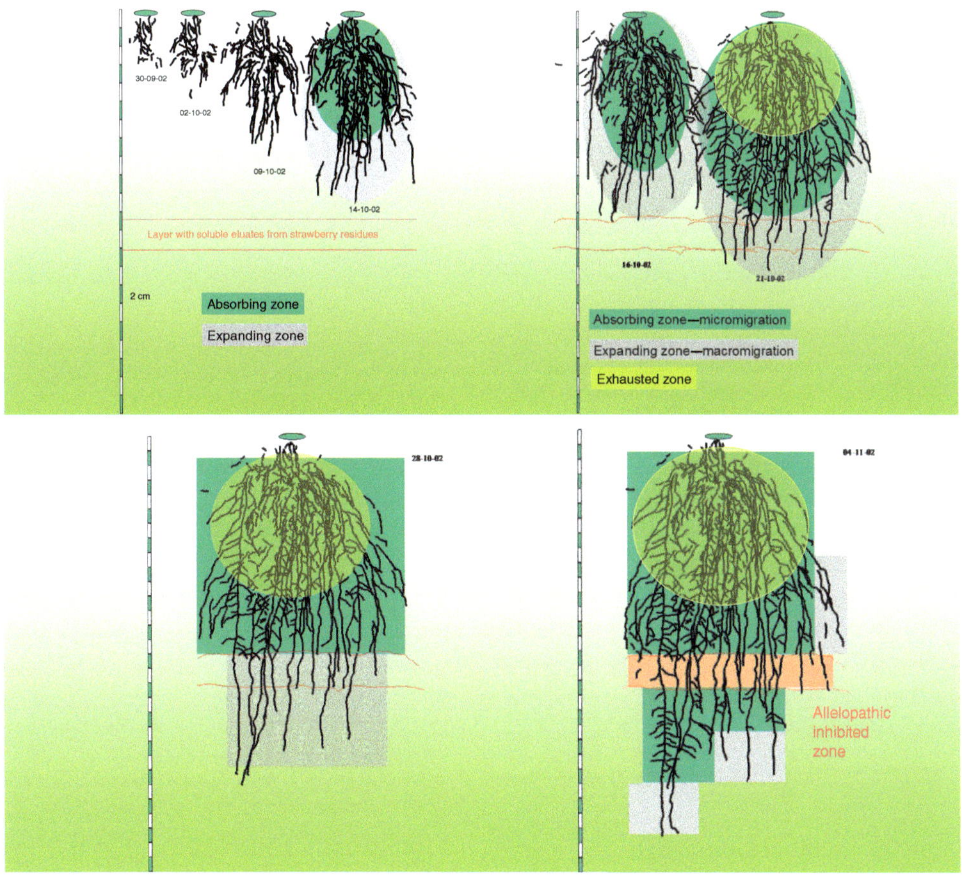

Fig. 8.7. Root growth of strawberry roots in bench rhizotrons over the time period indicated. The red lines indicate the layer in which the soluble eluates from strawberry residues were placed daily. (From Savini and Neri, 2004.)

Fig. 8.8. Strawberry root growth without (control) and with the addition of strawberry eluates.

layer, made of unsuberified cells, which functions as storage tissue, largely in the form of starch, and is not directly involved in absorption. After a certain period of growth, they form new laterals that are transient but very efficient in exploring fertile niches.

The roots are sensitive to their own allelopathic exudates, which, in addition to root competition (external function), may determine the spacing of the roots belonging to the same plant (internal function) and the avoidance of roots of neighbouring plants and other physical hindrances (external function) (Falik *et al.*, 2005).

An architectural model of strawberry root growth can be thus described (Figs 8.9 and 8.10) by the presence of:

1. Apical dominance, which allows root branching at a certain distance from the root apex (approximately 1–10 cm depending on the genotype and growing conditions).
2. Positive geotropism of all the root apices.
3. Hyponasty, which promotes an initial horizontal lateral growth.
4. Rapid root turnover, which determines root death in exploited niches.
5. Autopathy, which results in root distancing and the malfunctioning of aged roots and scarce root formation in exploited niches.

The growth of the new roots is very dynamic and is able to continuously replenish the absorbing root network. Rootlets actively absorb faster than the processes of diffusion or solubilization of mineral nutrients, and thus need a continuous substrate source. To overcome the reduction of nutrients, constant micromigration of the roots is necessary and thus root debris amasses in the old soil, resulting in intensive saprophytic activity in this area.

Root turnover requires 25–75% of the whole plant carbon budget depending on the soil fertility and plant species, although lower values can be found in cultivated conditions. Fertilization reduces root turnover, which is accelerated when there are alternate wet/dry conditions with a continuous shift of the roots. This dynamic micromigration is due to shedding of the obsolete root network and a strong interaction with self-produced residues and residues from other dispathic species nearby (Neri, 1998).

Roots tend to avoid returning to previously used niches because of allelopathic root repulsion (dispathy or autophobia) resulting

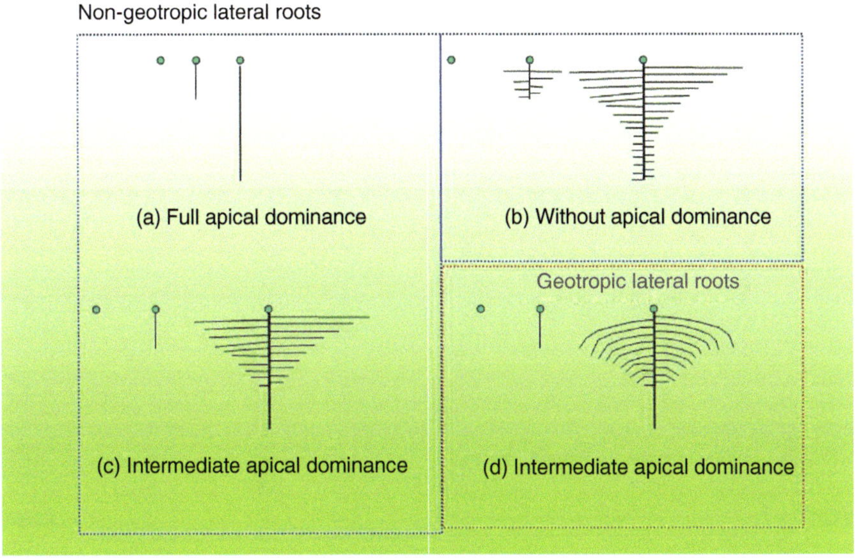

Fig. 8.9. Model of root growth of a strawberry plant showing a single non-geotropic root with no ramification because of full apical dominance (a), without apical dominance (b) and with intermediate apical dominance able to control lateral formation close to the apex (c), and a geotropic root with intermediate apical dominance plus a gravitropic response of laterals (d).

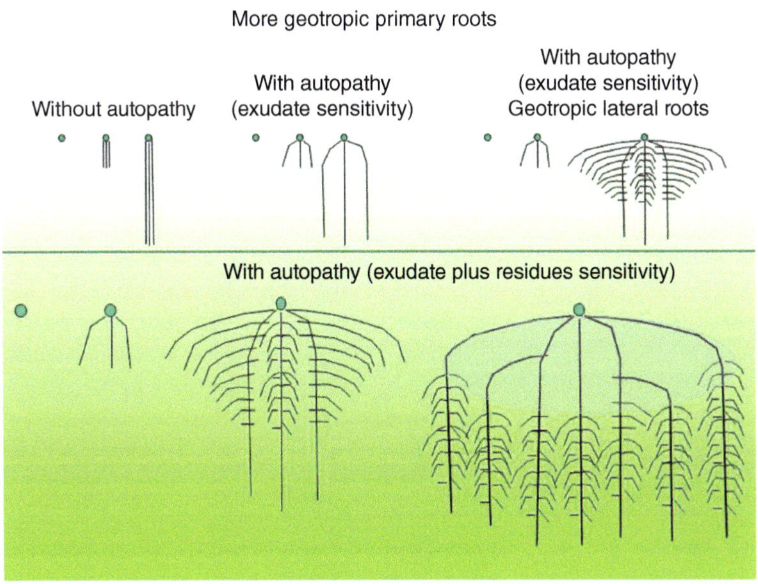

Fig. 8.10. Upper panel: a primary root with geotropic behaviour and without autopathy (left), with geotropic behaviour and with autopathy (middle), and with autopathy, apical dominance and geotropism of laterals (right). Lower panel: diagram showing the ageing and dying of older roots within the network.

from the production of its own residues in these niches. This dispathy is also determined by organic residues produced by microorganisms in the soil. It is worth noting that the organic residues from particular species may be tolerated by other species, and can even be stimulating, accounting for the success of specific crop rotations. Thus, trophism (meaning the capacity of the plant to obtain and use nutrients) needs dynamic control, not simply the presence of nutrients (Neri, 1998, 2004).

In the search for new niches, new roots follow the signals excreted by existing roots (primary allelopathic substances) and from previous root debris (secondary allelopathic substances). In this way, roots can rapidly find free niches, which is crucial for young juvenile plants growing in formerly occupied soils. It is also crucial when the roots of neighbourhood plants are very close, and the plant needs to produce more transmigrating roots (macromigrating roots) in the quest for new available niches. The macromigrating roots are poorly absorbing roots in comparison with micromigrant and lateral roots, but are very effective in the pursuit of new niches in the boundary layer of the exploited root zone. Residues accumulated from diverse species rotating in the exploited territory result in a more efficient humification (Zucconi, 2003; Zucconi *et al.*, 2003).

8.3. The Strawberry Root During the Growth Cycle and Production

Root growth can be manipulated during plant propagation by drip irrigation and fertigation, which renews the available water and nutrients, possibly diluting strawberry residues and probably stimulating the roots to grow at a higher density. The substrate and the pot (dimensions and form) are also crucial for the quality of plants propagated in a tray. In short rotation systems, it is also necessary that the organic residues from a previous strawberry crop are mineralized (biological, chemical or physical conversion), or completely humified to allow growth of the strawberry in the same field (Neri and Savini, 2006).

8.3.1. Stolon and runner relationships with the mother plant

The stoloniferous plants produce long sympodial stolons able to create ramets (rooted rosettes) at their terminal buds, which allow the daughter plant to explore with its own root system very different soil territories far from the mother plant root system. These peculiar shoots in strawberry are produced by the meristem at the leaf axil during vigorous growth phases; they have two nodes and are vegetative. Each ramet (a new individual clonal plant) shows the capacity to immediately produce a stolon at the axil of the basal bract, even before root formation. Over time, strawberries thus create a chain of ramets connected by stolons (runner chain). When each ramet starts to grow as a new plant, it may produce other stolons from axillary meristems depending on the rosette-growing conditions. Therefore, if the ramet has its own root system and is in fertile soil, the growth will be rapid and new ramets will be formed, but if the ramet has poor growing conditions, there will only be the runner chain, with the growth of a single stolon that is preformed inside each ramet (Fig. 8.11).

A stolon is a type of strawberry shoot and may be very elongated but always has two nodes and is vegetative and able to support the ramet (or rooted tip) until it becomes self-sufficient with its own root system (Fig. 8.12). In a single runner chain, the reciprocal capacity of the ramets to sustain growth and share locally abundant resources or to tolerate a local stress via differential root activity is still under debate. However, in the case of heterogeneity of resources such as water and nitrogen in the soil, with a reciprocal transport between ramets through the stolons, the connected plants have a higher total biomass, indicating a reciprocal transport capacity.

Thus, bare-root strawberry 'Camarosa' ramets, joined in pairs by their own stolons,

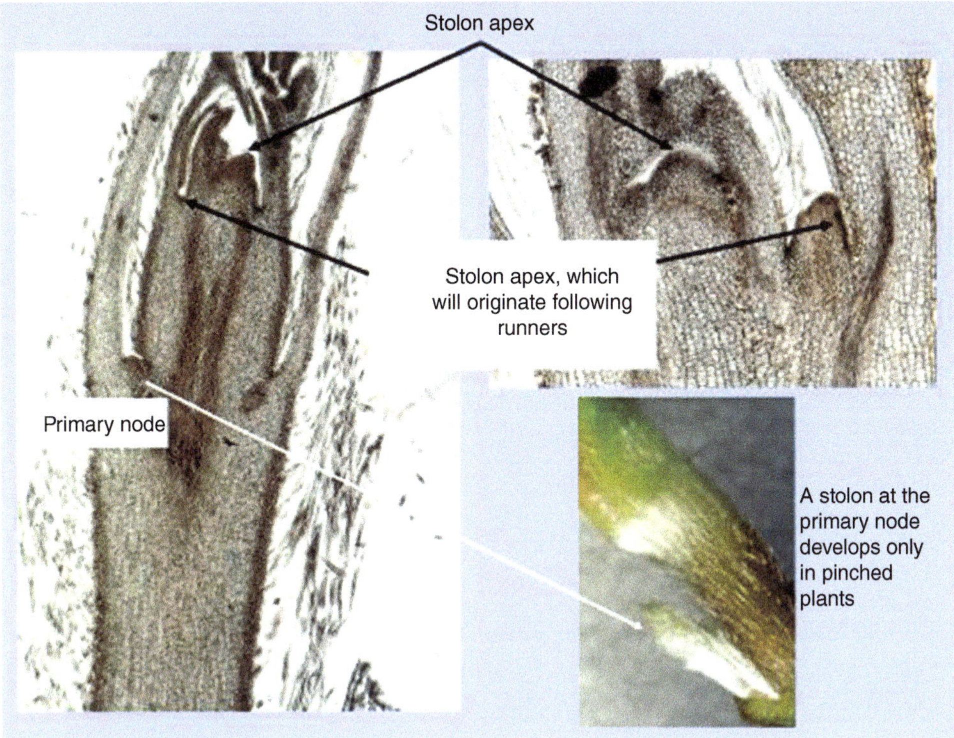

Fig. 8.11. Asymmetric stolon formation in a ramet meristem.

Fig. 8.12. Root formation in a stolon plant, showing, on the left of the stolon, the formation of a new runner chain. (Picture drawn by F. Massetani.)

were studied in a two-pot factorial experiment with bark decortication along the stolon, removal of the root system or glyphosate application to one of the two ramets (Savini *et al.*, 2008). The mother plant was considered to be the older ramet and the daughter plant was the younger. The mother plant did not show a strong hierarchic predominance, and when the roots of one ramet were cut, leaf number and chlorophyll content were decreased slightly in both plants. The decortication did not interfere with the water supply, but it limited the carbon allocation towards the daughter plant when the mother's roots were cut (but not vice versa) (Savini *et al.*, 2008). Glyphosate action was localized in the sprayed ramet: the chlorophyll content was reduced within 2 days and the plants expired after 4 days. The experiment specifically disabled phloem transportation (by decortication or girdling) without affecting water supply in response to localized stress (root cut, herbicide spray) applied to one ramet of the interconnected pair of strawberry plants, indicating that the xylem is able to transport water and nutrients in both directions (Savini *et al.*, 2008).

8.3.2. Tray plant growth

The fruit-growing season has changed fundamentally in recent decades. In the past, in north Europe the typical strawberry season was in the spring–summer period (from June to July). The season has now been expanded from the beginning of spring (April) to the end of autumn (November). The harvest season may be even longer, but becomes uneconomical in comparison with winter–spring production (from December to May) in the Mediterranean climate regions, which now account for half of the production in Europe (Massetani *et al.*, 2011).

Strawberry producers have therefore changed their supply of plants for the extended season of production. There is now a significant demand for strawberry 'tray' plants to be used in such programmed production (Savini *et al.*, 2006; Kehoe *et al.*, 2009). The plants in trays are propagated fresh and eventually can be cold stored (potted frigo plants), as was done for bare-root plants. Tray plants have gained popularity among growers in the Netherlands, Belgium and surrounding countries such as France, the UK and northern Italy (Lieten, 2002).

Tray plants are increasingly replacing the frigo 'waiting-bed' and A+ bare-root plants. They offer many advantages in comparison with soil-grown 'bare-root' plants. Runners and cuttings are grown on substrates, reducing the risk of infection by root diseases and enabling better control of plant nutrition. The physiological control of root growth and axillary meristems improves plant storage and establishment after cold storage, and enhances fruit production (Savini *et al.*, 2006). These programmed plants with well-developed inflorescences and efficient plant architecture and root systems can support heavy early production in the greenhouse just 60 days after planting from the cold room. In programmed forced production systems in continental climates, strawberry tray plants (also called plug plants, Fig. 8.13) are thus rapidly substituting the traditional bare-root systems.

In such conditions, the roots play an important role in flower induction with a strong indirect action. Poor root activity (induced by small dimensions of the pot and temporary low nitrogen content or water stress) is more flower inducing. The root quality also directly influences the plant quality, because a well-developed root system with young roots is helpful in overcoming the long storage period and the stress of transplanting in the field.

The plant quality is more controlled by shoot-to-root equilibrium and the total amount of the reservoir than by the dimension of the crown. The average fruit weight is higher for plants propagated from large runners and from plants grown in a more fertile substrate (Kehoe *et al.*, 2009). However, when large runners were propagated later in the season, the average fruit weight was reduced. Flower induction was regular during the season and the flower development was similar for the different runner categories (Kehoe *et al.*, 2009).

Fig. 8.13. Tray plant root systems at the proper time for cold storage. The root is well developed in the centre of the pot (top left), being developed mainly in the external part of the substrate with an empty centre (top right). Compare this plant with a poorly developed one (bottom left). A well-developed root system in the centre of the tray substrate is shown after a transplant from a small plug (bottom right).

8.3.3. Root–shoot interaction and plant collapse

The use of intensive planting systems with up to 70,000 plants per hectare, both in the soil or using artificial substrates in containers, reduces the volume of soil or substrate in the containers available for root growth because of the high-density planting. Production rates can now be more than 1 kg of fruit per plant per season and there is a real risk that the photosynthetic potential is not sufficient to feed the growing fruits and simultaneously provide the roots with the amount of energy they need to function (Fig. 8.14).

The rotation of different species is often not convenient, and thus strawberry plants are frequently replanted in the same soil, reducing the root efficiency. In these conditions, an excess of fruit production can be a problem (Fig. 8.15). The level of carbohydrates produced by the leaves and allocated to the fruits may become 'starved', which blocks water absorption. The result is a sudden leaf wilting with the possible collapse of the whole plant. Differential pruning, as shown in Fig. 8.15, with reduction of the number of fruits, may restore the shoot-to-root equilibrium, resulting in a healthy plant (Neri *et al.*, 1998).

8.4. Allelopathic Interactions and Replanting Problems

8.4.1. The ambivalent relationship of strawberry with organic residues

As described above, there is mounting evidence that roots can sense (and respond to) not only the richness of soil resources but also the occurrence (and even the identity) of other roots in explored niches, and the existence of inert objects, such as the wall of a soil container, before they even make contact. Detection mechanisms appear to involve non-toxic signals and allelochemicals (Schenk, 2006).

The accumulation of residues (self-produced or supplemented) in strawberry

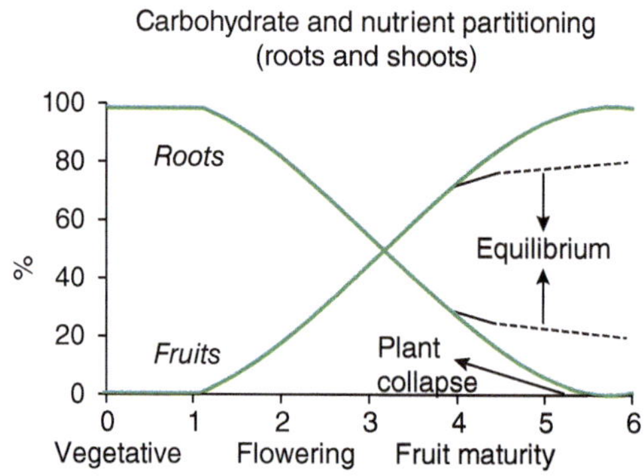

Fig. 8.14. Plant collapse due to replanting conditions may occur when the allocation of carbohydrates is directed mainly towards the fruits and thus root starvation occurs, reducing their absorption capacity and transpiration. If the environmental conditions are characterized by high irradiation and temperatures and by vapour pressure deficit, the plants are more prone to collapse. The collapse can appear after a few harvests (there are generally two to three harvests per week over a 4-week period) and occurs at the peak of production at the fourth or fifth harvest (indicated by the dotted line). The *y*-axis indicates the percentage of carbohydrates and nutrients partitioned into the roots and shoots. Growth stages are indicated by 0–6.

Fig. 8.15. Plant collapse due to poor replanting conditions and the results of differential pruning in strawberry 'Marmolada'. Removal of 50% of the leaves induced a worsening of the condition, while removal of 50% of the fruit enabled the plants to regain a healthy status. (From Neri *et al.*, 1998.)

was studied using bench rhizotrons. The strawberry roots were very sensitive to allelopathic substances (Figs 8.7 and 8.8), applied locally to the substrate as solid or liquid eluates from strawberry residues. In the treated area, the roots showed a specific behaviour and early reddening without the formation of root hairs or laterals (Neri *et al.*,

2005). This strongly suggests that allelopathic control of root growth guides the roots in terms of niche occupation and distancing. In fact, in the treated layer along the substrate profile, the roots grew in groups and grew deeper than usual, in a sort of avoidance mechanism to limit contact with the excess of allelopathic substances.

Supplementation of organic debris from other compatible species such as barley and broad bean limited the adverse action of strawberry residues (Fig. 8.16). These findings appear to support the practice of barley and broad bean green manuring prior to replanting strawberries in the same field (Neri and Savini, 2006).

8.4.2. Strategies to overcome replanting problems

In temperate climates, soil fertility is firmly associated with the presence of humified organic matter; the formation of vegetal soil is the primary means of expanding biomass production per unit area. In the wild, nature accomplishes this process efficiently, but in cultivated fields, this process is partially neglected, leading to an impoverishment of soil quality (Neri, 1998; Zucconi, 2003).

The organic residues from a single crop disturb the humification process, inducing abnormal decompositions, which may limit the maintenance of humus content and stimulate the release of toxic metabolites. These toxic metabolites are the source of specific allelopathic effects (dispathy) accounting for 'soil sickness' in replanting conditions. The toxins from the previous crop residues may hamper root absorption and promote dystrophic problems. The application of soil fumigants to rapidly create favourable replanting conditions is particularly aggressive towards soil biological equilibrium, in particular against mycorrhizae (Branzanti *et al.*, 1998; Baruzzi *et al.*,

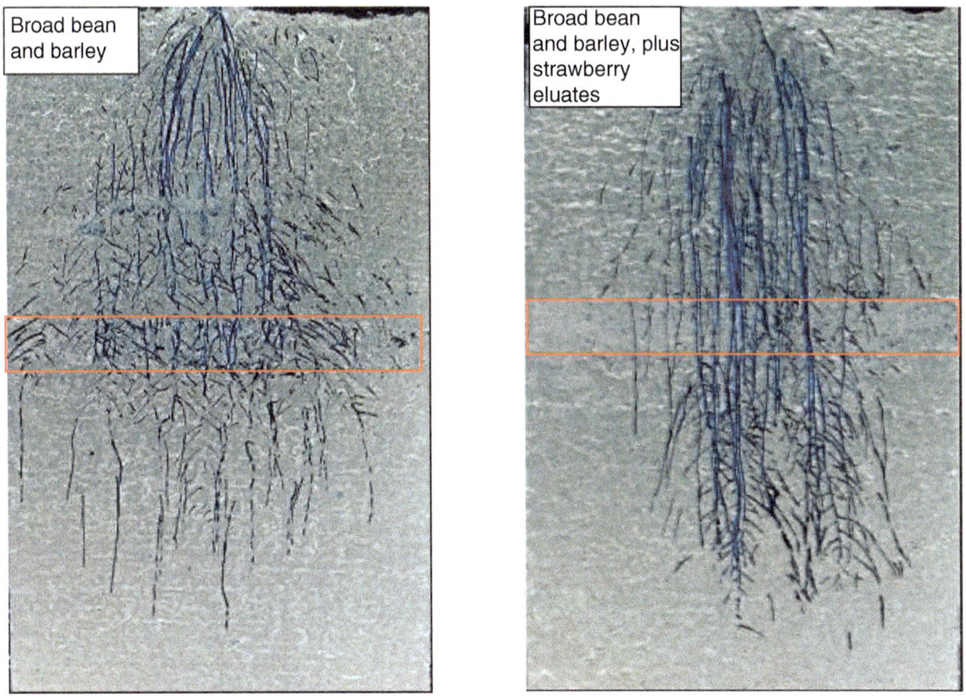

Fig. 8.16. Effect of supplementation of organic debris from compatible species. The impact of a layer of broad bean and barley residues 16 days after transplanting was extremely positive on strawberry root growth (left) and was partially able to counteract the negative effect of strawberry residue eluates (right).

2000). Integrated fruit production systems and/or organic/ecological farming systems do not allow this technique, but conventional farmers often request it when no virgin land is available inside the farm for new plantings.

Soil decline due to excess cultivation, insufficient organic matter production and disturbance of humification cannot continue. The worse the soil decline, the greater is the impact of abiotic and biotic stresses due to climate change on plant architecture (Neri *et al.*, 2012). The durability of an agricultural system can be significantly improved through enhanced control of soil organic matter by forcing humification, a practice that requires biodiversity, crop rotation, use of organic soil additions and reduction of pesticides, fertilizers and soil tillage to become truly effective (Zucconi, 2003).

Some cover crops fix nitrogen, and organic mulching often provides significant amounts of available phosphorus and potassium, enriching the soil microflora and stimulating mycorrhizae, thus helping to transform nutrients into forms that can be assimilated more easily by the growing strawberry plants. Before planting in temperate climates, compost or manure can be applied at a rate of up to 20 t ha^{-1} if the carbon:nitrogen (C:N) ratio is high (>20), or at a rate of less than 10 t ha^{-1} with lower C:N ratio to limit nitrogen leaching. This approach increases the levels of humus and nutrient accessibility, and aids the development of the cover crops in the following year. Mulching with straw between the rows helps control weeds, reducing competition for soil nutrients. All these practices limit environmental impoverishment by stopping soil erosion and nitrogen leaching, enhancing diversity and decreasing the amount of chemicals applied to the cultivation (Neri *et al.*, 1998). Defending the landscape by establishing farming systems that are more sustainable improves the resort importance of the land and can promote agrotourism, which is a rapidly growing sector of developed economies.

Coenotrophy is the concerted interaction of diverse microflora acting on a polygenic substrate under microaerobic conditions and is fundamental for humification (Zucconi, 2003). Under coenotrophic conditions, the production of humic complexes is highly effective, with limited loss of carbon. Humification includes the early disruption of organic polymers into soluble molecules, which rapidly undergo polymerization and polycondensation (Zucconi *et al.*, 2003). Strawberry roots like to grow in the humic substrate and with other complex molecules, but they do like fresh organic residues that are rapidly mineralizing.

8.4.3. How to improve soil quality

Humic complexes are very important stable organic components of the soil with colloidal properties. They improve the physical, chemical and biological characteristics of the substrate, and help control negative allelopathic interactions with other species and with self-produced organic residues. Conventional agriculture not only disrupts the humification process but also promotes the mineralization of humic substances in the soil and the subsequent nutrient assimilation by crop plants. This reduces durability and soil fertility because the humus matter decays over time. The nutrient and energy budgets of the different conventional crop systems usually do not take this problem into account.

Higher erosion and mineralization, limited humification and poor soil correction with external organic compost have caused an intense reduction in soil humus matter in several Mediterranean, tropical and temperate countries. This decrease has aggravated the spread of desertification. In fewer than 40 years, several Italian regions that practice intensive agriculture have lost 1% of soil organic matter. In the upper 50 cm of the soil profile, about 65 t ha^{-1} of organic matter has been mineralized, releasing a total of 2 t of nitrogen, or 50 kg ha^{-1} of N per year. This loss of humus means that more external organic residues need to be added to the soil to maintain the equilibrium. Humus creates healthy soil, which promotes plant growth, high fruit quality and natural tolerance to diseases and pests. Humus is one of the key factors for sustainable agriculture that must be recovered in integrated or organic crop production.

Although organic residues are subject to rapid mineralization, humus mineralizes at a reduced rate, 1–2% per year in temperate conditions, and thus can last in the soil for decades. 'Soil sickness' refers to the progressive soil quality reduction as a consequence of the repeated culture of a sole crop. It mainly affects the sole crop that has been repeatedly cultivated, and affects other species less, particularly if they are unrelated, but it must be considered the first symptom of soil degradation (Zucconi 2003). Monoculture of a sole species fosters the accumulation of noxious materials (dispathy) in the soil, which disrupt coenotrophy. This situation decreases the transformation of organic matter into humus and depresses soil fertility. This less fertile soil is progressively unable to sustain the growth of the sole species cultivated year after year, with pathogen and parasite outbreaks becoming more frequent, and nutrient deficiencies become more common. The actual origins of these problems are disruption of the humification process, loss of suppressivity and the phytotoxic effects of monospecific residues (Neri and Savini, 2005; Neri *et al.*, 2005; Zucconi *et al.*, 2003; Bonanomi *et al.*, 2006), which are able to rapidly modify strawberry root functioning and lateral root formation (Neri *et al.*, 2011).

Again, sustainability can be restored by increasing the amount of soil organic matter, which closely imitates the process of humification, by enhancing coenoses and organic additions, by reducing the use of pesticides and fertilizers, and by limiting soil tillage. Humification will augment suppressivity, which limits the increase of soilborne pathogens and pests. Better availability of nutrients due to high humus levels in the soil results in plants that are healthier and more tolerant to pathogens and parasites, decreasing the requirements for pesticides (Gangatharan and Neri, 2012). Sustainability can thus be improved depending on the type of cultivation system (Neri *et al.*, 1996; Zucconi, 2003).

Thus, it is evident that the presence of humic substrates may improve absorption by strawberry roots (Fig. 8.17), although the presence of organic residues that are not

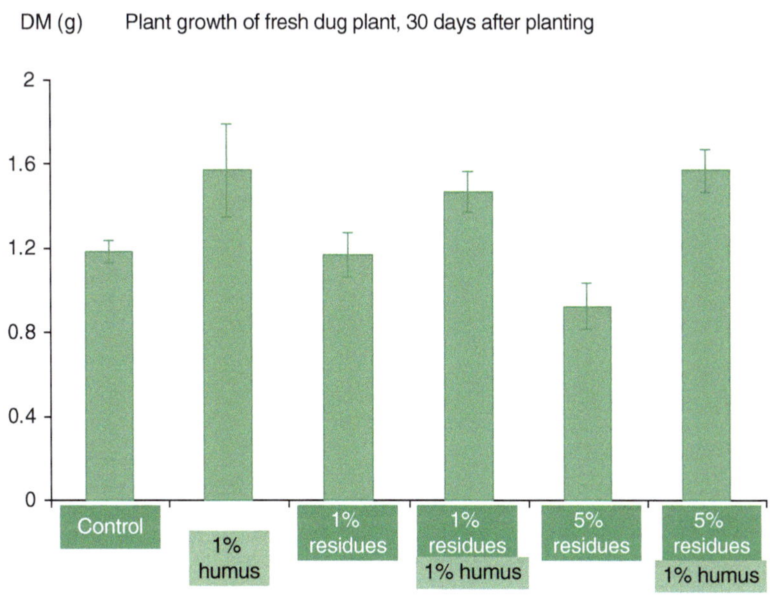

Fig. 8.17. Dry matter (DM) of the young strawberry plant with or without strawberry residues and humus. (From Savini and Neri, 2005.)

completely mineralized or humified can have a negative effect on strawberry root growth.

Thus, the soil surrounding plants is organized into discrete niches used by a single root (Zucconi, 2003). The presence of primary allelopathic factors inhibits new roots from entering an occupied niche. These factors improve the soil, favour absorption and produce a favourable rhizosphere, but induce the need for root transmigration to other favourable niches. This allows soil to be used very efficiently by the resident root in the short term and by the coenoses in the long term. When the root of the resident species ages and dies off, a new root of a compatible plant will be able to occupy the now-empty niche space.

8.5. Conclusions

Strawberry root growth is influenced by environmental and cultural factors, nutrient availability, internal physiological conditions such as carbon supply from the leaves (carbon allocation and partitioning, and shoot-to-root ratio) and allelopathic interactions with residues from previous cultures, and exudates, leachates and debris left by root growth and turn over. Roots show discontinuous periods of elongation and ramification, associated with shoot and root functioning and their reciprocal interaction. New crown roots have a positive geotropic behaviour, and grow a few millimetres apart, a response that is probably due to self-produced allelochemicals. This chapter has discussed an architectural model of strawberry root growth showing that, if temperature and water availability are at optimal levels, the proliferation of roots in favourable soil patches is rapid, as is the depletion of nutrients and the accumulation of allelochemicals. It is suggested that, in sustainable low-input strawberry field production systems, it is essential to introduce crop rotation over several years so that other residues can accumulate in the soil, and also to improve humification (or stabilization) of organic residues with the addition of diversified organic additions in order to recover soil fertility and natural suppressiveness against pathogens.

Acknowledgement

The author acknowledges the Polytechnic University of Marche for supporting this research work.

References

Baruzzi, G., Branzanti, B., Faedi, W., Gentili, M., Lucchi, P. and Neri, D. (2000) Effetto della micorrizazione e della sostanza organica su piante di fragola. *Frutticoltura* 12, 83–88.

Bonanomi, G., Giorgi, V., Del Sorbo, G., Neri, D. and Scala, F. (2006) Olive mill residues affect saprophytic growth and disease incidence of foliar and soilborne plant fungal pathogens. *Agriculture, Ecosystems & Environment* 115, 194–200.

Branzanti, M., Gentili, M. and Neri, D. (1998) Controllo del marciume del colletto in fragola mediante funghi micorrizici arbuscolari. *Rivista di Frutticoltura* 12, 77–80.

Chiatante, D., Di Iorio, A. and Scippa, G.S. (2007a) Modification of root architecture in woody plants is possible for the presence of two different mechanisms of lateral root production: the effect of slope in *Spartium junceum* L. seedlings. *Plant Biosystems* 141, 502–511.

Chiatante, D., Scippa, G.S., Di Iorio, A., De Micco, V. and Sarnataro, M. (2007b) Lateral root emission in woody taproots of *Fraxinus ornus* L. *Plant Biosystems* 141, 204–213.

Cline, M. (1997) Concepts and terminology of apical dominance. *American Journal of Botany* 84, 1064–1069.

Endeshaw, S.T., Lodolini, E.M. and Neri, D. (2015a) Effects of untreated two-phase olive mill pomace on potted olive plantlets. *Annals of Applied Biology* 166, 508–519

Endeshaw, S.T., Lodolini, E.M. and Neri, D. (2015b) Effects of olive shoot residues on shoot and root growth of potted olive plantlets. *Scientia Horticulturae* 182, 31–40.
Falik, O., Reides, P., Gersani, M. and Novoplansky, A. (2005) Root navigation by self inhibition. *Plant, Cell and Environment* 28, 562–569.
Gangatharan, R. and Neri, D. (2012) Can biodiversity improve soil fertility in agroecosytems? *New Medit* 4, 11–18.
Hancock, J.F. (1999) *Strawberries*. Crop Production Science in Horticulture Series, CABI Publishing, Wallingford, UK.
Hodge, A. (2009) Root decisions. *Plant, Cell and Environment* 32, 628–640.
Inderjit, K.M. and Foy, C.L. (2001) On the significance of field studies in allelopathy. *Weed Technology* 15, 792–797.
Kehoe, E., Savini, G. and Neri, D. (2009) The effects of runner grade, harvest date and peat growing media on strawberry tray plant fruit production. *Acta Horticulturae* 842, 699–702.
Kobayashi, K. (2004) Factors affecting phytotoxic activity of allelochemicals in soil. *Weed Biology and Management* 4, 1–7.
Lakshmanan, V., Selvaraj, G. and Bais, H.P. (2014) Functional soil microbiome: belowground solutions to an aboveground problem. *Plant Physiology* 166, 689–700.
Laskowski, M.J., Williams, M.E., Nusbaum, H.C. and Sussex, I.M. (1995) Formation of lateral root meristems is a two-stage process. *Development* 121, 3303–3310.
Lieten, P. (2002) The use of cold stored plant material in central Europe. *Acta Horticulturae* 567, 553–560.
Massetani, F., Gangatharan, R. and Neri, D. (2011). Plant architecture of strawberry in relation to abiotic stress, nutrient application and type of propagation system. *Genes, Genomes and Genomics* 5 (Special Issue 1), 1–12.
Massetani, F., Savini, G., Neri, D. (2014) Effect of rooting time, pot size and fertigation technique on strawberry plant architecture. *Journal of Berry Research* 4, 217–224.
McGraw-Hill (2012) *McGraw-Hill Encyclopedia of Science & Technology*, 11th edn. McGraw-Hill Education, NY.
Neri, D. (1998) Soil organic matter in sustainable agriculture. In: Borsari, B. and Vidrine, M.F. (eds) *Low Input Agriculture: Feasible Alternatives to Conventional Agricultural Practices*. Eunice, Lousiana, pp. 4–13.
Neri, D. (2004) Low-input apple production in central Italy: tree and soil management. *Journal of Fruit and Ornamental Plant Research* 12, 69–76.
Neri, D. and Savini, G. (2005) Root growth and architecture in strawberry. *Acta Horticulturae* 708, 39–44.
Neri, D. and Savini, G. (2006) Crescita e struttura radicale della fragola: impatto dei residui organici. *Italus Hortus* 13, 60–64.
Neri, D., Madia, T., Zucconi, F. and Guardigli, G. (1996) Root growth of rice seedlings in relation to crop-residues metabolism. In: Ito, O., Johansen, C., Adu-Gyamfi, J.J., Katayama, K., Kumar-Rao, J.V.D.K. and Rego, T.J. (eds) *Roots and Nitrogen in Cropping Systems of the Semi-arid Tropics*. Japan International Research Center for Agricultural Sciences, Tsukuba, Japan, pp. 389–399.
Neri, D., Bonanomi, G., Cozzolino, E. and Zucconi, F. (1998) Studi sugli apporti di sostanza organica nel fragoleto. *Frutticoltura* 5, 47–54.
Neri, D., Lee, O.N., Savini, G., Kurokura, T. and Sugiyama, N. (2011) Lateral root formation in the strawberry *Fragaria* × *ananassa* Duch. revealed by histone H4 *in situ* hybridisation. *Plant Biosystems* 145, 406–410.
Neri, D., Sugiyama, N. and Inujima, A. (2002) Effects of organic residues on strawberry root growth. *Acta Horticulturae* 567, 503–506.
Neri, D., Sugiyama, N., Iwama, T., Akagi, H., (2003) Effect of apical pinching on the growth of branch crowns and stolons in strawberry plants. *Journal of the Japanese Society for Horticultural Science* 72, 389–392.
Neri, D., Sugiyama, N. and Inujima, A. (2005) Effects of organic residues on strawberry root growth. *International Journal of Fruit Science* 5, 127–137.
Neri, D., Baruzzi, G., Massetani, F. and Faedi, W. (2012) Strawberry production in forced and protected culture in Europe as a response to climate change. *Canadian Journal of Plant Science* 92, 1021–1036.
Polverigiani, S., Kelderer, M. and Neri, D. (2014). Growth of 'M9' apple root in five Central Europe replanted soils. *Plant Root* 8, 55–63.
Savini, G. and Neri, D. (2004) Strawberry architectural model. *Acta Horticulturae* 649, 169–176.
Savini, G. and Neri, D. (2005) Acclimatamento radicale e fotosintesi nella fragola. *Frutticoltura* 4, 49–52.
Savini, G., Letouzé, A., Sabbadini, C. and Neri, D. (2006) Evaluation of tray-plant quality in the propagation phase. *Acta Horticulturae* 708, 231–236.
Savini, G., Giorgi, V., Scarano, E. and Neri, D. (2008) Strawberry plant relationship through the stolon. *Physiologia Plantarum* 134, 421–429.

Schenk, H.J. (2006) Root competition: beyond resource depletion. *Journal of Ecology* 94, 725–739.

Scheres, B., van den Toorn, H. and Heidstra, R. (2004) Root genomics: towards digital *in situ* hybridization. *Genome Biology* 5, 227.

Sugiyama, N., Iwama, T., Inaba, Y., Neri, D. (2004) Varietal differences in the formation of branch crowns in strawberry plants. *Journal of the Japanese Society for Horticultural Science* 73, 216–220.

Zucconi, F. (1996) Root dynamics in natural and agricultural plants, and the making of domestication. In: Ito, O., Johansen, C., Adu-Gyamfi, J.J., Katayama, K., Kumar-Rao, J.V.D.K. and Rego, T.J. (eds) *Roots and Nitrogen in Cropping Systems of the Semi-arid Tropics.* Japan International Research Center for Agricultural Sciences, Tsukuba, Japan, pp. 103–128.

Zucconi, F. (2003) *Declino del Suolo e Stanchezza del Terreno*. Pitagora, Bologna, Italy.

Zucconi, F., Bonanomi, G., Cozzolino, E., Neri, D. (2003) Il ruolo della sostanza organica nella coltura della fragola. In: *Atti Convegno Nazionale SOI Sulla Fragola,* 10–14 April 2000, Metaponto, Italy, pp. 237–250.

9 Plant Propagation Techniques and Types of Plants

Marina Gambardella[1]*, Francesca Massetani[2,3] and Davide Neri[3,4]
[1]*Departamento de Fruticultura y Enología – Pontificia Universidad Católica de Chile, Santiago, Chile;* [2]*HORT Soc. Coop., Università Politecnica delle Marche, Ancona, Italy;* [3]*Dipartimento di Scienze Agrarie, Alimentari ed Ambientali – Università Politecnica delle Marche, Ancona, Italy;* [4]*Centro di Ricerca per la Frutticoltura – Consiglio per la Ricerca in Agricoltura e l'Analisi dell'Economia Agraria, Roma, Italy*

9.1. Introduction

Although the strawberry is a perennial species, the main producing countries use an annual or biennial crop system, so farmers must replant year after year in planting densities ranging from 52,000 to 60,000 plants ha^{-1}. This continuous replacement of plants gives rise to the nursery industry, which has dynamism that is as much or more spectacular than fruit production. The area designated for cultivation in each season must be supplied with millions of plants.

Strawberry fruit production is now year round with winter production in Mediterranean climates, spring/summer production in colder regions, and programmed out-of-season production with protecting and forcing systems (Neri *et al.*, 2012). Each planting system needs appropriate plant types and planting techniques. Thus, at the time of establishing a strawberry orchard, plants must have physiological, health and development characteristics in agreement with the farmer's needs, together with a high content of reserve carbohydrates, and buds should have a proper

* mgambardella@uc.cl

degree of floral differentiation. These features are crucial to the success of the crop.

From a botanical point of view, *Fragaria* × *ananassa* is a hybrid species that can naturally propagate by seed (gamic) or runners (vegetatively, agamic) (Savini *et al.*, 2005). Propagation by seed is almost exclusively used in breeding for genetic improvement, in order to obtain new genotypes from which the breeders can select new varieties. In some cases, the seed is also used as a propagation system to produce fruit, for example with some varieties of *Fragaria vesca*, or with varieties adapted to garden cultivation.

However, the great industry of strawberry plants is based on their ability to produce stolons and runner chains capable of developing new self-rooted plants.

Runners grow from axillary buds located in the crown, primarily in response to high vigour with rapid growth, generally during long days (photoperiods) and with high temperatures and optimal water and nutrient availability. When the runners form a new bud, a new crown of a daughter plant is produced, which receives water and nutrients through the runner's vascular system (Savini *et al.*, 2008). According to Strand (2008), a mother plant could produce about 100 daughter plants during the season, but this depends on the variety, nursery location and agricultural works carried out. This author also stated that short-day varieties usually produce more runners than day-neutral varieties.

9.2. Plant Production Systems

Most strawberry nurseries are located in areas with very specific characteristics. These facilities require large flat surfaces, whose soils are deep, very sandy and well drained, with a pH between 6 and 7, and with a good water supply. Nurseries must be established in isolated areas, away from fruit production, in order to promote sanitary conditions. Altitude and latitude of the nursery should also be taken into account, as these elements determine the time when the plant's floral differentiation and physiological maturity occur.

The production of runners is favoured in areas whose geographical location – north or south – ensures an adequate photoperiod and temperatures between 25 and 30°C in summer. Autumn temperatures should decrease quickly, so that plants receive an accumulation of chilling hours (below 7°C), which in some cases is achieved in high-altitude areas (over 1000 m above sea level) or in more extreme latitudes. In Europe, nurseries are mainly located in Spain (provinces of Avila, Segovia and Valladolid), northern Italy (Veneto and Emilia-Romagna regions) and France (Aquitaine and Midi Pyrenées). Countries such as Poland and Romania have also developed an interesting nursery industry.

In North America, the main area for nursery production is northern California, south-east of Oregon and North Carolina in the USA and Ontario in Canada. In South America, Chile and Argentina produce strawberry plants to meet their own demand and to supply countries such as Mexico, Brazil, Peru, Bolivia and Colombia, among others.

It should be noted that the strawberry plant production process requires laboratory technologies such as *in vitro* culture, techniques of sanitation and detection of viral diseases, and prolonged periods of cold storage, among others, which imply specialization and a high technological level. For this reason, in order to obtain plants of adequate genetic and phytosanitary quality, the plant production process should be performed in specialized nurseries, following a methodology that has been proven and internationally accepted.

The production system basically consists of the following four stages.

9.2.1. Nuclear stock

The first stage of the propagation system is to establish a collection of mother plants from the varieties to be propagated, i.e. varieties of commercial interest. In the case of starting production with plants that have remained in the field for some time and whose health condition is not certain, they must undergo a sanitation process (thermotherapy or meristem culture) and verification of their genetic identity.

Under a production scheme of certified plants, once sanitation is completed, mother plants must be kept in individual containers, inside greenhouses with anti-aphid mesh, observing isolation conditions specified by regulations. The plants should be isolated from the ground, and a sterile, previously disinfected substrate must be used.

9.2.2. Propagation stock I

The material to establish the propagation stock is obtained from the nuclear stock plants. Two methods can be used to accomplish this: (i) micropropagation; and (ii) propagation by runners.

Micropropagation involves multiplication entirely *in vitro*, beginning with meristems, apical tips or axillary buds from nuclear stock plants. According to the methodology developed by Boxus (1992), the meristem should be between 0.2 and 0.4 mm in size, to ensure that it is free of viruses. This system of meristem culture is also used to obtain plants free from fungi, bacteria and other diseases, and to reinvigorate plants. Meristem plants produce many more daughter plants than are produced by conventionally propagated plants. The meristem, under *in vitro* culture conditions, gives rise to a plant that can be micropropagated if regulations in the country allow it, or taken directly to acclimatization under a scheme of 'one meristem, one mother plant'. In the case of using micropropagation to increase the number of plants at this stage and thus reduce costs, a limit of ten multiplication cycles has been established in order to avoid problems of somaclonal variation. At this stage, pre-base or F_0 material is obtained.

For propagation by runners, runner tips from nuclear stock plants are pinned down in separate pots of sterilized growing medium. The pots in which the runner tips are rooted are kept at a higher level than the nuclear stock pots to avoid transmission of soil or root pathogens through watering. When the runners have rooted, they are separated from the parent nuclear stock plants.

Plants originating from micropropagation or by runners, so-called F_0 material, are transferred to the substrate and acclimated to establish the propagation stock I, in structures protected with anti-aphid mesh to minimize risks of infection with disease vector insects. Each plant is set in an individual container (1 × 1 m bins) with a sterile substrate. Under these conditions, propagation is performed through runners at a rate ranging from 60 to 100 plants per mother plant. The process takes one season, about 8–9 months, and the plant, described as F_1, is finally obtained.

9.2.3. Propagation stock II

Plants are harvested in winter, when they are in vegetative recess. They are kept until the spring and are established directly in the field. Although this stage is carried out

Fig. 9.1. Mother plants (a) and propagation stock I (b, c) in a greenhouse with double doors, a footbath and insect mesh.

without isolating structures, it is still necessary to maintain distance from other strawberry plants, either in nurseries or in fruit production orchards. In these conditions, one can estimate a yield of approxumately 80 plants per mother plant. The plants obtained in this stage are called F_2.

9.2.4. Commercial nursery

F_2 plants are used to establish the commercial nursery in previously disinfected sandy soil that is highly loose and homogeneous. To achieve proper development of mother plants and favour runner production, planting is done in rows arranged directly on the surface, without ridges. The recommended planting distance is 1.6–1.8 m between rows and 0.30–0.6 m within rows, with a density of approximately 10,000–18,000 plants per hectare, depending on varietal vigour and the type of propagated plant. For example, with the cultivar 'Gariguette', the distances are 1.6 × 0.30 m and production is based on stolon tips for making the misted tip plants and/or tray plants in the summer. Usually, a sprinkler irrigation system is employed, which facilitates runner rooting, although care should be taken not to compact the nursery's surface. Weeds may be a major problem in strawberry commercial nurseries, especially with the removal of methyl bromide, as alternative fumigants do not efficiently control various weed species. Frequently, soil-active herbicides are applied before planting in order to reduce weeds. Once the nursery is established, herbicides should not be applied, and weeds must be controlled manually or with a rotary tiller.

After the plantation is established, the following agricultural tasks are carried out: removal of all flowers and leaves to promote vegetative growth; burying of runners as they are formed to stimulate rooting; guiding of runners so they distribute homogeneously in the field; raking of the soil between plantation rows to favour runner rooting and root aeration; and strict control of pests and diseases.

Depending on the desired type of plant, harvest takes place in autumn or winter, when the plants have reached the winter dormancy period (some cultivars, such 'Gariguette', will show a reddish colour at this point, as seen in the middle picture of Fig. 9.2.).

9.3. Plant Types and Cycles

The advances in increasingly specialized cultivation techniques, together with the expansion of production areas to more extreme climate zones, have forced nurseries to produce different types of specific plants adapted to modern cultivation requirements. Plant types provided by the nurseries through various propagation techniques are the starting point for a wide range of producing cycles aiming towards expansion of the fruiting season and cultivation in different areas. In fact, the cultivation technique of strawberry has undergone a constant evolution over the last 20 years. The most traditional plant cycle is the annual

Fig. 9.2. A commercial nursery showing (a) the establishment of mother plants spring, (b) the plants in recess before harvest and (c) the harvesting process.

crop cycle with a single fruiting season in the spring (using June-bearing, short-day varieties) following the summer planting. In many areas, other innovative growing techniques have been introduced and the harvest of strawberries now covers the whole year.

There are two traditional types of strawberry plants: 'fresh' and 'frigo' (cold stored). There also exist six other plant types (Table 9.1) that are capable of being adapted to more specialized cultivation systems, such as programmed cultivation, protected cultivation and off-soil cultivation.

9.3.1. Bare-root plants

Fresh plants and frigo plants are lifted from the nursery field and cleaned of the soil before storage and planting; therefore, their root system is free of the substrate.

Fresh and green plants are dug plants harvested in early autumn, before vegetative rest. They can be distinguished into 'fresh plants' when defoliated (mechanically before digging) and 'green plants' when the leaves are retained. They were used initially in various cultivation areas, but were later abandoned in northern environments because of unsatisfactory agronomic results compared with cold-stored material. Dug plants are ready for transplanting after their summer growth in the nursery. In cold climates, planting can take place from mid-July until early August, while in Mediterranean climates the planting is at the end of September to the beginning of October because earlier summer temperatures are too hot. Nevertheless, fresh plants should be taken to the field as soon as possible to avoid dehydration and rot problems once they are harvested. Thus, they are used for autumn plantings, when they are removed from the nursery and brought directly to the fields already prepared for the planting. If the operation cannot be done quickly enough, plants may be cold stored at 2°C for up to 1 week. They generally do not bear differentiated inflorescences, and flower induction occurs during the autumn months in the field, while the plant is still growing. Fruiting takes place from May to June with a duration of 1 month in northern climates, but the plants can also be used for early forcing production in spring (Lieten, 2005) or for winter production in Mediterranean climates with plastic covers.

Nurseries that produce high-altitude freshly dug plants are located in areas with a high accumulation of chilling hours in early autumn (high zones, as in the case of the province of Avila in Spain or northern California, or northern European countries such as Poland). These types of plants are well adapted to early production, because when they are transferred to the field in autumn, they have effectively been induced to flower, and an adequate transformation of root sugars to more complex carbohydrates has taken place, thus favouring the transplanting process. Due to the environmental conditions under which they are rooted at high altitude in the summer, at planting the plants already have at least one differentiated inflorescence. The fruit production occurs early in the months of February–March, with a duration of 2–3 months, until June. Early winter production is possible from January.

Frigo plants, on the other hand, are starting from stolons rooted early in the nursery, which are harvested when they have already initiated their winter vegetative rest period (January–February in the northern hemisphere, and July–August in the southern hemisphere). Once harvested, the plants undergo a selection and classification process, and are stored in transparent plastic bags inside wooden or plastic boxes, where they remain for 5–6 months at a temperature of −2°C. At this temperature, respiratory activity is strongly reduced to avoid consumption of reserve substances and to tolerate long storage periods, but no tissue damage is produced. Prior to cold storage, the plants are disinfected with fungicides and moistened in order to prevent dehydration inside cold-storage rooms. If the plants are cold stored for too long, they will suffer during transplanting in September–October and will not grow well and differentiate in order to start production from December to January in mild climates (typical conditions of the Mediterranean and southern Italy).

Table 9.1. Nursery plant classification. For interpretation of the diagrams, see Chapter 7. (Modified from Neri *et al.*, 2009.)

Commercial name	Physiological status	Starting material	Root free of substrate (bare roots)	Pots	Cold storage	No. of crowns and inflorescences (architectural model) at transplanting	Season and duration of production	Crown diameter (mm)	Nodes on the crown (*n*)	Side flower buds (*n*)	Flowers per plant (*n*)	Plant architecture
Freshly dug plant: fresh plant = without leaves; green plant = with leaves	Freshly dug plant; not flower differentiated	Runner	Yes	No	No	1 crown, 0 inf.	Traditional crop, spring–summer/ 1 month	9–13	9–14	0	0	
Freshly dug plant high altitude (two categories as above)	Freshly dug plant; flower differentiated	Runner	Yes	No	No	1 crown, 1 inf.	Winter–spring/2–3 months	9–13	9–14	0	5–10	

Frigo plant (A, A+, AA+ plant size)	Dug plant; cold stored; differentiated	Runner	Yes	No	Yes, 7 months	1–2 crowns, 3–4 inf.	Spring–summer/1 month	A–, 6–8; A, 8–12; A+, 12–15; AA+, 15–18	A, 7–11; A+, 9–13; AA+,12–17	A, 3–5; A+, 4–6; AA+, 6–8	A, 15–25; A+, 25–35; AA+, 35–45	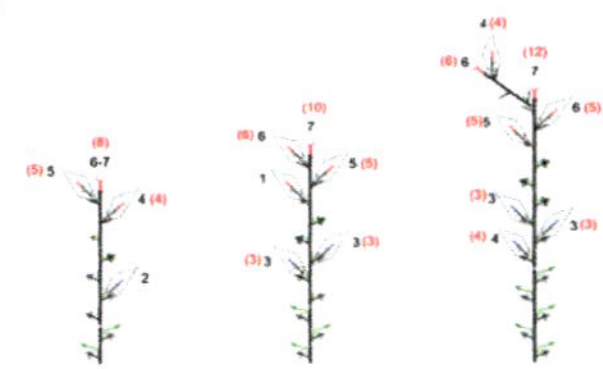
Waiting-bed plant	Large dug plant; cold stored; flower differentiated	Freshly dug plant; frigo plant; rooted runner	Yes	No	Yes, 3–9 months	3–4 crowns, 4–6 inf.	Out-of-season, spring–autumn/ 1 month	13–19	8–17	6–10	40–70	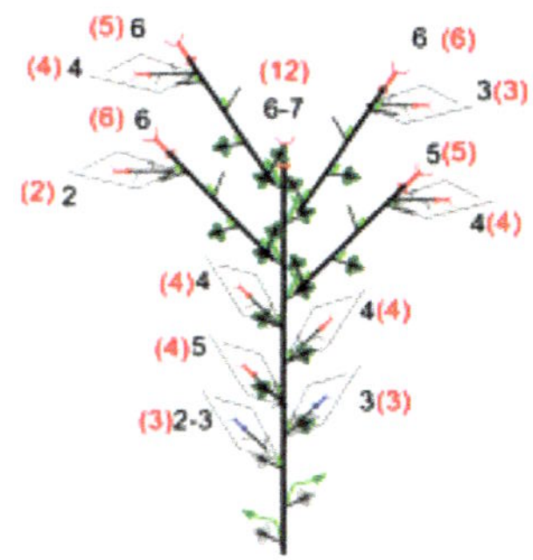
Rooted runner (misted tip or plug plants)	Fresh plant in pot; not flower differentiated	Runner	No	Yes	No	1 crown, 0 inf.	Spring/ 1 month	8–10	5–8	0	0	

Continued

Table 9.1. Continued.

Commercial name	Physiological status	Starting material	Root free of substrate (bare roots)	Pots	Cold storage	No. of crowns and inflorescences (architectural model) at transplanting	Season and duration of production	Crown diameter (mm)	Nodes on the crown (*n*)	Side flower buds (*n*)	Flowers per plant (*n*)	Plant architecture
Rooted runner (misted tip)	Fresh plant in pot; flower differentiated	Runner	No	Yes	No	1 crown, 1 inf.	Winter–spring/ 3 months	8–10	5–8	0–3	15–20	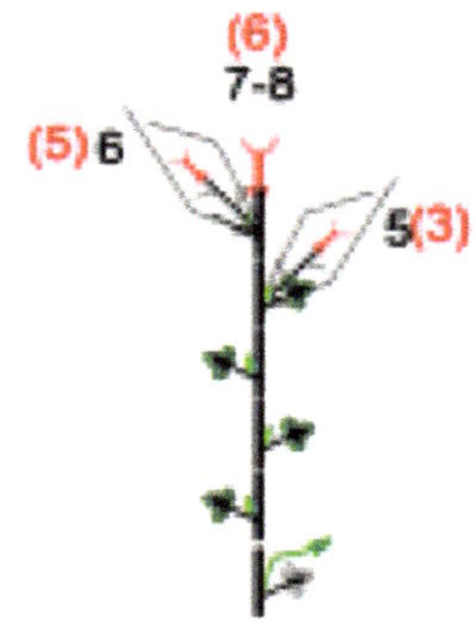
Tray plant	Pot plant; cold stored; flower differentiated	Runner	No	Yes	Yes, 3–9 months	1 crown, 2–3 inf.	Out-of-season, spring–autumn /1–2 months	12–15	10–16	4–5	20–40	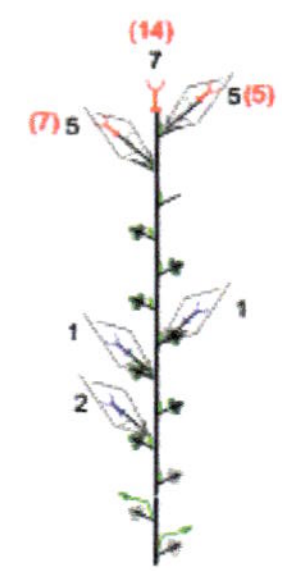

Tray plant from frigo plants	Large plant in pot; cold stored; flower differ-entiat-ed	Frigo plant	No	Yes	Yes, 4–9 months	2–3 crowns, 3–6 inf.	Out-of-season, spring–autumn/ 1 month	14–18	8–17	5–10	30–60	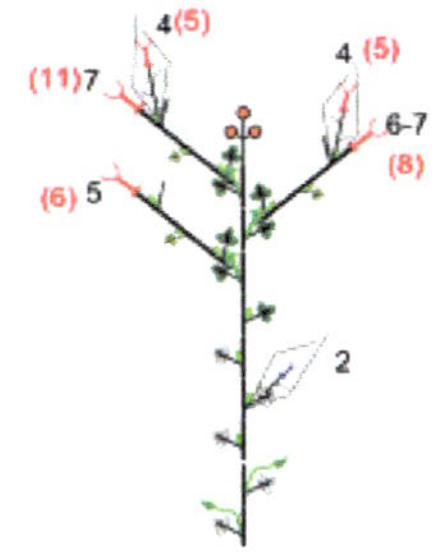
Minitray plant	Plant in pot; cold stored; flower differ-entiat-ed	Runner	No	Yes	Yes	1 crown, 2–3 inf.	Out-of-season, spring–autumn /1–2 months	10–13	8–12	4–5	15–35	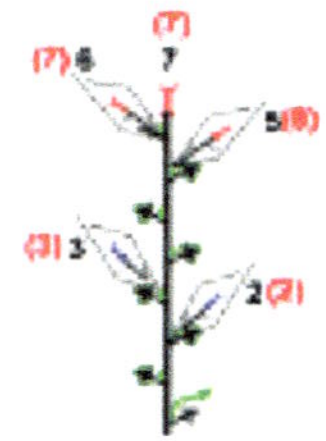

Flower induction takes place in the nursery field, under the photoperiod and temperature conditions of the autumn and winter before digging; therefore, at the beginning of cold storage the plants are already flower induced and have one to two crowns and three to four inflorescences. However, these plants in most cases do not have the ability to obtain good fruits from the formed inflorescences. The common technique involves removing the first single poor-quality inflorescence by hand as soon as the flowering stem is evident. After cold storage, the plants are ready to be planted in different seasons, commonly during the summer (July–August) in the field for annual production systems where fruiting occurs in spring–summer, lasting 1 month. In northern mountain environments, planting generally begins in early June at the highest altitudes and continues throughout the month moving to lower altitude. After transplantation, the growth at the end of the summer induces the development of several side shoots or crowns (Savini *et al.*, 2005) during the autumn that undergo flower formation and provide fruits in the following spring.

During the selection and packaging process, the excess soil and sand that remains in the roots is removed and the foliage is eliminated, leaving only two to three young leaves close to the apex. In both types of plants, classification is performed based mainly on crown diameter and root length, as these are some of the main indicators of plant quality. In Europe, according to these parameters, plants are classified into:

- Extra quality or A+: 12–15 mm in diameter, 12 cm in length, 250–300 plants per box.
- First quality or A: 8–12 mm in diameter, 10 cm in length, 600–700 plants per box.
- Second quality or A–: 6–8 mm in diameter, 8 cm in length, 900–1000 plants per box.

A+ plants are used when fruit production immediately after planting is required, taking advantage of the flower buds differentiated in the nursery. Therefore, A+ plants are used for programmed in-soil and soil-less crops for autumn fruiting (in the Italian Verona region), where planting starts in the last week of August until the first week of September and double fruiting is provided: the first crop takes place in the autumn following the planting from the flower buds differentiated in the nursery, and the second crop starts in the following spring due to plastic tunnels applied in January to anticipate the production. They are also used for the strawberry in-soil and soil-less out-of-season production during the summer (in the Italian Trentino region). Only large plants (A++ or AA+) have a good chance of providing significant programmed fruiting after transplanting, which is usually carried out in May–June, for the summer harvest season (early July–late August) (Faby, 1997).

In comparison with cold-stored plants, fresh plants show an increase in early ripening and also increased fruit quality, along with the possibility of delaying planting, with the corresponding savings in irrigation water. The use of fresh plants produced in high-altitude nurseries or in northern latitudes (Poland, Spain) allows flower differentiation in early September with a favourable thermophotoperiod. Forty days after transplanting, the plants reach full bloom, and after a further 40 days, the first fruits are ripe. It is worth noting that the use of underdeveloped plants from nurseries with a high planting density causes the growth of thin plants and may prolong the vegetative stage, delaying and reducing flower bud differentiation, which must already have been initiated in the nursery.

Waiting-bed plants were developed initially in the Netherlands in the early 1970s to achieve early production, making it possible to produce fruit when there is low supply and high prices in the markets (Kirschbaum *et al.*, 2000). They are cold-stored, large-sized plants between 18 and 22 mm in diameter (100–150 plants per box), capable of beginning production 5–8 weeks after planting, depending on variety, planting date and environmental conditions (Kirschbaum *et al.*, 2000). Their use is intended primarily for programmed crops: they are planted from April until July (Lieten, 2005) and in a short time they must ensure high production and high-quality fruits with a uniform size from spring (from early July until late August in open field) to autumn (under plastic tunnels), for about 1 month. The plants can be obtained from frigo plants, dug plants or rooted runners (the plants from fresh material are almost 1 year younger). At the nursery, they follow the same procedure as the above two types, the difference being that at harvest they are not immediately delivered to producers, but are brought to a 'bed' or fallow area with substrate, in order to thicken the crown, thus allowing increased earliness and productivity. For this reason, in Europe, waiting-bed plants are used mainly for off-the-ground cultivation and early planting, as they are more productive than type A+ plants, given that they grow so that they bear more than one flower-differentiated crown per plant and have better root and crown reserves. Before cold storage, the plants show some four to six inflorescences distributed in three to four crown shoots at digging time, along with many flower-differentiated large buds (up to 10) bearing 30–70 flowers, although they can contain more than 100 flowers.

9.3.2. Potted plants

Some years ago, in various parts of the world, a new way of producing strawberry plants was developed, the so-called potted plants, which are produced by generating unrooted runners from mother plants, and placing them in seedbeds, trays or speedlings with substrate. Afterward, they are subjected to a highly humid environment, while they develop their root system, in order to produce new plants within a short period of time, approximately 4 weeks (Barclay Poling and Maas, 1998). These plants have a higher production cost, which influences their low commercial availability; however, this method has already spread in Europe and USA. It is worth noting that in North Carolina there was an increase from 1 million to 8 million plants between 1992 and 1998 (Durner *et al.*, 2002). The growing interest in producing these types of plants may be explained by the fact that they have several advantages compared with bare-root plants, such as excellent early vigour (Rowley *et al.*, 2010), easy transplanting, low mortality (Durner *et al.*, 2002) and, according to Bish *et al.* (2002), a considerably lower water requirement during establishment. Furthermore, according to Giménez *et al.* (2009), when transplanted with an intact root system, their initial growth is better just after plantation is performed, leading to earlier and better quality flowering and harvesting.

Bare-root plants, on the other hand, suffer serious root damage, whether physical (during planting), through the attack of pathogens during the storage period or as a result of transplanting stress. Consequently, many plants die during the first week after plantation, or have poor vegetative growth, which causes delayed fruiting and reduced yield (Durner *et al.*, 2002).

In recent years, a special interest emerged in these potted plants, when the process of eliminating methyl bromide as a soil fumigant was initiated. This product was essential in preparing the soil for strawberry nurseries, because it allowed control of diseases, nematodes and weeds (García-Sinovas *et al.*, 2008). Potted plants, on the other hand, are established in a sterile substrate mixture, significantly reducing the possibility of any soil disease (Bish *et al.*, 2002). Substrate sterilization can be achieved by various treatments, such as water vapour (steaming).

There is some confusion regarding the designation of the different types of potted plants. First, there are les mottes, cime radicate or plug plants, in the French, Italian and English literature, respectively. These correspond to plants obtained through the rooting of runner tips ('rooted runners') in seedbeds with cells of 10–15 cm in diameter, under a misting system (therefore they are also called 'misted tip' plants). This type of plant is small in size, as it remains for less time at the nursery (3–4 weeks), and it has partially replaced the traditional bare-root fresh plant, showing increased earliness, better fruit quality, and a superior survival rate and cultivation homogeneity.

These plug plants at transplanting time show a single crown, but the inflorescences are not yet formed. In late summer, the misted tip plants are planted and the flower differentiation begins. Fruit production in cold winter areas (northern Italy) takes place in the spring months, from April to June, lasting 1 month. In mild winters (Mediterranean climate) with plastic tunnels, production starts in January and reaches a peak in February and March, with a prolonged spring production until the temperature become too high.

Differentiated rooted runners are plug misted tips already have differentiated flowers at transplantation, which occurs from late summer to the beginning of autumn. Fruiting, which begins early in the winter months, lasts for more than 3 months. More precocious (probably before Christmas time, which has the highest prices) and long-lasting productions can be obtained.

In addition, there are tray plants, which are of a larger size, as they are placed in cells of 7–8 cm and remain for at least 3–4 months in the nursery, where the crown is thickened during autumn under sprinkler fertigation. Subsequently, they remain cold stored (with their 'loaf' of new roots, and younger leaves), and the time frame depends on the variety and the needs of the producer, as with this type of plant it is possible to obtain early production if they are placed in heated nurseries off the ground. Currently, for out-of-season production, tray plants are used as they are less susceptible to problems associated with adverse winter conditions. Tray plants are produced from stolons or rooted runners placed in trays with a peat substrate during the summer (August), and then lifted from the nursery and cold stored during the winter. They are usually kept at –1.5 to –2.0°C, to allow preservation for 6-month periods. Due to their high cost (approximately €0.38–0.40 per plant, two to three times more expensive than frigo plants), they are used almost exclusively for high technology, off-the-ground cultivation. After the cold storage, planting is possible in different seasons in soil-less cultivation, allowing late out-of-season harvest, lasting 1–2 months, between summer and autumn.

Cultivation techniques of tray plants in the autumn must provide the appropriate conditions for flower initiation in order to obtain optimum production and quality of fruit during the winter and spring. They benefit from environments with good sunshine and mild temperatures from September to November. They develop one crown with two to three differentiated inflorescences before transplantation. Most profitable inflorescences are located in the apical portion of the crown, at the top of two to three extension crowns that will grow during the cultivation allowing the enlargement of the plant and providing good crop potential. A further two to six flower-differentiated buds may form along the axis of the plant, but generally they are not able to grow and provide fruits (Bosc *et al.*, 2012), because of bud mortality during cold storage or because they remain dormant. At lifting, the organs of the flowers have almost completely appeared. Therefore, after transplanting, only a few steps of flower development still have to take place (distension and colouring), leading to blooming 30 days later and first harvest after a further 30 days. Longer permanence of the plant under promoting conditions (mild temperature) in the nursery allows the formation of a higher number of flowers inside the inflorescence, with the total ranging from 20 to 40 flowers per plant. It is important to evaluate before cold storage the number of inflorescences and the number of flowers per inflorescence, plus their position along the stem of the crowns to estimate the production potential and the length of the fruiting period (harvest season). For this reason, in Europe there are several companies that sell tray plants with a certificate called flower mapping (or plant architecture).

Tray plants can be produced from small-sized cold-stored (frigo) plants. In the first phase (until early September), the plants grow in the same mountain environment where they are able to achieve an optimal vegetative growth, due to the cooler and milder weather. Subsequently, at temperature lowering, the plants are transferred to milder areas (e.g. Verona, Po Valley, Italy) where flower bud differentiation can continue, potentially obtaining more productive plants to be planted in the following summer. The frigo plants rapidly produce two to three new shoots growing in the pots, and thus at the beginning of autumn they already have two to three crowns, bearing three to six inflorescences. Under the early chilling autumn conditions, during dormancy, they are cold stored in fridges at –1 to –2°C for a variable period, depending on the timing of planting. Transplanting can take place from early spring in March to late summer or November, providing out-of-season production for 1 month. The advantage of these plants is their large size, which allows an abundant production in a short time. These plants are suitable for forcing to obtain programmed production in greenhouses. In France, for some high-quality productions, planting takes place in mid-December in heated greenhouses for productions from February to April. In this case, at the end of November the plants are placed in cold rooms at 3–4°C for 3 weeks (to satisfy the chilling requirements) and then planted in soil-less systems. Because of the high producing costs, these plants are used solely for soil-less and field programmed cultivation, providing better fruit quality compared with waiting-bed plants.

The same procedure applies to minitray plants, which differ only in being rooted later in smaller pots (5–6 cm in diameter), and for their lower cost, close to €0.22 per plant. They usually provide two to three good inflorescences with a total of 15–30 flowers per plant. Possibly, with this type of plant, flower induction is earlier than with tray plants and the plants are more equilibrated (fruit:leaf ratio), although they are less productive in programmed cycles.

The first stage to obtain potted plants or plug plants is the production of runners, which grow from the mother plants. Runners should be harvested when the first rootlets begin to appear and at least two trifoliate leaves have sprouted. According to Durner *et al.* (2002), these leaves must not exceed 10 cm in length from the crown and must have a minimum of 6.5 cm; moreover, the diameter of the crown must range between 20 and 50 mm. When cutting, it is necessary to leave at least 1 cm of the runner, so that it serves as an anchor at the moment of planting (Fig. 9.3) (Rowley *et al.*, 2010). In the case of tray plants, runners are harvested when they have a visible root system, i.e. the roots have a minimum length of 0.5 cm.

Mother plants should be established in mid-spring on light textured soils for runner production, as their growth is favoured by high temperatures (over 24°C) and long photoperiods, greater than 16 h (Rowley *et al.*, 2010). This may be performed under field conditions, or inside a greenhouse or tunnel. Contact of runners with the ground must be avoided, so if plants are in the field, it is necessary to use a mulch over the beds (Rowley *et al.*, 2010). In a greenhouse, an off-the-ground cultivation system is employed, i.e. at height. In this way, runner harvest is facilitated and production is cleaner, as soil fumigation with methyl bromide is avoided (Durner *et al.*, 2002).

Fig. 9.3. (a, b) A newly cut runner plant (a), showing 1 cm of runner left at its base for anchorage to the substrate (b). (c) The runner planted in a container.

When cutting the runners, they should be quickly planted and brought to conditions of high relative humidity (95%) in order to avoid dehydration. The storage time for runners may be up to 1 week at 0°C, but ideally planting should be carried out as soon as possible after harvest (Durner *et al.*, 2002). However, Lieten (2005) was more categorical and indicated they may be stored for up to 2 days.

Speedlings can be made of various materials, such as polystyrene, polyethylene or polypropylene, which differ in their useful life and cost. In addition, there are numerous sizes of seedbeds specially designed for strawberries, for example, from 100 to 200 cm^3 per cell for plug plants; nevertheless, for tray plants, a volume of 280 cm^3 should be sufficient to obtain adequate root development (Lieten, 2005).

The tray's colour has an effect on the temperature of the peat, which in turn may influence root growth and development, although Lieten (2005) stated that colour does not significantly affect yield or size of the crown.

9.3.3. Cycles

Many different growing cycles (Fig. 9.4) are possible using the available nursery material and by modulating the planting strategies and the growing technique. The planting time is fundamental for good results and varies depending on the area, the variety and the

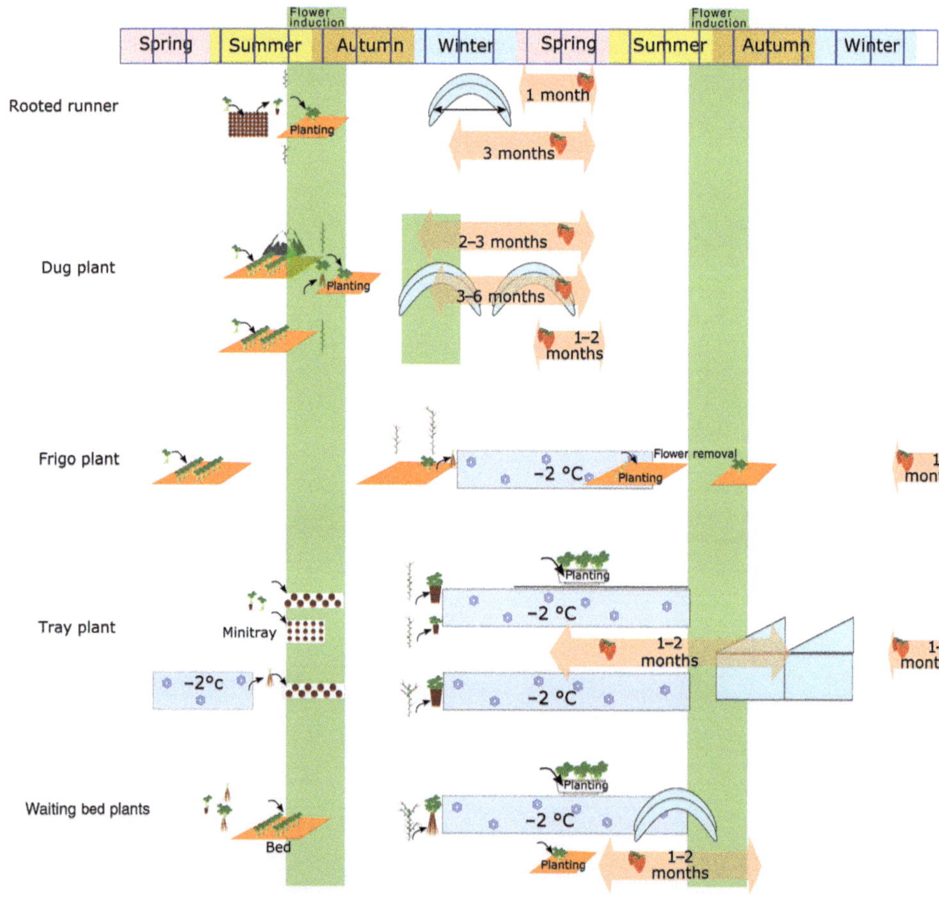

Fig. 9.4. Different growing cycles according to the available nursery material, the planting strategies and the growing technique.

cultivation technique. In Europe, strawberry production starts in the south of Spain and Italy beginning in December in protected cultivation in Sicily, followed by protected productions in Campania and Basilicata. Planting begins in mid-August using cold-stored plants and continues up to the first 10 days of October using rooted tips and bare-root fresh plants. The now widespread use of fresh plants, planted in early autumn and grown under protection (in tunnels), allows an early and prolonged production starting about 2 months after planting until June, but cultivars with a low chilling requirement are needed. In mid-April, the harvest begins in protected cultivations in Veneto, Emilia-Romagna and Marche, followed after about 15 days by the productions in the open field using frigo plants. In mid-June, programmed crops begin in the mountain areas in Piedmont, and later in Trentino, and in June and July production is possible in central European countries and even later in Scandinavian countries using tray plants. The remontant varieties are used to get multiple out-of-season cropping during the summer–autumn season (from the beginning of July until October; Lieten, 2005), generally planting cold-stored plants in the field during the spring (March–May). In the autumn, crops are started in the Verona area where spring production is also provided under plastic tunnels. Protecting and forcing systems allow strawberry fruiting during most parts of the year, from spring until the beginning of the winter season in the Netherlands and Belgium (Lieten, 2005), using tray plants and waiting-bed plants in programmed soil-less crops where the harvest season derives from the planting time. The nursery production of strawberry plants has become established, even in areas less suited to this crop, due to the spread of plant cold storage, digging from nurseries during the winter when they are in full rest and then using these plants for summer planting. Recently, for cultivation in the field, the so-called '60-day' programmed culture has gained importance, using cold-stored flower-differentiated plants (waiting bed, A+ frigo and tray plants) for summer planting and late summer–autumn production. Overall, central European regions are able to provide a long fruiting season.

9.4. Plant Certification Systems

Overall, the certification scheme consists of a propagation process, which is controlled by official bodies, based on precise technical standards that ensure plant health in every step of the process. Each country has its own certification rules, although some general aspects are present in every country. Nevertheless, current regulations must be continuously reviewed, as the nursery industry has undergone dramatic changes in the last decade, forcing them to adapt these regulations to the different types of plants that are now in the market.

Plant certification rules must be met at each stage of the propagation process, regarding various technical aspects such as: use and treatment of base material, isolation, infrastructure availability, use of substrates and soil treatments, tool handling, plant identification and labelling, and specific phytosanitary restrictions. Specialist inspectors, through regular visits, monitor all these technical considerations to check proper compliance with all regulations. The main considerations of the regulations required by each of the propagation stages identified in the previous item are presented in Table 9.2.

With respect to isolation, each stage of the certified plant programme must meet specific requirements, as outlined in Table 9.2. Moreover, the environment of the land on which propagation plants are located must be free of weeds and at least 10 m away from plants, trees and living fences, whose pests and diseases could be transmitted to nursery or mother plants.

Greenhouses that are used in the nuclear stock and propagation stock I stages must be completely isolated, with double anti-aphid mesh of 20/10 densities, a double door, foot disinfection systems at the entrance, and a paved or isolated floor. The roof should protect the inside of the

Table 9.2. Considerations of the regulations required by each of the propagation stages.

Stage	Main characteristics	Isolation conditions	Intensity of phytosanitary and genetic identity control	Plant category and labelling
Nuclear stock	Mother plants cultivated under screen house in individual containers	100 m away from any strawberry cultivation area	100% of plants at this stage undergo phyto-sanitary, visual and laboratory controls, together with visual genetic identity controls	'Pre-base' category, white label with purple stripe
Propagation stock I	Mother plants cultivated under screen house in bin-type large-sized individual containers	100 m away from any strawberry cultivation area	Laboratory phytopatho-logical analysis: 2% of plants in the case of bacteria; 30% of plants in the case of fungi. Genetic identity: 100%	'Base' category, white label
Propagation stock II	Mother plants cultivated under screen house tunnel, or directly in the field, on fumigated land, with rotations of at least 5 years	500 m away from any strawberry cultivation area	Laboratory analysis: 0.2% of plants in the case of viruses and phytoplasmas; 30% of plants in the case of fungi. Genetic identity: 2%	'Base' category, white label
Certified material	Cultivation in open field, on land without strawberry for at least 4 years (or 2 years, when soil is fumigated)	250 m away from any strawberry cultivation area, and 5 m away from other nurseries	Laboratory analysis: 0.02%. Genetic identity: 0.02%	'Certified' category, blue label

greenhouse from rainwater. There must also be a hall area prior to the entrance, to store the implements used for plant management and for staff robes.

Soils used in stages III (propagation stock II) and IV (commercial nursery) must have a controlled rotation of at least 5 years without strawberry plants, with species that are not affected by the same pathogens that attack strawberries. Furthermore, substrates and soils must be fumigated with either water vapour or authorized chemicals.

Tools used in any stage of the propagation process must be used exclusively in that stage, and must be disinfected between one plant and the other with 3% sodium hypochlorite. Tools should also be properly identified.

The specific phytosanitary restrictions are different in each of the propagation stages. Becasue it is essential that the nuclear stock comprises material that is completely free of pathogens and diseases, the intensity of control must be increased during this stage. To this end, careful monitoring of possible organisms present in individuals must be carried out. For viruses and phytoplasmas, laboratory tests are necessary, taking into consideration the restrictive regulations according to the current legislation of each country. In these cases, different mechanisms for microorganism detection are available, the selection of which depends directly on the pathogen species involved. Table 9.3 shows the methods recommended by the European and Mediterranean Plant Protection Organization (EPPO) for detection and identification of viruses, viruses-like organisms and phytoplasmas that may be present in *Fragaria* × *ananassa*.

Table 9.3. Recommended methods for virus, virus-like and phytoplasma detection. (From OEPP/EPPO 2008.)

Pathogen	Symptoms in cultivars	Mechanical inoculation to herbaceus host	Indicators for leaf graft transmission	Alternative tests
Virus and virus-like agents occurring in the EPPO region and which are tested for in this scheme				
Aphid-borne (SCV, SMYEV, SMoV, SVBV)			+	+
Leafhopper-borne	+			
Nematode-borne (ArMV, RpRSV, SLRV, TBRV)		+		
Virus and virus-like agents not present in the EPPO region or of minor importance that may optionally be tested				
SPMYEV			+	+
Strawberry latent C			+	
Leafhopper-borne virus	+			+
Nematode-borne (ToRSV)		+	+	
StCFV			+	+
Leaf roll, witches' broom	+		+	
Multiplier plant	+			
Feather-leaf	+		+	
SPaV			+	+
SNSV		+	+	+
BPYV, FClCV, ClLV, ApMV				+

SCV, strawberry crinkle virus; SMYEV, strawberry mild yellow-edge virus; SMoV, strawberry mottle virus; SVBV, strawberry vein banding virus; ArMV, Arabis mosaic virus; RpRSV, raspberry ringspot virus; SLRV, strawberry latent ringspot virus; TBRV, tomato black ring virus; SPMYEV, Strawberry pseudo mild yellow-edge virus; ToRSV, tomato ringspot virus; StCFV, chlorotic fleck virus; SPaV, strawberry pallidosis-associated virus; SNSV, strawberry necrotic shock virus; BPYV, beet pseudo yellows virus; FClCV, *Fragaria chiloensis* cryptic virus; ClLV, *Fragaria chiloensis* latent virus; ApMV, apple mosaic virus.

During the different steps of certification, it is necessary to perform visual inspections of the plants to detect any pest or disease that is affecting them. Moreover, tolerances at different stages are suggested (Table 9.4). It is important to also include the bacteria, oomycetes, fungi, arthropods and nematodes listed in Table 9.4.

9.5. Conclusions

Strawberry plant production in the nursery is based on the ability of the plants to produce stolons, which form new buds, resulting in new crowns of daughter plants. The production of runners is favoured in nurseries where adequate photoperiods and temperatures are ensured. The production system for certified material consists of four basic stages from the collection of healthy mother plants (nuclear stock) of verified varieties from which protected propagation stock I is established through to micropropagation or the production of runners, until the propagation stock II is set in the field to propagate mother plants of the commercial nursery. In the commercial nursery, various propagation techniques lead to the production of different plant types (bare-root or potted) that are the appropriate specific material to carry out a wide range of producing cycles. At the time of establishing a strawberry orchard, plants must have physiological (proper floral differentiation and content of carbohydrates), development and certified health characteristics. The certification system follows the specific certification rules of each country, based on technical standards that ensure plant health at every step of the propagation process.

Table 9.4. Suggested tolerances in visual inspection for strawberry pests at different stages of certification (From OEPP/EPPO, 1994.)

Microorganism		Incidence (%)			
		Nuclear stock	Propagation stock I	Propagation stock II	Commercial nursery
Bacteria	*Xanthomonas fragariae*	0	0	0	0
Oomycetes	*Phytophthora cactorum*	0	0	0	1
	Phytophthora fragariae var. *fragariae*	0	0	0	0
Fungi	*Colletotrichum acutatum*	0	0	0	0
	Verticillium dahliae and V. albo-atrum	0	0	0	2
	Rhizoctonia fragariae	0	0	0	1
Arthropods	*Chaetosiphon fragaefolii*	0	0	1	1
	Phytonemus pallidus fragariae	0	0	0	0.1
Nematodes	*Aphelenchoides* spp.	0	0	0	0
	Ditylenchus dipsaci	0	0	0	0

References

Barclay Poling, E. and Maas, J.L. (1998). Strawberry plug transplant technology. *Acta Horticulturae* 513, 393–402.

Bish, E.B., Cantliffe, D.J. and Chandler, C.K. (2002) Temperature conditioning and container size affect early season fruit yield of strawberry plug plants in a winter, annual hill production system. *Hortscience* 37, 762–764.

Bosc, J.P., Neri, D., Massetani, F. and Bardet, A. (2012). Relationship between plant architecture and fruit production of the short-day strawberry cultivar 'Gariguette'. *Acta Horticulturae* 926, 229–236.

Boxus, P. (1992) Mass production of strawberry and new alternatives for some horticultural crops. In: Kurata, K. and Kozai, T. (eds) *Transplant Production Systems*. Kluwer Academic Publishers, The Netherlands, pp. 151–162.

Durner, E.F., Poling, E.B. and Maas, J.L. (2002) Recent advances in strawberry plug transplant technology. *HortTechnology* 12, 545–550.

Faby, R. (1997) The productivity of graded 'Elsanta' frigo plants from different origins. *Acta Horticulturae* 439, 449–456.

García-Sinovas, D., García-Méndez, E., Andrade, M.A., Becerril, M., De Cal, A., Melgarejo, P., Salto, T., Martínez-Beringola, M.L., Redondo, C., Martínez-Treceño, A., Medina J.J., Soria, C. and López-Aranda, J.M. (2008). Strawberry nurseries in Spain: alternatives to MB, 2007 Results. In: *Proceedings of the Annual International Research Conference on Methyl Bromide Alternatives and Emissions Reductions*, Orlando, Florida.

Giménez, G., Andriolo, J.L., Janisch, D., Cocco, C. and Picio, M.D. (2009) Cell size in trays for the production of strawberry plug transplants. *Pesquisa Agropecuaria Brasileira* 44, 726–729.

Kirschbaum, D.S., Honorato, J. and Cantliffe, D.J. (2000) Strawberry waiting bed plants: a valid alternative to increase early and total yields in sub-tropical regions. *Scientia Horticulturae* 84, 83–90.

Lieten P. (2005) Strawberry production in central Europe. *International Journal of Fruit Science* 5, 91–105.

Neri, D., Savini, G., Massetani, F. and Bertoldi, M. (2009) Tipologie e qualità del materiale vivaistico. In: *Proceedings of the VIIth National Conference 'La Fragola presente e futuro'*, Marsala, Italy, pp. 180–189.

Neri, D., Baruzzi, G., Massetani, F. and Faedi, W. (2012) Strawberry production in forced and protected culture in Europe as a response to climate change. *Canadian Journal of Plant Science* 92, 1021–1036.

OEPP/EPPO (2008) Certification scheme for strawberry. *Bulletin OEPP/EPPO Bulletin* 38, 430–437.

Rowley, D., Black, B. and Drost, D. (2010) *Strawberry Plug Plant Production*. USU Extension Publication Horticulture/High Tunnels/2010-02pr.

Savini, G., Neri, D., Zucconi, F. and Sugiyama, N. (2005) Strawberry growth and flowering. *International Journal of Fruit Science* 5, 29–50.

Savini, G., Giorgi, V., Scarano, E. and Neri, D. (2008) Strawberry plant relationship through the stolon. *Physiologia Plantarum* 134, 421–429.

Strand, L.L. (2008) *Integrated Pest Management for Strawberries*, 2nd edn. UCANR Publications, Oakland, California.

10 Fungal Diseases of Strawberry and their Diagnosis

Carlos Garrido[1], Victoria E. González-Rodríguez[1], María Carbú[1], Amjad M. Husaini[2] and Jesús M. Cantoral[1]*

[1]Departamento de Biomedicina, Biotecnología y Salud Pública, Facultad de Ciencias del Mar y Ambientales, Instituto Universitario de Investigación Vitivinícola y Agroalimentaria (IVAGRO), Universidad de Cádiz, Puerto Real, Spain; [2]Centre for Plant Biotechnology, Sher-e-Kashmir University of Agricultural Sciences and Technology of Kashmir, Shalimar, Jammu and Kashmir, India

10.1. Introduction

10.1.1. Modern strawberry, a domesticated species for production

Over the centuries, men have cultivated different species of the genus *Fragaria* for strawberry production. The genus *Fragaria* belongs to the family Rosaceae and consists of more than 28 species, including several subspecies (Garrido *et al.*, 2011). More than 300 years ago, species such as *Fragaria chiloensis*, *Fragaria virginiana*, *Fragaria vesca* and *Fragaria moschata* were

* jesusmanuel.cantoral@uca.es

cultivated across the world for strawberry production. Progressively, these species were subjected to a 'domestication' process and became adapted to different climatic regions, developing resistance to regional pests and leading to special fruit characteristics, resulting in the development of new species and subspecies (http://www.juntadeandalucia.es/organismos/agriculturapescaydesarrollorural.html). The modern strawberry species *Fragaria* × *ananassa* is the most evident result of these processes. This species was developed in France around the middle of the 18th century, by crossing of *Fragaria virginiana* Duchesne (from North America) and *Fragaria chiloensis* Duchesne (from South America). After several rounds of crossing between hybrids, a plant similar to the modern *Fragaria* × *ananassa* was obtained. This new species, a hybrid, developed larger, more fragrant and tastier red berries, and showed a great capacity for regional adaptation (Maas, 1998). Due to these characteristics, this new species replaced the old ones, especially after it was introduced to America and the rest of Europe. Over several decades, *F.* × *ananassa* became the most important species commercially and is now predominantly cultivated for strawberry production worldwide (Garrido *et al.*, 2011).

Fragaria × *ananassa* shows a high level of regional adaptation due to the phenotypic plasticity of its plants. As the original hybrid plants were widely cultivated in different continents, and in more than 50 countries, the regional environmental pressures led to the development of new cultivars. Photoperiod, temperature, humidity, composition of the substratum, ease of obtaining nutrients and the presence of pathogens were some of the different environmental characteristics that contributed to the appearance of different cultivars. Although it is difficult to know the exact number of cultivars present in the world, only a dozen of them constitute the most common varieties used for world strawberry production (Garrido *et al.*, 2009a).

10.1.2. Economic importance of world strawberry production

The Statistical Office of The Food and Agriculture Organization (FAO) has released estimates of crop production in the world, classified by commodities × country or by countries × commodity (http://faostat.fao.org). The FAO estimated that the annual world production of strawberry exceeded 7,700,000 t in 2014. This data represented an increase of almost 1,200,000 t in comparison with 2010 (FAO, 2010–2014). *Fragaria* × *ananassa* is cultivated in more than 70 countries, with Asia and the Americas being the major producers of strawberry, accounting for 49.7 and 25.2%, respectively (FAO, 2010–2014). The European Union (EU) is third, with 19.2% of world strawberry production. Between 2010 and 2014, China increased its strawberry production, and is currently the country with the highest level of production, with an average of 2,600,000 t per year, relegating the USA, traditionally the most important producer, to second place with only half the production level of China at 1,300,000 t. In the USA, production is concentrated mainly in three states, California, Florida and Oregon, in decreasing order of importance. In the EU, Spain is the most important country, with an average production between 2010 and 2014 of 285,000 t. Spanish production is concentrated in the provinces of Huelva and Cádiz in the south-west of Spain, with a total cultivated area of approximately 7000 ha (FAO, 2010–2014).

10.1.3. Pathogen incidence in strawberry production

Strawberry plants, like other commercial crops, can be damaged by environmental, genetic and biological factors, either directly or by interactions between these factors (Garrido *et al.*, 2011). These strawberry health problems lead to significant economic losses to the growers and producer countries. Environmental factors causing damage to the cultivars can be minimized by creating good agricultural practice facilities. Strawberry plants are affected by arthropods, nematodes, fungi, bacteria and viruses. These cause considerable financial losses, running into hundreds of millions of dollars/euros per year (FAO, 2008–2013) (http://www.juntadeandalucia.es/organismos/agriculturaypesca.html).

There are several governmental agencies dedicated to preventing the spread of pathogens during international trade. The two major global agencies are the FAO and the European and Mediterranean Plant Protection Organization (EPPO). The EPPO is an intergovernmental organization, with 50 member countries, responsible for cooperation in plant protection in Europe and the Mediterranean region (http://www.eppo.org). This organization has the aim of helping its member countries prevent the entry or spread of dangerous pests. For this purpose, EPPO identifies pests, makes proposals on the phytosanitary measures that could be taken, and makes available several standards on Pests Risk Analysis databases. Since the 1970s, this organization has periodically published several pest lists that include: 'List of pests recommended for regulation as quarantine pests' (A1 and A2 list), 'Action List', 'Alert List' and 'List of invasive alien plants' (Capote *et al.*, 2012). In response to the lists and reports published by this organization, the member countries adopt various different phytosanitary measures and plant protection programmes. We recommend consulting the relevant one at the EPPO website (http://www.eppo.int/QUARANTINE/quarantine.htm).

More than 50 different species of phytopathogenic fungi can infect *F.* × *ananassa* cultivars, but not all of them have the same commercial significance (Table 10.1). In relation to quarantine pest lists, in recent years several genera of strawberry fungal pathogens, such as *Colletotrichum acutatum*, *Botrytis cinerea* and *Phytophthora* spp. have been included in the EPPO A2 list (the A2 list includes pests that are locally present in the EPPO region but not widely distributed), but in the last public A2 list from September 2015, only *Phytophthora fragariae* (specific for strawberry), *Verticillium dahliae*, *Verticillium albo-atrum* and *Fusarium oxysporum* were included (EPPO, 2015). This is not good news, as it implies that *Colletotrichum acutatum* and *B. cinerea* are currently widespread throughout Europe. It has therefore become more important to conduct studies on these fungi, as well as making an effort to control the spread of *Phytophthora* spp. and *Verticillium* spp. in the EPPO region.

10.2. The Main Fungal Pathogens of Strawberry

Strawberry plants are affected by a large number of diseases caused by fungi, bacteria, viruses, nematodes and arthropods. These pathogens cause damage on the leaves, roots, crowns and fruits. Susceptibility of the host plant, the particular pathogen and favourable environmental conditions (temperature, moisture) are the three main factors necessary for the initiation and development of plant disease. In the cultivated strawberry (*F.* × *ananassa*), each cultivar differs in its susceptibility to different pathogens and to their races or pathotypes. These races are not evenly distributed around the world, and are often present only in specific strawberry crop regions. Therefore, the damage caused in strawberry crops and the economic losses generated are of different scales for each region (Maas, 2004).

The American Phytopathological Society (APS) is the premier society dedicated to the study and control of plant diseases. The APS has published significant breakthroughs in plant pathology, mycology, virology, bacteriology, nematology and related disciplines for more than 100 years. APS members have been responsible for publishing a list of the most common types of damage affecting more than 100 crops, including strawberry cultivation (http://www.apsnet.org/publications/commonnames/Pages/Strawberry.aspx). This list includes more than 67 species of fungal pathogens (Table 10.1), of which 47 species belong to the phylum

Table 10.1. Strawberry fungal pathogens.

Genus	Species	Strawberry disease	Reference[a]
Alternaria	*alternata*	Black leaf spot	Miyamoto *et al.* (2009)
	tenuissima	*Alternaria* fruit rot	Shafique *et al.* (2009)
Armillaria	*mellea*	*Armillaria* crown rot and root rot	Prodorutti *et al.* (2009)
Aspergillus	*niger*	*Aspergillus* fruit rot	Chiotta *et al.* (2009)
Botrytis	*cinerea*	*Botrytis* rot fruit; grey mould	Vallejo *et al.* (2002)
Cladosporium	spp.	*Cladosporium* fruit rot	Ruiz-Moyano *et al.* (2009)
Cercospora	*fragariae*	*Cercospora* leaf spot	Maas *et al.* (1998)
	vexans	*Cercospora* leaf spot	Maas *et al.* (1998)
Colletotrichum	*acutatum*	Anthracnose on leaves and fruit, crown rot and black spot	Garrido *et al.* (2009a)
	gloeosporioides	Anthracnose on leaves and fruit, crown rot and black leaf spot	Chung *et al.* (2010)
	fragariae	Anthracnose on leaves and fruit; crown rot and black spot	Ortega-Morales *et al.* (2009)
Coniothyrium	*fragariae*	Black root rot	Douglas Gubler and Converse (1993)
	fuckelli	Black root rot	Pertot *et al.* (2012)
Cylindrocarpon	*destructans*	Root rot	Martin *et al.* (2002)
Diplocarpon	*earlianum*	Leaf scorch	Whitaker *et al.* (2009)
Fusarium	*oxysporum*	*Fusarium* wilt	Avis *et al.* (2009)
	sambucinum	Fruit blotch	Hunter *et al.* (1974)
Gnomonia	*comari*	Leaf blotch; stem end rot	Morocko *et al.* (2007)
Hainesia	*lythri*	Black root rot; *Hainesia* leaf spot	Douglas Gubler and Converse (1993)
Idriella	*lunata*	*Idriella* root rot	Maas *et al.* (1998)
Macrophomina	*phaseolina*	*Macrophomina* leaf blight	Maas *et al.* (1998)
		Macrophomina root rot	Javaid *et al.* (2009)
Mucor	*hiemalis*	*Mucor* fruit rot	Hauke *et al.* (2004)
	mucedo	*Mucor* fruit rot	Hauke *et al.* (2004)
	piriformis	*Mucor* fruit rot	Hauke *et al.* (2004)
Mycosphaerella	*fragariae*	Purple leaf spot; black seed disease	Ehsani-Moghaddam *et al.* (2006)
	louisianae	Purple leaf spot	Maas *et al.* (1998)
Olpidium	*brassicae*	*Olpidium* root infection	Douglas Gubler and Converse (1993)
Pestalotia	*longisetula*	*Pestalotia* fruit rot	Maas *et al.* (1998)
Penicillium	*cyclopium*	*Penicillium* fruit rot	Gutierrez *et al.* (2009)
	expansum	*Penicillium* fruit rot	Liu *et al.* (2007)
	frequentans	*Penicillium* fruit rot	Redondo *et al.* (2009)
	purpurogenum	Fruit rot; fruit blotch	Redondo *et al.* (2009)
Peronospora	*potentillae*	Downy mildew; fruit blotch	Choi *et al.* (2009)
Phytophthora	*bisheria*	Root rot	Abad *et al.* (2008)
	cactorum	Leather rot of fruit; *Phytophthora* crown and root rot	Nicastro *et al.* (2009)
	citricola	*Phytophthora* crown and root rot	Haesler *et al.* (2008)
	citrophthora	Leather rot; *Phytophthora* crown and root rot	Kong *et al.* (2009)
	cyptogera	Root rot	Abad *et al.* (2008)
	fragariae	Red stele, red core and root rot	Nicastro *et al.* (2009)
	megasperma	Crown rot	Abad *et al.* (2008)
	nicotianae	Leather rot; *Phytophthora* crown and root rot	Böszörményi *et al.* (2009)

Continued

Table 10.1. Continued.

Genus	Species	Strawberry disease	Reference[a]
Phoma	*lycopersici*	Leaf stalk rot	Maas *et al.* (1998)
	terrestris	Grey sterile fungus root rot	Douglas Gubler and Converse (1993)
Phomopsis	*obscurans*	*Phomopsis* leaf blight	Nita *et al.* (2003)
Pichia	*membranefaciens*	Postharvest rot	Douglas Gubler and Converse (1993)
	subpelliculosa	Postharvest rot	Douglas Gubler and Converse (1993)
Pythium	*ultimum*	Black root rot	Triky-Dotan *et al.* (2009)
Rhizoctonia	*fragariae*	Black root rot; anther and pistil blight;	Lamondia *et al.* (2005)
	solani	Hard brown rot; *Rhizoctonia* leaf blight	Chiba *et al.* (2009); Liu *et al.* (2009)
Rhizopus	*stolonifer*	*Rhizopus* rot	Maas *et al.* (1998)
	sexualis	*Rhizopus* rot	Maas *et al.* (1998)
Rosellinia	*necatrix*	Dematophora crown and root rot (white root rot)	Pliego *et al.* (2012)
Saccharomyces	*cerevisiae*	Postharvest rot	Douglas Gubler and Converse (1993)
	kluyveri	Postharvest rot	Douglas Gubler and Converse (1993)
Schizoparme	*straminea*	Fruit blotch	Douglas Gubler and Converse (1993)
Sclerotinia	*sclerotiorum*	*Sclerotinia* crown and fruit rot	Ren *et al.* (2010)
Sclerotium	*rolfsii*	*Sclerotium* rot; southern blight; fruit blotch	Errakhi *et al.* (2009)
	fragariae	*Septoria* hard rot and leaf spot	Maas *et al.* (1998)
Septoria	*aciculosa*	*Septoria* leaf spot	Maas *et al.* (1998)
	fragariaecola	*Septoria* leaf spot	Maas *et al.* (1998)
Sphaeropsis	*malorum*	Fruit blotch	Douglas Gubler and Converse (1993)
Sphaerotheca	*macularis*	Powdery mildew	Davik *et al.* (2005)
Stagonospora	*fragariae*	*Stagonospora* hard rot	Maas *et al.* (1998)
Verticillium	*albo-atrum*	*Verticillium* wilt	Larsen *et al.* (2007)
	dahliae	*Verticillium* wilt	Costa *et al.* (2007)
Zygosaccharomyces	*bailli*	Postharvest rot	Douglas Gubler and Converse (1993)
	florentinus	Postharvest rot	Douglas Gubler and Converse (1993)

[a]Relevant publications from authors working with the pathogen in the plant pathology field.

Ascomycota, ten to the Oomycota, four to the Basidiomycota, one to the Chytridiomycota and five to the Zygomycota. Several species of fungi are capable of causing damage to more than one part of the strawberry plant (leaf, root, crown and/or fruit), whereas other species affect only a specific part of the plant; for example, *Colletotrichum acutatum* causes anthracnose on leaves and fruits, crown rot and black spot, while *B. cinerea* causes damage to the fruit only (Table 10.1).

Various fungi invade the primary roots and attack/destroy secondary roots, limiting the plant's ability to take up water and nutrients. Leaf pathogens cause various injuries or overwinter in dead leaves and petioles, and form spores. These spores are disseminated by wind and/or rain and irrigation water, and initiate new infections. Fruit can be affected by a large number of genera of fungi, with *B. cinerea*, *Phytophthora cactorum* and *Colletotrichum* spp. being the major strawberry fruit disease problems worldwide, resulting in the greatest fruit losses. These also account for the highest amounts of fungicide use for strawberry fruit protection.

10.2.1. *Botrytis cinerea*

B. cinerea Pers.: Fr. is a phytopathogenic ascomycete that causes grey mould in more than 200 crop species worldwide, without any apparent host specificity. In strawberry, this pathogen is the cause of the disease known as '*Botrytis* rot fruit' (Table 10.1), which causes tremendous losses in the field (expected 80–90% loss of both flowers and strawberries) during rainy and cloudy periods, just before or during harvest and storage. This necrotrophic fungus attacks different organs, such as shoots, leaves, flowers and fruit, and is more destructive on mature or senescent tissue, but usually gains entry to such tissues at the flowering stage. Flowers are usually infected during blossoming, and the pathogen then enters the young fruits at a very early stage of their development. It remains latent for a considerable period before rapidly decomposing the tissues when environmental factors, such as relative humidity, and fruit physiology are optimal. The fungus infects the fruit, causing it to become deformed, dried out, dark and rapidly covered with a powdery layer of spores, which gives a grey appearance. In general, strawberries that are in contact with the ground or with another rotten strawberry or dead leaves in dense foliage are the ones that are commonly affected by this phytopathogenic fungus.

The long latency period of the pathogen from early infection until symptoms appear makes grey mould control very difficult. Therefore, the most common method used to control the infection and its spread has been the regular application of fungicides throughout flowering. Another alternative is the use of cultivars resistant to grey mould. The genetic resistance in strawberry to *Botrytis* infection appears to be multigenic and has a very low general combining ability. There has been little success in breeding and selecting cultivars resistant to this disease (Maas, 2004). A recent study determined the resistance of new cultivars, currently grown in Florida, and advanced selections from the University of Florida's breeding programme to *Botrytis* fruit rot (Seijo *et al.*, 2008). The results demonstrated that the cultivars 'Camarosa', 'Florida Radiance', 'Florida Elyana' and advanced selections 99-117 and 99-164 showed good levels of resistance, whereas 'Camino Real', 'Ventana', 'Treasure', 'Candonga', 'Strawberry Festival' and 'Sweet Charlie' were more susceptible. The major difficulty encountered in the fight against this disease is the lack of natural genetic resistance to grey mould in strawberry germplasm, making unsuccessful all attempts to incorporate tolerance to this disease into strawberry lines.

The development of biotechnology has provided new opportunities to enhance disease resistance for strawberry breeding. Vellicce *et al.* (2006) obtained transformed strawberry plants (cultivar 'Pájaro') using three defence-related genes: *ch5B*, encoding a chitinase from *Phaseolus vulgaris*; and *gln2* and *ap24*, encoding a glucanase and a thaumatin-like protein, respectively, both from *Nicotiana tabacum*. The results showed that constitutive expression of the bean *ch5B* gene in this strawberry cultivar was an effective strategy to provide protection against *B. cinerea*. The authors concluded that the same methodology could be used to introduce resistance into *F.* × *ananassa* germplasm. Other studies on transformation of strawberry plants using *Agrobacterium*-mediated transformation have resulted in the development of transgenic lines expressing glucose oxidase (Jin *et al.*, 2005) and thaumatin-like proteins (Schestibratov and Dolgov, 2005). Both of these transgenic strawberry lines showed a significantly higher level of resistance to grey mould.

10.2.2. *Colletotrichum* spp.

Colletotrichum spp. comprises a diverse range of important phytopathogenic fungi that cause pre- and postharvest crop losses worldwide. Three species have been reported as causal agents of strawberry (*F.* × *ananassa*) anthracnose: *Colletotrichum acutatum*, *C. fragariae* and *C. gloeosporioides* (telemorph, *Glomerella cingulata*) (Denoyes-Rothan *et al.*, 2003) (Table 10.1), which can infect strawberry plants in nurseries, field production, or during harvest and storage of fruits. All three species can be found on all parts of the plant (Denoyes-Rothan *et al.*, 2003), although *C. fragariae* and *C. gloeosporioides* are the principal causal agents of anthracnose crown rot, whereas *Colletotrichum acutatum* has been well documented as the principal causal agent of anthracnose fruit rot, also known as black spot (Mertely and Legard, 2004).

The worldwide distribution of the three species may vary. In the USA, *C. gloeosporioides* and *C. fragariae* have been described as causing severe losses, mainly in the south-east region (Florida), while *Colletotrichum acutatum* is the major species present in the south-west (California). However, in Europe, anthracnose is most often caused by *Colletotrichum acutatum*. The strawberry anthracnose symptoms produced are similar and occur most commonly on runners, flowers and fruits. In the fruit, the fungus causes circular lesions, which are firm and sunken and become black spots on mature fruit (Garrido *et al.*, 2008). Within the crown tissue, *Colletotrichum* spp. cause red-brown discoloration and necrosis of the fruits (Garrido *et al.*, 2008; Smith, 2008), or even the wilting of infected plants during periods of moisture stress, such as early afternoon in the summer. Under environmental conditions that favour infection, this process may continue for several days until the crown infection is extensive and causes the entire plant to wilt and die.

To colonize the host, these fungi develop many specialized infection structures, including germ tubes, appresoria, intracellular hyphae and secondary necrotrophic hyphae. The infection begins when the conidia adhere to plant surfaces, produce germ tubes and then continue forming appresoria, which penetrate the cuticle directly (Curry *et al.*, 2002). The pathogen grows beneath the cuticle by forming a subcuticular intramural network of hyphae before spreading throughout the tissue. In the case of *C. gloeosporioides*, the process continues with the formation of secondary necrotrophic hyphae, which kill plant cells by using cellular debris as nutrients (Bailey *et al.*, 1992).

Control of anthracnose can be difficult, as few available fungicides are effective against anthracnose of strawberry; once an epidemic of anthracnose fruit rot begins in a susceptible cultivar, it is nearly impossible to control (Maas, 2004). The prospects for developing cultivars resistant to anthracnose are relatively good. Although the genetics of anthracnose resistance are complicated, gains from recurrent selection have been possible because of the high broad-sense heritability estimates for resistance (Maas, 2004). Currently, only a few resistant cultivars are available. In the work performed by Seijo *et al.* (2008), the cultivars 'Sweet Charlie', 'Ruby Gem', 'Florida Elyana' and 'Florida Radiance' proved to be the most resistant; 'Strawberry Festival' and advanced selection 99-117 were intermediate in susceptibility; and 'Albion', 'Camarosa', 'Camino Real', 'Ventana', 'Candonga' and 'Treasure' were susceptible/highly susceptible. Transformation of strawberry plants using *Agrobacterium*-mediated transformation has shown that expression of a β-1,3-glucanase gene, isolated from the antagonist soil fungus *Trichoderma harzianum*, in strawberry enhanced anthracnose resistance (Mercado *et al.*, 2007).

10.2.3. *Fusarium oxysporium*

Within the genus *Fusarium*, two species have been described, *Fusarium oxysporium* and *Fusarium sambucinum* (Table 10.1), that are capable of causing significant damage to strawberry plants. *Fusarium oxysporium* is the main cause of the disease called *Fusarium* wilt, while *Fusarium sambucinum* is responsible for the disease called fruit blotch and the root rots or storage rot. However, it is *Fusarium oxysporum* that is the major cause of economic loss in the cultivation of strawberries, and therefore there have been a greater number of studies about its biological and morphological description, mode of infection and pathogen control.

The ascomycete *Fusarium oxysporium* causes disease on many economically important crop species. This cosmopolitan soilborne fungus is considered a normal constituent of the fungal rhizosphere community of plants (Fravel *et al.*, 2003). Propagules of the pathogen can survive for long periods in soil organic matter or the rhizosphere of many plant species (Fravel *et al.*, 2003), which may serve as the primary inoculum for other susceptible hosts. The fungus penetrates strawberry plants through the roots, colonizes the root cortex and then grows into the xylem, disrupting water transport throughout the plant. This leads to the typical symptoms of reddish-brown discoloration on the crown, drying and early senescence of mature leaves, followed by stunting and wilting of the entire plant. In addition, transplants are infected through runners from infected mother plants (Nam *et al.*, 2009).

Wilt-inducing isolates of *Fusarium oxysporium* are host-specific fungal pathogens and have been divided into more than 120 different *formae speciales* (f. sp.) based on their host specificity (Fravel *et al.*, 2003). Among these, *Fusarium oxysporium* f. sp. *fragariae* is the specific pathogen responsible for the wilt of strawberry. *Fusarium oxysporium* f. sp. *fragariae* has been reported from strawberry-producing regions throughout the world, including Australia, Korea, China, Spain and the USA (Fang *et al.*, 2012). Traditionally, management of *Fusarium* wilt on strawberry plants has been mainly through chemical soil fumigation; however, the most cost-effective and environmentally sustainable strategy is the use of resistant cultivars. The successful development and deployment of resistant cultivars requires an understanding of the defence responses of strawberry against *Fusarium oxysporium* f. sp. *fragariae*, which remain poorly understood (Fang *et al.*, 2013). Major gene resistance has provided effective control of *Fusarium* wilt in many crops, although the durability of resistance has been variable, with the emergence of new pathogenic resistant races (Islas, 2012). Fang *et al.* (2012) has reported that the strawberry cultivar 'Camarosa' is the most susceptible cultivar to *Fusarium oxysporium* f. sp. *fragariae*, while 'Festival' is the most resistant cultivar. The development of strawberry cultivars resistant to *Fusarium* wilt is currently one main focus area in the University of California strawberry breeding programme (Islas, 2012).

10.2.4. *Phytophthora* spp.

The oomycete genus *Phytophthora* comprises fungus-like organisms; it includes more than 80 species and harbours a group of important plant pathogens, capable of causing considerable economic losses to food crops and ornamentals. Of these, at least eight species (Table 10.1) have been isolated from the roots, crowns, stolons and fruits of affected strawberry plants in different regions of the world (Abad *et al.*, 2008). Species of *Phytophthora* associated with strawberry include *Phytophthora bisheria*, *Phytophthora cactorum*, *Phytophthora citricola*, *Phytophthora citrophthora*, *Phytophthora cryptogea*, *Phytophthora fragariae* var. *fragariae*, *Phytophthora megasperma* and *Phytophthora nicotianae*. The species *Phytophthora cactorum* and *Phytophthora fragariae* var. *fragariae* are the most important pathogens and occur in almost all countries where strawberry is cultivated.

Phytophthora cactorum has been reported as the causal agent of more than 200 plant species diseases from over 60 different families (Eikemo *et al.*, 2004). In strawberry, it causes leather rot of the fruit, stem and crown, and root rot (Abad *et al.*, 2008). These diseases have been described in the USA and other temperate to subtropical regions. In general, *Phytophthora cactorum* has a wide host range; however, not all strains are able to infect all host species. *Phytophthora cactorum* isolated from strawberry crowns has been shown to be genetically very uniform and has been suggested to originate from a single clone, at least within the EU (Chen *et al.*, 2011). The analysis of microsatellites of *Phytophthora cactorum* from strawberry showed that leather rot of strawberry fruit and crown rot are not caused by genetically different strains of the species (Hantula *et al.*, 2000). Eikemo *et al.* (2004) discovered than none of the *Phytophthora cactorum* isolates from other hosts could cause crown rot symptoms in strawberry. Therefore, the causal agent of crown rot is often referred to as a distinct pathotype of *Phytophthora cactorum*, as it cannot be distinguished morphologically from leather rot (Hantula *et al.*, 2000). Infection usually occurs during warm periods with prolonged wetness. Symptoms typically develop during early to mid-summer. The youngest leaves turn bluish green and often wilt suddenly; wilting quickly spreads to the entire plant, which collapses and dies, typically within a few days.

Red stele root, caused by *Phytophthora fragariae* var *fragariae*, was first observed in Scotland and is generally most severe in areas with cool and moist climates. There are no known natural hosts other than strawberry and loganberry, although artificial infection of other hosts has been possible. Symptoms usually appear on the upper parts of plants that come under stress in late spring or early summer, especially in low-lying wet areas. They may die just before fruiting or may produce a few small fruits. Younger leaves can have a blue-green coloration, while older ones turn yellow or red. In the root, there is a reddish discoloration of the steles, which occurs when the soil is cool, and the root later starts to rot from the tip upwards. Plants with severe root rot are often stunted and may wilt in hot weather.

The control of diseases caused by *Phytophthora* spp. is often difficult due to the release into the soil of resistant perennating structures, oospores and/or chlamydospores. Traditionally, *Phytophthora* spp. have been controlled by pre-plant soil fumigation or applying fungicides or mixtures of the two. Soil fumigation may reduce the inoculum in the soil but may not eradicate the fungus, and the use of fungicides has led to development of resistant strains (Maas, 2004). The use of resistant cultivars is by far the most reliable way to avoid root rot and crown rot problems. Many cultivars are available in North America and Europe that are resistant to this disease. In the case of cultivars resistant to *Phytophthora fragariae* var *fragariae*, there exists a race-specific resistance to the simpler races. Thus, most strawberry breeding programmes have concentrated on incorporating resistance against multiple races and have extended the range of useful cultivars that can be planted in sites infested with multiple races of this pathogen. This has been complicated in the past because the nature of genetic resistance was largely unknown and crossing two red stele-resistant parents often gave unpredictable results. However, recent studies have shown that the control of resistance and pathogenicity involves at least five genes for resistance (Maas, 2004).

10.2.5. *Verticillium* spp.

Verticillium is a genus of fungi belonging to the phylum Ascomycota, and causes wilt of over 300 plant species. Two *Verticillium* spp., *Verticillium dahliae* and *Verticillium albo-atrum* (Table 10.1) cause enormous economic losses in the cultivation of strawberries because they cause vascular diseases in the plants (Bhat and Subbarao, 1999). These soilborne plant pathogens are distributed throughout the world. *Verticillium* fungi overwinter in soil and plant debris as dormant mycelium or microsclerotia. When the conditions are favourable, the microsclerotia germinate and the hyphae may penetrate the root hairs directly, but infection is aided by breaks or wounds in the rootlets. Once inside the root, the fungus invades and destroys the xylem, resulting in reduced water uptake by the plant, causing the plants to wilt and wither. The symptoms produced by this pathogen vary depending on the susceptibility of the cultivar and cannot easily be differentiated from those for red stele or black root rot. The initial symptoms appear rapidly, when a sudden onset of high temperatures, high light intensity or drought interrupt mild conditions. *Verticillium* wilt tends to be most severe in plants that are already fruiting, and symptoms may continue throughout the summer and autumn. When the infection is established, the leaves droop, wilt, turn dry and become reddish-brown or dark yellow at the edges and between the veins. The strawberry plants are stunted and flattened with small yellowish leaves due to lack of water. Brown-bluish black streaks or smears may appear on runners and leaf petioles. The new roots growing from the crown often become overshadowed with blackened tips and the inside of the crown may appear to have brownish stripes.

Excellent long-range control of *Verticillium* wilt has been obtained by pre-plant soil fumigation or by the use of resistant strawberry cultivars. Several old and new cultivars are available that are moderately to highly resistant to *Verticillium* wilt. The transformation of strawberry cultivars using *Agrobacterium*-mediated transformation protocol has also been successful. Chalavi *et al.* (2003) carried out the transformation of 'Joliette' strawberry with a chitinase gene (*pcht*28) from *Lycopersicon chilense*. In growth chamber studies, they observed that the transgenic strawberry plants that expressed *pcht*28 had significantly higher resistance to *V. dahliae*.

10.2.6. Other genera

As well as the above-discussed fungal pathogens, there are other types of fungi capable of causing damage to the fruit, crown and root of strawberry plants whose incidence in the field is much smaller, and therefore the economic losses in strawberry cultivation are lower. Within the genera of fungi responsible for damage to the fruit are *Rhizoctonia fragariae* (anther and pistil blight), *Mycosphaerella fragariae* (black seed disease) and *Sclerotinia sclerotiorum* (sclerotinia fruit rot). This group of fungi vary widely in severity of disease, and no specific control measures have been developed against them, although pre-plant soil treatment and cultivation practices, such as mulching and removal of plant debris, may help to minimize their incidence. Others, such as *Gnomonia comari* (stem end rot), *Pestalotia longisetula* (*Pestalotia* fruit rot), *Alternaria* spp. (*Alternaria* rot) and *Cladosporium* spp. (*Cladosporium* rot), *Aspergillus niger* (*Aspergillus* rot), and *Stagonospora fragariae* (*Stagonospora* hard rot) are of less commercial importance. Lastly, there is one group of filamentous fungi that are especially important because they cause postharvest losses to strawberry in storage. This group includes *Rhizopus stolonifer* and *Rhizopus sexualis* (*Rhizopus* rot); *Mucor mucedo*, *Mucor piriformis* and *Mucor hiemalis* (*Mucor* fruit rot); *Penicillium cyclopium*, *Penicillium frequentans*, *Penicillium expansum* and *Penicillium purpugenum* (*Penicillium* fruit rot). Additionally, some yeasts also cause damage at postharvest, such as *Saccharomyces cerevisiae*, *Saccharomyces kluyveri*, *Pichia membranifaciens*, *Pichia subpelliculosa*, *Zygosaccharomyces bailii* and *Zygosaccharomyces florentinus* (Douglas Gubler and Converse, 1993). The importance of these pathogens in causing postharvest rot has been substantially reduced by modern storage and shipping methods (Maas, 1998).

The crowns and roots of strawberry plants may be also attacked by other fungi (Table 10.1), with less serious commercial consequences. These include *Armillaria mellea* (*Armillaria* root and crown rot), *Coniothyrium fragariae* and *Coniothyrium fuckelli* (black root rot), *Cylindrocarpon destructans* (root rot), *Dematophora necatrix* (*Dematophora* root and crown rot), *Hainesia lythri* (black root rot), *Idriella lunata* (*Idriella* root rot), *Macrophomina phaseolina* (*Macrophomina* root rot), *Phoma lycopersici* (leaf stalk rot) and *Phoma terrestris* (grey sterile fungus root rot), *Rosellinia necatrix* (white root rot) and *Sclerotinia sclerotiorum* (*Sclerotinia* crown rot) (Douglas Gubler and Converse, 1993; Agrios, 2011)

The genus *Pythium* spp., especially the species *Pythium ultimum*, is the most widespread strawberry root pathogen favoured by cool climates. This species has been shown to be a major cause of black root rot disease. *Pythium* spp. are not only strawberry pathogens; they can attack many other crops, both annual and perennial. In strawberry, *Pythium* spp. destroy juvenile root tissue, such as feeder rootlets. Control of *Pythium* spp. is effective with the application of chloropicrin mixtures.

Two other species that are able to kill structural roots as well as feeder rootlets of strawberries are *Rhizoctonia solani* and *Rhizoctonia fragariae*. Lesions on young roots are reddish-brown at first, and darken with age. The infected crowns show internal brown discoloration of basal tissues and frequently collapse and die. Traditionally, these diseases have been controlled by fumigating the soil prior to planting (Garrido *et al.*, 2010).

The leaf of the plant is also an organ that can be affected by fungal pathogens. There are at least 19 species that cause leaf damage, some only occasionally, depending on the susceptibility of cultivar and environmental conditions (Table 10.1). When enough leaf tissue has been destroyed by disease, the plant is weakened, and in such cases, the plant is more subject to winter injury. Additionally, the pathogens that cause these diseases on leaves can infect the berries, causing quality problems or even loss of fruits. *Alternaria alternata* (*Alternaria* black leaf spot), *Colletotrichum acutatum*, *Colletotrichum gloeosporioides* and *Colletotrichum fragariae* (anthracnose leaf spot and irregular leaf spot); *Diplocarpon earlianum* (leaf scorch), *Mycosphaerella fragariae* (leaf spot), *Phomopsis obscurans* (*Phomopsis* leaf blight), *Rhizoctonia solani* (*Rhizoctonia* leaf blight) and *Sphaerotheca macularis* (powdery mildew) are fungi causing lesions on the leaves. These diseases occur on strawberry plants in all areas of the world, from temperate climates to subtropical and tropical regions. The major effect of these pathogens is the progressive destruction of the foliage, which may weaken plants and reduce yields. Some pathogens cause very distinctive leaf symptoms; for example, *Sphaerotecha macularis* forms white patches of mycelium on the abaxial surface of the leaf. Other species cause similar, even identical symptoms, from circular spots with grey centres and dark margins to irregular purplish red or brown areas with dark reddish purple margins. It is usual in the literature to find incorrect identifications, and many citations confirm that symptoms are often confused among pathogens; for example, leaf scorch, caused by *Diplocarpon earlianum* is often confused with the symptoms of leaf spot caused by *Mycosphaerella fragariae*; and the symptoms produced by the latter are also wrongly identified as *Phomopsis* leaf blight caused by *Phomopsis obscurans* (Maas, 1998).

Leaves can be also attacked by fungi other than those discussed above, but their commercial importance is minor since they are limited to a few particular regions or countries and they occur only occasionally in strawberry. These other minor diseases of leaves include *Gnominia comari* (leaf blotch); *Mycophaerella louisianae* (purple leaf spot); *Septoria fragariae*, *S. aciculosa* and *S. fragariaecola* (septoria leaf spots); *Macrophomina phaseolina* (macrophomina leaf blight); *Cercospora fragariae* and *C. vexans* (cercospora leaf spots); *Sclerotium rolfsii* (sclerotium rot); and *Phoma lycopersici* (leaf stalk rot) (Maas, 1998).

10.3. Diagnosis Methods and Field Monitoring of Strawberry Diseases

10.3.1. Molecular methods applied to phytopathogenic fungi

Phytopathogenic fungi are able to infect any tissue at any stage of strawberry plant growth. These fungi show complex life cycles, including both sexual and asexual reproduction stages (Agrios, 2011). The biological variability of pathogenic races gives them the advantage of adaptation in climatologically different environments, ranging from dry and desert zones to wet and hot regions. This ability to adapt allows them to cause disease in almost all *Fragaria* × *ananassa* cultivars.

The development of molecular methods has led to the generation of much information about these diseases (Garrido *et al.*, 2009b). Traditionally, identification and characterization of fungi were focused on classical microbiological studies, but molecular methods have added to our understanding of these microorganisms. Molecular methods allow a more accurate characterization and differentiation of phytopathogenic fungi. These methods include iso-enzyme comparisons, restriction fragment length polymorphism (RFLP) analyses of mitochondrial DNA, AT-rich analyses, random amplified polymorphic DNA (RAPD) analyses, genus- and species-specific polymerase chain reaction (PCR) methods and enzyme-linked immunosorbent assays (ELISAs) (Garrido *et al.*, 2009a, 2012).

In recent years, many changes in the cultural practices have taken place in the cultivation of crops in general, and strawberry in particular. The use of toxic chemical compounds such as methyl bromide to control and prevent the appearance of pathogens has been progressively replaced by new biological, cultural and physical methods. These methods have led to the discovery of alternative approaches to disease control. They have reduced the massive fumigant use for soil and plant treatment, replacing them with biocontrol agents (microorganisms), biofumigants and alternative non-persistent chemical compounds of biological origin.

10.3.2. Past and present diagnosis methods

The first step in implementing appropriate strategies for disease management and appropriate control measures requires the unambiguous identification of the organism(s) responsible for the disease of the strawberry crop (Garrido *et al.*, 2011). Because many fungal pathogens of strawberry produce similar symptoms, it is important to be able to distinguish between different species. After accurate identification of the pathogen, it is essential to design a correct programme for its management. Many pathogens are subjected to special regulation through quarantine programmes agreed among producer countries. Therefore, pathogen identification is crucial to all aspects of fungal diagnostics and epidemiology in the field of plant pathology, and also in medical science, environmental studies and biological control (McCartney *et al.*, 2003; Atkins and Clark, 2004). The best way forward for pathogen identification is to use faster methods, making it possible to detect diseases even before symptoms appear (Debode *et al.*, 2009; Garrido *et al.*, 2009a).

Fungal pathogens cause diseases in strawberry using different reproductive structures, including asexual and sexual life cycles. Morphological characterization of the structures developed by fungal pathogens has been the basis used by researchers for identifying organisms to the genus/species level, and for classifying these pathogens into families, orders and classes. This classical method has been used for the detection and identification of fungal pathogens, including visual interpretation of plant symptoms, characterization of fungal structures and biochemical/chemical analyses. These studies have contributed much information, increasing our biological knowledge of these fungal species, but they have many limitations pertaining to accuracy and reliability in the detection/diagnosis. They are often time-consuming and laborious methods, and the organisms themselves must be capable of being cultured in the laboratory. This fact is a significant handicap, because less than 1% of the microorganisms present in an environmental sample can be cultured (Lievens *et al.*, 2005). Such classical analyses also require experienced and skilled laboratory staff, who need to have extensive knowledge of taxonomy (McCartney *et al.*, 2003).

In the past two decades, new methodologies based on molecular and immunological strategies have been used including among others: RFLP analyses of mitochondrial DNA (Sreenivasaprasad *et al.*, 1992; Garrido *et al.*, 2008), amplified fragment length polymorphism (AFLP), AT-rich analyses (Freeman *et al.*, 2000), RAPD (Whitelaw-Weckert *et al.*, 2007), genus- and species-specific PCR analysis (Mills *et al.*, 1992; Sreenivasaprasad *et al.*, 1992; Martinez-Culebras *et al.*, 2003; Garrido *et al.*, 2008; Capote *et al.* 2012), real-time PCR studies (Garrido *et al.*, 2009a) and ELISA assays (Hughes *et al.*, 1997). These techniques minimize the time to diagnosis and increase accuracy in identification of the microorganisms (Atkins and Clark, 2004; Capote *et al.* 2012).

10.3.3. PCR alternatives applied to fungal diagnosis

PCR is the most important and sensitive technique presently available for the detection of plant pathogens. Advances in PCR technology have opened up alternative approaches to the detection and identification of strawberry fungal pathogens, even on-site alternatives under field conditions, away from the laboratory, are now available, providing results in a short time (Capote *et al.* 2012). The evolution of PCR towards real-time PCR allows faster and more accurate detection and quantification of plant pathogens in an automated reaction. The advantages of PCR techniques are the high sensitivity, high specificity and reliability. Moreover, it is not necessary to isolate the pathogen from the infected material, reducing the diagnosis time from weeks to hours, and allowing the detection and identification of non-culturable pathogens. This characteristic has been especially useful in the analysis of symptomless plants. However, the frequent presence of PCR inhibitors in plant tissues or soil can reduce considerably the sensitivity of the reactions and may even result in false-negative results.

The optimization of an experimental PCR set-up focuses on three fundamental steps: (i) extraction of total community DNA/RNA from the environmental sample; (ii) selection of a specific target region of the DNA/RNA to identify the fungus; (iii) identification of the presence of the target DNA/RNA region in the sample (Atkins and Clark, 2004; Capote *et al.* 2012). In recent years, many research studies have been published reporting improvements to each of the fundamental steps described above, and working on some of the most serious fungal pathogens of strawberry, such as *B. cinerea* (Suarez *et al.*, 2005), *Colletotrichum acutatum* (Debode *et al.*, 2009; Garrido *et al.*, 2009a), *Colletotrichum gloeosporioides*, *Colletotrichum* spp. (Garrido *et al.*, 2009a), *Fusarium oxysporum* (Lievens *et al.*, 2003), *Verticillium albo-atrum* (Larsen *et al.*, 2007) and *Verticillium dahliae* (Atallah *et al.*, 2007).

The first step in PCR diagnostic analysis consists of collection of the samples from the plant material. This can be symptomatic or non-symptomatic tissue. An efficient and reliable DNA extraction method is then used. Commercial kits are now available for the extraction of fungal DNA from environmental samples such as plant tissue or soil, supplied by several companies (e.g. Dynabeads® DNA Direct from ThermoFisher Scientific; Soil DNA Isolation kit from Mo Bio Laboratories), but these kits can be costly and are not always totally reliable with respect to co-extraction of PCR inhibitors. Garrido *et al.* (2009a) optimized a DNA extraction protocol that can be used for samples of strawberry plant material directly, or from fungal colonies removed from an agar plate. This method uses sample material physically ground using a grinding machine, in the presence of CTAB (cetyl trimethyl ammonium bromide) lysis buffer. Garrido *et al.* (2009a) demonstrated that this method is very reliable for extracting DNA from any strawberry plant material. The sensitivity and accuracy of PCR protocols depends mainly on the instrumentation and technique used (i.e. conventional PCR versus real-time PCR), but in a high proportion of cases, this sensitivity depends on the quality of the total community DNA/RNA extracted from the environmental samples.

One disadvantage of PCR-based methods is the inability to discriminate viable from non-viable fungi or fungal structures. The detection of mRNA by reverse transcription (RT)-PCR is considered an accurate indicator of cell viability (Sheridan *et al.*, 1998). In RT-PCR, the RNA is reverse transcribed using the enzyme reverse transcriptase. The resulting cDNA is then amplified using a PCR-based method. The most frequent application of this technique in phytopathology is the analysis of plant and fungal gene expression during disease development (Yang *et al.*, 2010). Many PCR variations have been developed to improve the sensitivity, specificity, speed and throughput, and to allow the quantification of the fungal pathogen in the plant and the environment.

A second disadvantage is the selection of the target DNA/RNA to amplify. Known conserved genes with enough sequence variation are selected and a PCR diagnostic assay can then be designed to perform phylogenetic analysis. The ribosomal RNA (rRNA) genes are the regions that were used by Debode *et al.* (2009) and Garrido *et al.* (2009a) to develop protocols for identification of members of the genus *Colletotrichum*, and to distinguish between the three species causing anthracnose in strawberry: *Colletotrichum acutatum, Colletotrichum gloeosporioides and Colletotrichum fragariae.* These regions are the most commonly used DNA regions targeted to design primers for PCR-based identification and detection of fungal plant pathogens, because of the highly variable sequences of the internal transcribed spacers, ITS1 and ITS2, which separate the 18S/5.8S and 5.8S/28S rRNA genes, respectively (Garrido *et al.*, 2009a). As recent examples, PCR methods for identification of *Sclerotium rolfsii* (Jeeva *et al.*, 2010) and *Colletotrichum capsici* (Torres-Calzada *et al.*, 2011) were developed based on specific sequences of the ITS region. Another highly variable region of the rRNA genes is the intergenic spacer region, which separates the 28S/18S rRNA genes (Suarez *et al.*, 2005), but these regions have been less used than ITSs. Other regions are becoming more widely studied, for example, sequences of the β-tubulin gene (Suarez *et al.*, 2005; Atallah *et al.*, 2007; Debode *et al.*, 2009), translation elongation factor 1α (Geiser *et al.*, 2004; Knutsen *et al.*, 2004; Kristensen *et al.*, 2005), calmodulin (Mulè *et al.*, 2004), avirulence genes (Lievens *et al.*, 2007, 2008) and mitochondrial genes such as the multicopy *cox I* and *cox II* and their intergenic region (Martin and Tooley, 2003; Seifert *et al.*, 2007; Nguyen and Seifert, 2008).

The third step in effective screening protocol is the detection step. Many variants of conventional PCR such as nested PCR and co-operational amplification (co-PCR), multiplex PCR, PCR-ELISA, RT-PCR, PCR-denaturing gradient gel electrophoresis (PCR-DGGE) and real-time-PCR have been developed to improve the sensitivity, specificity, speed and performance, and to permit quantification of the plant pathogenic fungus. Co-PCR is a method that enhances sensitivity and minimizes contamination risks. In co-PCR, a single reaction containing four primers, one pair internal to the other, enhances the production of the longest fragment by the co-operational action of all amplicons (Olmos *et al.*, 2002). Co-PCR is usually coupled with dot blot hybridization by using a specific probe to enhance the specificity of the detection and provides a sensitivity level similar to nested PCR. Nested PCR consists of two consecutive rounds of amplification in which two external primers amplify a large amplicon, which is then used as a target for a second round of amplification using two internal primers (Porter-Jordan *et al.*, 1990). This method has been widely used for detection and/or further characterization of numerous fungi (Hong *et al.*, 2010; Meng and Wang, 2010; Qin *et al.*, 2011; Wu *et al.*, 2011). In both nested and co-PCR, the use of external primers can be used for generic amplification and the internal primers for further and more specific characterization of the amplified product at the species or strain level.

Real-time PCR (also known as quantitative PCR or qPCR) is currently considered the best method for detection of plant pathogens. There are many advantages of real-time PCR over conventional PCR, including the fact that this system does not require the use of post-PCR processing (e.g. electrophoresis, hybridization), avoiding the risk of carryover contamination and reducing assay labour and material costs. Real-time PCR is more sensitive, more accurate and less time consuming than conventional end-point qPCR (Lievens *et al.*, 2005). It allows monitoring of the reaction during the amplification process by use of a fluorescent signal that increases proportionally with the number of amplicons generated and with the number of targets present in the sample. For the fluorescent signal, different chemistries can be used, such as TaqMan®, SYBR® Green I, molecular beacons, Scorpions® or a new approach, developed by Invitrogen, called LUX. The TaqMan® system consists of a fluorogenic probe specific to the DNA target, which anneals to the target between the PCR primers.

The TaqMan® system tends to be the most sensitive of the methods but the new approach, LUX, is achieving the same level of sensitivity while further simplifying the techniques, as well as reducing the cost. This new system uses two primers, one of which has a hairpin-loop structure with a fluorophore (Suarez *et al.*, 2005). Although this method is replacing TaqMan® probes in real-time PCR technology, with the objective of providing cheaper, reliable methods with the specificity of TaqMan® probes without some of the constraints, to date TaqMan® technology continues to be used in the majority of research projects working with strawberry fungal pathogens. SYBR® Green I is a fluorescence intercalating dye with a high affinity for double-stranded DNA. The overall fluorescent signal from a reaction is proportional to the amount of double-stranded DNA present in the sample. Molecular beacons (Tyagi and Kramer, 1996; Capote *et al.*, 2012) are specific oligonucleotide probes (15–40-mers) flanked by two complementary 5–7-mer arm sequences, with a fluorescent dye covalently attached to the 5′ end and a quencher dye at the 3′ end. When the probe hybridizes to the target sequence, this allows the emission of a fluorescent signal. Molecular beacons have allowed real-time specific quantification of *Fusarium equiseti* (Maciá-Vicente *et al.*, 2009). Scorpions® are bifunctional molecules in which an upstream hairpin probe is covalently linked to a downstream primer sequence (Whitcombe *et al.*, 1999). The hairpin probe contains a fluorophore at the 5′ end and a quencher at the 3′ end. The loop portion of the scorpion probe is complementary to the target sequence.

The level of specificity and sensitivity of real-time PCR-based methods with some of the main strawberry fungal pathogens reported in several publications vouch for the results offered by this technology. Suarez *et al.* (2005) designed three TaqMan® probe/primer sets based on the ribosomal intergenic spacer, the β-tubulin gene and the sequence-characterized amplified region (SCAR) marker of *B. cinerea* published by Rigotti *et al.* (2002).

For the detection of *Colletotrichum* spp. and for monitoring strawberry anthracnose using real-time PCR, new protocols were published by Debode *et al.* (2009) and Garrido *et al.* (2009a). Both groups mainly used the ITS regions to design the sets of probe/primers. Garrido *et al.* (2009a) tested the specificity of all assays using DNA from isolates of six species of *Colletotrichum* and from DNA of another nine fungal species commonly found associated with strawberry material. Additionally, they checked that samples did not contain PCR inhibitors co-extracted during the DNA extractions using one universal pair of primers for the 5.8S rRNA gene by PCR SYBR® Green I amplifications. These assays were highly specific for *Colletotrichum* spp., *Colletotrichum acutatum* and *Colletotrichum gloeosporioides*, and no cross-reactions were observed with either related plant pathogens or healthy strawberry plant material. The real-time PCR assay detected the equivalent of 7.2 conidia per plant inoculated with a serial dilution of *Colletotrichum acutatum* spores, demonstrating the high degree of sensitivity (Garrido *et al.*, 2009a). The sensitivity of the new real-time PCR assays was compared with that of previously published conventional PCR assays; they were confirmed to be 100 times more sensitive than the latter.

Bilodeau *et al.* (2011) developed a new assay for rapid detection and quantification of *V. dahliae*, responsible for strawberry wilt, which can cause significant crop loss. To provide a faster means for estimating pathogen populations, a multiplexed TaqMan® real-time PCR assay based on the rRNA gene intergenic spacer was developed for *V. dahliae*. Variation in copy number of the rRNA gene was also evaluated among isolates by SYBR® Green I real-time PCR amplification of the *V. dahlia*-specific amplicon, compared with amplification of several single-copy genes, and was estimated to range from approximately 24 to 73 copies per haploid genome, which translated into possible differences in results among isolates of around 1.8 cycle thresholds. For the assay, they used an internal control for valuation of inhibition due to the presence of PCR inhibitors in DNA extracted from soil samples. This method provides an accurate and rapid means for quantification of *V. dahliae* over a wide range of inoculum densities. The species-specific marker for *V. dahliae* described can be used for pathogen identification and quantification in plant tissue, but in this case, another internal control must be used that will allow comparison of the quantity of pathogen relative to the plant DNA. Likewise, due to the unique sequence of the internal control and the ease with which it can be modified to be amplified with other primers, the internal control from the *V. dahliae* assay could also be used in other molecular quantification assays. A good correlation was observed in regression analysis (R^2 = 0.96) between real-time PCR results and inoculum densities determined by soil plating in a range of field soils with pathogen densities as low as one to two microsclerotia g^{-1} soil.

Although real-time PCR is currently considered the best standard method for detection of plant pathogens, other assay formats are applied to plant pathogens, such as multiplex PCR, which is based on the use of several PCR primer pairs in the same reaction, allowing the simultaneous and sensitive detection of different DNA targets, reducing both time and cost. This method is useful in plant pathology, as plants are usually infected by more than one pathogen. This needs an accurate and careful design of primers, and optimization of their relative concentrations is required to obtain equilibrate detection of all target fungi. Multiplex PCR has been used for differentiating two pathotypes of *Verticilliun albo-atrum* infecting hop (Radišek *et al.*, 2004). Another multiplexing method that allows simultaneous detection and identification of multiple oomycetes and fungi in complex plant or environmental samples is the use of the ligation detection system using padlock probes. Padlock probes are long oligonucleotide probes containing asymmetric target-complementary regions at their 5′ and 3′ ends and also incorporate a desthiobiotin moiety for specific capture and release, an internal endonuclease IV cleavage site for linearization and a unique sequence identifier for standardized microarray hybridization. Padlock probes have been used for the simultaneous detection of *Phytophthora cactorum*, *Phytophthora nicotianae*, *Pythium ultimum*, *Pythium aphanidermatum*, *Pythium undulatum*, *Rhizoctonia solani*, *Fusarium oxysporum* f. sp. *radicis-lycopersici*, *Fusarium solani*, *Myrothecium roridum*, *Myrothecium verrucaria*, *V. dahliae* and *V. albo-atrum* in samples collected from horticultural water circulation systems in a single assay (van Doorn *et al.*, 2009).

10.3.4. Other molecular techniques used to study fungal pathogens

A combination of PCR and ELISA has been used for the detection of several species of *Phytophthora* and *Pythium*. This method is as sensitive as nested PCR and can easily be automated, and hence is very suitable for routine diagnostic purposes. It is based on the use of forward and reverse primers carrying biotin and an antigenic group at their 5′ ends. PCR-amplified DNA can be immobilized microtiter plates (e.g. on avidin- or streptavidin-coated plates) via the biotin moiety of the forward primer and then can be quantified by an ELISA specific for the antigenic group of the reverse primer.

A PCR-DGGE detection tool based on the amplification of the ITS region has been applied recently to detect multiple species of *Phytophthora* from plant material and environmental samples (Lilja *et al.*, 2008; Capote *et al.*, 2012). In PCR-DGGE, target DNA from plant or environmental samples is first amplified by PCR and is then subjected to denaturing electrophoresis. Sequence variants of particular fragments migrate to different positions in the denaturing gradient gel, allowing very sensitive detection of polymorphisms in DNA sequences. The bands obtained in the gel can be extracted, cloned or reamplified, and sequenced for identification. These techniques are suitable for the identification of novel or unknown organisms, and the most abundant species can readily be detected. However, this method has some problems because it is time consuming, is poorly reproducible and the analysis of complex communities of microorganisms may be difficult due to the large number of bands obtained.

Fingerprinting analyses have also been useful for identifying specific sequences used for the detection of fungi at very low taxonomic level, and even to differentiate strains of the same species with different host ranges, virulence, compatibility groups or mating types. The most prominent of these techniques are RFLP, RAPD, AFLP and the use of microsatellites.

RFLP involves restriction enzyme digestion of the pathogen DNA, followed by separation of the fragments by electrophoresis in agarose or polyacrylamide gels to detect differences in the size of DNA fragments (Capote *et al.*, 2012). These differences are used to distinguish fungal species. This early technique has been progressively supplanted by PCR-RFLP, which combines the amplification of a target region with the further digestion of the PCR products. PCR-RFLP analysis of the ITS region demonstrated the presence of different anastomosis groups within isolates of *Rhizoctonia solani* (Panecoucque and Höfte, 2009; Capote *et al.*, 2012); PCR primers specific to members of the genus *Phytophthora* were used to amplify and further digest the resulting amplicons, yielding a specific restriction pattern of 27 different *Phytophthora* spp. (Drenth *et al.*, 2006).

In some cases, when data obtained using specific sequences from databases are inadequate because a specific primer cannot be designed, screening of arbitrary regions of the genome is often the next step. This strategy consists of an initial RAPD screening and subsequent analyses of the products with the object of developing a SCAR marker. This protocol was used by Larsen *et al.* (2007) to develop SCAR markers for *Verticillium albo-atrum*. Rigotti *et al.* (2002), studying *B. cinerea* isolates, identified SCAR markers for the specific identification of this pathogen in *Fragaria* × *ananassa*. Three years later, Suarez *et al.* (2005) compared their primers designed in the intergenic spacer regions and showed that this new assay was more sensitive than all the previous assays for *B. cinerea*. RAPD assays have been used to analyse the genetic diversity among different species and races of *Fusarium* spp. (Drenth *et al.*, 2006; Lievens *et al.*, 2007). SCAR primers have been used to distinguish among several *Fusarium oxysporum* f. sp. (Lievens *et al.*, 2008). The results obtained from RAPD profiles are easy to interpret. This is a rapid and inexpensive technique, but high-quality DNA is needed, although only in low quantity (Chandra *et al.*, 2011). These methods have a problem with many basidiomycetes and oomycetes that are heterokaryons, diploids or polyploids (Fourie *et al.*, 2011) because RAPD is a dominant marker and so cannot measure the genetic diversity affected by the number of alleles at a locus, or differentiate homozygote and heterozygote individuals.

AFLP technology has the capability to amplify between 50 and 100 fragments at one time and to detect various polymorphisms in different genomic regions simultaneously. This technique consists of the use of restriction enzymes to digest total genomic DNA, followed by ligation of restriction half-site specific adaptors to all restriction fragments. A selective amplification of these restriction fragments is performed with PCR primers that have within their 3′ end the corresponding adaptor sequence and selective bases. Depending on the primers used and the reaction conditions, amplification of fungal genomes produces genetic polymorphisms specific to the genus, species or strain (Liu *et al.*, 2009; Capote *et al.*, 2012). This technique is reproducible and highly sensitive, but a disadvantage is that it requires a larger quantity of DNA. AFLP has been used to differentiate fungal isolates at several taxonomic levels and to separate non-pathogenic strains of *Fusarium oxysporum* (Stewart *et al.*, 2006). AFLP profiles have also been widely used for the phylogenetic analysis of *Fusarium oxysporum* complexes (Baayen *et al.*, 2000; Groenewald *et al.*, 2006; Fourie *et al.*, 2011; Capote *et al.*, 2012).

Microsatellites, also known as simple sequence repeats or short tandem repeats, can differ in repeat number among individuals, and their distribution in the genome is almost random. They have been used for studies of the genetic diversity of plant pathogenic fungi within species, for example in *Sclerotinia sclerotiorum* (Winton and Hansen, 2001). Recently, they have also been utilized for the development of a system for detection and differentiation of phytopathogenic fungi using DNA hybridization technology. Currently, this is one of the most suitable techniques for detecting and quantifying multiple pathogens present in a sample (plant, soil or water) in a single assay. A DNA array is a collection of species-specific oligonucleotides or cDNAs (known as probes) immobilized on a solid support that is subjected to hybridization with a labelled target DNA. This technology has been applied to the detection of oomycete plant pathogens using specific oligonucleotides designed on the ITS region (Anderson *et al.*, 2006; Izzo and Mazzola, 2009). Lievens *et al.* (2005) developed a DNA array for the specific detection and identification, within 24 h, of the strawberry fungal pathogen *Fusarium oxysporium* and the quarantine pathogens *V. albo-atrum* and *V. dahliae*. This assay has proved to be highly sensitive, detecting 2.5 pg of DNA for *V. dahliae*, 0.35 pg of DNA for *V. albo-atrum* and 0.5 pg of DNA for *F. oxysporium*. Using a *cox I* high-density oligonucleotide microarray, Chen *et al.* (2009) could identify *Penicillium* spp. Moreover, Lievens *et al.* (2007) could detect and differentiate *F. oxysporum f. sp. cucumerinum and F. oxysporum f. sp. radicis-cucumerinum* pathogens using a DNA array containing genus-, species- and f. sp.-specific oligonucleotides. This array-based detection procedure for plant pathogens has been shown to be relatively cost-effective, and may lead to a comprehensive pathogen assessment method for detecting and quantifying all known pathogens (fungi, bacteria and nematodes, as well as viruses) (Capote *et al.*, 2012).

In 2011, Huang *et al.* (2011) demonstrated the usefulness of the heteroduplex mobility assay for phylogenetic analysis and for the discrimination of closely related *Colletotrichum* spp. This assay is a sequence-dependent method that has been widely used to determine sequence divergence and phylogenetic relationships of closely related organisms including fungi, viruses, phytoplasma, bacteria and protozoan parasites. It was demonstrated to detect single-base differences among sequences. For this reason, heteroduplex DNA patterns (HPs) using ITS regions were used as a tool to simplify the methodology for *Colletotrichum* spp. identification, as the unique HPs observed resemble a 'barcode' (White *et al.*, 1990). In this study, 29 monoconidial strains of *Colletotrichum* spp., including 18 *Colletotrichum gloeosporioides*, two *Colletotrichum capsici*, two *Colletotrichum graminicola*, one *Colletotrichum acutatum*, one *Colletotrichum musae*, one *Colletotrichum dematium*, one *Colletotrichum lindemuthianum* and three other unidentified species of *Colletotrichum*, were used. The fragments for HP analysis were prepared by PCR amplification of the ITS regions from 29 *Colletotrichum* strains using primers ITS1 and ITS4 (White *et al.*, 1990). The fragments amplified from six *Colletotrichum* strains, named Cg1, Cg2, Cg16, Cgr1, Cl1 and Cm1, were used as references. The barcode-like HPs obtained by cross-pairing HP analyses of these six reference strains appeared to create a model system for species/strain identification. The analysis required low-cost electrophoresis equipment and could discriminate single-based differences in sequences. The unique HPs obtained indicated a possibility of using this method as a barcode-like DNA system for discriminating economically important anthracnose fungi and possibly for the detection of newly evolved strains. HP analysis using cross-pairing of ITS fragments from taxonomically defined strains was suggested as a cost-effective and reliable method for discriminating closely related species of *Colletotrichum*.

10.3.5. Web-based decision support systems

Identificator is a new tool proposed by Pertot *et al.* (2012), which is based on the use of the web for visual identification of plant diseases, and, although it can be used in any crop, initially it was developed using strawberry as the crop type. The system is based on a multiaccess key of identification and specifically on the selection of pictures by the user, and can be used remotely from a desktop as well as from a smart phone or personal digital assistant. The system was developed following a simple approach: visual identification where images and/or short descriptions are used to uniquely identify diseases where possible and to suggest refining the visual identification process in cases of ambiguous identification. The identification procedure follows a four-step process: (i) the user is requested to identify the infected parts of the plant; (ii) the user is presented with a set of images and descriptions of possible symptoms that may appear on these plant parts; (iii) the user selects all the images/descriptions that match the plant at hand; (iv) the system presents all pictures of the disease (even ones that the user did not select, possibly on other parts of the plant), together with a short description of the disease. Sometimes, a laboratory analysis is required, which may be indicated at this step by the system. There are two aspects that make Identificator unique: (i) its simplicity and (ii) its flexibility, as the system can be adapted to any crop in any language, because the information on diseases and their characteristics can be edited by the user directly into the system tables. In fact, the internet opens up new opportunities for knowledge dissemination in the form of web-based expert/ decision support systems.

Scientific journals publish research from research groups across the world. This data is compiled by several associations or agencies for the scientific community, who are producing very useful databases. We will discuss three of these: the Centre for Agriculture and Biosciences International (CABI), Rothamsted Research, and the Broad Institute of Massachusetts Institute of Technology (MIT) and Harvard, for their contribution to the scientific and crop producer community.

CABI (http://www.cabi.org) is a not-for-profit international organization that has a special relevance in the agriculture community because of its aim of improving people's lives by providing information and applying scientific expertise to solve problems. Knowledge transfer from the laboratory to the field is the mission of this organization, using scientific publishing, development projects, and research and microbial services.

Rothamsted Research (http://www.rothamsted.ac.uk/index.php) is the longest running agricultural research station in the world, providing science advances and innovation since its foundation in 1843. This station was founded by the Biotechnology and Biological Sciences Research Council (BBSRC; http://www.bbsrc.ac.uk/home/home.aspx), and their mission is to deliver knowledge and new practices to increase crop productivity and quality and to develop environmentally sustainable solutions for food production and bioenergy. Rothamsted Research manages a very useful database, the Pathogen–Host Interaction base (PHI-base). PHI-base is an open access internet resource that provides information on pathogenicity, virulence and effector genes from different pathogens. This information is only added to the database when it is clearly demonstrated that the research group has experimentally tested the role of each gene.

The third institution is the Broad Institute of MIT and Harvard. This is a younger institution, founded less than 10 years ago, and supports many research projects and compiles useful information that includes fungal pathogens of agriculture and crop relevance. The information available in this database will be described in more detail below.

The last updated version of PHI-base was released in February 2013 and includes a total of 2421 genes from 107 pathogens. Nine species of strawberry fungal pathogens are included in the updated list version: *Alternaria alternata*, *B. cinerea*, *Cladosporium fulvum*, *Colletotrichum acutatum*, *Colletotrichum gloeosporioides*, *Fusarium oxysporum*, *Phytophthora cactorum*, *Sclerotinia sclerotiorum* and *V. dahliae*. Between them, a total of 143 genes are described in the databases (Table 10.2), with *B. cinerea* and *Fusarium oxysporum* having the highest number of characterized genes, 66 and 38, respectively. This database is updated with the information that research groups

Table 10.2. List of studied genes from strawberry fungal pathogens included in the Pathogen–Host Interaction database (PHI-base) 2013.

Pathogen species	Gene_name	Phenotype of mutant	PHI_ID	Database	Accession no.	Reference
Alternaria alternata	*AKT1*	Loss of pathogenicity	PHI:133	EMBL	BAA36588	Tanaka *et al.* (1999)
	AKT2	Loss of pathogenicity	PHI:134	EMBL	BAA36589	Tanaka *et al.* (1999)
	AMT	Loss of pathogenicity	PHI:160	EMBL	AAF01762	Johnson *et al.* (2000)
	AaFUS3	Loss of pathogenicity	PHI:2303	EMBL	GQ414506	Lin *et al.* (2009)
	aapk1	Reduced virulence	PHI:2317	Uniprot	B0LW64	–
	RLAP1	Loss of pathogenicity	PHI:2320	Uniprot	C0LD25	–
	YAP1	Loss of pathogenicity	PHI:2373	Uniprot	B9V258	Lin *et al.* (2009)
	AaGa1	Unaffected pathogenicity	PHI:2414	Uniprot	D2X3B5	Wang *et al.* (2010)
	ACTT2	Reduced virulence	PHI:2431	Uniprot	C9K1M7	Miyamoto *et al.* (2008)
	AaNoxA	Reduced virulence	PHI:2448	Uniprot	H9AWW9	Yang *et al.* (2012)
	AFT1	Loss of pathogenicity	PHI:508	EMBL	BAB69076	Hatta *et al.* (2002)
	AFT3	Loss of pathogenicity	PHI:509	EMBL	BAB69078	Hatta *et al.* (2002)
	AaSLT2	Reduced virulence	PHI:2318	Uniprot	D3J127	Yago *et al.* (2011)
Botrytis cinerea	*bcnoxA*	Reduced virulence	PHI:1023	EMBL	CAP12516	Segmüller *et al.* (2008)
	bcnoxB	Reduced virulence	PHI:1024	EMBL	CAP12517	Segmüller *et al.* (2008)
	bcnoxR	Reduced virulence	PHI:1025	EMBL	CAP12326	Segmüller *et al.* (2008)
	bcplc1	Reduced virulence	PHI:1026	EMBL	AAB39564	Schumacher *et al.* (2008)
	bcpg1	Reduced virulence	PHI:1027	EMBL	AAC64374	–
	bcpme1	Reduced virulence	PHI:1028	EMBL	CAC29255	Valette-Collet *et al.* (2003)
	bcpme2	Unaffected pathogenicity	PHI:1029	EMBL	CAD21438	–
	BCPGA1	Reduced virulence	PHI:103	EMBL	AAC64374	ten Have *et al.* (1998)
	bcatrA	Unaffected pathogenicity	PHI:1030	EMBL	CAA93142	Del Sorbo *et al.* (2008)
	bcSAK1	Reduced virulence	PHI:1031	EMBL	CAJ85638	Segmüller *et al.* (2007)
	Bmp3	Reduced virulence	PHI:1032	EMBL	ABJ51957	–
	BcatrB	Sensitive to chemical	PHI:1160	EMBL	CAB52402	Schoonbeek *et al.* (2001)
	BMP1	Loss of pathogenicity	PHI:161	EMBL	AAG23132	Zheng *et al.* (2000)
	BcatrB	Reduced virulence	PHI:202	EMBL	CAB52402	Schoonbeek *et al.* (2001)
	BCG1	Reduced virulence	PHI:203	EMBL	CAC19871	Gronover *et al.* (2001)
	BcBOA2	Reduced virulence	PHI:2289	Uniprot	JQ665433	Dalmais *et al.* (2011)
	BcBOA6	Reduced virulence	PHI:2290	Uniprot	B1GVX7	Dalmais *et al.* (2011)
	BcSpl1	Reduced virulence	PHI:2291	Uniprot	A6RNZ1	Frías *et al.* (2011)
	BcFRP1	Unaffected pathogenicity	PHI:2295	Uniprot	A6SB12	Jonkers *et al.* (2011)
	BcCdc42	Reduced virulence	PHI:2298	Uniprot	A6SSN8	Kokkelink *et al.* (2011)

BCFHG1	Unaffected pathogenicity	PHI:2304	EMBL	CAP74387	Turrion-Gomez *et al.* (2010)
BcFKBP12	Increased virulence	PHI:2305	EMBL	ABA25866	Melendez *et al.* (2009)
BcCRZ1	Reduced virulence	PHI:2308	Uniprot	A6RI22	Schumacher *et al.* (2008)
BcatrB	Reduced virulence	PHI:2309	Uniprot	Q9UW03	Stefanato *et al.* (2009)
bos5	Loss of pathogenicity	PHI:2319	Uniprot	A6S644	Yan *et al.* (2010)
BcNma	Reduced virulence	PHI:2334	Uniprot	E2GL14	Finkelshtein *et al.* (2011)
BcPIE3	Reduced virulence	PHI:2335	Genbank	DQ140394	Gioti *et al.* (2008)
PGIP2	Effector[a]	PHI:2347	Uniprot	Q5TIP4	Ferrari *et al.* (2008)
Bcchs3a	Reduced virulence	PHI:2359	Uniprot	Q8TG14	Arbelet *et al.* (2010)
bac	Reduced virulence	PHI:2368	Uniprot	Q9P880	Schumacher *et al.* (2008)
bcpka2	Unaffected pathogenicity	PHI:2369	Uniprot	A6S414	Schumacher *et al.* (2008)
bcpka1	Reduced virulence	PHI:2370	Uniprot	C0H5W3	Schumacher *et al.* (2008)
bcpkaR	Reduced virulence	PHI:2371	Uniprot	C0H5W5	Schumacher *et al.* (2008)
bcras2	Reduced virulence	PHI:2372	Uniprot	A6RUJ6	Schumacher *et al.* (2008)
Bap1	Unaffected pathogenicity	PHI:2374	Uniprot	D2T177	Temme *et al.* (2009)
BAC	Reduced virulence	PHI:240	EMBL	CAB77164	–
BcAtf1	Increased virulence	PHI:2447	Broad	B0510_6765	Temme *et al.* (2012)
Bcchs1	Reduced virulence	PHI:276	EMBL	CAA54909	Soulié *et al.* (2003)
BCP1	Reduced virulence	PHI:277	EMBL	AAQ16572	Viaud *et al.* (2003)
BCPME1	Reduced virulence	PHI:278	EMBL	CAC29255	Valette-Collet *et al.* (2003)
BcPLS1	Loss of pathogenicity	PHI:329	EMBL	CAD43406	Gourgues *et al.* (2004)
BcSOD1	Reduced virulence	PHI:330	EMBL	CAD88591	–
BcBOT1 (*CND5*)	Reduced virulence	PHI:438	EMBL	AAQ16576	Siewers *et al.* (2005)
BTP1	Reduced virulence	PHI:441	EMBL	CAE55153	–
BcPG2	Reduced virulence	PHI:492	EMBL	AAV84614	Kars *et al.* (2005)
FRT1	Unaffected pathogenicity	PHI:538	EMBL	AAU87358	Doehlemann *et al.* (2005)
BCPME2	Unaffected pathogenicity	PHI:540	EMBL	CAD21438	–
LIP1	Unaffected pathogenicity	PHI:541	EMBL	AAU87359	–
BCCAT2	Unaffected pathogenicity	PHI:542	EMBL	AAK77951	–
BCATRD	Unaffected pathogenicity	PHI:543	EMBL	CAC41639	–
BCMFS1	Unaffected pathogenicity	PHI:544	EMBL	AAF64435	Hayashi *et al.* (2002)
BMP1	Loss of pathogenicity	PHI:545	EMBL	AAG23132	Doehlemann *et al.* (2006a)
XYN11A	Reduced virulence	PHI:546	EMBL	AAZ03776	Brito *et al.* (2006)
CEL5A	Unaffected pathogenicity	PHI:547	EMBL	AAT40313	–

Continued

Table 10.2. Continued.

Pathogen species	Gene_name	Phenotype of mutant	PHI_ID	Database	Accession no.	Reference
	BcPIC5	Reduced virulence	PHI:548	EMBL	ABA25866	Gioti *et al.* (2006)
	BCCHS3a	Reduced virulence	PHI:549	EMBL	AAM14606	Soulié *et al.* (2006)
	BOS1	Reduced virulence	PHI:550	EMBL	AAL37947	Viaud *et al.* (2006)
	BcLCC2	Unaffected pathogenicity	PHI:552	EMBL	AAK77953	Schouten *et al.* (2002)
	BCG3	Reduced virulence	PHI:574	EMBL	BAD93277	Doehlemann *et al.* (2006a)
	TPS1	Unaffected pathogenicity	PHI:650	EMBL	ABG25558	Doehlemann *et al.* (2006b)
	TRE1	Unaffected pathogenicity	PHI:651	EMBL	ABG25559	Doehlemann *et al.* (2006b)
	CUTA	Unaffected pathogenicity	PHI:69	EMBL	CAA93255	–
	os-1/bos1/barA	Resistant to chemical	PHI:837	EMBL	AAL30826	Cui *et al.* (2002)
	BcCCC2	Loss of pathogenicity	PHI:2483	Uniprot	A6SEF3	Saitoh *et al.* (2010)
	Ste11	Loss of pathogenicity	PHI:2484	Uniprot	B6VCT9	Schamber *et al.* (2010)
	Ste7	Loss of pathogenicity	PHI:2485	Uniprot	A6RT73	Schamber *et al.* (2010)
	Ste50	Loss of pathogenicity	PHI:2486	Uniprot	A6S3X7	Schamber *et al.* (2010)
	ste12	Reduced virulence	PHI:2487	Uniprot	B6VCT8	Schamber *et al.* (2010)
Cladosporium fulvum	*AVR4*	Effector[a]	PHI:18	EMBL	CAA69643	Westerink *et al.* (2004)
	AOX1	Reduced virulence	PHI:199	EMBL	AAF82788	Segers *et al.* (2001)
	Avr2	Effector[a]	PHI:2344	Uniprot	Q8NID8	van Esse *et al.* (2008)
	AVR2	Effector[a]	PHI:472	EMBL	CAD16675	Rooney *et al.* (2005)
	AVR4E	Effector[a]	PHI:529	EMBL	AAT28197	Takken *et al.* (1999)
	AVR9	Effector	PHI:7	EMBL	CAA42824	Snoeijers *et al.* (2003)
	ECP1	Reduced virulence	PHI:70	EMBL	CAA78400	Lauge *et al.* (1997)
	ECP2	Effector	PHI:71	EMBL	CAA78401	Lauge *et al.* (1997)
Colletotrichum acutatum	*KLAP1*	Loss of pathogenicity	PHI:481	EMBL	AAX14039	Chen *et al.* (2005)
Colletotrichum gloeosporioides	*CgDN3*	Loss of pathogenicity	PHI:164	EMBL	AAB92221	Stephenson *et al.* (2000)
	CgMEK1 (*EMK1*)	Loss of pathogenicity	PHI:165	EMBL	AAD55385	Kim *et al.* (2000)
	CHIP2	Unaffected pathogenicity	PHI:166	EMBL	AAD53262	Kim *et al.* (2000)
	CHIP3	Unaffected pathogenicity	PHI:167	EMBL	AAF00024	Kim *et al.* (2000)
	PELB	Reduced virulence	PHI:222	EMBL	AAD09857	Yacoby *et al.* (2001)
	CHIP6	Reduced virulence	PHI:243	EMBL	AAD00894	Kim *et al.* (2002)
	CAP20	Loss of pathogenicity	PHI:27	EMBL	AAA77678	Hwang *et al.* (1995)

	CgDN24	Unaffected pathogenicity	PHI:521	EMBL	AAB92223	Stephenson *et al.* (2005)
Fusarium oxysporum	*XlnR*	Unaffected pathogenicity	PHI:1020	Uniprot	A8QJI7	Calero-Nieto *et al.* (2007)
	CTF1	Unaffected pathogenicity	PHI:1021	EMBL	ABR12478	–
	XLNR	Unaffected pathogenicity	PHI:1022	EMBL	ABN41464	–
	FOXG_00076	Unaffected pathogenicity	PHI:1101	Uniprot	J9MB32	López-Berges *et al.* (2009)
	FOXG_02277	Unaffected pathogenicity	PHI:1103	Uniprot	J9MHC1	López-Berges *et al.* (2009)
	FOXG_08661	Unaffected pathogenicity	PHI:1106	Uniprot	J9N0G7	López-Berges *et al.* (2009)
	FOXG_00016	Reduced virulence	PHI:1107	Uniprot	J9MAX2	López-Berges *et al.* (2009)
	PG1	Unaffected pathogenicity	PHI:114	EMBL	AAC05015	–
	PGX1	Unaffected pathogenicity	PHI:181	EMBL	AAK81847	–
	ARG1	Reduced virulence	PHI:200	EMBL	BAB40769	Namiki *et al.* (2001)
	FMK1	Loss of pathogenicity	PHI:215	EMBL	AAG01162	Di Pietro *et al.* (2001)
	Fhk1	Reduced virulence	PHI:2252	Genbank	GQ871928	Rispail and Di Pietro (2010)
	Fmk1	Reduced virulence	PHI:2253	Genbank	AF286533	Rispail and Di Pietro (2010)
	Putative zinc finger transcription factor	Unaffected pathogenicity	PHI:2277	Uniprot	J9MB32	López-Berges *et al.* (2009)
	MeaB	Unaffected pathogenicity	PHI:2278	Uniprot	J9MHC1	López-Berges *et al.* (2009)
	Conserved hypothetical protein	Unaffected pathogenicity	PHI:2279	Uniprot	J9N0G7	López-Berges *et al.* (2009)
	Velvet protein family	Reduced virulence	PHI:2280	Uniprot	J9MAX2	López-Berges *et al.* (2009)
	MeaB	Increased virulence	PHI:2282	Uniprot	J9MHC1	López-Berges *et al.* (2010)
	AreA	Reduced virulence	PHI:2283	Uniprot	J9MJU9	López-Berges *et al.* (2010)
	Msb2	Loss of pathogenicity	PHI:2284	Uniprot	J9N257	Pérez-Nadales and Di Pietro (2011)
	Fmk1	Loss of pathogenicity	PHI:2285	Genbank	AF286533	Pérez-Nadales and Di Pietro (2011)
	ftf1	Unaffected pathogenicity	PHI:2307	Uniprot	Q2V9X3	Ramos *et al.* (2007)
	FGA1	Reduced virulence	PHI:251	EMBL	BAB69488	Jain *et al.* (2002)
	FOW1	Reduced virulence	PHI:254	EMBL	BAB85760	Inoue *et al.* (2002)
	CHSV	Increased virulence	PHI:285	EMBL	AAO49384	Ortoneda *et al.* (2004)
	FGB1	Reduced virulence	PHI:300	EMBL	AAO91808	Jain *et al.* (2003)
	foSNF1	Reduced virulence	PHI:301	EMBL	AAN32715	Ospina-Giraldo *et al.* (2003)
	PacC	Increased virulence	PHI:315	EMBL	AAM95700	Caracuel *et al.* (2003)
	CHS2	Reduced virulence	PHI:336	EMBL	AAT77182	Martín-Udíroz *et al.* (2004)
	CHS7	Reduced virulence	PHI:337	EMBL	AAT77184	Martín-Udíroz *et al.* (2004)
	REN1	Unaffected pathogenicity	PHI:375	EMBL	BAC55015	Ohara *et al.* (2004)

Continued

Table 10.2. Continued.

Pathogen species	Gene_name	Phenotype of mutant	PHI_ID	Database	Accession no.	Reference
	SIX1	Effector[a]	PHI:379	EMBL	CAE55866	Rep *et al.* (2004)
	FGA2	Loss of pathogenicity	PHI:395	EMBL	BAD44729	Jain *et al.* (2005)
	FRP1	Loss of pathogenicity	PHI:490	EMBL	AAT85969	Duyvesteijn *et al.* (2005)
	GAS1	Reduced virulence	PHI:522	EMBL	AAX78216	Caracuel *et al.* (2005)
	XYL3	Unaffected pathogenicity	PHI:569	EMBL	AAC06239	Gómez-Gómez *et al.* (2002)
	XYL4	Unaffected pathogenicity	PHI:570	EMBL	AAK27975	Gómez-Gómez *et al.* (2002)
	FOW2	Loss of pathogenicity	PHI:734	EMBL	BAE98264	Imazaki *et al.* (2007)
Phytophthora cactorum	*PcF*	Effector[a]	PHI:665	EMBL	AAK63068	Orsomando *et al.* (2001)
Sclerotinia sclerotiorum	*Sssod1*	Reduced virulence	PHI:2314	Uniprot	A7E5X4	
	cna1	Reduced virulence	PHI:2361	Uniprot	Q0H744	Harel *et al.* (2006)
	Ss-ggt1	Reduced virulence	PHI:2411	Uniprot	A7F946	Li *et al.* (2012)
	PAC1	Reduced virulence	PHI:314	EMBL	AAF93178	Rollins (2003)
Verticillium dahliae	*VGB*	Reduced virulence	PHI:2310	Genbank	JQ665433	Tzima *et al.* (2012)
	Ave1	Effector[a]	PHI:2331	Uniprot	H9DUR1	de Jonge *et al.* (2012)
	VMK1	Loss of pathogenicity	PHI:483	EMBL	AAW71477	Rauyaree *et al.* (2005)

[a]Plant avirulence determinant.

provide to the organization after an experiment has been validated and the role of the gene has been demonstrated sufficiently. The organization database includes not only pathogenicity factors, but also 'loss of pathogenicity', 'reduced virulence', 'increased virulence' or 'unaffected pathogenicity'.

A total of 25 new genes have been included in Table 10.2 since 2010. There have been 13 new *B. cinerea* genes. Seven genes (*BcBOA2*, *BcBOA6*, *BcSpl1*, *BcCdc42*, *BcNma*, *Bcchs3a* and *Ste12*) reduce virulence in *B. cinerea*. Four new pathogenicity factors, *BcCCC2*, *Ste11*, *Ste7* and *Ste50* have been described by Saitoh *et al.* (2010) and Schamber *et al.* (2010). More recently, one gene catalogued as having 'increased virulence' has been described by Temme *et al.* (2012) in this fungus.

In the case of *B. cinerea*, the genomes of two strains have been sequenced and published, and are available in the Broad Institute database. *B. cinerea* has a total genome size of approximately 42 Mb, and more than 16,000 of its genes have been sequenced (Table 10.3 and Fig. 10.1). Currently, the genome is being annotated by different groups, due to the genomics information obtained from molecular studies of the fungus, including knock-out mutants, pathogenicity tests, metabolic studies and morphological characterization of mutants.

There were a total of six new entries for *Fusarium oxysporum* in the PHI-base during 2010–2013. These included two genes that reduced the virulence, *Fhk1* (Rispail and Di Pietro, 2010) and *AreA* (López-Berges *et al.*, 2009); two new pathogenicity factors, *Msb2* and *Fmk1* (Pérez-Nadales and Di Pietro, 2011); and one gene that increases the virulence level, *MeaB* (López-Berges *et al.*, 2010). This fungus is very important because of a very wide host range of crops and is included as quarantine pathogen in Europe for the EPPO agency. The sequence of the total genome is also available in the Broad Institute database, with a 61 Mb genome size and 17,708 sequenced genes, together with the sequences of two other species: *Fusarium verticillioides*, with a total genome size of 41 Mb and 14,169 sequenced genes; and *Fusarium graminearum*, with a smaller genome size of 36 Mb and 13,321 sequenced genes.

Another six new genes were included in the PHI-base during the 2010–2013. Yang *et al.* (2010) and Yago *et al.* (2011) described the role of the *AaNoxA* and *AsSLT2* genes, respectively, catalogued as 'reduced virulence' in *Alternaria alternata*. This fungus can cause disease in other important crops such as pear, apple, calamondin and tobacco, among others. More than 60% of its genes, included in the PHI-base, are annotated as 'loss of pathogenicity' (Table 10.2). *Sclerotinia sclerotiorum* and *V. dahliae* must also be mentioned because, during 2012, three genes were included in PHI-base, and their genomes have also been completely sequenced and are available in the Broad Institute genome project database. They have a total genome size of 38 Mb (14,503 genes) for *Sclerotinia sclerotiorum* and 33 Mb (10,535 genes) for *V. dahliae* (Table 10.2 and Fig. 10.1).

10.3.6. Proteomics advances in strawberry fungal pathogens

Proteomics studies have made considerable progress in recent years, accumulating relevant information about the biology of fungal pathogens. Proteomic techniques have been useful for obtaining extensive and decisive biological information about the life cycles of many organisms (Garrido *et al.*, 2010). However, these techniques are dependent on advances in the genetic information accumulated in databases, as the result of protein sequencing must be compared with genome databases. In other cases, proteins can be identified by *de novo* peptide sequencing and sequence alignment (Garrido *et al.*, 2010). Another important limitation of the techniques is the need to have an effective protein extraction method to obtain a good and representative biological sample, and the majority of proteins spots then need to be identified by two-dimensional electrophoresis for a complete proteomics study. All these limitations are progressively being overcome; we strongly recommended the specific reviews by Garrido

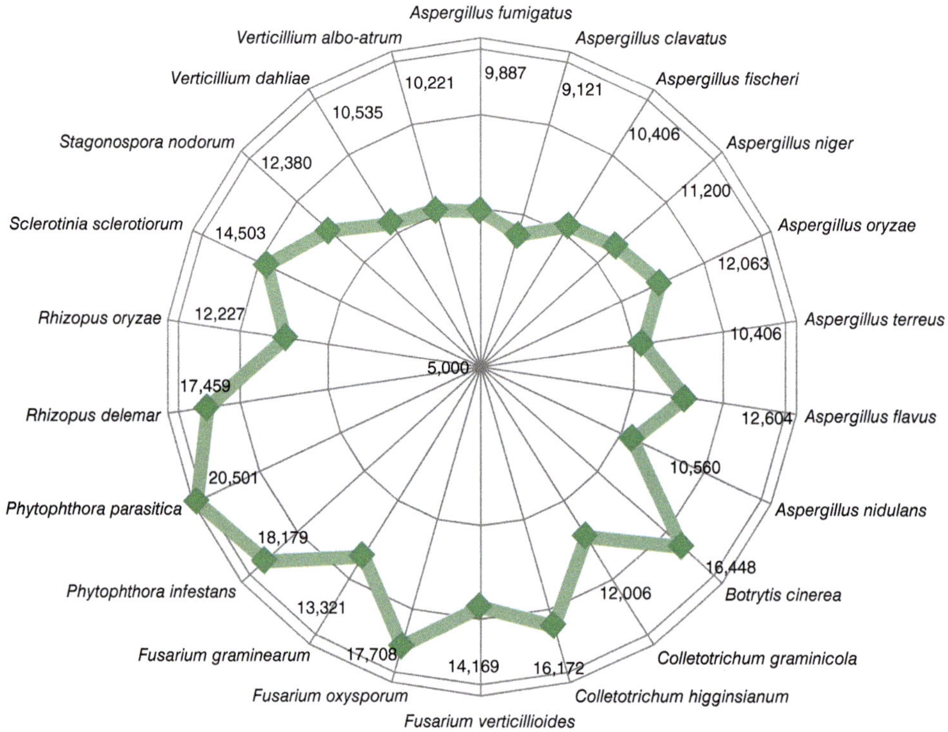

Fig. 10.1. Number of genes sequenced from each genome sequencing project of fungal strawberry genera available in the Broad Institute of MIT and Harvard database.

Table 10.3. Genome sequence projects of fungal strawberry genera available in the Broad Institute of MIT and Harvard database.

Genus	Species	Strain sequenced	Genome size (Mb)
Aspergillus	*fumigatus*	AF293	29.4
	clavatus	NRRL1	
	fischeri	NRRL 181	
	niger	ATCC 1015	
	oryzae	RIB40/ATCC 42149	
	terreus	NIH 264	
	flavus	NRRL3357	
	nidulans	FGSC A4	
Botrytis	*cinerea*	B05.10/T4	42.66/41.61
Colletotrichum	*graminicola*	M1.001	51.60
	higginsianum	IMI349063	49.08
Fusarium	*verticillioides*	7600	41.78
	oxysporum	4287	61.36
	graminearum	PH-1	36.45
Phytophthora	*infestans*	T30-4	228.54
	parasitica	INRA-310 (V2)	82.39
Rhizopus	*delemar*	RA 99-880	46.09
	oryzae	1006PhL	36.35
Sclerotinia	*sclerotiorum*	ATCC18683	38.33
Stagonospora	*nodorum*	SN15	37.24
Verticillium	*dahliae*	VdLs.17	33.83
	albo-atrum	VaMs.102	32.83

et al. (2010) and González-Fernández *et al.* (2010) for more information regarding the experimental design of proteomics studies.

The biocontrol agent *Trichoderma* spp. is one of the microorganisms, in relation to strawberry crops, about which much biological information has accumulated using proteomics techniques. This fungus provides an abundance of biotechnologically valuable proteins and secondary metabolites. The *Trichoderma harzianum* and *Trichoderma atroviride* proteomes were first described by Grinyer *et al.* (2004) and subsequently by Woo *et al.* (2006). They concluded that several resistance genes are upregulated or activated, which allows the plant to recognize them (Woo *et al.*, 2006). Proteomics studies have demonstrated the presence of large numbers of effective membrane pumps in *Trichoderma* spp., allowing their adaptation to environmental and biochemical stresses (Woo *et al.*, 2006). *Trichoderma* spp. establishes a strong molecular-based communication/interaction with infected plants. The fungus induces resistance mechanisms. Three types of elicitor are produced by the fungus: (i) proteins, such as proteases, xylanases, chitinases and glucanases; (ii) Avr-like proteins; and (iii) cell wall degradation enzymes (Woo *et al.*, 2006).

In relation to phytopathogenic fungi, *B. cinerea* is one of the most studied fungi from the proteomics point of view. Several research groups have determined much biological information about this pathogen (Fernández-Acero *et al.*, 2006, 2007; Shah *et al.*, 2009a,b; Fernández and Novo, 2010). Fernández-Acero *et al.* (2006) reported the first proteome analysis of *B. cinerea*, comparing two strains with different levels of virulence (Fernández-Acero *et al.*, 2007). Malate dehydrogenase proteins were identified and found to be overexpressed in the more virulent strain (*B. cinerea* CECT-2100). The role of malate dehydrogenase as a pathogenicity factor had been suggested previously, as this enzyme catalyses the reversible conversion of oxalacetate and malate; oxalacetate has been shown to be an oxalic acid precursor, which has been described as a pathogenicity factor in *B. cinerea* (Lyon *et al.*, 2007).

In a subsequent study, Fernández-Acero *et al.* (2009) analysed the proteins produced by *B. cinerea* during cellulose degradation. A total of 267 spots were identified, and proteins were found that could play a significant role in plant infection, specifically *B. cinerea* peptidyl-prolyl *cis*-trans isomerase and glyceraldehyde 3-phosphate dehydrogenase (Fernández-Acero *et al.*, 2009). Shah *et al.* (2009a) published an approach to the secretome of *B. cinerea* using a high-throughput liquid chromatography tandem mass spectrometry (MS/MS) approach. *B. cinerea*-secreted proteins were identified as transport proteins, carbohydrate metabolism proteins, peptidases and oxidation/reduction proteins used by *B. cinerea* for plant infection and colonization (Shah *et al.*, 2009a). In 2010, our group developed a proteomics approach for studying the secretome of *B. cinerea* using two-dimensional electrophoresis combined with MS/MS analysis. More than 70 protein spots were identified, and although many of these have not yet been functionally assigned, their regulation under the conditions assayed demonstrated their functional significance in the mechanisms of infection used by *B. cinerea*, which will advance our functional understanding of *Botrytis* pathogenesis (Fernández-Acero *et al.*, 2010). For further details of the specific case of *B. cinerea*, we recommend the review by González-Fernández *et al.* (2010).

A proteomic study of strawberry was published by Fang *et al.* (2012). This work tried to elucidate the defence mechanisms used by strawberry against anthracnose causal agents. The authors inoculated strawberry leaves with *Colletotrichum fragariae* and characterized at different times the changes in protein synthesis by the plant. They showed a dynamic overview of the altered protein expression involved in metabolism, photosynthesis, energy production, antioxidant activity and chaperone activity. The results demonstrated that there exists a protein network consisting of several functional components in the plants, including a dynamic balance between reactive oxygen species production and scavenging, accelerated biosynthesis of heat-shock proteins, impaired glycolysis and enhanced cell wall lignin formation, in response to infection by *Colletotrichum fragariae*. The authors indicated the possible existence of a systematic *Colletotrichum fragariae* resistance mechanism in strawberry leaves. An understanding of how plants and pathogens recognize each other and establish a successful or unsuccessful relationship is crucial for exploring the defence mechanisms of plants (Fang *et al.*, 2012).

10.4. Conclusions

More than 50 genera of fungal pathogens are known to cause diseases in *Fragaria* × *ananassa*. This crop is one of the most commercially important fruit crops in the world and is continually subjected to the risk of fungal disease, with subsequent economic losses for growers and producer countries. In recent years, many studies have focused on the development of new molecular methods, leading to a better understanding of fungal pathogens and improving the strategies available for disease control. This chapter has discussed the latest results, based on genomics, transcriptomics and proteomics approaches, for the study of strawberry fungal pathogens. We have described the most important diseases of strawberry crops caused by phytopathogenic fungi, and have discussed the most recent advances in the development and optimization of molecular methods and better tools for detection, identification, control and understanding of these pathogens, and their application in strawberry production fields.

Acknowledgements

We would like to thank the Spanish Government, the Ministerio de Ciencia y Tecnología, the Junta de Andalucía (Regional Government), and the EU (Fondos Europeos de Desarrollo Regional) for financing of most of the work of our group mentioned in this chapter: Spanish DGICYT AGL2006-13401-C02-02/AGR and AGL2009-13359-C02-02; AGL2012-39798-C02 y AGL2015-65684-C2-2R; Junta de Andalucía, PO7-FQM-02689; and by the University of Cádiz, project UCA 18DGUE1102 V.E. González-Rodríguez was financed by grant FPU of the Ministerio de Educación, Government of Spain, ref. AP2009-1309.

References

Abad, Z.G., Abad, J.A., Coffey, M.D., Oudemans, P.V., Man in 't Veld, W.A., de Gruyter, H., Cunnington, J. and Louws, F.J. (2008) *Phytophthora bisheria* sp. nov., a new species identified in isolates from the Rosaceous raspberry, rose and strawberry in three continents. *Mycologia* 100, 99–110.

Agrios, G.N. (2011) *Plant Pathology*, 5th edn. Academic Press, London.

Anderson, N., Szemes, M., O'Brien, P., De Weerdt, M., Schoen, C., Boender, P. and Bonants, P. (2006) Use of hybridization melting kinetics for detecting *Phytophthora* species using three-dimensional microarrays: demonstration of a novel concept for the differentiation of detection targets. *Mycological Research* 110, 664–671.

Arbelet, D., Malfatti, P., Simond-Côte, E., Fontaine, T., Desquilbet, L., Expert, D., Kunz, C., Soulié, M.C. (2010) Disruption of the *Bcchs3a* chitin synthase gene in *Botrytis cinerea* is responsible for altered adhesion and overstimulation of host plant immunity. *Molecular Plant–Microbe Interactions* 23, 1324–1334.

Atallah, Z.K., Bae, J., Jansky, S.H., Rouse, D.I. and Stevenson, W.R. (2007) Multiplex real-time quantitative PCR to detect and quantify *Verticillium dahliae* colonization in potato lines that differ in response to *Verticillium* wilt. *Phytopathology* 97, 865–872.

Atkins, S.D. and Clark, I.M. (2004) Fungal molecular diagnostics: a mini review. *Journal of Applied Genetics* 45, 3–15.

Avis, T.J., Rioux, D., Simard, M., Michaud, M. and Tweddell, R.J. (2009) Ultrastructural alterations in *Fusarium sambucinum* and *Heterobasidion annosum* treated with aluminum chloride and sodium metabisulfite. *Phytopathology* 99, 167–175.

Baayen, R.P., O'Donnell, K., Bonants, P.J.M., Cigelnik, E., Kroon, L.P.N.M., Roebroeck, E.J.A. and Waalwijk, C. (2000) Gene genealogies and AFLP analyses in the *Fusarium oxysporum* complex identify monophyletic and nonmonophyletic formae speciales causing wilt and rot disease. *Phytopathology* 90, 891–900.

Bailey, J.A., O'Connell, R.J., Pring, R.J., Nash, C. and Jeger, M.J. (1992) Infection strategies of *Colletotrichum* species. In: Bailey, J.A. and Jeger, M.J. (eds) *Colletotrichum: Biology, Pathology and Control*. CAB International, Wallingford, UK, pp. 88–120.

Bhat, R.G. and Subbarao, K.V. (1999) Host range specificity in *Verticillium dahliae*. *Phytopathology* 89, 1218–1225.

Bilodeau, G.J., Koike, S.T., Uribe, P. and Martin, F.N. (2011) Development of an assay for rapid detection and quantification of *Verticillium dahliae* in soil. *Phytopathology* 102, 331–343.

Böszörményi, E., Ersek, T., Fodor, A., Fodor, A.M., Földes, L.S., Hevesi, M., Hogan, J.S., Katona, Z., Klein, M.G., Kormány, A., Pekár, S., Szentirmai, A., Sztaricskai, F. and Taylor, R.A. (2009) Isolation and activity of *Xenorhabdus* antimicrobial compounds against the plant pathogens *Erwinia amylovora* and *Phytophthora nicotianae*. *Journal of Applied Microbiology* 107, 746–759.

Brito, N., Espino, J.J., González, C. (2006) The endo-β-1,4-xylanase Xyn11A is required for virulence in *Botrytis cinerea*. *Molecular Plant–Microbe Interactions* 19, 25–32.

Calero-Nieto, F., Di Pietro, A., Roncero, M.I. and Hera, C. (2007) Role of the transcriptional activator XlnR of *Fusarium oxysporum* in regulation of xylanase genes and virulence. *Molecular Plant–Microbe Interactions* 20, 977–985.

Capote, N., Pastrana, A.M., Aguado, A., and Sánchez-Torres, P. (2012). Molecular tools for detection of plant pathogenic fungi and fungicide resistance In: Cumagun, C.J. (ed.) *Plant Pathology*. InTech. Available from: http://www.intechopen.com/books/plant-pathology/molecular-tools-for-detection-of-plant-pathogenic-fungi-and-fungicide-resistance.

Caracuel Z., Roncero M.I., Espeso EA., González-Verdejo, C.I., García-Maceira, F.I. and Di Pietro, A. (2003) The pH signalling transcription factor PacC controls virulence in the plant pathogen *Fusarium oxysporum*. *Molecular Microbiology* 48, 765–779.

Caracuel, Z., Martínez-Rocha, A.L., Di Pietro, A., Madrid, M.P. and Roncero, M.I. (2005) *Fusarium oxysporum gas1* encodes a putative β-1,3-glucanosyltransferase required for virulence on tomato plants. *Molecular Plant–Microbe Interactions* 18, 1140–1147.

Chalavi, V., Tabaeizadeh, Z. and Thibodeau, P. (2003) Enhanced resistance to *Verticillium dahliae* in transgenic strawberry plants expressing a *Lycopersicon chilense* chitinase gene. *Journal of the American Society for Horticultural Science* 128, 747–753.

Chandra, N., Wulff, E.G. Udayashankar, A.C., Nandini, B.P., Niranjana, S.R., Mortensen, C.N. and Prakash, H.S. (2011) Prospects of molecular markers in *Fusarium* species diversity. *Applied Microbiology and Biotechnology* 90, 1625–1639.

Chen, H.Q., Cao, L. Dekkers, K.L., Rollins, J.A., Ko, N.J., Timmer, L.W., and Chung, K.R. (2005) A gene with domains related to transcription regulation is required for pathogenicity in *Colletotrichum acutatum* causing Key lime anthracnose. *Molecular Plant Pathology* 6, 513–525.

Chen, W., Seifert, K.A. and Lévesque, C.A. (2009) A high density *COX1* barcode oligonucleotide array for identification and detection of species of *Penicillium* subgenus *Penicillium*. *Molecular Ecology Resources* 9, 114–129.

Chen, X., Klemsdal, S.S. and Brurberg, M.B. (2011) Identification and analysis of *Phytophthora cactorum* genes up-regulated during cyst germination and strawberry infection. *Current Genetics* 57, 297–315.

Chiba, S., Salaipeth, L., Lin, Y.H., Sasaki, A., Kanematsu, S. and Suzuki, N. (2009) A novel bipartite double-stranded RNA mycovirus from the white root rot fungus *Rosellinia necatrix*: molecular and biological characterization, taxonomic considerations, and potential for biological control. *Journal of Virology* 83, 12801–12812.

Chiotta, M.L., Ponsone, M.L., Combina, M., Torres, A.M. and Chulze, S.N. (2009) *Aspergillus* section Nigri species isolated from different wine-grape growing regions in Argentina. *International Journal of Food Microbiology* 136, 137–141.

Choi, Y.J., Shin, H.D., Thines, M. (2009) Two novel *Peronospora* species are associated with recent reports of downy mildew on sages. *Mycological Research* 113, 1340–1350.

Chung, W.H., Chung, W.C., Peng, M.T., Yang, H.R. and Huang, J.W. (2010) Specific detection of benzimidazole resistance in *Colletotrichum gloeosporioides* from fruit crops by PCR-RFLP. *New Biotechnology* 27, 17–24.

Costa, R., Gomes, N.C., Krögerrecklenfort, E., Opelt, K., Berg, G. and Smalla, K. (2007) *Pseudomonas* community structure and antagonistic potential in the rhizosphere: insights gained by combining phylogenetic and functional gene-based analyses. *Environmental Microbiology* 9, 2260–2273.

Cui, W., Beever, R.E., Parkes, S.L., Weeds, P.L., Templeton, M.D. (2002) An osmosensing histidine kinase mediates dicarboximide fungicide resistance in *Botryotinia fuckeliana* (*Botrytis cinerea*). *Fungal Genetics and Biology* 36, 187–198.

Curry, K.J., Abril, M., Avant, J.B. and Smith, B.J. (2002) Strawberry anthracnose: histopathology of *Colletotrichum acutatum* and *C. fragariae*. *Phytopathology* 92, 1055–1063.
Dalmais, B., Schumacher, J., Moraga, J., Pêcheur, P., Tudzynski, B., Collado, I.G., Viaud, M. (2011) The *Botrytis cinerea* phytotoxin botcinic acid requires two polyketide synthases for production and has a redundant role in virulence with botrydial. *Molecular Plant Pathology* 12, 564–579.
Davik, J. and Honne, B.I. (2005) Genetic variance and breeding values for resistance to a wind-borne disease *Sphaerotheca macularis* (Wallr. ex Fr.) in strawberry (*Fragaria* × *ananassa* Duch.) estimated by exploring mixed and spatial models and pedigree information. *Theoretical and Applied* Genetics 111, 256–264.
Debode, J., van Hemelrijck, W., Baeyen, S., Creemers, P., Heungens, K. and Maes, M. (2009) Quantitative detection and monitoring of *Colletotrichum acutatum* in strawberry leaves using real-time PCR. *Plant Pathology* 58, 504–514.
de Jonge, R., van Esse, H.P., Maruthachalam, K., Bolton, M.D., Santhanam, P., Saber, M.K., Zhang, Z., Usami, T., Lievens, B., Subbarao, K.V. and Thomma, B.P. (2012) Tomato immune receptor Ve1 recognizes effector of multiple fungal pathogens uncovered by genome and RNA sequencing. *Proceedings of the National Academy of Sciences USA* 109, 5110–5115.
Del Sorbo, G., Ruocco, M., Schoonbeek, H.J., Scala, F., Pane, C., Vinale, F. and De Waard, M.A. (2008) Cloning and functional characterization of *BcatrA*, a gene encoding an ABC transporter of the plant pathogenic fungus *Botryotinia fuckeliana* (*Botrytis cinerea*) *Mycological Research* 112, 737–746.
Denoyes-Rothan, B., Guérin, G., Délye, C., Smith, B., Minz, D., Maymon, M. and Freeman, S. (2003) Genetic diversity and pathogenic variability among isolates of *Colletotrichum* species from strawberry. *Phytopathology* 93, 219–228.
Di Pietro, A., García-MacEira, F.I., Méglecz, E. and Roncero, M.I. (2001) A MAP kinase of the vascular wilt fungus *Fusarium oxysporum* is essential for root penetration and pathogenesis. *Molecular Microbiology* 39, 1140–1152.
Doehlemann, G., Molitor, F. and Hahn, M. (2005) Molecular and functional characterization of a fructose specific transporter from the gray mold fungus *Botrytis cinerea*. *Fungal Genetics and Biology* 42, 601–610.
Doehlemann, G., Berndt, P. and Hahn, M. (2006a) Different signalling pathways involving a Galpha protein, cAMP and a MAP kinase control germination of *Botrytis cinerea* conidia. *Molecular Microbiology* 59, 821–835.
Doehlemann, G., Berndt, P. and Hahn, M. (2006b) Trehalose metabolism is important for heat stress tolerance and spore germination of *Botrytis cinerea*. *Microbiology* 152, 2625–2634.
Douglas Gubler, W. and Converse, R.H. (1993) *Disease of Strawberry*. American Phytopathological Society, St Paul, Minnesota.
Drenth, A., Wagels, G., Smith, B., Sendall, B., O'Dwyer, C., Irvine, G. and Irwin, J.A.G. (2006) Development of a DNA-based method for detection and identification of *Phytophthora* species. *Australasian Plant Pathology* 35, 147–159.
Duyvesteijn, R.G., van Wijk, R., Boer, Y., Rep, M., Cornelissen, B.J. and Haring, M.A. (2005) Frp1 is a *Fusarium oxysporum* F-box protein required for pathogenicity on tomato. *Molecular Microbiology* 57, 1051–1063.
Ehsani-Moghaddam, B., Charles, M.T., Carisse, O. and Khanizadeh, S. (2006) Superoxide dismutase responses of strawberry cultivars to infection by *Mycosphaerella fragariae*. *Journal of Plant Physiology* 163, 147–153.
Eikemo, H., Klemsdal, S.S., Riisberg, I., Bonants, P., Stensvand, A. and Tronsmo, A.M. (2004) Genetic variation between *Phytophthora cactorum* isolates differing in their ability to cause crown rot in strawberry. *Mycological Research* 108, 317–324.
Errakhi, R., Lebrihi, A. and Barakate, M. (2009) In vitro and in vivo antagonism of actinomycetes isolated from Moroccan rhizospherical soils against *Sclerotium rolfsii*: a causal agent of root rot on sugar beet (*Beta vulgaris* L.). *Journal of Applied Microbiology* 107, 672–681.
EPPO (2015) *Quarantine Pests for Europe*. European and Mediterranean Plant Protection Organization, Paris, France.
Fang, X., Kuo, J., You, M.P., Finnegan, P.M. and Barbetti, M.J. (2012) Comparative root colonisation of strawberry cultivars Camarosa and Festival by *Fusarium oxysporum* f. sp. *fragariae*. *Plant and Soil* 358, 75–89.
Fang, X., Jost, R., Finnegan, P.M. and Barbetti, M.J. (2013) Comparative proteome analysis of the strawberry-*Fusarium oxysporum* f. sp. *fragariae* pathosystem reveals early activation of defense responses as a crucial determinant of host resistance. *Journal of Proteome Research* 12, 1772–1788.
FAO (2010–2014) *Statistical Databases*. Food and Agriculture Organization of the United Nations, Rome, Italy.
Fernández, R.G. and Novo, J.V.J. (2010) Proteomics of fungal plant pathogens: the case of *Botrytis cinerea*. In: Méndez-Vilas, A. (ed.) *Current Research, Technology and Education Topics in Applied Microbiology and Microbial Biotechnology*. Formatex, Spain, pp. 205–217.

Fernández-Acero, F.J., Jorge, I., Calvo, E., Vallejo, I., Carbú, M., Camafeita, E., López, J.A., Cantoral, J.M. and Jorrín, J. (2006) Two-dimensional electrophoresis protein profile of the phytopathogenic fungus *Botrytis cinerea*. *Proteomics* 6, S88–S96.

Fernández-Acero, F.J., Jorge, I., Calvo, E., Vallejo, I., Carbú, M., Camafeita, E., Garrido, C., López, J.A., Jorrin, J. and Cantoral, J.M. (2007) Proteomic analysis of phytopathogenic fungus *Botrytis cinerea* as a potential tool for identifying pathogenicity factors, therapeutic targets and for basic research. *Archives of Microbiology* 187, 207–215.

Fernández-Acero, F.J., Colby, T., Harzen, A., Cantoral, J.M. and Schmidt, J. (2009) Proteomic analysis of the phytopathogenic fungus *Botrytis cinerea* during cellulose degradation. *Proteomics* 9, 2892–2902.

Fernández-Acero, F.J., Colby, T., Harzen, A., Carbú, M., Wieneke, U., Cantoral, J.M. and Schmidt, J. (2010) 2-DE proteomic approach to the *Botrytis cinerea* secretome induced with different carbon sources and plant-based elicitors. *Proteomics* 10, 2270–2280.

Ferrari, S., Galletti, R., Pontiggia, D., Manfredini, C., Lionetti, V., Bellincampi, D., Cervone, F. and De Lorenzo, G. (2008) Transgenic expression of a fungal endo-polygalacturonase increases plant resistance to pathogens and reduces auxin sensitivity. *Plant Physiology* 146, 669–681.

Fourie, G., Steenkamp, E.T., Ploetz, R.C., Gordon, T.R. and Viljoen, A. (2011) Current status of the taxonomic position of *Fusarium oxysporum* formae specialis *cubense* within the *Fusarium oxysporum* complex. *Infection, Genetics and Evolution* 11, 533–542.

Fravel, D., Olivain, C. and Alabouvette, C. (2003) *Fusarium oxysporum* and its biocontrol. *New Phytologist* 157, 493–502.

Freeman, S., Minz, D., Jurkevitch, E., Maymon, M. and Shabi, E. (2000) Molecular analyses of *Colletotrichum* species from almond and other fruits. *Phytopathology* 90, 608–614.

Garrido, C., Carbú, M., Fernández-Acero, F., Budge, G., Vallejo, I., Colyer, A. and Cantoral, J. (2008) Isolation and pathogenicity of *Colletotrichum* spp. causing anthracnose of strawberry in south west Spain. *European Journal of Plant Pathology* 120, 409–415.

Garrido, C., Carbú, M., Fernández-Acero, F.J., Boonham, N., Colyer, A., Cantoral, J.M. and Budge, G. (2009a) Development of protocols for detection of *Colletotrichum acutatum* and monitoring of strawberry anthracnose using real-time PCR. *Plant Pathology* 58, 43–51.

Garrido, C., Carbú, M., Javier Fernández-Acero, F., Vallejo, I. and Manuel Cantoral, J. (2009b) Phylogenetic relationships and genome organisation of *Colletotrichum acutatum* causing anthracnose in strawberry. *European Journal of Plant Pathology* 125, 397–411.

Garrido, C., Cantoral, J.M., Carbu, M., Gonzalez-Rodriguez V.E. and Javier Fernandez-Acero, F. (2010) New proteomic approaches to plant pathogenic fungi. *Current Proteomics* 7, 306–315.

Garrido, C., Carbú, M., Fernández-Acero, F.J., González-Rodríguez, V.E. and Cantoral, J.M. (2011) New insight in the study of strawberry fungal pathogens. In: Husaini, A.M. and Mercado, J.A. (eds) *Genomics, Transgenics, Molecular Breeding and Biotechnology of Strawberry*. Global Science Books, Japan/UK, pp. 24–39.

Garrido, C., Fernández-Acero, F.J., Carbú, M., González-Rodríguez, V.E., Liñeiro, E. and Cantoral, J.M. (2012) Molecular microbiology applied to the study of phytopathogenic fungi. In: Magdeldin, S. (ed.) *Gel Electrophoresis – Advanced Techniques*. InTech, pp. 139–156. Available at: http://www.intechopen.com/books/gel-electrophoresis-advanced-techniques/molecular-microbiology-applied-to-the-study-of-phytopathogenic-fungi.

Geiser, D.M., del Mar Jiménez-Gasco, M., Kang, S., Makalowska, I., Veeraraghavan, N., Ward, T.J., Zhang, N., Kuldau, G.A. and O'Donnell, K. (2004) FUSARIUM-ID v. 1.0: a DNA sequence database for identifying *Fusarium*. In: Mulè, G., Bailey, J.A., Cooke, B.M. and Logrieco, A. (eds) *Molecular Diversity and PCR-detection of Toxigenic Fusarium Species and Ochratoxigenic Fungi*. Springer, The Netherlands, pp. 473–479.

Gioti, A., Simon, A., Le Pêcheur, P., Giraud, C., Pradier, J.M. and Viaud, M., Levis, C. (2006) Expression profiling of *Botrytis cinerea* genes identifies three patterns of up-regulation in planta and an FKBP12 protein affecting pathogenicity. *Journal of Molecular Biology* 358, 372–386.

Gioti, A., Pradier, J.M., Fournier, E., Le Pêcheur, P., Giraud, C., Debieu, D., Bach, J., Leroux, P. and Levis, C. (2008) A *Botrytis cinerea* emopamil binding domain protein, required for full virulence, belongs to a eukaryotic superfamily which has expanded in Euascomycetes. *Eukaryotic Cell* 7, 368–378.

Gómez-Gómez, E., Ruíz-Roldán, M.C., Di Pietro, A., Roncero, M.I. and Hera, C. (2002) Role in pathogenesis of two endo-β-1,4-xylanase genes from the vascular wilt fungus *Fusarium oxysporum*. *Fungal Genetics and Biology* 35, 213–222.

González-Fernández, R., Prats, E. and Jorrín-Novo, J.V. (2010) Proteomics of plant pathogenic fungi. *Journal of Biomedicine and Biotechnology* 2010, 932527.

Gourgues, M., Brunet-Simon, A., Lebrun, M.H. and Levis, C. (2004) The tetraspanin BcPls1 is required for appressorium-mediated penetration of *Botrytis cinerea* into host plant leaves. *Molecular Microbiology* 51, 619–629.

Grinyer, J., McKay, M., Nevalainen, H. and Herbert, B. (2004) Fungal proteomics: initial mapping of biological control strain *Trichoderma harzianum*. *Current Genetics* 45, 163–169.

Groenewald, S., van den Berg, N., Marasas, W.F.O. and Viljoen, A. (2006) The application of high-throughput AFLPs in assessing genetic diversity in *Fusarium oxysporum* f. sp. *cubense*. *Mycological Research* 110, 297–305.

Gutiérrez, L., Escudero, A., Batlle, R. and Nerín, C. (2009) Effect of mixed antimicrobial agents and flavors in active packaging films. *Journal of Agricultural and Food Chemistry* 57, 8564–8571.

Haesler, F., Hagn, A., Frommberger, M., Hertkorn, N., Schmitt-Kopplin, P., Munch, J.C. and Schloter, M. (2008) In vitro antagonism of an actinobacterial *Kitasatospora* isolate against the plant pathogen *Phytophthora citricola* as elucidated with ultrahigh resolution mass spectrometry. *Journal of Microbiological Methods* 75, 188–195.

Hantula, J., Lilja, A., Nuorteva, H., Parikka, P. and Werres, S. (2000) Pathogenicity, morphology and genetic variation of *Phytophthora cactorum* from strawberry, apple, rhododendron, and silver birch. *Mycological Research* 104, 1062–1068.

Harel, A., Bercovich, S. and Yarden, O. (2006) Calcineurin is required for sclerotial development and pathogenicity of *Sclerotinia sclerotiorum* in an oxalic acid-independent manner. *Molecular Plant–Microbe Interactions* 19, 682–693.

Hatta, R., Ito, K., Hosaki, Y., Tanaka, T., Tanaka, A., Yamamoto, M., Akimitsu, K. and Tsuge, T. (2002) A conditionally dispensable chromosome controls host-specific pathogenicity in the fungal plant pathogen *Alternaria alternata*. *Genetics* 161, 59–70.

Hauke, K., Creemers, P., Brugmans, W. and van Laer, S. (2004) Signum, a new fungicide with interesting properties in resistance management of fungal diseases in strawberries. *Communication in Agricultural and Applied Biological Sciences* 69, 743–755.

Hayashi, K., Schoonbeek, H.J. and de Waard, M.A. (2002) Bcmfs1, a novel major facilitator superfamily transporter from *Botrytis cinerea*, provides tolerance towards the natural toxic compounds camptothecin and cercosporin and towards fungicides. *Applied and Environmental Microbiology* 68, 4996–5004.

Hong, S.-Y., Kang, M.-R., Cho, E.-J., Kim, H.-K. and Yun, S.-H. (2010) Specific PCR detection of four quarantine *Fusarium* species in Korea. *Plant Pathology Journal* 26, 409–416.

Huang, T.-P., Yeh, Y. and Tzeng, D.D.-S. (2011) Barcode-like heteroduplex DNA pattern as an aid for rapid identification of anthracnose fungi. *New Biotechnology* 28, 72–78.

Hughes, K.J.D., Lane, C.R. and Cook, R.T.A. (1997) Development of a rapid method for the detection and identification of *Colletotrichum Acutatum*. In: Dehne, H.W., Adam, G., Diekmann, M., Frahm, J., Mauler-Machnik, A. and van Halteren, P. (eds) *Diagnosis and Identification of Plant Pathogens*. Springer, The Netherlands, pp. 113–116.

Hunter, T. and Jordan, V.W.L. (1974) Fruit blotch of strawberry caused by *Fusarium sambucinum* Fuckel. *Plant Pathology* 23, 118.

Hwang, C.S., Flaishman, M.A. and Kolattukudy, P.E. (1995) Cloning of a gene expressed during appressorium formation by *Colletotrichum gloeosporioides* and a marked decrease in virulence by disruption of this gene. *The Plant Cell* 7, 183–193.

Imazaki, I., Kurahashi, M., Iida, Y. and Tsuge, T. (2007) Fow2, a Zn(II)2Cys6-type transcription regulator, controls plant infection of the vascular wilt fungus *Fusarium oxysporum*. *Molecular Microbiology* 63, 737–753.

Inoue, I., Namiki, F. and Tsuge, T. (2002) Plant colonization by the vascular wilt fungus *Fusarium oxysporum* requires FOW1, a gene encoding a mitochondrial protein. *The Plant Cell* 14, 1869–1883.

Islas, C.M. (2012) Fusarium wilt caused by *Fusarium oxysporum* f. sp. *fragariae*: an emerging disease of strawberry in California. Thesis, University of California, Davis, California.

Izzo, A.D. and Mazzola, M. (2009) Hybridization of an ITS-based macroarray with ITS community probes for characterization of complex communities of fungi and fungal-like protists. *Mycological Research* 113, 802–812.

Jain, S., Akiyama, K., Mae, K., Ohguchi, T. and Takata, R. (2002) Targeted disruption of a G protein alpha subunit gene results in reduced pathogenicity in *Fusarium oxysporum*. *Current Genetics* 41, 407–413.

Jain, S., Akiyama, K., Takata, R. and Ohguchi, T. (2005) Signaling via the G protein α subunit FGA2 is necessary for pathogenesis in *Fusarium oxysporum*. *FEMS Microbiology Letters* 243, 165–172.

Javaid, A. and Amin, M. (2009) Antifungal activity of methanol and *n*-hexane extracts of three *Chenopodium* species against *Macrophomina phaseolina*. *Natural Products Research* 23, 1120–1127.

Jeeva, M.L., Mishra, A.K., Vidyadharan, P., Misra, R.S. and Hegde, V. (2010) A species-specific polymerase chain reaction assay for rapid and sensitive detection of *Sclerotium rolfsii*. *Australasian Plant Pathology* 39, 517–523.

Jin, W., Yin, S., Lu, R., Yuan, W., Dong, J. and Wang, G. (2005) GO gene transformation of strawberry and its resistance to gray mold. *Molecular Plant Breeding* 3, 797–800.

Johnson, R.D., Johnson, L., Itoh, Y., Kodama, M., Otani, H. and Kohmoto, K. (2000) Cloning and characterization of a cyclic peptide synthetase gene from *Alternaria alternata* apple pathotype whose product is involved in AM-toxin synthesis and pathogenicity. *Molecular Plant–Microbe Interactions* 13, 742–753.

Jonkers, W., van Kan, J.A., Tijm, P., Lee, Y.W., Tudzynski, P., Rep, M. and Michielse, C.B. (2011) The FRP1 F-box gene has different functions in sexuality, pathogenicity and metabolism in three fungal pathogens. *Molecular Plant Pathology* 12, 548–563.

Kars, I., Krooshof, G.H., Wagemakers, L., Joosten, R., Benen, J.A. and van Kan, J.A. (2005) Necrotizing activity of five *Botrytis cinerea* endopolygalacturonases produced in *Pichia pastoris*. *The Plant Journal* 43, 213–225.

Kim, Y.K., Kawano, T., Li, D. and Kolattukudy, P.E. (2000) A mitogen-activated protein kinase kinase required for induction of cytokinesis and appressorium formation by host signals in the conidia of *Colletotrichum gloeosporioides*. *The Plant Cell* 12, 1331–1343.

Kim, Y.K., Liu, Z.M., Li, D. and Kolattukudy, P.E. (2000) Two novel genes induced by hard-surface contact of *Colletotrichum gloeosporioides* conidia. *Journal of Bacteriology* 182, 4688–4695.

Knutsen, A.K., Torp, M. and Holst-Jensen, A. (2004) Phylogenetic analyses of the *Fusarium poae*, *Fusarium sporotrichioides* and *Fusarium langsethiae* species complex based on partial sequences of the translation elongation factor-1α gene. *International Journal of Food Microbiology* 95, 287–295.

Kokkelink, L., Minz, A., Al-Masri, M., Giesbert, S., Barakat, R., Sharon, A. and Tudzynski, P. (2011) The small GTPase BcCdc42 affects nuclear division, germination and virulence of the gray mold fungus *Botrytis cinerea*. *Fungal Genetics and Biology* 48, 1012–1019.

Kong, P., Moorman, G.W., Lea-Cox, J.D., Ross, D.S., Richardson, P.A. and Hong, C. (2009) Zoosporic tolerance to pH stress and its implications for *Phytophthora* species in aquatic ecosystems. *Applied and Environmental Microbiology* 75, 4307–4314.

Kristensen, R., Torp, M., Kosiak, B. and Holst-Jensen, A. (2005) Phylogeny and toxigenic potential is correlated in *Fusarium* species as revealed by partial translation elongation factor 1α gene sequences. *Mycological Research* 109, 173–186.

Lamondia, J.A. and Cowles, R.S. (2005) Comparison of *Pratylenchus penetrans* infection and *Maladera castanea* feeding on strawberry root rot. *Journal of Nematology* 37, 131–135.

Larsen, R.C., Vandemark, G.J., Hughes, T.J. and Grau, C.R. (2007) Development of a real-time polymerase chain reaction assay for quantifying *Verticillium albo-atrum* DNA in resistant and susceptible alfalfa. *Phytopathology* 97, 1519–1525.

Laugé, R. (1997) The in planta-produced extracellular proteins ECP1 and ECP2 of *Cladosporium fulvum* are virulence factors. *Molecular Plant–Microbe Interactions* 10, 725–734.

Li, M., Liang, X., Rollins, J.A. (2012) *Sclerotinia sclerotiorum* γ-glutamyl transpeptidase (Ss-Ggt1) is required for regulating glutathione accumulation and development of sclerotia and compound appressoria. *Molecular Plant–Microbe Interactions* 25, 412–420.

Lievens, B., Brouwer, M., Vanachter, A.C.R.C., Lévesque, C.A., Cammue, B.P.A. and Thomma, B.P.H.J. (2003) Design and development of a DNA array for rapid detection and identification of multiple tomato vascular wilt pathogens. *FEMS Microbiology Letters* 223, 113–122.

Lievens, B., Brouwer, M., Vanachter, A.C.R.C., Lévesque, C.A., Cammue, B. and Thomma, B.P.H.J. (2005) Quantitative assessment of phytopathogenic fungi in various substrates using a DNA macroarray. *Environmental Microbiology* 7, 1698–1710.

Lievens, B., Claes, L., Vakalounakis, D.J., Vanachter, A.C.R.C. and Thomma, B.P.H.J. (2007) A robust identification and detection assay to discriminate the cucumber pathogens *Fusarium oxysporum* f. sp. *cucumerinum* and f. sp. *radicis-cucumerinum*. *Environmental Microbiology* 9, 2145–2161.

Lievens, B., Rep, M. and Thomma, B.P.H.J. (2008) Recent developments in the molecular discrimination of formae speciales of *Fusarium oxysporum*. *Pest Management Science* 64, 781–788.

Lilja, A., Rytköken, A. and Hantula, J. (2008) Introduced pathogens found on ornamentals, strawberry and trees in Finland over the past 20 years. *Agricultural and Food Science* 20, 74–85.

Lin, C.H., Yang, S.L., Wang, N.Y. and Chung, K.R. (2009) The FUS3 MAPK signaling pathway of the citrus pathogen *Alternaria alternata* functions independently or cooperatively with the fungal redox-responsive AP1 regulator for diverse developmental, physiological and pathogenic processes. *Fungal Genetics and Biology* 47, 381–391.

Liu, B., Wasilwa, L.A., Morelock, T.E., O'Neill, N.R. and Correll, J.C. (2007) Comparison of *Colletotrichum orbiculare* and several allied *Colletotrichum* spp. for mtDNA RFLPs, intron RFLP and sequence variation, vegetative compatibility, and host specificity. *Phytopathology* 97, 1305–1314.

Liu, L.N., Zhang, J.Z. and Xu, T. (2009) Histopathological studies of sclerotia of Rhizoctonia solani parasitized by the EGFP transformant of Trichoderma virens. *Letters in Applied Microbiology* 49, 745–750.

López-Berges, M.S., Di Pietro, A., Daboussi, M.-J., Wahab, H.A., Vasnier, C., Roncero, M.I.G., Dufresne, M. and Hera, C. (2009) Identification of virulence genes in *Fusarium oxysporum* f. sp. *lycopersici* by large-scale transposon tagging. *Molecular Plant Pathology* 10, 95–107.

López-Berges, M.S., Rispail, N., Prados-Rosales, R.C. and Di Pietro, A. (2010) A nitrogen response pathway regulates virulence functions in *Fusarium oxysporum* via the protein kinase TOR and the bZIP protein MeaB. *The Plant Cell* 22, 2459–2475.

Lyon, G.D., Goodman, B.A. and Williamson, B. (2007) *Botrytis cinerea* perturbs redox processes as an attack strategy in plants. In: Elad, Y., Williamson, B., Tudzynski, P. and Delen, N. (eds) *Botrytis: Biology, Pathology and Control.* Springer, The Netherlands, pp. 119–141.

Maas, J.L. (1998) *Compendium of Strawberry Diseases*, 2nd edn. American Phytopathological Society, St Paul Minnesota.

Maas, J.L. (2004) Strawberry disease management. In: Naqvi, S.A.M.H. (ed.) *Diseases of Fruits and Vegetables: Volume II.* Springer, The Netherlands, pp. 441–483.

Maciá-Vicente, J.G., Jansson, H.-B., Talbot, N.J. and Lopez-Llorca, L.V. (2009) Real-time PCR quantification and live-cell imaging of endophytic colonization of barley (*Hordeum vulgare*) roots by *Fusarium equiseti* and *Pochonia chlamydosporia*. *New Phytologist* 182, 213–228.

Martin, F.N. and Bull, C.T. (2002) Biological approaches for control of root pathogens of strawberry. *Phytopathology*, 92, 1356–1362.

Martin, F.N. and Tooley, P.W. (2003) Phylogenetic relationships among *Phytophthora* species inferred from sequence analysis of mitochondrially encoded cytochrome oxidase I and II genes. *Mycologia* 95, 269–284.

Martinez-Culebras, P.V., Querol, A., Suarez-Fernandez, M.B., Garcia-Lopez, M.D. and Barrio, E. (2003) Phylogenetic relationships among *Colletotrichum* pathogens of strawberry and design of PCR primers for their identification. *Journal of Phytopathology* 151, 135–143.

Martín-Udíroz, M., Madrid, M.P. and Roncero, M.I. (2004) Role of chitin synthase genes in *Fusarium oxysporum*. *Microbiology* 150, 3175–3187.

McCartney, H.A., Foster, S.J., Fraaije, B.A. and Ward, E. (2003) Molecular diagnostics for fungal plant pathogens. *Pest Management Science* 59, 129–142.

Meléndez, H.G., Billon-Grand, G., Fèvre, M. and Mey, G. (2009) Role of the *Botrytis cinerea* FKBP12 ortholog in pathogenic development and in sulfur regulation. *Fungal Genetics and Biology* 46, 308–320.

Meng, J. and Wang, Y. (2010) Rapid detection of *Phytophthora nicotianae* in infected tobacco tissues and soil samples based on its *Ypt1* gene. *Journal of Phytopathology* 158, 1–7.

Mercado, J.A., Martín-Pizarro, C., Pascual, L., Quesada, M.A., Pliego-Alfaro, F., de los Santos, B., Romero, F., Galvez, J., Rey, M., de la Viña, G., Llobell, A., Yubero-Serrano, E.M., Muñoz-Blanco, J. and Caballero, J.L. (2007) Evaluation of tolerance to *Colletotrichum acutatum* in strawberry plants transformed with *Trichoderma*-derived genes. *Acta Horticulturae* 738, 383–388.

Mertely, J.C. and Legard, D.E. (2004) Detection, isolation, and pathogenicity of *Colletotrichum* spp. from strawberry petioles. *Plant Disease* 88, 407–412.

Mills, P.R., Sreenivasaprasad, S. and Brown, A.E. (1992) Detection and differentiation of *Colletotrichum gloeosporioides* isolates using PCR. *FEMS Microbiology Letters* 98, 137–143.

Miyamoto, Y., Masunaka, A., Tsuge, T., Yamamoto, M., Ohtani, K., Fukumoto, T., Gomi, K., Peever, T.L. and Akimitsu, K. (2008) Functional analysis of a multicopy host-selective ACT-toxin biosynthesis gene in the tangerine pathotype of *Alternaria alternate* using RNA silencing. *Molecular Plant–Microbe Interactions* 21, 1591–1599.

Miyamoto, Y., Ishii, Y., Honda, A., Masunaka, A., Tsuge, T., Yamamoto, M., Ohtani, K., Fukumoto, T., Gomi, K., Peever, T.L. and Akimitsu, K. (2009) Function of genes encoding acyl-CoA synthetase and enoyl-CoA hydratase for host-selective act-toxin biosynthesis in the tangerine pathotype of *Alternaria alternate*. *Phytopathology* 99, 369–377.

Morocko, I. and Fatehi, J. (2007) Molecular characterization of strawberry pathogen *Gnomonia fragariae* and its genetic relatedness to other *Gnomonia* species and members of Diaporthales. *Mycological Research* 111, 603–614.

Mulè, G., Susca, A., Stea, G. and Moretti, A. (2004) Specific detection of the toxigenic species *Fusarium proliferatum* and *F. oxysporum* from asparagus plants using primers based on calmodulin gene sequences. *FEMS Microbiology Letters* 230, 235–240.

Nam, M.H., Park, M.S., Kim, H.G. and Yoo, S.J. (2009) Biological control of strawberry *Fusarium* wilt caused by *Fusarium oxysporum* f. sp. *fragariae* using *Bacillus velezensis* BS87 and RK1 formulation. *Journal of Microbiology and Biotechnology* 19, 520–524.

Namiki, F., Matsunaga, M., Okuda, M., Inoue, I., Nishi, K., Fujita, Y., Tsuge, T. (2001) Mutation of an arginine biosynthesis gene causes reduced pathogenicity in *Fusarium oxysporum* f. sp. *melonis*. *Molecular Plant–Microbe Interactions* 14, 580–584.

Nguyen, H.D.T. and Seifert, K.A. (2008) Description and DNA barcoding of three new species of *Leohumicola* from South Africa and the United States. *Persoonia: Molecular Phylogeny and Evolution of Fungi* 21, 57–69.

Nicastro, G., Orsomando, G., Ferrari, E., Manconi, L., Desario, F., Amici, A., Naso, A., Carpaneto, A., Pertinhez, T.A., Ruggieri, S. and Spisni, A. (2009) Solution structure of the phytotoxic protein PcF: the first characterized member of the *Phytophthora* PcF toxin family. *Protein Science* 18, 1786–1791.

Nita, M., Ellis, M.A. and Madden, L.V. (2003) Reliability and accuracy of visual estimation of phomopsis leaf blight of strawberry. *Phytopathology* 93, 995–1005.

Ohara, T., Inoue, I., Namiki, F., Kunoh, H. and Tsuge, T. (2004) REN1 is required for development of microconidia and macroconidia, but not of chlamydospores, in the plant pathogenic fungus *Fusarium oxysporum*. *Genetics* 166, 113–124.

Olmos, A., Bertolini, E. and Cambra, M. (2002) Simultaneous and co-operational amplification (co-PCR): a new concept for detection of plant viruses. *Journal of Virological Methods* 106, 51–59.

Orsomando, G., Lorenzi, M., Raffaelli, N., Dalla Rizza, M., Mezzetti, B. and Ruggieri, S. (2001) Phytotoxic protein PcF, purification, characterization, and cDNA sequencing of a novel hydroxyproline-containing factor secreted by the strawberry pathogen *Phytophthora cactorum*. *Journal of Biological Chemistry* 276, 21578–21584.

Ospina-Giraldo, M.D., Mullins, E. and Kang, S. (2003) Loss of function of the *Fusarium oxysporum SNF1* gene reduces virulence on cabbage and *Arabidopsis*. *Current Genetics* 44, 49–57.

Ortega-Morales, B.O., Ortega-Morales, F.N., Lara-Reyna, J., De la Rosa-García, S.C., Martínez-Hernández, A. and Montero-M, J. (2009) Antagonism of *Bacillus* spp. isolated from marine biofilms against terrestrial phytopathogenic fungi. *Marine Biotechnology* 11, 375–383.

Ortoneda, M., Guarro, J., Madrid, M.P., Caracuel, Z., Roncero, M.I., Mayayo, E. and Di Pietro, A. (2004) *Fusarium oxysporum* as a multihost model for the genetic dissection of fungal virulence in plants and mammals. *Infection and Immunity* 72, 1760–1766.

Pannecoucque, J. and Höfte, M. (2009) Detection of rDNA ITS polymorphism in *Rhizoctonia solani* AG 2-1 isolates. *Mycologia* 101, 26–33.

Pérez-Nadales, E. and Di Pietro, A. (2011) The membrane mucin Msb2 regulates invasive growth and plant infection in *Fusarium oxysporum*. *The Plant Cell* 23, 1171–1185.

Pertot, I., Kuflik, T., Gordon, I., Freeman, S. and Elad, Y. (2012) Identificator: A web-based tool for visual plant disease identification, a proof of concept with a case study on strawberry. *Computers and Electronics in Agriculture* 84, 144–154.

Pliego, C., López-Herrera, C., Ramos, C., Cazorla, F.M. (2012) Developing tools to unravel the biological secrets of *Rosellinia necatrix*, an emergent threat to woody crops. *Molecular Plant Pathology*, 13, 226–239.

Porter-Jordan, K., Rosenberg, E.I., Keiser, J.F., Gross, J.D., Ross, A.M., Nasim, S. and Garrett, C.T. (1990) Nested polymerase chain reaction assay for the detection of cytomegalovirus overcomes false positives caused by contamination with fragmented DNA. *Journal of Medical Virology* 30, 85–91.

Prodorutti, D., Vanblaere, T., Gobbin, D., Pellegrini, A., Gessler, C. and Pertot, I. (2009) Genetic diversity of *Armillaria* spp. infecting highbush blueberry in northern Italy (Trentino region). *Phytopathology* 99, 651–658.

Qin, L., Fu, Y., Xie, J., Cheng, J., Jiang, D., Li, G. and Huang, J. (2011) A nested-PCR method for rapid detection of *Sclerotinia sclerotiorum* on petals of oilseed rape (*Brassica napus*). *Plant Pathology* 60, 271–277.

Radišek, S., Jakše, J. and Javornik, B. (2004) Development of pathotype-specific SCAR markers for detection of *Verticillium albo-atrum* isolates from hop. *Plant Disease* 88, 1115–1122.

Ramos, B., Alves-Santos, F.M., García-Sánchez, M.A., Martín-Rodrigues, N., Eslava, A.P. and Díaz-Mínguez, J.M. (2007) The gene coding for a new transcription factor (*ftf1*) of *Fusarium oxysporum* is only expressed during infection of common bean. *Fungal Genetics and Biology* 44, 864–876.

Rauyaree, P., Ospina-Giraldo, M.D., Kang, S., Bhat, R.G., Subbarao, K.V., Grant, S.J. and Dobinson, K.F. (2005) Mutations in VMK1, a mitogen-activated protein kinase gene, affect microsclerotia formation and pathogenicity in *Verticillium dahliae*. *Current Genetics* 48, 109–116.

Redondo, C., Cubero, J. and Melgarejo, P. (2009) Characterization of *Penicillium* species by ribosomal DNA sequencing and BOX, ERIC and REP-PCR analysis. *Mycopathologia* 168, 11–22.

Ren, L., Li, G. and Jiang, D. (2010) Characterization of some culture factors affecting oxalate degradation by the mycoparasite *Coniothyrium minitans*. *Journal of Applied Microbiology* 108, 173–180.

Rep, M., van der Does, H.C., Meijer, M., van Wijk, R., Houterman, P.M., Dekker, H.L., de Koster, C.G. and Cornelissen, B.J. (2004) A small, cysteine-rich protein secreted by *Fusarium oxysporum* during colonization of xylem vessels is required for I-3-mediated resistance in tomato. *Molecular Microbiology* 53, 1373–1383.

Rigotti, S., Gindro, K., Richter, H. and Viret, O. (2002) Characterization of molecular markers for specific and sensitive detection of *Botrytis cinerea* Pers.: Fr. in strawberry (*Fragaria* × *ananassa* Duch.) using PCR. *FEMS Microbiology Letters* 209, 169–174.

Rispail, N. and Di Pietro, A. (2010) The two-component histidine kinase Fhk1 controls stress adaptation and virulence of *Fusarium oxysporum*. *Molecular Plant Pathology* 11, 395–407.

Rollins, J.A. (2003) The *Sclerotinia sclerotiorum* pac1 gene is required for sclerotial development and virulence. *Molecular Plant–Microbe Interactions* 16, 785–795.

Rooney, H.C., Van't Klooster, J.W., van der Hoorn, R.A., Joosten, M.H., Jones, J.D. and de Wit, P.J. (2005) *Cladosporium* Avr2 inhibits tomato Rcr3 protease required for Cf-2-dependent disease resistance. *Science* 308, 1783–1786.

Ruiz-Moyano, S., Benito, M.J., Martín, A., Aranda, E., Hernández, A. and Córdoba, M.G. (2009) Characterization of molds isolated from smoked paprika by PCR-RFLP and micellar electrokinetic capillary electrophoresis. *Food Microbiology* 26, 776–782.

Saitoh, Y., Izumitsu, K., Morita, A. and Tanaka, C. (2010) A copper-transporting ATPase BcCCC2 is necessary for pathogenicity of *Botrytis cinerea*. *Molecular Genetics and Genomics* 284, 33–43.

Schamber, A., Leroch, M., Diwo, J., Mendgen, K. and Hahn, M. (2010) The role of mitogen-activated protein (MAP) kinase signalling components and the Ste12 transcription factor in germination and pathogenicity of *Botrytis cinerea*. *Molecular Plant Pathology* 11, 105–119.

Schestibratov, K.A. and Dolgov, S.V. (2005) Transgenic strawberry plants expressing a thaumatin II gene demonstrate enhanced resistance to *Botrytis cinerea*. *Scientia Horticulturae* 106, 177–189.

Schoonbeek, H., Del Sorbo, G. and de Waard, M.A. (2001) The ABC transporter BcatrB affects the sensitivity of *Botrytis cinerea* to the phytoalexin resveratrol and the fungicide fenpiclonil. *Molecular Plant–Microbe Interactions* 14, 562–571.

Schouten, A., Wagemakers, L., Stefanato, F.L., van der Kaaij, R.M. and van Kan, J.A. (2002) Resveratrol acts as a natural profungicide and induces self-intoxication by a specific laccase. *Molecular Microbiology* 43, 883–894.

Schumacher, J., Kokkelink, L., Huesmann, C., Jimenez-Teja, D., Collado, I.G., Barakat, R., Tudzynski, P. and Tudzynski, B. (2008) The cAMP-dependent signaling pathway and its role in conidial germination, growth, and virulence of the gray mold *Botrytis cinerea*. *Molecular Plant–Microbe Interactions* 21, 1443–1459.

Segers, G., Bradshaw, N., Archer, D., Blissett, K. and Oliver, R.P. (2001) Alcohol oxidase is a novel pathogenicity factor for *Cladosporium fulvum*, but aldehyde dehydrogenase is dispensable. *Molecular Plant–Microbe Interactions* 14, 367–377.

Segmüller, N., Ellendorf, U., Tudzynski, B. and Tudzynski, P. (2007) BcSAK1, a stress-activated mitogen-activated protein kinase, is involved in vegetative differentiation and pathogenicity in *Botrytis cinerea*. *Eukaryotic Cell* 6, 211–221.

Segmüller, N., Kokkelink, L., Giesbert, S., Odinius, D., van Kan, J. and Tudzynski, P. (2008) NADPH oxidases are involved in differentiation and pathogenicity in *Botrytis cinerea*. *Molecular Plant–Microbe Interactions* 21, 808–819.

Seifert, K.A., Samson, R.A., Houbraken, J., Lévesque, C.A., Moncalvo, J.-M., Louis-Seize, G. and Hebert, P.D. (2007) Prospects for fungus identification using *CO1* DNA barcodes, with *Penicillium* as a test case. *Proceedings of the National Academy of Sciences USA* 104, 3901–3906.

Seijo, T.E., Chandler, C.K., Mertely, J.C., Moyer, C. and Peres, N.A. (2008) Resistance of strawberry cultivars and advanced selections to anthracnose and *Botrytis* fruit rots. *Proceedings of the Florida State Horticultural Society* 121, 246–248.

Shafique, S., Bajwa, R. and Shafique, S. (2009) Mutation of *Alternaria tenuissima* FCBP-252 for hyper-active α-amylase. *Indian Journal of Experimental Biology* 47, 591–596.

Shah, P., Atwood, J.A., Orlando, R., El Mubarek, H., Podila, G.K. and Davis, M.R. (2009a) Comparative proteomic analysis of *Botrytis cinerea* secretome. *Journal of Proteome Research* 8, 1123–1130.

Shah, P., Gutierrez-Sanchez, G., Orlando, R. and Bergmann, C. (2009b) A proteomic study of pectin-degrading enzymes secreted by *Botrytis cinerea* grown in liquid culture. *Proteomics* 9, 3126–3135.

Sheridan, G.E.C., Masters, C.I., Shallcross, J.A. and Mackey, B.M. (1998) Detection of mRNA by reverse transcription-PCR as an indicator of viability in *Escherichia coli* cells. *Applied and Environmental Microbiology* 64, 1313–1318.

Siewers, V., Viaud, M., Jimenez-Teja, D., Collado, I.G., Gronover, C.S., Pradier, J.M., Tudzynski, B. and Tudzynski, P. (2005) Functional analysis of the cytochrome P450 monooxygenase gene bcbot1 of *Botrytis cinerea* indicates that botrydial is a strain-specific virulence factor. *Molecular Plant–Microbe Interactions* 18, 602–612.

Smith, B.J. (2008) Epidemiology and pathology of strawberry anthracnose: a North American perspective. *HortScience* 43, 69–73.

Snoeijers, S.S., Pérez-García, A., Goosen, T. and de Wit, P.J. (2003) Promoter analysis of the avirulence gene Avr9 of the fungal tomato pathogen *Cladosporium fulvum* in the model filamentous fungus *Aspergillus nidulans*. *Current Genetics* 43, 96–102.

Soulié, M.C., Piffeteau, A., Choquer, M., Boccara, M. and Vidal-Cros A (2003) Disruption of *Botrytis cinerea* class I chitin synthase gene *Bcchs1* results in cell wall weakening and reduced virulence. *Fungal Genetics and Biology* 40, 38–46.

Soulié, M.C., Perino, C., Piffeteau, A., Choquer, M., Malfatti, P., Cimerman, A., Kunz, C., Boccara, M. and Vidal-Cros, A. (2006) *Botrytis cinerea* virulence is drastically reduced after disruption of chitin synthase class III gene (*Bcchs3a*) *Cell Microbiology* 8, 1310–1321.

Sreenivasaprasad, S., Brown, A.E. and Mills, P.R. (1992) DNA sequence variation and interrelationships among *Colletotrichum* species causing strawberry anthracnose. *Physiological and Molecular Plant Pathology* 41, 265–281.

Stefanato, F.L., Abou-Mansour, E., Buchala, A., Kretschmer, M., Mosbach, A., Hahn, M., Bochet, C.G., Métraux, J.P., Schoonbeek, H.J. (2009) The ABC transporter BcatrB from *Botrytis cinerea* exports camalexin and is a virulence factor on *Arabidopsis thaliana*. *The Plant Journal* 58, 499–510.

Stephenson, S.A., Hatfield, J., Rusu, A.G., Maclean, D.J., and Manners, J.M. (2000) CgDN3: an essential pathogenicity gene of *Colletotrichum gloeosporioides* necessary to avert a hypersensitive-like response in the host *Stylosanthes guianensis*. *Molecular Plant–Microbe Interactions* 13, 929–941.

Stephenson, S.A., Stephens, C.M., Maclean, D.J. and Manners, J.M. (2005) *CgDN24*: a gene involved in hyphal development in the fungal phytopathogen *Colletotrichum gloeosporioides*. *Microbiol Research* 160, 389–397.

Stewart, J.E., Kim, M.-S., James, R.L., Dumroese, R.K. and Klopfenstein, N.B. (2006) Molecular characterization of *Fusarium oxysporum* and *Fusarium commune* isolates from a conifer nursery. *Phytopathology* 96, 1124–1133.

Suarez, M.B., Walsh, K., Boonham, N., O'Neill, T., Pearson, S. and Barker, I. (2005) Development of real-time PCR (TaqMan®) assays for the detection and quantification of *Botrytis cinerea* in planta. *Plant Physiology and Biochemistry* 43, 890–899.

Takken, F.L., Thomas, C.M., Joosten, M.H., Golstein, C., Westerink, N., Hille, J., Nijkamp, H.J., de Wit, P.J. and Jones, J.D. (1999) A second gene at the tomato Cf-4 locus confers resistance to *Cladosporium fulvum* through recognition of a novel avirulence determinant. *The Plant Journal* 20, 279–288.

Tanaka, A., Shiotani, H., Yamamoto, M. and Tsuge, T. (1999) Insertional mutagenesis and cloning of the genes required for biosynthesis of the host-specific AK-toxin in the Japanese pear pathotype of *Alternaria alternata*. *Molecular Plant–Microbe Interactions* 12, 691–702.

Temme, N. and Tudzynski, P. (2009) Does botrytis cinerea ignore H_2O_2-induced oxidative stress during infection? Characterization of *Botrytis* activator protein 1. *Molecular Plant–Microbe Interactions* 22, 987–998.

Temme, N., Oeser, B., Massaroli, M., Heller, J., Simon, A., González Collado, I., Viaud, M. and Tudzynski, P. (2012) BcAtf1, a global regulator, controls various differentiation processes and phytotoxin production in *Botrytis cinerea*. *Molecular Plant Pathology* 13, 704–718.

ten Have, A., Mulder, W., Visser, J. and van Kan, J.A. (1998) The endopolygalacturonase gene *Bcpg1* is required for full virulence of *Botrytis cinerea*. *Molecular Plant Microbe Interactions* 11, 1009–1016.

Torres-Calzada, C., Tapia-Tussell, R., Quijano-Ramayo, A., Martin-Mex, R., Rojas-Herrera, R., Higuera-Ciapara, I. and Perez-Brito, D. (2011) A species-specific polymerase chain reaction assay for rapid and sensitive detection of *Colletotrichum capsici*. *Molecular Biotechnology* 49, 48–55.

Triky-Dotan, S., Austerweil, M., Steiner, B., Peretz-Alon, Y., Katan J. and Gamliel A (2009) Accelerated degradation of metam-sodium in soil and consequences for root-disease management. *Phytopathology* 99, 362–368.

Turrion-Gomez, J.L., Eslava, A.P. and Benito, E.P. (2010) The flavohemoglobin BCFHG1 is the main NO detoxification system and confers protection against nitrosative conditions but is not a virulence factor in the fungal necrotroph *Botrytis cinerea*. *Fungal Genetics and Biology* 47, 484–496.

Tyagi, S. and Kramer, F.R. (1996) Molecular beacons: probes that fluoresce upon hybridization. *Nature Biotechnology* 14, 303–308.
Tzima, A.K., Paplomatas, E.J., Tsitsigiannis, D.I. and Kang, S. (2012) The G protein β subunit controls virulence and multiple growth- and development-related traits in *Verticillium dahliae*. *Fungal Genetics and Biology* 49, 271–283.
Valette-Collet, O., Cimerman, A., Reignault, P., Levis, C. and Boccara, M. (2003) Disruption of *Botrytis cinerea* pectin methylesterase gene Bcpme1 reduces virulence on several host plants. *Molecular Plant–Microbe Interactions* 16, 360–367.
Vallejo, I., Carbú, M., Muñoz, F., Rebordinos, L. and Cantoral, J.M. (2002) Inheritance of chromosome-length polymorphisms in the phytopathogenic ascomycete *Botryotinia fuckeliana* (anam. *Botrytis cinerea*) *Mycological Research* 106, 1075–1085.
van Doorn, R., Sławiak, M., Szemes, M., Dullemans, A.M., Bonants, P., Kowalchuk, G.A. and Schoen, C.D. (2009) Robust detection and identification of multiple oomycetes and fungi in environmental samples by using a novel cleavable padlock probe-based ligation detection assay. *Applied and Environmental Microbiology* 75, 4185–4193.
van Esse, H.P., van't Klooster, J.W., Bolton, M.D., Yadeta, K.A., van Baarlen, P., Boeren, S., Vervoort, J., de Wit, P.J. and Thomma, B.P. (2008) The *Cladosporium fulvum* virulence protein Avr2 inhibits host proteases required for basal defense. *The Plant Cell* 20, 1948–1963.
Vellicce, G., Ricci, J.D., Hernández, L. and Castagnaro, A. (2006) Enhanced resistance to *Botrytis cinerea* mediated by the transgenic expression of the chitinase gene *ch5B* in strawberry. *Transgenic Research* 15, 57–68.
Viaud, M., Brunet-Simon, A., Brygoo, Y., Pradier, J.M. and Levis, C. (2003) Cyclophilin A and calcineurin functions investigated by gene inactivation, cyclosporin A inhibition and cDNA arrays approaches in the phytopathogenic fungus *Botrytis cinerea*. *Molecular Microbiology* 50, 1451–1465.
Viaud, M., Fillinger, S., Liu, W., Polepalli, J.S., Le Pêcheur, P., Kunduru, A.R., Leroux, P. and Legendre, L. (2006) A class III histidine kinase acts as a novel virulence factor in *Botrytis cinerea*. *Molecular Plant–Microbe Interactions* 19, 1042–1050.
Wang, N.Y., Lin, C.H. and Chung, K.R. (2010) A Gα subunit gene is essential for conidiation and potassium efflux but dispensable for pathogenicity of *Alternaria alternata* on citrus. *Current Genetics* 56, 43–51.
Westerink, N., Brandwagt, B.F., de Wit, P.J. and Joosten, M.H. (2004) *Cladosporium fulvum* circumvents the second functional resistance gene homologue at the *Cf-4* locus (*Hcr9-4E*) by secretion of a stable avr4E isoform. *Molecular Microbiology* 54, 533–545.
Whitaker, V.M., Bradeen, J.M., Debener, T., Biber, A. and Hokanson, S.C. (2009) Rdr3, a novel locus conferring black spot disease resistance in tetraploid rose: genetic analysis, LRR profiling, and SCAR marker development. *Theoretical and Applied Genetics* 120, 573–585.
Whitcombe, D., Kelly, S., Mann, J., Theaker, J., Jones, C. and Little, S. (1999) Scorpions™ primers – a novel method for use in single-tube genotyping. *American Journal of Human Genetics* 65, 2333.
White, T.J., Bruns, S.L. and Taylor, J.W. (1990) Amplification and direct sequencing of fungal ribosomal RNA genes for phylogenetics. In: Innis, M.A., Gelfand, D.H., Sninsky, J.J. and White, T.J. (eds) *PCR Protocols: a Guide to Methods and Applications*. Academic Press, San Diego, pp. 315–322.
Whitelaw-Weckert, M.A., Curtin, S.J., Huang, R., Steel, C.C., Blanchard, C.L. and Roffey, P.E. (2007) Phylogenetic relationships and pathogenicity of *Colletotrichum acutatum* isolates from grape in subtropical Australia. *Plant Pathology* 56, 448–463.
Winton, L.M. and Hansen, E.M. (2001) Molecular diagnosis of *Phytophthora lateralis* in trees, water, and foliage baits using multiplex polymerase chain reaction. *Forest Pathology* 31, 275–283.
Woo, S.L., Scala, F., Ruocco, M. and Lorito, M. (2006) The molecular biology of the interactions between *Trichoderma* spp., phytopathogenic fungi, and plants. *Phytopathology* 96, 181–185.
Wu, C.P., Chen, G.Y., Li, B., Su, H., An, Y.L., Zhen, S.Z. and Ye, J.R. (2011) Rapid and accurate detection of *Ceratocystis fagacearum* from stained wood and soil by nested and real-time PCR. *Forest Pathology* 41, 15–21.
Yago, J.I., Lin, C.H. and Chung, K.R. (2011) The SLT2 mitogen-activated protein kinase-mediated signalling pathway governs conidiation, morphogenesis, fungal virulence and production of toxin and melanin in the tangerine pathotype of *Alternaria alternata*. *Molecular Plant Pathology* 12, 653–665.
Yakoby, N., Beno-Moualem, D., Keen, N.T., Dinoor, A., Pines, O. and Prusky, D. (2001) *Colletotrichum gloeosporioides* pelB is an important virulence factor in avocado fruit–fungus interaction. *Molecular Plant–Microbe Interactions* 14, 988–995.

Yang, S.L. and Chung, K.R. (2012) The NADPH oxidase-mediated production of hydrogen peroxide (H_2O_2) and resistance to oxidative stress in the necrotrophic pathogen *Alternaria alternata* of citrus. *Molecular Plant–Microbe Interactions* 13, 900–914.
Yang, Y., Hou, S., Cui, G., Chen, S., Wei, J. and Huang, L. (2010) Characterization of reference genes for quantitative real-time PCR analysis in various tissues of *Salvia miltiorrhiza*. *Molecular Biology Reports* 37, 507–513.
Yan, L., Yang, Q., Sundin, G.W., Li, H. and Ma, Z. (2010) The mitogen-activated protein kinase BOS5 is involved in regulating vegetative differentiation and virulence in *Botrytis cinerea*. *Fungal Genetics and Biology* 47, 753–760.
Zheng, L., Campbell, M., Murphy, J., Lam, S. and Xu, J.R. (2000) The *BMP1* gene is essential for pathogenicity in the gray mold fungus *Botrytis cinerea*. *Molecular Plant–Microbe Interactions* 13, 724–732.

11 New Biocontrol Strategies for Strawberry Fungal Pathogens

María Carbú[1], Victoria E. González-Rodríguez[1], Carlos Garrido[1], Amjad M. Husaini[2] and Jesús M. Cantoral[1]*

[1]*Departamento de Biomedicina, Biotecnología y Salud Pública, Facultad de Ciencias del Mar y Ambientales, Instituto Universitario de Investigación Vitivinícola y Agroalimentaria (IVAGRO), Universidad de Cádiz, Puerto Real, Spain;* [2]*Centre for Plant Biotechnology, Sher-e-Kashmir University of Agricultural Sciences and Technology of Kashmir, Shalimar, Srinagar, Jammu and Kashmir, India*

11.1. Introduction

Strawberry (*Fragaria* × *ananassa*) is an important high-value commercial crop produced both conventionally and organically in open fields and in plastic-covered tunnels. The strawberry plants are susceptible to damage caused by environmental factors such as dehydration of the fruit or physiological disorders, and can be infected by different pathogens such as arthropods, nematodes and various fungi, bacteria and viruses, among other pests, causing enormous economic losses to strawberry growers.

The group of phytopathogenic fungi that attack this crop is especially extensive. More than 60 species of fungus can affect *Fragaria* × *ananassa* cultivars, but not all of them have the same commercial importance. Damage caused by these pathogens represents economic losses to the agricultural sector valued at hundreds of millions of dollars/euros per year (FAO, 2008–2013). Among the most significant fungal genera are *Botrytis*, *Colletotrichum*, *Phytophthora*, *Fusarium*, *Verticillium* and *Rhizopus*. These pathogens are capable of infecting both the strawberry plant and the fruit postharvest,

* jesusmanuel.cantoral@uca.es

and cause significant damage to more than 100 other crops cultivated around the world. *Colletotrichum* spp., and more specifically *Colletotrichum acutatum*, *Colletotrichum gloeosporioides* and *Colletotrichum fragariae* (anthracnose leaf spot and irregular leaf spot) and the species *Rhizoctonia solani* (*Rhizoctonia* leaf blight) cause destruction of the foliage, which can weaken the plants and thus reduce crop yields (Garrido *et al.*, 2011). These diseases often occur in temperate areas of the tropical and subtropical regions where strawberry is cultivated.

Botrytis cinerea is considered the main causative agent of fruit rot, a disease known as *Botrytis* fruit rot and also as grey mould. However, it is known that other fungi such as *Colletotrichum* spp., *Phytophthora cactorum* and *Rhizopus stolonifer* are able to attack the fruit in the field and postharvest (Washington *et al.*, 1992). These rots start in the field during cultivation and continue throughout the processes of collection, cold storage and exposure at room temperature in the markets. The percentage of each of these genera of fungi in the total number of phytopathogenic fungi varies depending on the environmental conditions of each marketing year, the locality and even the different times when fruit collection is made (Sánchez *et al.*, 2003). In either case, damage to the fruit represents the most direct and visible business loss.

The roots and crown of the strawberry plants are also attacked by various fungi, such as *Phytophthora* (*P. fragariae* and *P. cactorum*), *Colletotrichum* (*C. acutatum*, *C. fragariae* and *C. gloesporoides*), *Verticillium* (*V. albo-atrum* and *V. dahliae*) and *Rhizopus* (*R. fragariae* and *R. solani*) (Garrido *et al.*, 2011). These fungi may attack and destroy secondary roots and invade primary roots, limiting the plant's ability to take up water and nutrients, resulting in the fruit drying and undergoing poor development, with wilting and plant collapse (Maas, 2004). Crown infection may derive from root infection from or direct infection of the crown itself; in either case, the infected strawberry plants fail to grow successfully.

Traditionally, the control of fungi and bacteria in crops was based on the use of chemical compounds such as fungicides and bactericides. According to the Food and Agriculture Organization, the use of fungicide decreased from 355,000 t in 2005 to 215,600 t in 2013 (FAO, 2005–2013).

Figure 11.1 shows the use of fungicides across the different continents, and shows that the USA and Europe consumed the highest quantities of fungicides, with maximum values in 2010 and 2011, respectively. After these time points, they reduced the use of fungicides, with the USA showing the greatest reduction. Asia, Africa and Oceania maintained the same level of use during 2005–2010, and these levels were lower than those in Europe and America. A possible explanation for the decrease in the use of fungicide in the USA and Europe could be the stricter regulations that these countries have applied for protection of the environment and consumer health. According to data available from the Business Association for Plant Protection (AEPLA), the plant protection market worldwide represented an expense of US$47,255 million in 2012. The expense in Europe due to the use of phytosanitary measures was estimated at €8,977,038 million in 2011, of which €679 million was the cost for Spanish farmers (AEPLA, 2012).

Worldwide, the most prevalent strawberry disease management practice is the application of chemical treatments, in conjunction with other practices for managing plant health, including, for example, the reduction of overcrowding and overwatering of plants, rapid removal of diseased plants and plant debris, and constant monitoring of plants to ensure healthy production stocks (Maas, 2004). Soil fumigation has been of paramount importance for strawberry fruit and nursery-plant production in recent decades, and has enabled growers to develop systems for high-value annual crop production. However, increasing the use of biological control agents for the control of pathogens of fruits and vegetables is becoming more common.

The use of biological methods for the control of plant pathogens has been studied for more than 70 years and appears to be a feasible option. However, and in view of the results obtained by Martin and Bull (2002), there is still a long way to go to develop an effective system of strawberry production

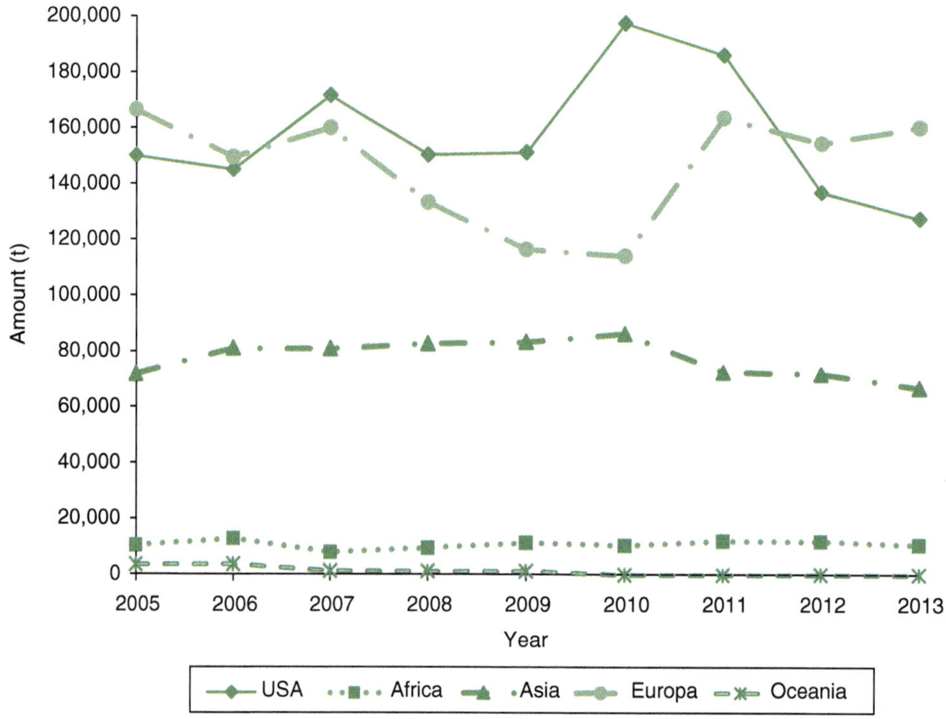

Fig. 11.1. The amount (t) of fungicides and bactericides used in different continents between 2005 and 2013.

based on the use of biological control methods with results similar to those obtained with soil fumigation by methyl bromide. In order to achieve successful control by these methods, it is necessary to have extensive knowledge about which pathogens are controlled, how they contribute to the growth and destruction of the field plantings, their distribution in the areas of production, the time of the year when they have the most significant impact on strawberry plants, how environmental parameters affect the damage, the influence of cropping practices on the severity of the damage and how biological control agents can exert their protective effects in plants (Martin and Bull, 2002).

11.2. Methyl Bromide and Other Fungicides

For the last 50 years, intensive strawberry cultivation in most regions of the world was done using pre-plant soil disinfestation with methyl bromide (MeBr) and chloropicrin, individually or in mixtures to control key soilborne pathogens, weeds and pests. Fumigation with MeBr causes: (i) an increased growth response in strawberries of about 30% in soils with low pathogen and weed pressures; (ii) a reduction in nitrogen fertilizer requirement of about 50%; and (iii) improved nutrient uptake (Wilhelm and Paulus, 1980; Porter *et al.*, 2006). Such benefits are due partly to the changes in soil biology and chemistry that fumigation induces, and the associated beneficial impacts in the rhizosphere of strawberries (Porter *et al.*, 2005). However, MeBr emissions can degrade stratospheric ozone (Butler, 1995), and therefore it was listed for phase-out for non-quarantine, soil disinfestation purposes under the *Montreal Protocol on Substances that Degrade the Ozone Layer* (Montreal Protocol). According to the Montreal Protocol, the import and manufacture of MeBr in the USA and other developed countries has been banned since 2005, after stepwise reductions in 1999, 2001 and 2003. Currently, there is no single registered alternative fumigant for all of the

MeBr uses and there is a need for environmentally sound and economically feasible alternatives. Initially, the efforts of the scientific community focused on finding new chemical compounds that had the same efficiency as that of MeBr for soil disinfestation and strawberry production. Some of the alternative fumigants to MeBr included metam sodium and 1,3-dichloropropene (1,3-D), but none have completely replaced MeBr. The fumigants 1,3-D and chloropicrin in combination with methyl isothiocyanate generators have shown to be the most promising alternatives to MeBr for strawberry production. Studies with the experimental fumigants methyl iodide and propargyl bromide have suggested that these compounds have higher reactivity than MeBr as stand-alone fumigants (Ajwa *et al.*, 2003). Samtani *et al.* (2011) evaluated the efficacy of steam disinfestations and other non-fumigant alternatives for soil disinfestation and strawberry production, and concluded that steam treatment applied pre-plant provided weed control similar to that of MeBr/chloropicrin and therefore this treatment could be considered as a viable replacement for MeBr.

However, the use of chemical compounds to control crop pathogens has many negative effects on the environment, and can cause the appearance of fungal strains resistant to these fungicides. Therefore, industries and regulatory authorities as well as the scientific community are focusing increasingly on the search for alternative methods for the control of strawberry diseases, including biological control, cultural measures and other measures that could be integrated into the production system of this crop.

The search for an alternative to MeBr is also important for nursery-plant production, and the transplantation of these plants in the field. If these plants are not pathogen free, the nursery industry may not be able to provide an appropriate number of transplants for commercial fruit production.

11.3. New Biological Control Agents

The biological control of plant disease was defined by Baker as 'the decrease in inoculum or the disease-producing activity of a pathogen accomplished through one or more organisms, including the host plant, but excluding man' (Baker, 1987). Based on this definition, a number of potential biological control methods have been developed, including: (i) plant-derived fungicides from secondary plant metabolites; (ii) the use of antagonistic microorganisms; and (iii) the manipulation of resistance responses in harvested commodities (Wilson *et al.*, 1991).

Tens of thousands of compounds that plants produce during secondary metabolism appear to be involved in plant fitness, because of their ability to 'repel or attract other organisms' (Wilson *et al.*, 1991). Several plant extracts (Grainge and Ahmed, 1988; Davidson and Parish, 1989; Qasem and Aau-Blan, 1996; Wilson *et al.*, 1997) and essential oils (Singh *et al.*, 1980; Kurita *et al.*, 1981; Wilson *et al.*, 1997) have shown antifungal activity against a wide range of fungi, such as *Botrytis cinerea*, *Rhizopus stolonifera* and *Monilinia fructicola*. Within the large reservoir of natural fungicides that exist in plants, it is reasonable to assume that these compounds may serve as a safe and effective alternative to synthetic fungicides. Such compounds, if properly formulated and applied, could be used directly or could serve as templates for synthetic analogues.

The biological control of fungal plant pathogens by microorganisms has been developed as an alternative to fungicide treatment and considerable success has been achieved in controlling both preharvest and postharvest diseases. These antagonistic microorganisms must interact with the crop plants, potential pathogens, environmental variables and indigenous organisms under the prevailing microclimatic conditions (Sutton, 1994). They must also be natural colonizers of strawberry fruits, leaves and roots, and be capable of interacting successfully with the plant, microbiological and other environmental conditions (Maas, 2004). Several mechanisms, including antibiosis, competition for limited nutrients and space, induction of plant defence responses, suppression of sporulation and mycoparasitism, have been proposed to be responsible for the biological control of fungal pathogens by these microbial agents. In strawberry, two main groups of microorganisms have been studied as biocontrol agents against strawberry fungal pathogens: (i) rhizobacteria, with *Pseudomonas* and *Streptomyces* spp. being the main members of this group; and (ii) members of the antagonistic fungal genus *Trichoderma*, the most effective fungal biocontrol agent identified to date (Berg *et al.*, 2000; Woo *et al.*, 2006; Garrido *et al.*, 2011).

11.3.1. Biofumigation

One of the biological approaches suggested by the United Nations Methyl Bromide Technical Options Committee as a possible component of soil disinfestation systems is biofumigation (Anon., 2006). Biofumigation refers to the agronomic practice based on incorporating soil amendment (fresh plant mass and manure) into the soil, which will release chemical substances (allelochemicals), known as isothiocyanates, able to suppress soilborne pests and diseases, plus a soil heater to enhance biological activities. These allelochemicals released from damaged tissues of members of the Brassicaceae and other minor families have the capacity to kill fungal, bacterial and nematode pathogens, and also weeds. Brassicaceae residues contain glucosinolates that yield isothiocyanates and other biologically active compounds when hydrolysed by the enzyme myrosinase. In addition to biofumigation, pest control attained in response to *Brassica* residue amendments has been attributed to altered soil biology (Cohen *et al.*, 2005; Mazzola *et al.*, 2007; Friberg *et al.*, 2009), at times acting through the modulation of host defence responses (Cohen and Mazzola, 2006). This practice involves growing selected Brassicaceae plants as biocidal green manure crops to amend the soil with a high level of organic matter and natural bioactive compounds. A preliminary study on the drying of some Brassicaceae plant organs to produce biocidal pellets to be used as organic treatments in addition to or as alternative to biocidal green manure was conducted by Lazzeri *et al.* (2004). The dried plants showed *in vitro* fungitoxic activity on two of the main agents of 'root rot', *Pythium irregulare and Rhizoctonia solani* (AG-5 strain) (Table 11.1). The authors concluded that, if tested on an industrial scale, these dried treatments might open up interesting applications as ecological alternatives to MeBr for the control of soilborne fungi, both in organic as well as in conventional agriculture.

The results obtained in laboratory tests by Mattner *et al.* (2008) showed that the *Brassica rapa/B. napus* fumigant crop had the capacity to suppress a diversity of soilborne fungal pathogens of strawberry (Table 11.1). At high doses, volatiles from the biofumigant crop were lethal to six different soilborne pathogens: *Alternaria alternata, Colletotrichum dematium, Cylindrocarpon destructans, Fusarium oxysporum, Pythium ultimum, Pythium cactorum* and *Rhizoctonia fragariae* (Table 11.1). They also suggested that allelochemicals other than isothiocyanates, such as nitriles, or other biological mechanisms might play a role in

Table 11.1. Biofumigants used as biological control in strawberry crops.

Plants	Allelochemical	Fungal pathogens	Reference
Brassicacea	Isothiocyanates	*Phythium irregular, Rhizoctonia solani*	Lazzeri *et al.* (2004)
Brassica rapa, Brassica napus	Isothiocyanates and nitriles	*Alternaria alternata, Colletotrichum dematium, Cylindrocarpon destructans, Fusarium oxysporum, Penicillium ultimum, P. cactorum, Rhizoctonia fragariae*	Mattner *et al.* (2008)
Lavandula angustifolia 'Maillette', *Lavandula intermedia* 'Grosso'	Monoterpenoids	*Verticillium dahliae*	Yohalem and Passey (2011)

biofumigation. Subsequent work by Yohalem and Passey (2011) has shown that various *Lavandula*-based materials (*Lavandula angustifolia* 'Maillette', *Lavandula intermedia* 'Grosso' and their waste material) can reduce the number of viable microsclerotia of *V. dahliae* in soil and inhibit wilt in strawberry (Table 11.1). The monoterpenoids associated with *Lavandula* spp. affect the viability of fungi by interference with membrane integrity. The presence and persistence of the monoterpenoids were correlated with a reduction in pathogen population. As such, these authors recommended the use of biocidal green manure in combination with other physical (solarization) or biological techniques with low environmental impact, and the use of genetically resistant strawberry varieties for effective control of several soil pests.

11.3.2. Antagonistic rhizobacteria

The rhizosphere is the site for interactions between plants, pathogenic microorganisms and antagonistic rhizobacteria/fungi. Antagonistic rhizobacteria have often been the focus of research in sustainable systems of agriculture because of their ability to suppress soilborne diseases. With the objective of isolating new biocontrol agents against strawberry *Verticillium* wilt, Berg *et al.* (2000) developed a strategy for effective selection and evaluation of these agents. A variety of different antagonistic bacteria was found for the two strawberry species under study (*Fragaria* × *ananassa* Thuill. 'Elanta' and *Fragaria viridis*). The bacterial species *Pseudomonas fluorescens*, *Streptomyces albidoflavus*, *Streptomyces rimosus* and *Streptomyces diastachromogenes* were found in rhizospheres of both strawberry species, while *Pseudomonas chlororaphis* was found only in the rhizosphere of *Fragaria* × *ananassa*, and *Streptomyces exfoliatus* only in the rhizosphere of *Fragaria viridis*. After *in vitro* assays, the authors concluded that all isolates inhibited the growth of *V. dahliae*. In biocontrol experiments in the greenhouse and in the field, bacterial treatment also resulted in efficient suppression of the pathogen, in addition to increased production of strawberries. It was found that *Pseudomonas* treatment enhanced the yield of strawberry fruits by as much as 344% (Berg *et al.*, 2000). The explanation of this effect lies in the interactions that take place between the host and the microorganisms; this association provides the plant with the nutrients and environmental conditions for enhanced development of its root systems. On the basis of all these results, the authors concluded that it was possible to patent three effective biocontrol agents (two isolates of *Pseudomonas fluorescens* and one *Streptomyces diastachromogenes* isolate) antagonistic to the *Verticillium* wilt pathogen. Anandhakumar and Zeller (2008) performed *in vitro* screening of more than 100 rhizobacteria to control *Phytophthora* disease. They found three bacteria of different genera, *Raoultella terrigena*, *Bacillus amyloliquefaciens* and *Pseudomonas fluorescens*, with inhibitory effects on the mycelial growth of *Phytophthora fragariae* var. *fragariae* and *Phytophthora cactorum*. Of these three species, *Raoultella terrigena* had the highest inhibitor activity. A parameter that plays an important role in the efficacy of the applied agent on the target pathogen is the method of its application or treatment of the biological agents. They can be applied prior to sowing, on transplants, on tissue culture plantlets or as drenches. In the study by Anandhakumar and Zeller (2008), the roots of strawberry plants were dipped in the bacterial cell solution before planting. The advantages of this method are that: (i) the roots are directly exposed to the bacterial inoculum free of soil; and (ii) the pre-treatment ensures higher levels of activity, reduces production costs, simplifies formulation and application, reduces extension costs and minimizes environmental impact. Anandhakumar and Zeller, (2008) also investigated whether a mixture of bacterial species gave better control against crown rot and red stele disease of strawberry than a single bacterial species alone. Under natural conditions, *Raoultella terrigena* and *Bacillus amyloliquefaciens* both showed a significant effect in the first season (37.5%), while in second season *Raoultella terrigena* showed more effectiveness (45.1%).

The development of novel molecular biology techniques has allowed a deeper understanding of the molecular mechanisms and genes involved in the antagonist capacity of rhizobacteria (Garrido *et al.*, 2011). Costa *et al.* (2007) employed a suite of tools to test whether antagonistic *Pseudomonas* spp. corresponded to dominant populations in the rhizosphere of *V. dahliae* host plants. They developed a novel polymerase chain reaction/electrophoresis denaturing gradient gel (PCR-DGGE) system to characterize *gacA* gene types within *Pseudomonas* spp. It is known that the *gacA* gene is required for the production of many secondary metabolites and exoenzymes in *Pseudomonas* spp. (Heeb and Haas, 2001) and that it promotes the synthesis of molecules involved in antagonistic interactions and disease suppression, such as antibiotics and biocides. A study carried out by Costa *et al.* (2007) suggested that this family of genes, in conjunction with the PCR-DGGE system, might be a suitable strategy for the simultaneous analysis of *Pseudomonas* community structure and function in soil studies.

In order to develop a biocontrol product for commercial strawberry production based on the use of the chitinolytic rhizobacterium *Serratia plymuthica*, Kurze *et al.* (2001) evaluated the potential of a strain isolated from the rhizosphere of oilseed rape for controlling *Verticillium* wilt and *Phytophthora* root rot of strawberry. Artificial inoculation with *Serratia plymuthica* controlled the wilt and root rot of the strawberry plants and even produced an increase in the number of buds, blossoms and fruit. On the basis of these results, they patented *Serratia plymuthica* (HRO-C48 strain) as a biological control (Berg *et al.*, 2002)

11.3.3. Plant growth-promoting bacteria

Another alternative for crop protection against pathogens is the biological control exerted by some plant growth-promoting bacteria (PGPB). These PGPB are beneficial for plant growth, yield and crop quality. This group of bacteria includes members of the genera *Acinetobacter*, *Alcaligenes*, *Arthrobacter*, *Azospirillum*, *Azotobacter*, *Bacillus*, *Beijerinckia*, *Burkholderia*, *Enterobacter*, *Erwinia*, *Flavobacterium*, *Rhizobium* and *Serratia* (Bashan and de Bashan, 2005). The mechanisms by which PGPB promote plant growth are not fully understood, but are thought to include: (i) the ability to produce plant hormones, such as auxins, cytokinins and gibberellins, and to inhibit ethylene production; (ii) asymbiotic nitrogen fixation; (iii) solubilization of inorganic phosphate and mineralization of organic phosphate and/or other nutrients; and (iv) antagonism against phytopathogenic microorganisms by production of siderophores, the synthesis of antibiotics, enzymes and/or fungicidal compounds, and competition with detrimental microorganisms. *Azospirillum* is a well-studied PGPB capable of naturally colonizing many plant species, including strawberry. Different *Azospirillum brasilense* strains isolated from strawberry plants show important characteristics within the PGPB group: (i) nitrogen-fixing activity; (ii) the production of siderophores and indoles; (iii) the promotion of growth of strawberry plants (Pedraza *et al.*, 2010; Tortora *et al.*, 2011, 2012); and (iv) protection of strawberry plants against anthracnose disease (Tortora *et al.*, 2012). Tortora *et al.* (2012) characterized the systemic resistance induced by *Azospirillum brasilense* on strawberry plants at the biochemical and molecular levels. Their results revealed the accumulation of salicylic acid and induction of defence-related genes, suggesting that this response is related to structural cell wall modifications as a consequence of the increase in phenolic compounds and callose deposition. They concluded that *Azospirillum brasilense* protected the strawberry plants against anthracnose disease caused by *Colletotrichum acutatum*.

In a later work, Guerrero-Molina *et al.* (2012) demonstrated the existence of an effective *Azospirillum brasilense* root association and further colonization through stolons from a single bacteria-inoculated strawberry mother plant to newborn daughter plants. As strawberry plants are commercially reproduced in nurseries by planting stolons into the soil, these results suggest the possibility of using a single inoculation with the selected PGPB to produce numerous plant generations in the nursery that are already inoculated and ready for planting in the field.

Another characteristic that has been studied is the growth-promoting effect of flower and foliar spraying and root inoculation of *Bacillus* and *Pseudomonas* on strawberry yield, growth and mineral content under organic growing conditions (Esitken *et al.*, 2010). The results of this study showed that root inoculation of *Bacillius* (M3 strain) and spraying of *Pseudomonas* (BA-8 strain) or *Bacillus* (OSU-142 strain) significantly affected fruit yield, vitamin C content and fruit number per plant, increased both the growth and phosphorus (P), iron (Fe), copper and zinc (Zn) content of the strawberry plant and increased the soil P, Fe, Zn, potassium and magnesium availability. The authors concluded that the bacteria tested in their study are a promising biofertilizer for fruit and vegetable production in sustainable and organic agricultural systems.

11.3.4. Antagonistic fungi

Biological control based on the use of antagonistic fungal agents has gained wide acceptance, next to antagonistic bacteria, primarily because of their broader spectrum in terms of disease control. In several assays performed on strawberry, table grape berries and kiwifruit, the yeast-like fungus *Aureobasidium pullulans* and the yeasts *Candida vanderwaltii* and *Candida oleophila* were the most effective antagonists of *Botrytis cinerea* and *Rhizopus stolonifer*. In 1997, *Aureobasidium pullulans* and *Candida oleophila* isolates were utilized in trials on strawberry grown in plastic tunnels (Lima *et al.*, 1997). Both antagonists showed activity against *Botrytis cinerea* and *Rhizopus stolonifer*, being more active when applied at full bloom and late petal fall than at fruit maturity. In both cases, competition for nutrients seemed to be the mode of action of these biocontrol agents. After microscopic examination of the *Candida oleophila* strain, the authors observed that there could also be a direct interaction between this yeast and *Botrytis cinerea*.

Another fungus antagonist that has proved capable of reducing the incidence of *Botrytis cinerea* on strawberry has been *Gliocladium roseum*. The mode of action of this ascomycete is by the reduction of inocula from infected leaves and protection of the flower parts from attack by the phytopathogen (Maas, 2004). Strangely, in some studies published in 1991, it was observed that inoculation of strawberry plants with non-pathogenic isolates of *Fusarium oxysporum* could be used to control wilt caused by *Fusarium* itself (Okayama, 1991; Tezuka and Makino, 1991).

When the biocontrol agent is studied under controlled environmental conditions, such as postharvest biological control, it is possible to modify and/or control the conditions so that the agent's activity is as optimal as possible. However, this control cannot be exercised when the antagonistic fungal agent is applied in the field, where fluctuations in environmental conditions may prevent the biocontrol agent from establishing, and may hamper its activity or survival in the phyllosphere. The fungus *Clonostachys rosea* (syn. *Gliocladium roseum*: teleomorph *Bionectria ochroleuca*) (Schroers *et al.*, 1999) is an antagonist that controls *Botrytis cinerea* efficiently on strawberry leaves, petals and stamens in experimental conditions (Peng and Sutton, 1991; Sutton and Peng, 1993; Sutton *et al.*, 1997). Based on these preliminary data, Cota *et al.* (2008) evaluated the efficiency of four *Clonostachys rosea* isolates in the biological control of strawberry grey mould under field conditions. In their experiments, applying *Clonostachys rosea* was as efficient as the use of fungicides (procymidone and captan) in reducing leaf colonization by *Botrytis cinerea*, as well as grey mould incidence in both flowers and fruits. The competition for nutrients and substrate is the major mechanism that *Clonostachys rosea* uses against *Botrytis cinerea* (Sutton *et al.*, 1997). The authors concluded that *Clonostachys rosea* reduced sporulation of *Botrytis cinerea* in leaves by competition with the saprophytic pathogen resulting in an effective reduction of the inoculum produced in crop debris, which contributed to the reduction in damage to flowers and fruits (Cota *et al.*, 2008). Furthermore, they observed that the effect was most striking when the isolates were applied twice a week, probably because the antagonist reached a larger number of flowers at several developmental stages. The application of *Clonostachys rosea* twice a week was more efficient in controlling grey mould than a weekly application of fungicides (procymidone and captan). More recently, it has been shown that *Clonostachys rosea* is highly susceptible to UV radiation and has reduced ability to antagonize a pathogen in the presence of sunlight (Morandi *et al.*, 2008; Costa *et al.*, 2012). Therefore, Costa *et al.* (2013) evaluated the ability of an isolate of *Clonostachys rosea*, selected previously for its tolerance to UV-B radiation, to control *Botrytis cinerea* on strawberry leaves in controlled experiments. The results indicated that UV-B radiation reduced the ability of *Clonostachys rosea* to control pathogens on strawberry leaves and that the applied inoculum of the biocontrol agent must therefore be higher than the effective dose to control *Botrytis cinerea*. They concluded that, for the development of products based on biological control agents, the effect of UV-B should be studied in efficacy studies (Costa *et al.*, 2013).

Another fungus that has been studied for its ability as a biocontrol agent against *Botrytis cinerea* is the ballistosporous yeast *Sporidiobolus pararoseus* belonging to Basidiomycota (Huang *et al.*, 2012). The authors evaluated the efficacy of the live yeast cells and their volatile organic compounds (VOCs) in the suppression of strawberry grey mould under controlled conditions. They found that, on strawberry fruits naturally infected with *Botrytis cinerea*, *Sporidiobolus pararoseus* not only suppressed this pathogen but also suppressed other fungi, including *Mucor* spp., *Penicillium* spp. and *Rhizopus* spp. Therefore, they suggested that *Sporidiobolus pararoseus* may have the potential to suppress a wide range of decay-inducing fungi on postharvest fruits. They also noticed that the VOCs produced by *Sporidiobolus pararoseus* could effectively inhibit both conidial germination and mycelial growth of *Botrytis cinerea* on agar medium and that it could suppress strawberry grey mould disease. Based on these results, Huang *et al.* (2012) suggested that the competition for limited nutrients and space and/or the production of antifungal VOCs could be the mechanism used by *Sporidiobolus pararoseus* for the suppression of *Botrytis cinerea* on postharvest strawberry fruits. The authors concluded that *Sporidiobolus pararoseus* has the potential to be developed as a biofumigant to control grey mould disease of strawberry fruits stored in cartons (Huang *et al.*, 2012).

Among the alternatives that have been described for the control of fungal diseases in crops, biological control is less effective than many of the commercial fungicides currently in use. Organic and inorganic additives have been applied in combination with biocontrol agents to obtain beneficial effects (El-Ghaouth *et al.*, 2000; Meng *et al.*, 2010). Phytic acid (PA) is a natural plant compound that has properties such as antioxidant or iron chelation. This compound is routinely added to foods to prevent product discoloration, increase nutritional quality and prolong shelf-life. However, there is no information concerning the effect of a combination of antagonistic yeast and PA on the control of postharvest diseases of strawberry. Against this background, Zhang *et al.* (2013) published a paper studying the effects of the yeast antagonist *Rhodotorula mucilaginosa*, used alone or in combination with PA, for controlling postharvest grey mould spoilage of strawberry and evaluated the efficacy of this yeast and PA, used alone or in combination, for controlling the natural spoilage of strawberry. They observed that *Rhodotorula mucilaginosa* as a stand-alone treatment had potential as a biocontrol agent for the control of postharvest grey mould decay of strawberry. Furthermore, they concluded that PA at a concentration of 4 μmol ml^{-1} could significantly enhance the biocontrol activity of *Rhodotorula mucilaginosa* against postharvest grey mould decay and natural decay development in strawberry. The mode of action of PA may be its direct inhibition of the growth of the pathogens, while promoting the growth of *Rhodotorula mucilaginosa* in strawberry and thus enhancing nutrient competition between them (Zhang *et al.*, 2013).

It has been observed that biological control of plant pathogens can be enhanced with the use of multiple-strain treatments as opposed to single-strain treatment (Elad *et al.*, 1998; Guetsky *et al.*, 2002a,b; Elad and Stewart, 2007), especially when applying biological control agents with different modes of action (Xu *et al.*, 2010). However, it is possible that such interactions between various biological control agents may cause a reduction in the effectiveness of biocontrol. Guetsky *et al.* (2001) investigated the use of a combination of the yeast *Pichia guilliermondii* and the bacterium *Bacillus mycoides* as an effective method to suppress *Botrytis cinerea* on strawberry leaves. Applied separately, the biological agents significantly inhibited spore germination, lesion formation and lesion development at most temperatures, relative humidities and spray-time combinations. However, the results were not totally reproducible under all conditions studied by the authors. However, when combined, the mixture of *Bacillus mycoides* and *Pichia guillermondii* suppressed *Botrytis cinerea* effectively under all conditions studied (Guetsky *et al.*, 2001).

The control of *Botrytis cinerea* on detached strawberry leaves was studied by Robinson-Boyer *et al.* (2009) and the efficiency of combinations of commercially available biological control agents was determined. Combinations were tested as a mixture or as sequential treatments before and after disease loading. The results demonstrated that combinations of biological control agents as mixtures resulted in less control (i.e. higher disease) than when the most effective biological control agent within the combination was applied alone, indicating possible antagonism between them. However, when two biological control agents were applied sequentially about 48 h apart, there was no strong evidence for antagonism between the agents. They concluded that there was usually no benefit in applying two biological control agents either together or sequentially, so in general they did not recommend a combination as the best method to control infections of strawberry leaves by *Botrytis cinerea*.

Control of powdery mildew, caused by *Podosphaera aphanis*, can be achieved by application of a single strain of antagonistic fungi as a biological control agent (Kiss, 2003). Sylla *et al.* (2013) developed a dual-culture method and leaf disc assays, and showed that there were various interactions *in vitro* between bacterial and fungal biocontrol strains against *Podosphaera aphanis*. For example, the combination of *Beauveria bassiana* and *Bacillus amyloliquefaciens* (FZB42 strain) presented the lowest inhibitory effects, while the combination of *Trichoderma harzianum* (strain T58) and *Bacillus subtilis* (strain FZB24), caused the greatest inhibition. On leaf discs, some of the multiple-strain treatments contributed to improved biological control (e.g. *Bacillus subtilis* + *Metarhizium anisopliae* (strain 43) or *Beauveria bassiana* + *Penicillium oxalicum*), while others did not show any significant effects (e.g. *Bacillus amyloliquefaciens* + *T. harzianum* (strain T22)). Moreover, it appeared that the entomopathogenic fungus *Metarhizium anisopliae* was a promising agent for multiple biological control agent treatments against *Podosphaera aphanis* conidiation in strawberry, while multiple treatments including *T. harzianum* and *Bacillus subtilis* was not. Sylla *et al.* (2013) concluded that, in order to consider treatment with multiple strains as an alternative for the control of powdery mildew in strawberry, it is necessary to have a greater knowledge of the microbial biogeography of strawberry plants/leaves, as well as the spectrum of exuded organic compounds from leaves and exudation sites (e.g. stomata, trichomes).

11.3.5. *Trichoderma* spp.

The genus *Trichoderma* has been the focus of many researchers and is the most effective fungal biocontrol agent found to date. *Trichoderma* spp. comprise almost 50% of the fungal biological control agents market, mostly as soil/growth enhancers, and this makes them interesting candidates for investigation. This genus has been known for a long time (since 1865), but its taxonomy and identification of the different species was not known until about 1969 (Rafai, 1969). The difficulty of distinguishing among the different species lay in the few morphological differences associated with the various species. However, the enormous progress made in the application of molecular techniques has enabled phylogenetic classification of *Trichoderma* spp. (Druzhinina *et al.*, 2006). Fungal species belonging to this genus are widely distributed throughout the world and can be isolated from soil, decaying wood and other forms of plant organic matter (Howell, 2003). Since the 1930s, it has been known that fungi of the genus *Trichoderma* have the potential to act as biocontrol agents of plant diseases. *Trichoderma* spp. have developed an antagonizing ability to interact, both parasitically and symbiotically, with different substrates and living organisms, including plants and other microbes (Woo *et al.*, 2006). There are currently a large number of publications demonstrating the ability of this fungus as a biocontrol agent; however, there is still a lack of knowledge about the molecular mechanisms that the fungus employs to carry out its antagonist activity. The reviews by Howell (2003), Woo *et al.* (2006) and Daguerre *et al.* (2014) summarize the main results obtained from genomic and proteomic studies of this fungus, with the aim of deepening our understanding of the mechanisms involved in the interactions that exist between *Trichoderma* spp., plants and fungal pathogens.

A number of studies have been conducted with the aim of selecting strains of *Trichoderma* spp. capable of carrying out effective control of infections caused by phytopathogenic fungi in strawberry production. Intensive research into biocontrol using *T. harzianum* has been carried out, and there have been some significant achievements in greenhouse crops and vineyards. The first biocontrol agent to be commercialized, registered and used was isolate T-39 of *T. harzianum* (TRICHODEX), which is effective against *Botrytis cinerea*, *Sclerotinia sclerotiorum*, *Cladosporium fluvum* and *Colletotrichum acutatum*. In 2004, Freeman *et al.* (2004) conducted a study that used various isolates of *Trichoderma*, including isolate T-39, to determine the influence of the use of different isolate combinations, timing applications and dosage rates on disease control. They concluded that all combinations of concentrations and application time intervals were effective in anthracnose disease reduction. However, it appeared that the time period between applications and concentrations are important parameters for grey mould control in strawberry. The isolates used in this study belonged to the species *T. harzianum*, *Trichoderma hamatum*, *Trichoderma longibrachiatum* and *Trichoderma atroviride*, indicating the taxonomic diversity of *Trichoderma* agents for biocontrol. Currently, more than 250 different *Trichoderma*-based agricultural products have been registered in the world, and are sold and applied to protect and improve yields of vegetables, ornamental plants and fruit trees (Woo *et al.*, 2014).

In recent decades, much effort has been made to understand the mechanisms used by *Trichoderma* spp. in their action as biocontrol agents against plant diseases. These fungi are characterized by: (i) the ability to use a variety of nutrient sources; (ii) being among the microorganisms that are most resistant to natural and man-made chemicals and toxins; (iii) the ability to effectively degrade some of these compounds, including hydrocarbons, chlorophenolic compounds, polysaccharides and xenobiotic pesticides (Harman *et al.*, 2004); (iv) their high versatility in adapting to many different environmental conditions; and (v) their large arsenal of molecular weapons to use against a particular microbe that they confront. Regarding the modes of action of members of the genus *Trichoderma*, several mechanisms have been described such as mycoparasitism, competition, the use of enzymes and antibiosis.

According to Punja and Utkhede (2003), *Trichoderma* spp. are the most widely studied mycoparasitic fungi. This fungal biocontrol agent is able to parasitize other fungi, and the mechanisms include coiling around the hyphae of the pathogen, penetration and subsequent dissolution of the host cytoplasm (Howell, 2003). It has been reported that the fungus is able to excrete various toxins that are toxic to *Rhizoctonia solani* and *Sclerotinia americana*, and are even strongly inhibitory for *Pythium ultimum* and *Phytophthora* spp.; two of the most important toxins excreted by *Trichoderma virens* are gliotoxin and gliovirin (Howell, 2003). In subsequent studies, a mutant of *T. virens* deficient for both mycoparasitism and toxin biosynthesis still retained biocontrol efficacy, indicating that neither mycoparasitism nor toxins are essential for biocontrol of these pathogens (Liu *et al.*, 2007). Another relevant feature of mycoparasitism is related to the production of extracellular enzymes (e.g. proteolytic enzymes, the glucanolytic system, chitinase) (Verma *et al.*, 2007). The role of these enzymes has been well documented and it has been observed that they are a key factor in pathogen cell wall lysis during mycoparasitism (Verma *et al.*, 2007; Daguerre *et al.*, 2014).

Trichoderma spp. are capable of growing rapidly and colonizing the root system of the treated plant, a condition necessary for the biocontrol agent to compete for space and nutrients efficiently. However, Howell (2003) described that strains of *Trichoderma koningii* were excellent root colonizers but exhibited little or no biocontrol activity against *Rhizopus solani* on cotton seedlings. In addition, *Trichoderma* spp. were shown to compete and sequester ions of iron by releasing siderophore-type compounds (Verma *et al.*, 2007), which are also used by PGPB, as previously. These compounds are capable of generating complex interactions between plants and pathogens, which may contribute to the plant's ability to reach a certain level of tolerance or resistance to damage. This mechanism is similar to that employed by *Pseudomonas* spp., in that it induces plant root and shoot growth, resistance to biotic and abiotic stresses, and improvements in the nutritional status of the plant. Harman (2000) showed that, after treatment of seed with *T. harzianum* (T-22 strain), corn planted in low-nitrogen soil produced plants that were greener and larger in the early part of the growing season. The authors reported a strong interaction between T-22 and the nitrogen-fixing bacterium *Bradyrhizobium japonicum*. Theoretically, the combination of a nitrogen-fixing bacterium and a fungus that enables the plant to utilize nitrogen more efficiently is directly related to a general beneficial growth effect on the root system, which thereby gives the plant more resistance against pathogens (Harman, 2000).

Antibiosis is the process of secretion of antimicrobial compounds by antagonistic fungi to suppress and/or kill pathogenic fungi in the vicinity of its growth area (Verma *et al.*, 2007). The elements that are probably related to the antibiosis of *Trichoderma* spp. are some antagonistic metabolites produced by the fungus. These metabolites have the characteristics of being linear, amphipathic polypeptides, namely peptaibols and peptaibiotics (Szekeres *et al.*, 2005). Szekeres *et al.* (2005) discussed the physico-chemical and biological properties of these antibiotic compounds, which included the disruption of lipid membranes, antimicrobial activities and induction of plant resistance.

Recent research into the possible mechanisms involved in biological control by *Trichoderma* spp. has revealed that enzymes such as chitinases, glucanases and proteases produced by the biocontrol agent are responsible for suppression of the plant pathogen (Howell, 2003). The first two function by breaking down the polysaccharides, chitin and β-glucans that are responsible for the rigidity of fungal cell walls, thereby destroying cell wall integrity In the case of proteases, the enzymes break down hydrolytic enzymes into peptide chains and/or their constituent amino acids and thereby destroy the capacity of the pathogen to act on plant cells. Nevertheless, when considering enzyme biosynthesis as a mechanism of biocontrol, one must take into account the synergistic effect between enzymes and antibiotics produced by *Trichoderma* spp. (Howell, 2003).

11.3.6. Summary

The possible strategies for biological control of phytopathogenic fungi that damage strawberry plants include the use of biofumigation, antagonistic rhizobacteria, PGPB and antagonistic fungal agents. The mechanisms used by these biocontrol agents are many and complex, and the use of a particular one will depend mainly on the type of interactions established between the plant host, the phytopathogenic fungus and the biocontrol agent. It is also important to consider that there are many factors that influence the effectiveness of the agent: soil type, pH, temperature, humidity, soil environment, rhizosphere, and method of application or treatment. Therefore, for effective biocontrol, there has to be a proper understanding of the combination of different mechanisms and factors that work synergistically in exercising adequate control of diseases of strawberry plants. It is therefore important and necessary to continue to deepen our knowledge of the mechanisms used by different biocontrol agents, including those already known and newly discovered agents.

11.4. Conclusion

Strawberry is a fruit of high commercial value, and its cultivation extends to numerous regions around the world. The strawberry plant is susceptible to many diseases including those caused by fungal pathogens, which are responsible for high economic losses to the strawberry sector. Traditionally, the control of these diseases is done by applying fungicides. Today, in almost all countries, application of fungicides is partially prohibited for the control of fungal rot during cultivation of strawberry and is completely prohibited after harvesting the fruit. Therefore, the use of biological control agents against these pathogens has become the most interesting alternative, and the scientific community is focusing its efforts in this area. In this chapter, we have discussed the use of MeBr and other fungicides to control fungal damage in strawberry cultivation. We also discussed the various alternatives that are being developed for biological control of strawberry crop diseases, as well as the mode of action of these biological control agents to control the various diseases of the crown, stem, leaves and strawberry fruit.

Acknowledgements

We would like to thank the Spanish Government, the Ministerio de Ciencia y Tecnología, the Junta de Andalucía (Regional Government) and the European Union (Fondos Europeos de Desarrollo Regional) for financing most of the work of our group given in this chapter: Spanish DGICYT AGL2006-13401-C02-02/AGR and AGL2009-13359-C02-02; AGL2012-39798-C02 y AGL2015-65684-C2-2R; Junta de Andalucía, PO7-FQM-02689; and by the University of Cádiz, project UCA 18DGUE1102. Victoria E. González-Rodríguez was financed by grant FPU of the Ministerio de Educación, Government of Spain, ref. AP2009-1309.

References

AEPLA (2012) *Statistical Databases*, Business Association for Plant Protection, Madrid, Spain.

Ajwa, H.A., Klose, S., Nelson, S.D., Minuto, A., Gullino, M.L., Lamberti, F. and Lopez-Aranda, J.M. (2003) Alternatives to methyl bromide in strawberry production in the United States of America and the Mediterranean region. *Phytopathologia Mediterranea* 42, 220–244.

Anandhakumar, J. and Zeller, W. (2008) Biological control of red stele (*Phytophthora fragariae* var. *fragariae*) and crown rot (*P. cactorum*) disease of strawberry with rhizobacteria. *Journal of Plant Diseases and Protection* 115, 49–56.

Anon. (2006) *Report of the Methyl Bromide Technical Options Committee*, United Nations Environment Programme, Nairobi, Kenya.

Baker, K.F. (1987) Evaluating concepts of biological control of plant pathogens. *Annual Reviews of Phytopathology* 25, 67–85.

Bashan, Y. and de Bashan, L. (2005) Bacteria/plant growth-promotion. In: Hillel, D. (ed.) *Encyclopedia of Soils in the Environment*. Elsevier, Oxford, UK, pp. 103–115.

Berg, G., Kurze, S., Buchner, A., Wellington, E.M. and Smalla, K. (2000) Successful strategy for the selection of new strawberry-associated rhizobacteria antagonistic to *Verticillium* wilt. *Canadian Journal of Microbiology* 46, 1128–1137.

Berg, G.D., Dahl, R. and Kurze, S. (2002) Strains of rhizobacteria for use against phytopathogenic soil fungi and methods of use thereof. Patent no. EP1021954 B1 Available at: http://www.google.es/patents/EP1021954B1?hl=es&cl=en.

Butler, J.H. (1995) Methyl bromide under scrutiny. *Nature* 376, 469–470.

Cohen, M.F. and Mazzola, M. (2006) Resident bacteria, nitric oxide emission and particle size modulate the effect of *Brassica napus* seed meal on disease incited by *Rhizoctonia* solani and *Pythium* spp. *Plant and Soil* 286, 75–86.

Cohen, M.F., Yamasaki, H. and Mazzola, M. (2005) *Brassica napus* seed meal soil amendment modifies microbial community structure, nitric oxide production and incidence of *Rhizoctonia* root rot. *Soil Biology and Biochemistry* 37, 1215–1227.

Costa, L., Rangel, D.N., Morandi, M.B. and Bettiol, W. (2012) Impact of UV-B radiation on *Clonostachys rosea* germination and growth. *World Journal of Microbiology and Biotechnology* 28, 2497–2504.

Costa, L.B., Rangel, D.E.N., Morandi, M.A.B. and Bettiol, W. (2013) Effects of UV-B radiation on the antagonistic ability of *Clonostachys rosea* to *Botrytis cinerea* on strawberry leaves. *Biological Control* 65, 95–100.

Costa, R., Gomes, N.C.M., Krögerrecklenfort, E., Opelt, K., Berg, G. and Smalla, K. (2007) *Pseudomonas* community structure and antagonistic potential in the rhizosphere: insights gained by combining phylogenetic and functional gene-based analyses. *Environmental Microbiology* 9, 2260–2273.

Cota, L.V., Maffia, L.A., Mizubuti, E.S.G., Macedo, P.E.F. and Antunes, R.F. (2008) Biological control of strawberry gray mold by *Clonostachys rosea* under field conditions. *Biological Control* 46, 515–522.

Daguerre, Y., Siegel, K., Edel-Hermann, V. and Steinberg, C. (2014) Fungal proteins and genes associated with biocontrol mechanisms of soil-borne pathogens: a review. *Fungal Biology Reviews* 28, 97–125.

Davidson P. M. and Parish M. E. (1989) Methods for testing the efficacy of food antimicrobials. *Food Technology* 43, 148–155.

Druzhinina, I.S., Kopchinskiy, A.G. and Kubicek, C.P. (2006) The first 100 *Trichoderma* species characterized by molecular data. *Mycoscience* 47, 55–64.

Elad, Y. and Stewart, A. (2007) Microbial control of *Botrytis* spp. In: Elad, Y., Williamson, B., Tudzynski, P. and Delen, N. (eds) *Botrytis: Biology, Pathology and Control*. Springer, The Netherlands, pp. 223–241.

Elad, Y., Kirshner, B., Yehuda, N. and Sztejnberg, A. (1998) Management of powdery mildew and gray mold of cucumber by *Trichoderma harzianum* T39 and *Ampelomyces quisqualis* AQ10. *BioControl* 43, 241–251.

El-Ghaouth, A., Smilanick, J.L., Wisniewski, M. and Wilson, C.L. (2000) Improved control of apple and citrus fruit decay with a combination of *Candida saitoana* and 2-deoxy-d-glucose. *Plant Disease* 84, 249–253.

Esitken, A., Yildiz, H.E., Ercisli, S., Figen Donmez, M., Turan, M. and Gunes, A. (2010) Effects of plant growth promoting bacteria (PGPB) on yield, growth and nutrient contents of organically grown strawberry. *Scientia Horticulturae* 124, 62–66.

FAO (2005–2013) *Statistical Databases*, Food and Agriculture Organization of the United Nations, Rome, Italy.

FAO (2008–2013) *Statistical Databases*, Food and Agriculture Organization of the United Nations, Rome, Italy.

Freeman, S., Minz, D., Kolesnik, I., Barbul, O., Zveibil, A., Maymon, M., Nitzani, Y., Kirshner, B., Rav-David, D., Bilu, A., Dag, A., Shafir, S. and Elad, Y. (2004) *Trichoderma* biocontrol of *Colletotrichum acutatum* and *Botrytis cinerea* and survival in strawberry. *European Journal of Plant Pathology* 110, 361–370.

Friberg, H., Edel-Hermann, V., Faivre, C., Gautheron, N., Fayolle, L., Faloya, V., Montfort, F. and Steinberg, C. (2009) Cause and duration of mustard incorporation effects on soil-borne plant pathogenic fungi. *Soil Biology and Biochemistry* 41, 2075–2084.

Garrido, C., Carbú, M., Fernández-Acero, F.J., González-Rodríguez, V.E. and Cantoral, J.M. (2011) New insight in the study of strawberry fungal pathogens. In: Husaini, A.M. and Mercado, J.A. (eds) *Genomics, Transgenics, Molecular Breeding and Biotechnology of Strawberry*. Global Science Books, Japan/UK, pp. 24–39.

Grainge, M. and Ahmed, S. (1988) *Handbook of Plants with Pest-Control Properties*. John Wiley & Sons, New York.

Guerrero-Molina, M.F., Winik, B.C. and Pedraza, R.O. (2012) More than rhizosphere colonization of strawberry plants by *Azospirillum brasilense*. *Applied Soil Ecology* 61, 205–212.

Guetsky, R., Shtienberg, D., Elad, Y. and Dinoor, A. (2001) Combining biocontrol agents to reduce the variability of biological control. *Phytopathology* 91, 621–627.

Guetsky, R., Shtienberg, D., Dinoor, A. and Elad, Y. (2002a) Establishment, survival and activity of the biocontrol agents *Pichia guilermondii* and *Bacillus mycoides* applied as a mixture on strawberry plants. *Biocontrol Science and Technology* 12, 705–714.

Guetsky, R., Shtienberg, D., Elad, Y., Fischer, E. and Dinoor, A. (2002b) Improving biological control by combining biocontrol agents each with several mechanisms of disease suppression. *Phytopathology* 92, 976–985.

Harman, G.E. (2000) Myths and dogmas of biocontrol: changes in perceptions derived from research on *Trichoderma harzinum* T-22. *Plant Disease* 84, 377–393.

Harman, G.E., Lorito, M. and Lynch, J.M. (2004) Uses of *Trichoderma* spp. to alleviate or remediate soil and water pollution. *Advances in Applied Microbiology* 56, 313–330.

Heeb, S. and Haas, D. (2001) Regulatory roles of the GacS/GacA two-component system in plant-associated and other gram negative bacteria. *Molecular Plant–Microbe Interactions* 14, 1351–1363.

Howell, C.R. (2003) Mechanisms employed by *Trichoderma* species in the biological control of plant diseases: the history and evolution of current concepts. *Plant Disease* 87, 4–10.

Huang, R., Che, H.J., Zhang, J., Yang, L., Jiang, D.H. and Li, G.Q. (2012) Evaluation of *Sporidiobolus pararoseus* strain YCXT3 as biocontrol agent of *Botrytis cinerea* on post-harvest strawberry fruits. *Biological Control* 62, 53–63.

Kiss, L. (2003) A review of fungal antagonists of powdery mildews and their potential as biocontrol agents. *Pest Management Science* 59, 475–483.

Kurita, N., Miyaji, M., Kurane, R. and Takahara, Y. (1981) Antifungal activity of components of essential oils. *Agricultural and Biological Chemistry* 45, 945–952.

Kurze, S., Bahl, H., Dahl, R. and Berg, G. (2001) Biological control of fungal strawberry diseases by *Serratia plymuthica* HRO-C48. *Plant Disease* 85, 529–534.

Lazzeri, L., Leoni, O. and Manici, L.M. (2004) Biocidal plant dried pellets for biofumigation. *Industrial Crops and Products* 20, 59–65.

Lima, G., Ippolito, A., Nigro, F. and Salerno, M. (1997) Effectiveness of *Aureobasidium pullulans* and *Candida oleophila* against postharvest strawberry rots. *Postharvest Biology and Technology* 10, 169–178.

Liu, X., Wang, J., Gou, P., Mao, C., Zhu, Z.-R. and Li, H. (2007) *In vitro* inhibition of postharvest pathogens of fruit and control of gray mold of strawberry and green mold of citrus by aureobasidin A. *International Journal of Food Microbiology* 119, 223–229.

Maas, J.L. (2004) Strawberry disease management. In: Naqvi, S.A.M.H. (ed.) *Diseases of Fruits and Vegetables: Volume II.* Springer, The Netherlands, pp. 441–483.

Martin, F.N. and Bull, C.T. (2002) Biological approaches for control of root pathogens of strawberry. *Phytopathology* 92, 1356–1362.

Mattner, S.W., Porter, I.J., Gounder, R.K., Shanks, A.L., Wren, D.J. and Allen, D. (2008) Factors that impact on the ability of biofumigants to suppress fungal pathogens and weeds of strawberry. *Crop Protection* 27, 1165–1173.

Mazzola, M., Brown, J., Izzo, A.D. and Cohen, M.F. (2007) Mechanism of action and efficacy of seed meal-induced pathogen suppression differ in a Brassicaceae species and time-dependent manner. *Phytopathology* 97, 454–460.

Meng, X.-H., Qin, G.-Z. and Tian, S.-P. (2010) Influences of preharvest spraying *Cryptococcus laurentii* combined with postharvest chitosan coating on postharvest diseases and quality of table grapes in storage. *LWT – Food Science and Technology* 43, 596–601.

Morandi, M.A.B., Mattos, L.P.V., Santos, E.R. and Bonugli, R.C. (2008) Influence of application time on the establishment, survival, and ability of *Clonostachys rosea* to suppress *Botrytis cinerea* sporulation on rose debris. *Crop Protection* 27, 77–83.

Okayama, K. (1991) Selection and effect of antagonistics on *Fusarium* wilt of strawberries. *Bulletin of the Nara-ken Agricultural Experiment Station* 22, 17–22.

Pedraza, R.O., Motok, J., Salazar, S.M., Ragout, A.L., Mentel, M., Tortora, M.L., Guerrero-Molina, M.F., Winik, B.C. and Díaz-Ricci, J.C. (2010) Growth-promotion of strawberry plants inoculated with *Azospirillum brasilense*. *World Journal of Microbiology and Biotechnology* 26, 265–272.

Peng, G. and Sutton, J.C. (1991) Evaluation of microorganisms for biocontrol of *Botrytis cinerea* in strawberry. *Canadian Journal of Plant Pathology* 13, 247–257.

Porter, I.J., Brett, R.W., Mattner, S.W. and Donohoe, H.E. (2005) Implications of the increased growth response after fumigation on future crop protection and crop production strategies. *Acta Horticulturae* 698, 229–238.

Porter, I.J., Trinder, L., Partington, D., Banks, J., Smith, S., Hannah, M. and Karavarsamis, N. (2006) *Validating the Yield Performance of Alternatives to Methyl Bromide for Pre-plant Fumigation.* UNEP, Nairobi, Kenya.

Punja, Z.K. and Utkhede, R.S. (2003) Using fungi and yeasts to manage vegetable crop diseases. *Trends in Biotechnology* 21, 400–407.

Qasem, J.R. and Aau-Blan, H.A. (1996) Fungicidal activity of some common weed extracts against different plant pathogenic fungi. *Journal of Phytopathology* 144, 157–161.
Rafai, M.A. (1969) A revision of the genus *Trichoderma*. *Mycological Papers* 116, 1–56.
Robinson-Boyer, L., Jeger, M.J., Xu, X.-M. and Jeffries, P. (2009) Management of strawberry grey mould using mixtures of biocontrol agents with different mechanisms of action. *Biocontrol Science and Technology* 19, 1051–1065.
Samtani, J.B., Ajwa, H.A., Weber, J.B., Browne, G.T., Klose, S., Hunzie, J. and Fennimore, S.A. (2011) Evaluation of non-fumigant alternatives to methyl bromide for weed control and crop yield in California strawberries (*Fragaria ananassa* L.). *Crop Protection* 30, 45–51.
Sánchez, J.P., Vega, J., Fernández, J.B., González, L. and Tubio, F.M. (2003) Efecto de los tratamientos fungicidas en pre-recolección en la producción de fresón y en las podredumbres post-recolección. *Boletín de Sanidad Vegetal – Plagas* 29, 441–452.
Schroers, H.-J., Samuels, G.J., Seifert, K.A. and Gams, W. (1999) Classification of the Mycoparasite *Gliocladium roseum* in *Clonostachys* as *C. rosea*, its relationship to *Bionectria ochroleuca*, and notes on other *Gliocladium*-like fungi. *Mycologia* 91, 365–385.
Singh, A.K., Dikshit, A., Sharma, M.L. and Dixit, S.N. (1980) Fungitoxic activity of some essential oils. *Economic Botany* 34, 186–190.
Sutton, J. (1994) Biological control of strawberry diseases. *Advances in Strawberry Research* 13, 1–12.
Sutton, J.C. and Peng, G. (1993) Biocontrol of *Botrytis cinerea* in strawberry leaves. *Phytopathology* 83, 615–621.
Sutton, J.C., Peng, G., Yu, H., Zhang, P.G. and Valdebenito-Sanhueza, R.M. (1997) *Glicocladium roseum* – a versatile adversary of *Botrytis cinerea* in crops. *Plant Disease* 81, 316–338.
Sylla, J., Alsanius, B.W., Krüger, E., Becker, D. and Wohanka, W. (2013) *In vitro* compatibility of microbial agents for simultaneous application to control strawberry powdery mildew (*Podosphaera aphanis*). *Crop Protection* 51, 40–47.
Szekeres, A., Leitgeb, B., Kredics, L., Antal, Z., Hatvani, L., Manczinger, L. and Vágvölgyi, C. (2005) Peptaibols and related peptaibiotics of *Trichoderma*. *Acta Microbiologica et Immunologica Hungarica* 52, 137–168.
Tezuka, N. and Makino, T. (1991) Biological control of *Fusarium* wilt of strawberry by nonpathogenic *Fusarium oxysporum* isolated from strawberry. *Annals of the Phytopathological Society of Japan* 57, 506–511.
Tortora, M.L., Díaz-Ricci, J.C. and Pedraza, R.O. (2011) *Azospirillum brasilense* siderophores with antifungal activity against *Colletotrichum acutatum*. *Archives of Microbiology* 193, 275–286.
Tortora, M.L., Díaz-Ricci, J.C. and Pedraza, R.O. (2012) Protection of strawberry plants (*Fragaria ananassa* Duch.) against anthracnose disease induced by *Azospirillum brasilense*. *Plant and Soil* 356, 279–290.
Verma, M., Brar, S.K., Tyagi, R.D., Surampalli, R.Y. and Valéro, J.R. (2007) Antagonistic fungi, *Trichoderma* spp.: panoply of biological control. *Biochemical Engineering Journal* 37, 1–20.
Washington, W., Shanmuganathgan, N. and Forbes, C. (1992) Fungicide control of strawberry fruit roots, and field occurrence of resistance of *Botrytis cinerea* to iprodiones, benomyl and dichlofluanid. *Crop Protection* 11, 355–360.
Wilhelm, S. and Paulus, A.O. (1980) How soil fumigation benefits the California strawberry industry. *Plant Disease* 64, 264–270.
Wilson, C.L., Wisniewski, M.E., Biles, C.L., McLaughlin, R., Chalutz, E. and Droby, S. (1991) Biological control of post-harvest diseases of fruits and vegetables: alternatives to synthetic fungicides. *Crop Protection* 10, 172–177.
Wilson, C.L., Solar, J.M., El Ghaouth, A. and Wisniewski, M.E. (1997) Rapid evaluation of plant extracts and essential oils for antifungal activity against *Botrytis cinerea*. *Plant Disease* 81, 204–210.
Woo, S.L., Scala, F., Ruocco, M. and Lorito, M. (2006) The molecular biology of the interactions between *Trichoderma* spp., phytopathogenic fungi, and plants. *Phytopathology* 96, 181–185.
Woo, S.L., Ruocco, M., Vinale, F., Nigro, M., Marra, R., Lombardi, N., Pascale, A., Lanzuise, S., Manganiello, G. and Lorito, M. (2014) *Trichoderma*-based products and their widespread use in agriculture. *Open Mycology Journal* 8, 71–126.
Xu, X.M., Salama, N., Jeffries, P. and Jeger, M.J. (2010) Numerical studies of biocontrol efficacies of foliar plant pathogens in relation to the characteristics of a biocontrol agent. *Phytopathology* 100, 814–821.
Yohalem, D. and Passey, T. (2011) Amendment of soils with fresh and post-extraction lavender (*Lavandula angustifolia*) and lavandin (*Lavandula* × *intermedia*) reduce inoculum of *Verticillium dahliae* and inhibit wilt in strawberry. *Applied Soil Ecology* 49, 187–196.
Zhang, H., Yang, Q., Lin, H., Ren, X., Zhao, L. and Hou, J. (2013) Phytic acid enhances biocontrol efficacy of *Rhodotorula mucilaginosa* against postharvest gray mold spoilage and natural spoilage of strawberries. *LWT – Food Science and Technology* 52, 110–115.

12 Emerging Diseases in Strawberry Crop: Charcoal Rot and *Fusarium* Wilt

Berta de los Santos[1]*, Manuel Chamorro[1], Juan J. Medina-Mínguez[2], Nieves Capote[1], Ana Aguado[1] and Fernando Romero[1]
[1]*Centro IFAPA Las Torres-Tomejil, CAPDER-Junta de Andalucía, Alcalá del Río, Sevilla, Spain;* [2]*Centro IFAPA Huelva, CAPDER-Junta de Andalucía, Huelva, Spain*

12.1. Introduction

Strawberry (*Fragaria* × *ananassa* Duch.) is one of the most widely consumed fruits throughout the world. World production has increased steadily over the years, surpassing 4.5 million t in 2013. Most of the production is located in the northern hemisphere, led by the USA with 30.3% (1,366,850 t) of worldwide production, followed by Mexico (360,426 t), Turkey (353,173 t), Spain (289,900 t), Egypt (242,297 t) and the Republic of Korea (231,803 t) in 2012 (http://faostat3.fao.org). Although data from the Food and Agriculture Organization of the United Nations (FAO) Statistics Division indicated that production in China was only 15,250 t in 2012, Zhang *et al.* (2012) stated that China was ranked first worldwide in terms of output of area of strawberry production in 2012, and the Office of Agricultural Affairs (Beijing, China) determined that China's fresh and frozen strawberry production was 2.1 million t and 150,000 t, respectively, in 2012. China only exports a small amount of fresh strawberries to neighbouring countries due to high shipping costs (http://gain.fas.usda.gov/Recent%20GAIN%20Publications/Strawberries_Beijing_China%20-%20Peoples%20Republic%20of_12-17-2012.pdf).

Fungal diseases cause numerous economic losses in all world crop areas, but mainly in those where the same soil is used year after year. In recent years, novel plant collapse problems have been detected (charcoal rot and *Fusarium* wilt), and outbreaks of crown

* berta.santos@juntadeandalucia.es

and root disease have impacted severely on strawberry production (Fang *et al.*, 2011b). The increase in the prevalence of crown and root rot may be related to the phase-out of methyl bromide (MeBr). In the mid-1990s, the Montreal Protocol marked the beginning of the end for MeBr utilization for soil disinfestation. Article 2 (developed) countries phased out MeBr use in 2005. Some of these countries are important strawberry fruit and plant producers (i.e. Australia, Canada, France, Germany, Italy, Israel, Japan, Poland, Republic of Korea, Spain, USA). In Article 5 (developing) countries, MeBr was phased out by 31 December 2014 (López-Aranda, 2014). Some authors have associated the changes in the fumigation products used and methods of applying them prior to planting strawberries with the establishment of *Macrophomina phaseolina* (causal agent of charcoal rot) (Mertely *et al.*, 2005; Zveibil and Freeman, 2005; Avilés *et al.*, 2008; Koike, 2008a, 2012). The development of *Fusarium* wilt (caused by *Fusarium oxysporum*) has also been associated with changes in the practices of pre-plant soil fumigation, but this has so far only been confirmed in southern California (Koike, 2012), although the establishment of *Fusarium oxysporum* is not usually associated with the use of fumigants (Phillips and Golzar, 2008). Furthermore, Sugimura *et al.* (2001) suggested that *Fusarium* wilt of strawberry could be controlled better by soil solarization than by fumigation with either MeBr or chloropicrin.

M. phaseolina and *F. oxysporum* cause similar symptoms: poor growth, plant stunting, dieback, plant collapse and crown discolour. Both pathogens are soilborne fungi, have a long persistence and are spread via soil (Koike, 2011).

12.2. *Fusarium* Wilt (*Fusarium* Yellows)

Fusarium spp. used to be considered a weak or sublethal pathogen in strawberries (Particka and Hancock, 2008; Gliessman and Muramoto, 2010) and had been reported as one of the soilborne pathogens involved in the black root rot disease complex (Miller, 1949; D'Ercole *et al.*, 1989; Kukkonen *et al.*, 2004). It was shown to interact with *Pratylenchus penetrans* in the occurrence of black root rot (Maas, 1998). In strawberry production fields, Albregts *et al.* (1996) and Locascio *et al.* (1999a) observed stunted and chlorotic strawberry plants. *Fusarium* spp. and *Rhizoctonia* were found on the brown roots of these small plants. However, pathogenicity tests with potential pathogens were not conducted, and thus the possible roles of any of the individual fungi in the disease were not elucidated (Locascio *et al.*, 1999a). Mortality due to lethal pathogens was rare in the USA or Spain, two of the major strawberry-producing countries, until the outbreak of charcoal rot and *Fusarium* wilt, although *Fusarium* spp. had been isolated from soils and plants (Albregts *et al.*, 1996; Fort *et al.*, 1996; de los Santos *et al.*, 2003; De Cal *et al.*, 2005).

Fusarium oxysporum is a cosmopolitan and common soilborne hemibiotrophic fungus, and is responsible for an enormous range of plant diseases, usually involving a vascular wilt syndrome, often as a result of the xylem vessels becoming blocked (Leslie and Summerell, 2006). The majority of isolates causing vascular wilts are specific strains that infect only a small number of host plants and that are differentiated on the basis of pathogenicity as *formae specialis* (f. sp.) (Leslie and Summerell, 2006). The strain that commonly attacks strawberry is *Fusarium oxysporum* Schlechtend.: Fr. f. sp. *fragariae* Winks & Williams, which causes *Fusarium* wilt, also known as *Fusarium* yellows (Maas, 1998).

Fusarium oxysporum f. sp. *fragariae* was first observed and described in Australia (Winks and Williams, 1965) and has been a problem primarily in this country and in Japan (Hancock, 1999). The disease has also been detected in: Argentina (Mena *et al.*, 1975; Baino *et al.*, 2010); Chile, where the fruit production has suffered important economic losses principally due to this disease (González *et al.*, 2005); China, where *Fusarium* wilt is considered a replant disease and occurs with *Verticillium* wilt, the major limitation for strawberry production in greenhouses (Zhang *et al.*, 2012); Japan (Tezuka and Makino, 1991); and Korea (Cho and Moon, 1984), where it is considered the most serious soilborne disease of nursery plants (Nam *et al.*, 2011a). In the USA, *Fusarium* wilt was first detected in California (Koike *et al.*, 2009a), and later in South Carolina (Williamson *et al.*, 2012). *F. oxysporum* survival in soil and pathogen-related plant mortality are under investigation in California (Daugovish and Fennimore, 2011; Koike *et al.*, 2013). In Spain, *F. oxysporum* was found to cause wilt and the death of strawberry plants in soil-less culture systems in the Huelva region (south-western Spain) (Arroyo *et al.*, 2009), although De Cal *et al.* (2005) had already detected the presence of *Fusarium* spp. in nursery soils. Later, *Fusarium* spp. was detected in soil and symptomatic plants in fruit production fields in Huelva (Miranda *et al.*, 2012). Redondo *et al.* (2012) found an increase of the pathogenic species of *Fusarium* in soil and plants in Spanish strawberry nurseries. More recently, *Fusarium solani* was detected in strawberry fruit production fields in south-western Spain, with symptoms comprising foliage wilt, plant stunting and drying, and death of older leaves (Pastrana *et al.*, 2013).

Symptoms of *Fusarium* wilt consist of wilting of foliage, drying and withering of older leaves, stunting of plants and reduced fruit production (Koike *et al.*, 2009b). As the disease progresses, only the central youngest leaves remain green (Koike, 2012). Chlorosis of the foliage (yellowing) rather than wilting may occur under cool conditions (Maas, 1998; Williamson *et al.*, 2012). The plants eventually collapse and die. The internal vascular and cortical tissues of the plant crowns show a brown-to-orange-brown discoloration (Fig. 12.1), and the lower crown tissues may decay extensively as the disease advances. Slimy, white or pink masses of micro- and macroconidia may be produced on infected plant parts (Maas, 1998; Koike *et al.*, 2009b; Williamson *et al.*, 2012). Although *F. oxysporum* can attack strawberry roots, causing discoloration and stripping, it is most frequently isolated from the crowns (Phillips and Golzar, 2008; Fang *et al.*, 2011a).

Symptoms of *Fusarium* wilt development in strawberry are favoured by high temperatures, which cause the infected leaves to

Fig. 12.1. A strawberry crown showing symptoms of *Fusarium* wilt.

wilt and die rapidly (Maas, 1998). With an increase in temperature from 17 to 27°C, the disease severity in plants inoculated with *F. oxysporum* increased significantly. While little disease developed at 17°C, in contrast, at 27°C, symptoms of *Fusarium* wilt were evident by 2 weeks after inoculation, with rapid wilt or death of the plants within 4 weeks, including severe disease symptoms on the crown, root and vascular tissues, along with poor development of the roots (Fang *et al.*, 2011a). If infected plants are subjected to stress, the disease progresses more rapidly and the symptoms are more severe. Such stress factors include environmental extremes (hot weather), underwatering, the presence of pests and physiological stress from heavy fruit loads (Koike, 2012). Other factors influencing the severity of *Fusarium* wilt are the inoculum level in the soil, environmental conditions and cultivar susceptibility (http://ceventura.ucdavis.edu/files/121665.pdf). *F. oxysporum* can be dispersed by many different means including wind and in soil, and through seeds or infected plant material (Leslie and Summerell, 2006).

In nurseries, *F. oxysporum* persists in the soil after the plants have been removed (Zhang *et al.*, 2012). *F. oxysporum* f. sp. *fragariae* penetrates strawberry plants through the roots, severely affecting the roots and crowns, and resulting in a rapid wilting and eventually death of the strawberry plants (Koike *et al.*, 2009b; Fang *et al.*, 2011a,b). Infection by *F. oxysporum* through runner propagation from infected mother plants of strawberry was shown in stolons and daughter plants hanging from raised beds (Nam *et al.*, 2011a). *Fusarium* spp. have also been isolated from strawberry plants coming from nurseries, so in strawberry fruit production fields, the inoculum could come from the soil and/or the mother plant (de los Santos, 2010).

Fusarium wilt of strawberry is caused by a *formae speciale* of *F. oxysporum* Schlechtendahl emend. Snyder & Hansen, which is host specific to strawberry and can only infect this crop (Snyder and Hansen, 1940; Koike, 2012). There are other *formae speciales* of *F. oxysporum*. Morphologically, these strains are very similar or identical. Identification has traditionally involved pathogenicity testing with sets of host differentials appropriate for the *formae speciales* in question (Leslie and Summerell, 2006). Molecular methods, such as polymerase chain reaction (PCR) have been described to resolve genetic variation among isolates within or between *formae speciales* of *F. oxysporum* (McDonald, 1997).

F. oxysporum has pathogenic, saprophytic and biocontrol strains that cannot be distinguished from each other by morphological analysis or sexual compatibility studies. The high genetic diversity among isolates has led to *F. oxysporum* being considered a complex of several different species (known as *Fusarium* spp. complex) (Baayen *et al.*, 2000).

The colony morphology of *F. oxysporum* on potato dextrose agar varies widely. As described by Arroyo *et al.* (2009) and Williamson *et al.* (2012), colonies of *F. oxysporum* isolated from strawberry tissues had a light purple mycelium and beige or orange reverse colony colours on acidified potato dextrose agar after 5 days of incubation at 25°C, whereas colonies on quarter-strength acidified agar were light purple (Fig. 12.2). The mycelia may be floccose, sparse or abundant, and range in colour from white to pale violet. Abundant pale orange or pale violet macroconidia are produced in a central spore mass in some isolates. Small pale brown, blue to blue-black or violet sclerotia may be produced abundantly by some isolates (Leslie and Summerell, 2006).

F. oxysporum has no known sexual stage but produces three types of asexual spores: microconidia, macroconidia and chlamydospores. The microconidia are abundant, generally single-celled, hyaline, and oval, elliptical or kidney shaped (5.9–9.2 × 2.1–3.4 μm) and are produced only in false heads. The macroconidia are abundant, of short or medium length (28.8–37.3 × 3.2–4.3 μm), three to five septate, thin-walled, straight to slightly curved and fusoid subulate with an attenuated apical cell and a foot-shaped basal cell (Nelson *et al.*, 1983; Leslie and Summerell, 2006; Arroyo *et al.*, 2009). The conidiophores are unbranched or branched monophialides (Nelson *et al.*, 1983). Chlamydospores are present and are usually formed

Fig. 12.2. Colony morphology of *Fusarium oxysporum* growing on potato dextrose agar.

singly or in pairs, but also may be found in clusters or in short chains. They can be either terminal or intercalary in aerial, submerged or surface hyphae. Chlamydospores are smooth or rough walled. In most isolates, they form readily and profusely in culture (Nelson *et al.*, 1983; Leslie and Summerell, 2006). These resilient structures allow the pathogen to survive in the soil for long periods of time (Koike, 2012). Thus, the most distinguishing characteristics of *F. oxysporum* are: the presence of chlamydospores, and microconidia borne in false heads on short monophialides (Nelson *et al.*, 1983).

Traditionally, the most prevalent techniques used to identify *F. oxysporum* relied on culture-based morphological approaches. However, these methods are often time consuming, laborious and require extensive knowledge of classical taxonomy. These limitations led to the development of molecular approaches based on the detection of pathogen DNA or RNA, with improved accuracy, rapidity and reliability. A wide variety of molecular methods have been used to detect, identify and quantify *Fusarium* spp. in plants and environmental samples.

PCR is the most prevalent method for *Fusarium* spp. detection. PCR allows the amplification of millions of copies of specific DNA sequences without the need to isolate the pathogen from the infected material, reducing the diagnosis time from weeks to hours. Conventional PCR for generic detection of *Fusarium* spp. can be performed using a set of primers based on the translation elongation factor 1α gene (TEF 1α: primer ef1, 5′-ATGGGTAAGGARGACAAGAC-3′, and primer ef2, 5′- GGARGTACCAGTSATCATGTT-3′) (O'Donnell *et al.*, 1998). For the specific detection by conventional PCR of *F. oxysporum* sensu lato, including saprophytic, biocontrol and pathogenic forms, several sets of primers have been reported: primers PFO2 (5′-CGGGGGATAAAGGCGG-3′) and PFO3 (5′-ACCGTGTAATACCCTGGG-3′), based in the 28S ribosomal RNA (rRNA) gene (Edel *et al.*, 2000), and primers FOF1 (5′-ACATACCACTTGTTGCCTCG-3′) and FOR1 (5′-CGCCAATCAATTTGAGGAACG-3′) designed within the internal transcribed spacer (ITS) region of the rRNA gene (Mishra *et al.*, 2003). Advances in PCR-based methods, such as real-time PCR, allow the rapid and accurate detection of *F. oxysporum* with improved sensitivity and specificity, as well as quantification of the pathogen in plants or environmental samples (Jiménez-Fernández *et al.*, 2010). Quantification allows the establishment of the damage threshold at which a pathogen can cause disease and the optimum time for applying preventative measures to avoid losses.

Different PCR-based protocols for specific detection of *F. oxysporum* belonging to different *formae speciales* or even races or pathotypes within a *forma specialis* have been developed for: *F. oxysporum* f. sp. *albedinis* (Fernandez *et al.*, 1998); *F. oxysporum* f. sp. *gladioli* (de Haan *et al.*, 2000); *F. oxysporum* f. sp. *phaseoli* (Alves-Santos *et al.*, 2002); *F. oxysporum* f. sp. *ciceris* (Jiménez-Gasco and Jiménez-Díaz, 2003); *F. oxysporum* f. sp. *asparragi* (Mulè *et al.*, 2004); *F. oxysporum* f. sp. *chrysanthemi* (Pasquali *et al.*, 2004); *F. oxysporum* f. sp. *basilici* (Pasquali *et al.*, 2006); *F. oxysporum* f. sp. *lactucae* (Pasquali *et al.*, 2007); *F. oxysporum* f. sp. *niveum* (Zhang *et al.*, 2005); *F. oxysporum* f. sp. *lycopersici* and *radicis-lycopersici* (Hirano and Arie, 2006; Inami *et al.*, 2010); *F. oxysporum* f. sp. *cucumerinum* and f. sp. *radicis-cucumerinum* (Lievens *et al.*, 2007); *F. oxysporum* f. sp. *vasinfectum* (Zambounis *et al.*, 2007); *F. oxysporum* f. sp. *cubense* (Lin *et al.*, 2009) and *F. oxysporum* f. sp. *melonis* (Luongo *et al.*, 2012). Biocontrol strains of *F. oxysporum* can also be identified by PCR-based methods (Cipriani *et al.*, 2009; Edel-Hermann *et al.*, 2011). As cited by Mirmajlessi *et al.* (2015), Suga *et al.* (2013) characterized and used some transposable elements in *F. oxysporum* f. sp. *fragariae* to design a specific set of PCR primers. The genomic region between *Han* and *Skippy* (as transposable elements) was amplified by an inter-retrotransposon-amplified polymorphism technique (IRAP-PCR), and specific primers were designed from this region. The developed PCR primers discriminated *F. oxysporum* f. sp. *fragariae* strains from non-pathogenic *F. oxysporum* strains and five other *formae speciales*. To the best of our knowledge, there are no other PCR-based techniques for strawberry *Fusarium* wilt diagnosis.

Several works have investigated the variability of *F. oxysporum* f. sp. *fragariae* related to vegetative compatibility groups (VCGs) and pathogenicity. Two strains belong to the same VCG if they are able to form a stable heterokaryon. These strains are said to be vegetatively compatible. The vegetative compatibility phenotype has a multigenic basis, as the number of alleles governing the trait is large. In fact, many *formae speciales* comprise strains that belong to multiple VCGs, suggesting a polyphyletic nature (independent origins) for most of them, including *F. oxysporum* f. sp. *fragariae*. Hyun *et al.* (1996) identified four major VCGs in *F. oxysporum* f. sp. *fragariae* isolates from Korea, by means of a nitrate reductase complementation test. These VCGs significantly correlated with four isozyme polymorphism groups with isozyme patterns of sterase, catalase, acid phosphate and leucine aminopeptidase. However, no correlation was found between VCGs and pathogenicity of the isolates. Using 22 pathogenic and non-pathogenic isolates of *F. oxysporum* f. sp. *fragariae* from the same country, Nagarajan *et al.* (2006) described three VCGs and one incompatible group. A high correlation between VCG groups and the three major clusters determined by random amplified polymorphic DNA (RAPD) markers analysis was established. In addition, VCG and RAPD patterns between pathogenic and non-pathogenic *F. oxysporum* isolates from strawberry were distinctly different. In an analysis of 27 pathogenic *F. oxysporum* f. sp. *fragariae* isolates from California, USA, two somatic compatibility groups were identified, each of which was associated with a different sequence of the TEF-1α gene (Islas *et al.*, 2011).

Molecular characterization of *Fusarium* spp. implies the use of identification strategies based on comparative sequences (Summerbell *et al.*, 2005). This method consists of PCR amplification of a selected region of the genomic DNA (target locus), followed by sequencing of the resulting amplicon and comparison against a database library such as GenBank (http://www.ncbi.nlm.nih.gov/), the European Molecular Biology Laboratory nucleotide sequence database (EMBL, http://www.ebi.ac.uk/embl/) or the DNA Data Bank of Japan (DDBJ, http://www.ddbj.nig.ac.jp/). GenBank is the most widely used database and contains a large number of sequences from different species and loci. However, a considerable percentage of the sequences reported in this database are erroneous or incomplete, or have been associated with misidentified organisms, leading to false identifications (Kang *et al.*, 2010). To overcome this problem, specific sequence databases for supporting strain identification and phylogenetic analyses of *Fusarium* have been developed: *Fusarium*-ID (http://isolate.fusariumdb.org/), *Fusarium* Comparative Genomics Platform (FCGP, http://genomics.fusariumdb.org/) and the *Fusarium* Community Platform (FCP, http://www.fusariumdb.org/wordpress/), an online research forum. Together, these platforms form the Cyber Infrastructure for *Fusarium* (CiF, http://www.fusariumdb.org/) (Park *et al.*, 2010). Different loci from *F. oxysporum* f. sp. *fragariae* have been sequenced and reported in GenBank, including a partial region of TEF-1α (O'Donnell *et al.*, 2009), β-tubulin, rRNA and the ITS region (O'Donnell *et al.*, 2009), exopolygalacturonase (*pgx1* and *pgx4*) and endopolygalacturonase (*pg1* and *pg5*) genes, the *mat1-1-1* gene for mating type protein, and trichothecene 3-*O*-acetyltransferase (all of which are non-referenced, published as direct submissions).

The choice of the appropriate locus is crucial for an accurate identification by sequencing. Among the loci used for identification and differentiation of *Fusarium* spp., the most informative is TEF1-α, which has a high level of interspecies variation combined with low levels of intraspecific variation. Other genes such as tubulin, calmodulin and the ITS region have lower levels of genetic variation and are moderately informative, while the ITS region of the rRNA gene constitutes the most problematic loci for *Fusarium* identification and differentiation because it fails to distinguish between closely related species (Balajee *et al.*, 2009).

Different PCR-based marker systems have been applied to studies of the genetic variability of *F. oxysporum* populations. A high level of genetic variation has been demonstrated among *F. oxysporum* f. sp. *fragariae* isolates from Korea. Analysis by RAPD and restriction fragment length polymorphism (RFLP) analyses separated the *F. oxysporum* f. sp. *fragariae* isolates into eight distinct groups (Nagarajan *et al.*, 2004, 2006).

12.2.1. *Fusarium* wilt management

Fusarium wilt is difficult to control due to the long viability of the resting structures (chlamydospores), the broad host range of the pathogen and the inability of fungicides to affect the pathogen (Zhang *et al.*, 2012). Recommended disease reduction measures include host plant resistance and numerous biological and chemical control methods.

Identifying and deploying resistant cultivars is considered to be the most cost-effective and environmentally sustainable strategy for control of *Fusarium* wilt (Fravel *et al.*, 2003; Particka and Hancock, 2005; MacKenzie *et al.*, 2006). Fang *et al.* (2012b) found that strawberry 'Camarosa' was the most susceptible cultivar to *F. oxysporum* f. sp. *fragariae*, while strawberry cultivars 'Festival', 'Aromas' and 'Camino Real' were relatively resistant to *F. oxysporum*, with 'Festival' being the most resistant cultivar. This resistant cultivar significantly impeded spore germination and penetration from 4 to 12 h post-inoculation and subsequent growth and colonization by this pathogen until 7 days post-inoculation compared with the susceptible 'Camarosa'. At 7 days post-inoculation, fungal colonization in the resistant 'Festival' remained confined mainly to the epidermal layer of the roots, while in the susceptible 'Camarosa', hyphae had heavily colonized the cortical tissue throughout and had also colonized the vascular tissues. The authors concluded that this study demonstrated for the first time that resistance of a strawberry cultivar to *F. oxysporum* f. sp. *fragariae* is a result of impedance of pathogen growth and colonization both on the plant surface and within host tissues (Fang *et al.*, 2012a).

Among the commonly used cultivars in the USA, 'Ventana', 'San Andreas' and 'Camino Real' are the most resistant to *Fusarium* wilt, whereas 'Albion', 'Monterey', 'Camarosa', 'Benicia' and 'Palomar' are the most susceptible (htpp://ceventura.ucdavis.edu/files/121665.pdf). Koike (2011) also reported the cultivars 'San Andreas' and 'Ventana' as resistant but not immune, and included 'Monterey' in this group, whereas 'Chandler' and 'Seascape' were considered as susceptible. Takahashi *et al.* (2003) studied the resistance to *F. oxysporum* in strawberry lines and cultivars used commonly in Japan.

Toyoda *et al.* (1991) used a system for plant regeneration from leaf-derived callus tissues of strawberry developed by Toyoda *et al.* (1990) and determined the resistance to *Fusarium* wilt following transplantation of regenerated plants to a pathogen-infested soil. They demonstrated the existence of somaclonal resistance against *F. oxysporum*.

Currently, the search for alternatives to the use of chemicals to control plant diseases is a major goal in the usage of biological methods. An environment-friendly alternative to protect roots against soilborne fungal pathogens is based on using naturally occurring rizhosphere-associated bacteria and fungi with antagonistic properties as biological control agents (Berg, 2007). The plant species or cultivar plays a key role in the composition of the rhizobacterial populations colonizing the roots (Smalla *et al.*, 2001). Antagonistic microorganisms found in the strawberry rhizosphere are typical for this crop (Berg *et al.*, 2002). The bacterial antagonistic community of strawberry is characterized by a high proportion of isolates with antifungal properties and a low diversity of species. Approximately 90% of antagonists isolated from the rhizosphere of field-grown strawberry plants belong to the *Pseudomonas* cluster. Other antagonistic species found in the strawberry rhizosphere are *Acinetobacter* spp., *Bacillus megaterium*, *Burkholderia cepacia*, *Comamonas acidovorans*, *Serratia grimesii*, and *Strenotrophomonas maltophilia* (Berg, 2007). Fluorescent *Pseudomonas* spp. were revealed to be highly abundant in strawberry rhizospheres in plants grown in Californian soils. Surprisingly, however, the fluorescent *Pseudomonas* population in soil was enhanced in fumigated fields in comparison with non-fumigated fields (Martin and Bull, 2002). De los Santos *et al.* (2003) studied the effects of repeated fumigation treatments on crop yields and *Trichoderma* soil populations over a period of 3 years, at two different places in the province of Huelva (south-western Spain). They concluded that the repeated applications of MeBr and other fumigants, such as chloropicrin, produce a significant increment in the soil population of *Trichoderma* (*Trichoderma inhamatum*, *Trichoderma viride* and *Trichoderma harzianum*) and in the crop yield, whereas in untreated soil, both *Trichoderma* populations and yield decreased year by year. Furthermore, the authors concluded that the results of their work showed a generalized presence of wild *Trichoderma* spp. in the soil in Huelva where strawberries are grown. *Trichoderma* spp. are among the most common saprophytic fungi and are well documented as effective biological control agents of plant disease caused by soilborne fungi. These species are often very fast growing and rapidly colonize substrates, thus excluding pathogens such as *Fusarium* spp. (Ozbay and Newman, 2004). Soil treatments with *T. harzianum* spores suppressed infestation with *F. oxysporum* f. sp. *vasinfectum* and *F. oxysporum* f. sp. *melonis*. Competition was a proposed mechanism, although it was not proven to be the main activity (Sivan and Chet, 1989). Perveen and Bokhari (2012) suggested that the secretion of diffusible non-volatile inhibitory substances by a *Trichoderma* isolate resulted in an inhibition zone in dual culture without any hyphae contact between *Trichoderma* and *F. oxysporum*. *T. harzianum* has been shown to occur along with a decreasing incidence of *Fusarium* wilt in strawberries (Moon *et al.*, 1995). However, despite a long history of using biological control for *Fusarium* wilt, most control methods are either ineffective or difficult to apply (Alabouvette *et al.*, 1998).

Non-pathogenic isolates of *Fusarium oxysporum* isolated from strawberry plants were found to control *Fusarium* wilt in field test when the plants were inoculated with the non-pathogenic isolates prior to planting (Okayama, 1991; Tezuka and Makino, 1991). These non-pathogenic isolates are common inhabitants of the strawberry rhizosphere and appear to be well-adapted to the ecological niche that otherwise would be colonized by the pathogenic species (Maas, 2004).

Induced systemic resistance to *Fusarium* wilt and changes in antioxidative abilities in mycorrizal strawberry plants have been investigated. Runner plants were inoculated with the arbuscular mycorrhizal fungus *Glomus mosseae* and analysed with a split root system. Six weeks after this inoculation, *F. oxysporum* f. sp. *fragariae* was inoculated. The disease incidence and severity of symptoms were lower in the shoots and root of the mycorrhizal plants, and induced resistance also appeared in the split root-treated mycorrhizal plants. After pathogen inoculation, the mycorrhizal plants showed higher values in the following parameters: superoxide dismutase activity, DPPH (2,2-diphenyl-1-picrylhydrazyl) radical (a stable organic nitrogen-centred free radical), scavenging activity, and total polyphenol and ascorbic acid contents. The authors concluded that plant growth enhancement and tolerance to *Fusarium* wilt including induced resistance had occurred in the mycorrhizal strawberry plants. In this case, the antioxidative abilities increased under pathogen-stressed conditions (Matsubara *et al.*, 2012). Bull *et al.* (2005) tested the effect of commercially prepared mycorrhizal inoculants on strawberry yield. None of the tested arbuscular mycorrhizae fungi inoculants influenced market yield and only one increased total yield. Additionally, the inoculants only sporadically increased the colonization of roots by the arbuscular mycorrhizae fungi. Although these inoculants were not effective under the conditions tested in this study, this does not rule out their usefulness with other cultivars or in other locations. However, these data suggest that naturally occurring inoculum from field soil and/or transplants was sufficient. Organic production fields might be the source for the inoculum colonizing non-inoculated plants, as these fields are not fumigated with MeBr and chloropicrin, in comparison with conventional production fields, although *Trichoderma* spp. and fluorescent *Pseudomonas* populations have been shown to increase after chemical fumigation (Martin and Bull, 2002; de los Santos *et al.*, 2003).

Fusarium wilt suppression has been attempted using antagonistic bacteria. Two isolates of *Bacillus velezensis* (BS87 and RK1), selected from soil in strawberry fields in Korea, showed high levels of antagonism towards *F. oxysporum* f. sp. *fragariae in vitro* (Nam *et al.*, 2009). The isolates were evaluated for control of *Fusarium* wilt in strawberries in pot trials and field trials. The antagonistic bacteria were applied using the root-dip method. In the pot trials, the optimum applied concentration of *Bacillus velezensis* for pre-plant root-dip application to control *Fusarium* wilt was 10^5–10^6 colony-forming units. However, in the field trials, the biological control efficacies of formulations of RK1 were similar to that of a conventional fungicide (copper hydroxide) when compared with a non-treated control. Thus, to increase the control of *Fusarium* wilt in strawberry it was suggested that antagonistic bacteria should be applied based on an optimal concentration. Zhang *et al.* (2012) isolated a strain of *Bacillus subtilis* that can control *Fusarium* and *Verticillium* wilts from replanted strawberry field soil.

In general, strawberries are planted as whole plants, or as bare-root plants without leaves (transplants) or runner plants. Prior to planting strawberries, use of the root-dip method is an excellent approach to introduce biological control agents (BCAs), as the almost-microorganism-free roots are more easily colonizable. For other plants, it is sometimes difficult to introduce BCAs into a stable microbial ecosystem such as the rhizosphere. However, because rhizosphere establishment is a key aspect of effective biological control, this makes strawberries an ideal host plant for BCAs. Another possibility for BCA application is via a drip-irrigation system (Berg, 2007). Some products on the market are suitable for controlling soilborne pathogens in strawberry.

These products are based on *Trichoderma* spp., *Bacillus subtilis* or *Bacillus amyloliquefaciens* (Berg, 2007; Nam *et al*., 2009).

12.3. Charcoal Rot

Macrophomina phaseolina (Tassi) Goidanich is a primarily soilborne pathogen with a wide distribution, varied host range, great longevity and high competitive saprophytic ability (Babu *et al*., 2010). About 500 plant diseases are caused by this fungus (Su *et al*., 2001), and in strawberry it is often termed 'charcoal rot' due to the discoloration or blackening of the crown and roots caused by the presence of the fungus in afflicted plants (Agostini, 2011) and the presence of small, black, macroscopically visible sclerotia that form in the shredded, parasitized host tissue and look like charcoal (Pratt, 2006).

Charcoal rot has emerged as an important disease in strawberry production systems that have abandoned soil fumigation with MeBr and that are in a period of transition in the implementation of alternatives (Mertely *et al*., 2005; Zveibil and Freeman, 2005; Avilés *et al*., 2008; Zveibil *et al*., 2012). Charcoal rot of strawberry has been reported in: Argentina, where the disease was observed in commercial farms in 2007, and where it is considered a potential threat for north-western Argentina's strawberry industry (Baino *et al*., 2011); Iran (Sharifi and Mahdavi, 2012); Italy (Angelini and nad Faedi, 2010); Spain (Avilés *et al*., 2008); Australia, where devastating outbreaks of crown and root diseases impacted severely on strawberry production in the 2007–2009 seasons (Golzar *et al*., 2007; Fang *et al*., 2011a); Greece, where in non-fumigated plots almost 25% of diseased plants are attributed mainly to mixed infection by *F. oxysporum* f. sp. *fragariae* and *M. phaseolina* (Tjamos *et al*., 2006); Turkey, where it is reported as *Macrophomina* spp. (Benlioğlu *et al*., 2004); and Israel, where one study suggested that the soil fumigation regimen for control of fungal pathogens, utilizing alternatives to MeBr, was not adequate to maintain healthy strawberry material at any of the production stages (Zveibil and Freeman, 2005). In the USA, *M. phaseolina* was first isolated from collapsed and dying plants from a commercial field in Florida (Mertely *et al*., 2005). The authors also suggested that *M. phaseolina* would be an emerging threat as the Florida strawberry industry transitioned from MeBr to other fumigants in 2005. Later, Koike (2008b) observed severely diseased strawberry plants in commercial fields in southern California, and noted that many of the Californian affected fields were not pre-plant fumigated with MeBr:chloropicrin fumigants. However, Sanchez *et al*. (2012) reported the presence of *M. phaseolina* in both fumigated (with MeBr) and non-fumigated fields, suggesting that the emergence of this pathogen in the Chilean strawberry crop was not related to the MeBr phase-out.

The first report of charcoal rot occurrence as a serious strawberry crown disease was in Illinois, USA (Tweedy and Powell, 1958). Other earlier records of *M. phaseolina* affecting strawberry were from India, France and Egypt. In Egypt, *M. phaseolina* was detected as causing a serious root rot of strawberries (Madkour and Aly, 1981). In 1989, in a few strawberry fields of Aquitania in south-western France, *M. phaseolina* was isolated from internal crown lesions of strawberry plants of several cultivars. The disease was frequently observed from fields where sunflowers were cultivated (Baudry and Morzieres, 1993). In India, *Macrophomina* leaf blight and dry crown rot, caused by *M. phaseolina* (syn. *Rhizoctonia bataticola* (Taubenhaus) E.J. Butler) and *Sclerotinia bataticola* Taubenhaus, was reported as a severe strawberry disease (Lele and Phatak, 1965). As described by Maas (1998), symptoms on leaves are more or less circular spots with ashy grey centres and dark purplish margins. They become oval to irregular shaped as they enlarge and the infection progresses. The lesions spread from the margin towards the centre of the leaflet. Puckering, leaf curling and defoliation are common in older plants. The runners and crowns of infected plants turn dark brown or black with irregularly shape lesions. The growth of new leaves or stolons is greatly suppressed. New roots are poorly developed, and parts of older ones may disintegrate.

Symptoms of charcoal rot in strawberry are marked by plant stunting, initial wilting of the foliage, drying and death of older leaves, and eventual collapse and death of the plants (Koike, 2008b) (Fig. 12.3). Similar to *Verticillium* wilt symptoms, the central leaves remain green and alive for a period of time. However, in contrast to most *Verticillium* infections, which show little to no discoloration of the crown when cut open, a plant infected with *Macrophomina* will show distinctive dark brown to orange-brown necrotic areas on the margins of the crown and along the woody vascular ring (Mertely *et al.*, 2005; Bolda and Koike, 2012) (Fig. 12.4). This disease also causes roots to become completely brown and rotted (Avilés *et al.*, 2008; Koike, 2012). Koike (2008b) found that fruiting bodies or other fungal structures were not observed directly on plant tissues, but dark, oblong-shaped sclerotia were observed in infected crown tissue by Avilés *et al.* (2008), and Mertely *et al.* (2005) occasionally observed the development of ostiolate pycnidia on the host tissues after 8–10 days of incubation.

Although *M. phaseolina* is regarded as a warm-climate pathogen and some of the earliest records of the pathogen affecting strawberry were from India and Egypt, where high soil temperatures (33–47°C) and low soil moisture conditions prevail, additional

Fig. 12.3. (a, b) Strawberry plants in the field showing symptoms of charcoal rot.

Fig. 12.4. (a, b) *Macrophomina phaseolina*-infected strawberry plants with dark brown necrotic areas in the crown.

outbreaks have been recorded from cooler, more temperate regions (soil temperatures ranging from 8 to 27°C) such as Spain and France (Zveibil *et al.*, 2012). Disease development is generally considered to be enhanced by some combination of heat stress, soil-water deficit, light-textured soil or the stress associated with host reproduction (Mihail, 1989). Under controlled conditions, *M. phaseolina* was most virulent and caused the most severe disease symptoms at 32°C, whereas plants inoculated with *M. phaseolina* did not show any disease symptoms at 17 or 22°C, presumably because the temperature was too low for disease development. This explains the dominance of this pathogen in the warmer months of the year and, in particular, the dominance of *M. phaseolina* in the late spring/early summer when temperatures often exceed 30°C and a large number of plants wilt and die rapidly in the field (Fang *et al.*, 2011a). In south-western Spain, symptoms of charcoal rot appeared in strawberry fields at the end of the season, when temperatures exceeded 25°C (Miranda *et al.*, 2012).

While various stresses favour symptom development and eventual mortality, high soil temperatures have been found to inhibit early root infection of cantaloupe (*Cucumis melo* L.) by this pathogen (Bruton *et al.*, 1987). In strawberry crops, the importance of elevated soil temperature, rather than moisture stress, on plant mortality caused by *M. phaseolina* was demonstrated by Zveibil *et al.* (2012), although only two levels of soil moisture were evaluated in the study. Sclerotia viability declined more rapidly in soil maintained at 25°C or at soil temperatures fluctuating from 18 to 32°C under greenhouse conditions, compared with sclerotia viability in soil kept at 30°C. After 30–40 weeks of exposure in soil, inocula maintained at 25 or 30°C, or at fluctuating temperatures in a greenhouse, declined to negligible levels. A significant increase in plant mortality was observed in infested soils maintained at 30°C versus 25°C, whereas water stress at 25 or 30°C did not affect plant mortality in *M. phaseolina*-infested soils (Zveibil *et al.*, 2012). Using *in vitro* assays, Khan (2007) demonstrated that the growth of warm-climate *M. phaseolina* was negligible at low temperatures and that it started growth at 15°C. The optimum temperature was 30°C, and its growth was strongly inhibited by temperatures of 40°C or above.

Soil moisture content has been considered to be a major factor affecting the survival of *M. phaseolina* in soil (Pratt, 2006). Soil moisture and water level have significant effects on the survival of sclerotia of *M. phaseolina*. Thus, in moist conditions, both high (>50°C) and low (–5 or 5°C) temperatures have been reported to adversely affect the survival and growth of *M. phaseolina* (Papavizas, 1977). As Zveibil *et al.* (2012) reported, high soil moisture and flooding caused a decline in sclerotia viability compared with dry soils, although the results of a study by Pratt (2006) did not clearly indicate that dry soils favoured the survival of sclerotia, as suggested by some authors, because optimal survival occurred in relatively moist soil at 50–100% moisture holding capacity. In this study, results and observations from experiments on soil moisture levels suggested that complex interactions between microorganisms and soil moisture could influence the survival of sclerotia. Olaya and Abawi (1996) reported that high temperatures and low soil-water potentials are important factors in causing this disease. In culture, this fungus grows at high temperatures (35°C) and at low osmotic potentials, especially if nutrients are available. The sclerotia germinate readily over a wide range of water osmotic potentials (Odvody and Dunkle, 1979). Soil humidity and, in particularly, the soil-water potential are of great importance in the ecology of *M. phaseolina* (Shokes *et al.*, 1977). Goudarzi *et al.* (2008) studied the effect of the osmotic and matric potential on sclerotia germination and mycelial growth of *M. phaseolina* and concluded that the matric potential was more important as a factor than the osmotic potential in promoting the vegetative growth of *M. phaseolina*. Factors that adversely affect the persistence of these propagules include repeated freezing and thawing of soil, low carbon:nitrogen ratios in soil and possibly high soil moisture (Short *et al.*, 1980).

The role of plant density in charcoal rot incidence has only been examined infrequently, but in studies with sunflower, peanuts and guayule (*Parthenium argentatum* A. Gray), disease incidence was inversely related to plant density (Mihail, 1989).

M. phaseolina survives as microsclerotia in the soil and on infected strawberry plant debris, when, at the end of the growing season in late spring, the plants are usually destroyed by mowing off the tops, ploughing the plants into the soil or applying herbicides. Alternatively, the plants may be removed individually by uprooting. In all cases, the debris of plants infected by soilborne pathogens may remain in the soil until the following growing season and thus, serve as a potential source of primary inoculum for new crops (Zveibil *et al.*, 2012). Some studies have suggested that the soil population of *M. phaseolina* microsclerotia increases as the host debris incorporated into the soil decomposes (Mihail, 1989). This may be one of the causes of the increase observed in the incidence of the disease in untreated soils where strawberry is grown year after year (de los Santos *et al.*, 2014). In contrast to *F. oxysporum*, at least in Spain *M. phaseolina* has not been detected in nurseries (Redondo *et al.*, 2012), so the strawberry plant is not considered a source of inoculum for *M. phaseolina*.

The microsclerotia serve as the primary source of inoculum and have been found to persist within the soil for up to 3 years (Dhingra and Sinclair, 1975), although other authors have described significant decreases in the number of sclerotia in soil and plant residues in the field within periods of one to several months (Pratt, 2006). Survival is greatly reduced in wet soils, with the microsclerotia surviving no more than 7–8 weeks and the mycelium no more than 7 days (http//www.cals.ncsu.edu/course/pp728/macrophomina/macrophominiap_haseolinia.HTM). The longevity of sclerotia is not known, but in south-western Spain, the *M. phaseolina* population has increased from year to year in non-fumigated soils where strawberry are grown since 2009 (Chamorro *et al.*, 2015). Short *et al.* (1980) reported that populations of germinable sclerotia in soil were directly related to the number of consecutive years of planting soybean and corn in the field and showed a twofold increase in 2 years. Furthermore, the severity and incidence of charcoal rot of soybean and strawberry is directly related to the population level of *M. phaseolina* in the soil, and soybean and strawberry yields are inversely related to the severity and incidence of charcoal rot (Short *et al.*, 1978; Miranda *et al.*, 2012). Wyllie (1988) reported that 15 microsclerotia g^{-1} soil is an acceptable level for the production of soybean.

The microsclerotia are produced in the host tissue and released into the soil as the infected plant decays. These multicelled structures allow persistence of the fungus under adverse conditions such as low soil nutrient levels and temperature above 30°C. Germination of the microsclerotia occurs when temperatures are between 28 and 35°C. Microsclerotia germinate on the root surface, and germ tubes form appresoria that penetrate the host epidermal cell walls by mechanical pressure and enzymatic digestion or through natural openings (Bowers and Russin, 1999). The hyphae first grow intercellularly in the cortex and then intracellularly through the xylem, colonizing the vascular tissue. Once in the vascular tissue, *M. phaseolina* spreads through the taproot and lower stem of the plant, producing microsclerotia that plug the vessels (Wyllie, 1988). The rate of infection increases with higher soil temperatures, and low soil moisture will further enhance disease severity, and the mechanical plugging of the xylem vessels by microsclerotia, toxin production, enzymatic action and mechanical pressure during penetration leads to disease development (http//www.cals.ncsu.edu/course/pp728/macrophomina/macrophominia_phaseolinia.HTM).

M. phaseolina is the type species of the genus *Macrophomina* and is also the name given to the coelomycete synanamorph of *Rhizoctonia bataticola* (Taubenh.) E.J. Butler (Crous *et al.*, 2006). *M. phaseolina* produces large numbers of sclerotia but no spores (Rossman and Palm-Hernández, 2008). For this reason, the fungus was considered to be related to other common sclerotia-forming fungi such as *Rhizoctonia solani* J.G. Kühn, a basidiomycete (Rossman and Palm-Hernández, 2008). According to Wheeler (1975), *M. phaseolina* belonged to the division Eumycota, subdivision Deuteromycotina, class Coelomycetes, order Sphaeropsidales (having pycnidial conidiomata), family Sphaerioidaceae and genus *Macrophomina*. Rossman and Palm-Hernández (2008) found that, in integrating the asexual species into an ascomycete phylogeny, some surprising relationships have become evident. One example of a relationship revealed using molecular sequence data (28S rRNA gene sequences) was the inclusion of *M. phaseolina*, the cause of charcoal rot and ashy stem blight, in the ascomycete family Botryosphaeriaceae, although the teleomorph is still unknown (Crous *et al.*, 2006). According to Index Fungorum (http://www.indexfungorum.org/), the new position of *M. phaseolina* in classification is: family Botryosphaeriaceae, order Botryosphaeriales, Incertae sedis, class Dothideomycetes, subdivision Pezizomycotina and division Ascomycota.

As reported by Crous *et al.* (2006), von Arx (1981) introduced the name *Tiarosporella phaseolina* (Tassi) van der Aa for *M. phaseolina*, and also reduced the genus *Macrophomina* to synonymy under *Tiarosporella* Höhn. This treatment has largely been ignored by the plant pathological and mycological community, although von Arx (1987) retained it. The genus *Tiarosporella* is characterized by having conidia formed from smooth, hyaline conidiogenous cells that lack periclinal thickenings and percurrent proliferations, and hyaline, subcylindrical to fusiform conidia that have irregular apical mucoid appendages (Nag Raj, 1993). Crous *et al.* (2006) were able to induce numerous strains of *M. phaseolina* to sporulate on sterile pine needles, and found that the conidia formed apical mucoid appendages. Thus, *M. phaseolina* (Tassi) Goid. (= *Tiarosporella phaseolina* (Tassi) Van der Aa) belongs to the anamorphic Ascomycetes and is characterized by the production of both pycnidia and sclerotia in host tissues and culture medium (Ndiaye, 2007). Pycnidial conidiomata are dark brown to black, solitary or gregarious, up to 200 μm in diameter and open by a central ostiole; the walls are multi-layered and the cells dark brown and thick-walled. The pycnidiospores are ellipsoid to obovoid, and measure 20–24 (range 16–32) × 7–9 (range 6–11) μm; immature conidia are hyaline. During sclerotia formation, 50–200 individual hyphal cells aggregate to form multicellular bodies named microsclerotia. The microsclerotia, which occur in host tissues or soil, are black, smooth and hard, and are variable in size depending on the available nutrients of the substrate on which the propagules are produced (Short *et al.*, 1978; Crous *et al.*, 2006; Ndiaye, 2007).

According to the first description by Maas (1998) of *M. phaseolina* isolated from strawberry plants with symptoms of leaf blight and dry crown rot, cultures from infested tissues on potato dextrose agar produced profuse fluffy, white mycelium. Within 3 days, numerous small, black, horny, fleshy, irregularly shaped sclerotia were produced, but sporulation has not been observed in connection with this disease in India, either in culture or on plant material. *M. phaseolina* isolated from strawberry plants showing symptoms of charcoal rot produces numerous dark, oblong or irregularly shaped sclerotia that are 55–67 to 170–190 μm long and 44–50 to 133–135 μm wide (Fig. 12.5a). When isolates were grown in 1.5% water agar with dried and sterilized wheat straw, dark, ostiolate pycnidia and hyaline, single-celled, cylindrical conidia were produced (Mertely *et al.*, 2005; Koike, 2008a; Baino *et al.*, 2011). Mertely *et al.* (2005) and Baino *et al.* (2011) reported that ostiolate pycnidia bearing relatively large, broadly ellipsoidal, hyaline conidia occasionally or rarely developed on the affected host tissue after 8–10 days of incubation. Colonies in culture range in colour from white to brown or grey, and darken with age (Fig. 12.5b). Hyphal branches generally form at right angles to parent hyphae, but branching is also common at acute angles.

Crous *et al.* (2006) determined the differences between *Tiarosporella* and *Macrophomina*, with the latter producing in the pycnidia percurrently proliferating conidiogenous cells and conidia that become dark brown and lose their apical

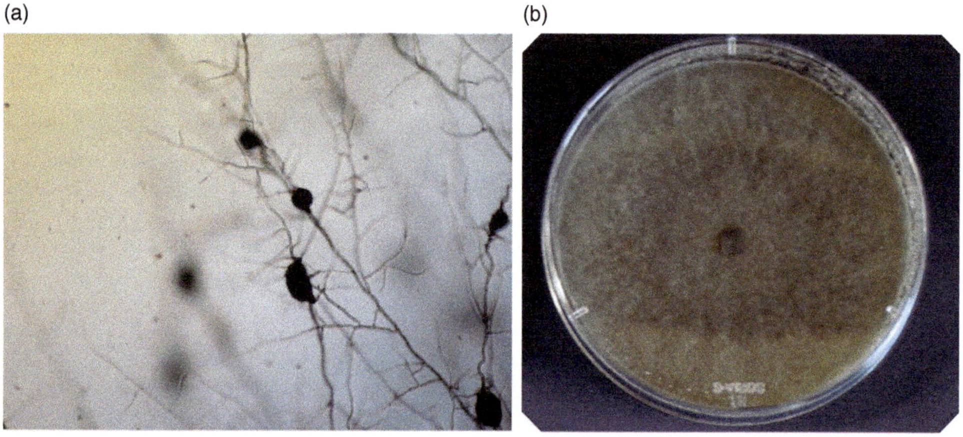

Fig. 12.5. (a) Sclerotia of *Macrophomina phaseolina* (magnification ×100). (b) Colony morphology on potato dextrose agar.

appendages at maturity. Based on these differences, the genus *Macrophomina* and the name *M. phaseolina* is retained. The emended description is:

> ***Macrophomina phaseolina*** (Tassi) Goid., Annali Sper. agr. N.S. 1: 457. 1947.
>
> *Basionym*: *Macrophoma phaseolina* Tassi, Bull. Lab. Ort bot. Siena IV: 9. 1901.
>
> = *Tiarosporella phaseoli* (Maubl.) Aa, Verh. Kon. Ned. Akad. Wetensch., Sectie 2, 68: 4. 1977.
>
> Additional synonyms listed by Holliday and Punithalingam (1988).
>
> Crous *et al.* (2006).

Although, only one species (*M. phaseolina*) is recognized within the genus *Macrophomina* (Walker, 1980), great variability in morphology and pathogenicity has been recognized among isolates from different hosts (Beas-Fernández *et al.*, 2006). No variations were observed among isolates from soybean, corn, sorghum or cotton root tissue and soil from fields cropped continuously with these species in terms of restriction patterns of DNA fragments amplified by PCR covering the ITS region, 5.8S rRNA gene and part of the 25S rRNA gene, suggesting that *M. phaseolina* constitutes a single species (Su *et al.*, 2001).

As reported by Rayatpanah *et al.* (2012), the genetic diversity of *M. phaseolina* could favour its survival and adaptation to variable environments, because significant morphological, physiological, pathogenic and genetic diversity has been reported. However, no clear evidence to suggest *formae speciale*, subspecies or physiological races has been reported. Furthermore, Muñoz-Cabañas *et al.* (2005) concluded that *M. phaseolina* showed certain pathogenic specialization but no different genetic races.

Su *et al.* (2001) also suggested host specialization in the genus. In this work, ten random primers were used to amplify the total DNA of 45 isolates, and banding patterns resulting from RAPD analysis were compared with the neighbour-joining method. Isolates from a given host were genetically similar to each other but distinctly different from those from other hosts. Moreover, chlorate-sensitive isolates showed genetic differences from chlorate-resistant isolates within a given host.

Jana *et al.* (2005) pointed out that the utility of DNA markers such as RFLP, RAPD, amplified fragment length polymorphism (AFLP), variable number tandem repeats, and simple sequence repeats (SSR) in detecting genetic variability is well established for many phytopathogenic fungi. Various molecular techniques have been used to develop methods for the genetic differentiation of populations of *M. phaseolina* using RFLPs of rRNA gene ITS regions (Su *et al.*, 2001), RAPD (Su *et al.*, 2001; Almeida *et al.*, 2003;

Jana *et al.*, 2003) and AFLP (Vandemark *et al.*, 2000; Mayék-Pérez *et al.*, 2001). However, none of these methods has so far been able to differentiate the isolates of *M. phaseolina* from specific hosts and geographical locations (Jana *et al.*, 2005). Single primers of SSR or microsatellite markers have been used for the characterization of genetic variability of different populations of *M. phaseolina* obtained from soybean and cotton, and host-specific DNA fingerprint groups were detected (Jana *et al.*, 2005). Baird *et al.* (2010) evaluated the genetic diversity of 109 isolates of *M. phaseolina* collected from different geographical regions and host species throughout the USA. The paired genetic distances matrix based on SSR data separated the *M. phaseolina* isolates into six groups. Isolates from strawberry (MPI81, MPI82 and MPI83) were included in group I, close to those of cotton, fir, pole bean and soybean. Based on these results, the authors concluded that parasexuality is unlikely and that mutations, seeds for planting or movement around the country of the pathogen on soil particles by farm equipment may be responsible for the variability.

Identification and detection of *M. phaseolina* is difficult because the isolates are morphologically very similar. Different scientists have adopted various methods to distinguish *M. phaseolina* isolates. Among the most applicable are: (i) morphological and cultural characterization; (ii) biochemical methods; and (iii) PCR-based molecular techniques (Babu *et al.*, 2010). Insufficient morphological variability within the genus has led some workers to partition this fungus on the basis of cultural characteristics. Furthermore, efforts to identify subspecies of *M. phaseolina* based on microsclerotia size, culture characteristics, changes in soil population in response to rotation and differences in pathogenicity have failed, usually because of the extreme variability within the species or difficulties in quantifying these characteristics (Su *et al.*, 2001). Utilization of chlorate has been used as a marker for identifying host-specific isolates in *M. phaseolina* (Babu *et al.*, 2010), but it was also observed that chlorate-sensitive isolates were distinct from chlorate-resistant isolates within a given host (Su *et al.*, 2001). Therefore, the chlorate phenotype might not be useful for studying host specialization in *M. phaseolina* (Babu *et al.*, 2010). There are few reports on the use of biochemical or serological techniques to detect and quantify *M. phaseolina* (Babu *et al.*, 2010). The identification of races of *M. phaseolina* based on differential hosts is time consuming and may be influenced by environmental factors (Jana *et al.*, 2005). Isolates of *M. phaseolina* could be exactly identified and detected, in both *in vitro* and *in vivo* conditions, by nuclear rRNA genes, particularly in the ITS region, which is a good target for phylogenetic analysis in fungi (Bruns *et al.*, 1991). Babu *et al.* (2007) noted that no such markers were available for specific detection of *M. phaseolina.* Their screening of GenBank for ITS sequences of *M. phaseolina* revealed the existence of very few sequences that showed any degree of variation among them. Different sequences from *M. phaseolina* isolates and other fungi from the EMBL and GenBank databases were aligned based on homologies among these isolates but not among other fungi, and were used to develop specific primers and an oligonucleotide probe (within the ITS region) by amplification of ITS region, followed by RFLP, sequencing and analysis of the ITS. Babu *et al.* (2007) evaluated the efficiency of these specific primers for identification/detection of *M. phaseolina* under *in vitro* conditions. Two primers, MpKFI (5′-CCGCCAGAGGACTATCAAAC-3′) and MpKRI (5′-CCGCCAG AGGACTATCAAAC-3′), were designed from the conserved region adjacent to the 5.8S gene. These primers were shown to be specific for *M. phaseolina.* Avilés *et al.* (2008), in their initial report on *M. phaseolina* as the cause of crown and root rot of strawberry in Spain, sequenced rRNA gene fragments from a single sclerotium (isolate CH 724) from the crowns of affected strawberry plants (Spanish Type Culture Collection, CECT 20715; GenBank accession no. AM410964) and the sequence showed a 99% identity with *M. phaseolina.*

Recently, Babu *et al.* (2011) developed a real-time quantitative PCR assay to detect and quantify *M. phaseolina* abundance in rhizosphere soil and plant tissue. They detected a specific sequence-characterized amplified region in the *M. phaseolina* genome and designed species-specific primers and a probe for real-time PCR. Both TaqMan and SYBR Green techniques were used to target an ~1 kb sequence-characterized amplified region of *M. phaseolina*, and two sets of specific primers were designed for SYBR Green (MpSyK forward/reverse primers) and TaqMan (MpTqK forward/reverse primers) assays. The minimum detection limit or sensitivity of the TaqMan assay was 30 fg ml^{-1} *M. phaseolina* DNA, and the limit of quantification of the *M. phaseolina* viable population was estimated as 0.66×10^5 colony-forming units g^{-1} soil equivalent to 10 pg ml^{-1} target DNA. This report was the first to demonstrate the use of real-time quantitative PCR assays with greater specificity and sensitivity to detect *M. phaseolina* population in soil and plant materials. This method seems to be a strong diagnostic tool for *M. phaseolina* (Mirmajlessi *et al.*, 2015).

12.3.1. Charcoal rot management

Reducing plant stress is one of the goals in charcoal management. Plants grown in conditions of high temperatures, drought or poor fertility (too high or too low) are most susceptible to the disease. Any cultural practices that minimize plant stress will reduce the risk of charcoal rot. This includes using moderate levels of plant populations and optimum fertility levels, especially in terms of phosphorus. Good fertility will not control charcoal rot but it can reduce the disease severity (http://fyi.uwex.edu/fieldcroppathology/soybean_pests_diseases/soybean_charcoal_rot/). The tolerance of *M. phaseolina* to reduced water potentials may be an important factor in the epidemic development of charcoal rot in the tropics characterized by high temperatures and droughts (Olaya and Abawi, 1996). Of the cultural practices that reduce disease severity, applying irrigation water seems to be the most effective (Goudarzi *et al.*, 2008). Water management has a significant effect on root colonization by *M. phaseolina* (Kendig *et al.*, 2000).

Other possible charcoal rot management steps could include applying post-plant fungicides (if effective; registered materials are available), rotating to non-host crops, and planting resistant or tolerant strawberry cultivars (Koike, 2008a).

The population of *M. phaseolina* in soil will increase when susceptible hosts (e.g. soybeans, sunflower, cotton, rice, cucurbits, wheat) are cropped in successive years and it is redistributed by tillage practices (Wyllie, 1988). Mihail (1989) reported that the effect of susceptible crops on inoculum levels and on inoculum dynamics was demonstrated in several studies with soybeans, with increases in the population in monoculture. Furthermore, the spatial pattern of microsclerotia in soil was found to be highly aggregated and to exhibit a gradient of low to high values across the experimental field (Mihail and Alcorn, 1987). In strawberry production, affected fields generally have small, limited patches with symptomatic plants. However, for locations where the disease has developed for more than one season, the patches can be quite large and appear to have spread from the initial problem area (Koike, 2008b; Chamorro *et al.*, 2015). The fungus is probably spread within and between fields mostly by the movement of soil during soil tillage and preparation operations (Koike, 2008a). Ploughing or other tillage does not readily destroy the charcoal rot organism. In the case of soybean and, in fields with a history of charcoal rot, rotating out of soybean production for 2 years was recommended. Small grains and corn can be planted during the 2-year interval between soybeans, because, although corn is a host for the pathogen, it generally supports lower populations of microsclerotia in soil than soybean (http://fyi.uwex.edu/fieldcroppathology/soybean_pests_diseases/soybean_charcoal_rot/). Fang *et al.* (2011b) pointed out that, in situations where there were periods of 3 or more years involving crop rotations between successive strawberry crops, the level of plant decline or death was lower, even in the absence of fumigation.

M. phaseolina has a broad host range and is reported as a pathogen of many vegetables, fruits and field crops. Host specificity was not evident when strawberry plants were inoculated with each of seven Israeli isolates of *M. phaseolina* obtained from six other plant species (spikenard (*Aralia* sp.), pincushion protea (*Leucospermum* sp.), protea (*Leucadendron* sp.), muskmelon (*Cucumis melo*), watermelon (*Citrullus lanatus*) and almond (*Amygdalus communis*)), suggesting the importance of keeping strawberry crops out of rotation with other host crops of the pathogen (Zveibil *et al.*, 2012). In Australia, serious losses to *M. phaseolina* have occurred following rotation with sorghum green manure crops, which are an important host and a potential source of inoculum of *M. phaseolina* (Hutton *et al.*, 2013). Bolda and Koike (2012) suggested *Rubus*, the genus of caneberries, as a candidate for the rotation, because *Rubus* has not been reported as a host, but this does not necessarily mean that such plants would never become infected. Furthermore, these authors pointed out that researchers have successfully infected strawberry with *M. phaseolina* from other hosts, but not vice versa. Furthermore, Mihail and Taylor (1995) suggested that isolates colonizing the Poaceae are more restricted in pathogenicity than the general population. Growing *Brassica* spp. as a part of crop rotation was reported to reduce soilborne disease and weed populations (Daugovish *et al.*, 2011a).

Zveibil *et al.* (2012) reported that survival of *M. phaseolina* in soil may also be affected by inoculum type, as bare sclerotia may be more vulnerable to the fungicidal effects of fumigation and solarization compared with inoculum within infected plant tissues such as strawberry stolons and crowns. In this regard, the results of Hutton *et al.* (2013) showed that a range of soil fumigants had the capacity to kill laboratory-grown microsclerotia of *M. phaseolina* in field soils. In contrast, only MeBr eradicated *M. phaseolina* in infected crowns buried in treated soils (Hutton, 2010). These results highlight the importance for growers to remove as much strawberry crop residue from infected fields as possible, prior to treatment with alternative fumigants (Hutton *et al.*, 2013). The results from Baird *et al.* (2003) indicated that farm practices that increase plant debris destruction immediately after harvest or those that enhance *Trichoderma* spp. populations may directly or indirectly lower the relative longevity of soilborne pathogens, including *M. phaseolina*. The authors suggested that it is possible that *Trichoderma* spp. were involved in direct tissue degradation and nutrient depletion of the root segments, or may have acted as a mycoparasite, reducing the survival of the pathogen. Furthermore, strains of *Trichoderma* have been tested against *M. phaseolina* from different crops. *T. harzianum* reduced root colonization by *M. phaseolina*, and the incidence of charcoal rot disease in soybean was less when the soils of the plots were protected by *Trichoderma*, indicating the possibility of biological control by *T. harzianum* in soybean disease management programmes (Sandoval and López, 2000). Isolates of *T. harzianum* Rifai strain inhibited linear growth and microsclerotia production in *M. phaseolina in vitro*, and coating melon seeds with *T. harzianum* conidia reduced disease incidence by 37.5–46.3%. *In vitro* growth of *T. harzianum* was affected less than that of the pathogen by applications of pentachloronitrobenzene or dazomet (Elad *et al.*, 1986). In strawberry field trials, soil fumigation with metam sodium or dazomet reduced *Trichoderma* populations to very low levels. Metam sodium and dazomet both generate methyl isothiocyanate in soil. The results suggest that this compound could be responsible for the decrease in *Trichoderma* populations observed after these treatments (de los Santos *et al.*, 2003). *Aspergillus* spp. and *Trichoderma* spp. can inhibit the growth and sclerotia production of *M. phaseolina*. Under natural field conditions, isolates of *Trichoderma* and *Aspergillus* were highly effective in reducing the sesame (*Sesamum indicum*) dead plant percentage resulting from *M. phaseolina infection* (Pineda and Gonnella, 1988). In field trials, biosolarization with chicken manure at 12,500 kg ha^{-1} + drip application of Tusal® (a commercial product containing *T. harzianum* and *T. viride*) was as effective as biosolarization with chicken manure at 25,000 kg/ha in controlling *M. phaseolina* soil populations and charcoal rot incidence. Mortality reached 43% in untreated soils, whereas in biosolarized soils + Tusal®, the percentage of dead plants was 13% (Chamorro *et al.*, 2015). Native strains of *Trichoderma* isolated from sorghum and common bean crop soils were investigated to assess their biocontrol potential over *M. phaseolina*, and two showed better hyperparasitic behaviour against *M. phaseolina* than the others (Larralde-Corona *et al.*, 2008). Volatile toxic substances produced by *T. harzianum* and *T. viride* diffuse easily through the pores of the soil and were found to inhibit the soilborne *M. phaseolina*, suppressing sclerotia formation in particular, so metabolites released by these *Trichoderma* spp. could be toxic and fungistatic to *M. phaseolina* (Sreedevi *et al.*, 2011). Soil application of talc-based formulations of *T. harzianum*, *T. polysporum* and *T. viride* effectively controlled the root rot of eggplant (*Solanum melongena* L.) caused by *Macrophomina* under field conditions (Ramezani, 2008).

High level of variation in *M. phaseolina*, its soil habitat and the good survival ability of its sclerotia make chemical control of this pathogen during the crop season difficult and uneconomical. Therefore, the most appropriate approach to combat this pathogen could be the use of resistant varieties (Khan, 2007). In preliminary tests, it appears that some strawberry cultivars differed in their susceptibility to this pathogen. In replicated shade-house experiments, cultivars such as 'Albion', 'Camarosa', 'Diamante' and 'Ventana' were quite susceptible. Inoculated plants showed wilting symptoms 2 weeks after exposure to the fungus, and by 4 weeks the plants had collapsed. In contrast, the cultivar 'Seascape' was highly tolerant and showed only minor dieback of the oldest leaves by the end of the experiment (Koike, 2008a). Later, Koike *et al.* (2009a) reported that, in non-fumigated soils, 'Camarosa' and 'Albion' had the greatest rates of decline due to *Macrophomina*, while 'Ventana' appeared to be the most tolerant. However, Fang *et al.* (2012b) reported that, under controlled conditions, against *M. phaseolina*, the cultivar 'Albion' was the most resistant, with 'Camarosa' being the most susceptible, while in the field, 'Camino Real' showed the greatest fruit yield in both fumigated and non-fumigated beds and was the most disease-resistant cultivar. Only 'Aromas' was resistant against *M. phaseolina* and *F. oxysporum*. Avilés *et al.* (2012) reported three susceptibility groups against *M. phaseolina*. The cultivars 'Carmela', 'Candonga', 'Antilla' and 'Florida Fortuna' comprised the most susceptible group, the second group, with moderate susceptibility, was composed of 'Camarosa' and 'Primoris', while the least-susceptible cultivar was 'Coral'. In a previous study, these authors found that isolate and (isolate × cultivar) interaction showed significant effects on charcoal rot severity. They concluded that their results indicate that characterized cultivars manifest different susceptibility to crown and root rot caused by *M. phaseolina*, depending on the isolate (Avilés *et al.*, 2009). The results of a study carried out with two isolates of *M. phaseolina*, one from soil and the other from a symptomatic strawberry plant, also indicated that the type of cultivar had a significant effect on charcoal rot severity, whereas (isolate × cultivar) interaction had a significant effect on the percentage of dead plants. 'Splendor' was the most resistant cultivar (Chamorro *et al.*, 2012).

12.4. Soil Disinfestation After the Phase-out of Methyl Bromide

After the phase-out of MeBr, a very effective soil fumigant, on 1 January 2005 in all industrialized countries, particularly in those characterized by an intensive use of this fumigant in the past, some phytopathological problems became difficult to manage. Special attention was given to the emergence of new diseases, such as charcoal rot or *Fusarium* wilt in strawberry, or the re-emergence of old ones on several crops (Colla *et al.*, 2012). One of the advantages of MeBr was that, when applied under a wide range of soil textures, moisture levels, temperatures and depths, it kept its relatively high efficacy towards soilborne pests. However, MeBr alternatives, such as 1,3-dichloropropene (1,3-D): chloropicrin (trichloronitromethane), are less flexible with regard to application methods and the conditions needed to enhance efficacy (Santos, 2007). Furthermore, alternative fumigants are more target pest specific and considerably less broad spectrum in terms of pest control activity than MeBr (Noling, 2002).

The first action for most vegetable growers was the replacement of MeBr with other fumigants, often applied in mixtures in order to achieve satisfactory results, but it was clearly understood from the beginning that such replacements would only be temporary (Colla *et al.*, 2012). As an alternative to MeBr, preferred chemicals for strawberry fruit production are 1,3-D:chloropicrin, metam sodium (sodium *N*-methyldithiocarbamate), dazomet and methyl iodide (Pinzon, 2011). Growers are adopting replacement fumigants to a greater extent each season. The most common forms of 1,3-D products are *cis-* or *trans*-isomers of 1,3-D, used with the fungicidal agent chloropicrin, at ratios such as 65% 1,3-D:35% chloropicrin (Nelson *et al.*, 2001). Chloropicrin is almost always applied in combination with other fumigants when used as a pathogen-specific treatment. Other chemicals initially foreseen as possible replacements for MeBr, such as methyl iodide, propargyl bromide and sodium azide, have not yet reached the market, while dimethyl disulfide (DMDS) is still under development in most countries (Gullino *et al.*, 2007). Although Paladin®, a fumigant containing the active ingredient DMDS, has already been registered in Israel and in the Florida Department of Agriculture and Consumer Services, this fumigant must be used with an approved virtually impermeable film and can be used for vegetables and fruit crops (Qin *et al.*, 2011; López-Aranda, 2014). In Europe, DMDS was submitted for registration in 2012 (F. Jiménez, Certis Europe, 2012, personal communication).

The re-evaluation of pesticides that took place under European Directive 91/414/EEC on Plant Protection Products, and later under the European Regulation No.1107/2009 concerning the placing of plant protection products on the market, dramatically reduced the arsenal of available fumigants in the European scenario. Dazomet was the first authorized soil fumigant included in Annex I, the positive list of European Directive 91/414/EEC, from 1 June 2011 for 10 years, followed by metam (including potassium and sodium), approved from 1 July 2012. Both soil fumigants may be authorized for application prior to planting, limited to one application every third year on the same field. Chloropicrin was withdrawn by 23 June 2012, although Member States could grant an additional period of grace expiring at the latest by 23 June 2013. 1,3-D authorization was withdrawn in 2008, and its grace period expired on 20 March 2009. A new application for inclusion in Annex I submitted by the producer was rejected by the Commission (Decision: 20 January 2011). To date, several Member States, notably Italy, Spain and Greece, have granted periods of 120-day authorizations for 1,3-D for emergency uses (Colla *et al.*, 2012).

In California, USA, the use of metam sodium increased rapidly from 1991 to 1995 and the use of metam potassium began in 2001. The decrease in metam sodium use was correlated with the increase in metam potassium use. Due to their relatively low cost, metam fumigants are widely used in many annual vegetable and fruit crops. Use of 1,3-D increased rapidly from 2000 to 2005, reflecting its importance in replacing MeBr. 1,3-D is an effective nematocide and has been used for many crops that are sensitive to nematode damage. Chloropicrin use has increased steadily since 2000. Chloropicrin is a good fungicide and is used in mixtures (sometimes as a warning agent) with MeBr and 1,3-D products. Strawberry production uses up to 70% chloropicrin (Gao and Wilhoit, 2011). 1,3-D and chloropicrin are restricted by township caps and buffer zone requirements because of their volatile organic compound emissions (Cabrera *et al.*, 2011).

Fumigants are ideally applied approximately 2 weeks prior to planting transplants but newer fumigant labels may increase this period to nearly a month. Fumigants are applied broadcast and as a soil mulch under polyethylene, or under a virtually (VIF) or totally (TIF) impermeable film, in order to reduce their emissions into the atmosphere

(Yates *et al.*, 2002). Using highly retentive mulches is an important component for improving fumigant activity against soil-borne pathogens. VIFs increase the duration under the mulch of relatively high fumigant concentrations, allowing more time for exposure of the pathogens to lethal rates and for lateral distribution in the soil (Santos *et al.*, 2007). Nowadays, fumigants are generally applied during the bedding process, and row middles are not treated so application rates can be reduced, promoting a reduction of dosages (Nelson *et al.*, 2001). The bed is then immediately mulched with black polyethylene or VIF. A recent evolution of this method is emulsifying fumigants in water for delivery into preformed beds through drip irrigation systems under plastic mulch (Duniway, 2002). Pest control efficacy of all the available fumigant alternatives to MeBr is highly dependent on uniform soil delivery and distribution, and uniformity of soil type and conditions within the field. Unlike MeBr, prevailing edaphic and climatic conditions during pre- and post-fumigant application are very important in dictating the treatment efficacy and crop response (Noling, 2002).

12.4.1. Chemical alternatives

Locascio *et al.* (1999b) claimed that studies of strawberry production indicated that marketable fruit yields of polyethylene-mulched strawberry without fumigation were as low as 40–80% of yields obtained with MeBr:chloropicrin fumigation. Locascio *et al.* (1999a) observed that plants in untreated control plots appeared to be stunted and chlorotic in comparison with plants grown in fumigated soils. Incidences of infection of plants were significantly higher for the following potential pathogens in plants from the untreated control plots compared with plots treated with MeBr:chloropicrin or chloropicrin alone: *Curvularia* spp., *F. solani* (36–78% of plants colonized), *M. phaseolina* (84–94% of plants colonized), *Pythium irregulare* and *Rhizoctonia* spp. The authors concluded that the high incidence of these pathogenic fungi isolated from plants in the untreated plots indicated that the soil treatments reduced pathogens that probably contributed to poor plant growth and lower yields in the control.

When comparing plants treated with low and high rates of drip-applied chloropicrin, methyl iodide, 1,3-D:choloropicrin (65:35), MeBr:chloropicrin with untreated controls, Daugovish *et al.* (2011c) did not observe significant mortality in any treatment (including untreated controls) until May. *M. phaseolina* was a primary pathogen isolated from the declining plants at the end of May, with highest mortality (11%) in low-dose chloropicrin treatment, which was significantly greater only when compared with methyl iodide (5%), which was slightly lower than all other treatments. This suggested a lack of efficacy of pre-plant fumigants against late-season decline associated with *M. phaseolina*. However, marketable yield was 28% less without fumigation. Yield decline in non-fumigated controls (compared with fumigated treatments) started to occur in March, but the pathogen-related mortality was not observed until May, suggesting that the negative effects of *M. phaseolina* and *F. oxysporum* on productivity may take place prior to plant collapse. Furthermore, when mortality was compared among the two central and two side rows of strawberry plants, it appeared that 53% more 'Camarosa' strawberry plants collapsed in side rows compared with central rows following chloropicrin treatment, while no differences were observed for the untreated controls. These observations suggested that fumigant distribution and, probably greater stress in the side rows, accelerated pathogen-induced mortality in these zones compared with the bed interior. They concluded that fumigants provide protection from these pathogens in the early and mid-season but not during May–June. However, in field trials carried out in south-western Spain, pre-plant treatments with chloropicrin alone, 1,3-D:chloropicrin, DMDS + chloropicrin, methyl iodide + chloropicrin or dazomet (incorporated into soil with Mix-Tiller®) not only reduced *M. phaseolina* population in soils, but also provided protection from the pathogen until the end of the season in May. Mortality reached 20% for 'Camarosa' strawberry plants in untreated plots, but in fumigated soils mortality did not reach 10%, and there were no dead plants in soils treated with dazomet (Miranda *et al.*, 2012).

Cabrera *et al.* (2011) evaluated the efficacy of reduced rates of 1,3-D:choloropicrin followed by a standard high-density polyethylene tarpaulin or a TIF on plant-parasitic nematodes, soilborne pathogens and weeds. The type of tarpaulin did not affect the efficacy of reduced 1,3-D and chloropicrin rates against nematodes, soilborne pathogens and weeds. However, higher soilborne pathogen incidence was encountered in the reduced rate treatments compared with full rates under both tarpaulin types. This suggested that fumigant concentrations at reduced rates may be sufficient for adequate nematode control in sandy soil conditions but are not sufficient for the control of some soilborne pathogens such as *Fusarium* spp. Field trials by Qin *et al.* (2011) demonstrated the excellent performance of TIFs on fumigant emission control and pest control efficacy. TIFs can effectively retain fumigants under the tarpaulin and in surface soils with improved uniform distribution. These findings suggested that a proper fumigant rate for ensuring good efficacy must be determined by considering the fumigant property and the soil type. An overdose of fumigant may delay tarpaulin-cutting time or cause phytotoxicity for crops, while too a low fumigant level may not be able to achieve the fumigation efficacy target.

In soil treated with methyl iodide, applied under a VIF, control of *F. oxysporum* was as effective as treatment with MeBr: the higher rate provided 100% control and mortality was not observed. Higher and moderate rates of methyl iodide produced yields equivalent to MeBr treatment. In soils treated with DMDS, plant mortality was not observed and the yield reduction was not significant compared with MeBr treatments. The authors concluded that, as methyl iodide was applied at 40% lower rates than MeBr, growers can use fewer chemicals while still sustaining production, and emission is minimized. In the USA, methyl iodide has been approved for use as a preplant biocide, and in Europe this compound is not registered. DMDS seems to be a promising and attractive alternative with high efficacy and potential reduction of buffer zones (Othman *et al.*, 2011). DMDS has been identified as one of the volatile compounds produced when soil is amended with cabbage and solarized, which leads to a reduction in fungal plant pathogens and nematodes. This material has zero ozone depletion potential and is reported to have a complex mode of action affecting mitochondrial function and causing inhibition of cytochrome oxidase (Rosskopf *et al.*, 2000).

Zveibil *et al.* (2012) studied the effect of several fumigants in control of *M. phaseolina*. The fumigants were each administered via drip irrigation through two plastic pipes per plot placed on the soil surface. Fumigation was conducted under impermeable polyethylene plastic sheets spread on the surface of the plots, including the non-fumigated control plots. The soils were previously inoculated with different types of *M. phaseolina* inoculum. MeBr and metam sodium consistently provided the most effective eradication of several types of inoculum of *M. phaseolina* (sclerotia and naturally infected crowns and stolons). However, fumigation with chloropicrin or a combination of 1,3-D:chloropicrin was not consistently as effective as MeBr or metam sodium fumigation at the two soil depths evaluated in this study. Similar results were obtained by Hutton *et al.* (2013). They carried out field trials to screen the efficacy of different soil fumigants against *M. phaseolina*. Fumigants were applied by shank injection (except for dazomet, which was incorporated into the soil with a rotary hoe) under low-density polyethylene to soils previously inoculated with sclerotia or infected crown tissues. MeBr:chloropicrin (50:50) was the only fumigant to eradicate *M. phaseolina* from buried infected crowns 7 days after fumigation. In contrast, methyl iodide:chloropicrin (30:70), 1,3-D:chloropicrin (65:35) and chloropicrin alone reduced the survival of *M. phaseolina* in infected crowns by up to 97% but did not eradicate it. All fumigants, including dazomet and methyl iodide:chloropicrin, reduced the viability of microsclerotia of *M. phaseolina* by 74–82%.

Chemicals have been used in combination with solarization. Yields with solarization combined with chloropicrin or metam sodium and 1,3-D were statistically similar to those achieved with the individual fumigants and with MeBr, whereas chemical treatments resulted in significantly higher yields than those obtained in the untreated control and with soil solarization alone (Rosskopf *et al.*, 2000). Solarization with dazomet (400 kg ha^{-1}) controlled strawberry diseases caused by *F. oxysporum* and *M. phaseolina* as effectively as MeBr (Yücel *et al.*, 2002). Combining solarization with a reduced dose of chemical fumigants such as metam sodium achieved similar strawberry yields to MeBr treatment (López-Aranda *et al.*, 2000), although this treatment did not control *M. phaseolina*. Plant mortality was higher than that observed in the untreated control (Medina-Mínguez *et al.*, 2012; Miranda *et al.*, 2012).

12.4.2. Non-chemical alternatives

Several approaches to the production of strawberry without MeBr or other chemicals have been studied, such as soil-less production, anaerobic soil disinfestation, heat disinfestation, solarization, and the use of biofumigants and herbicides. The soil-less system is a strawberry production method used frequently in Europe (Fennimore *et al.*, 2012), but in this production system, Arroyo *et al.* (2009) detected *F. oxysporum* affecting strawberries in south-western Spain. Legard (2011) pointed out that soil-less production systems were as productive as fumigant-based production system but were not economically competitive due primarily to the high cost and limited availability of the substrates (peat and coconut coir).

Soil solarization is a non-chemical approach to soil disinfestation, involving covering the soil with a clear plastic film to trap solar radiation and accumulate heat. In this way, soil temperatures can be raised to levels that are lethal to many plant pathogens (such as fungi, bacteria, weeds, nematodes and insects) (Katan, 1981). The drawbacks of solarization are the non-productive occupation of soil for a long period, the climate dependency and the mobility to control all pathogens (Ros *et al.*, 2008). The use is restricted in some areas by the length and intensity of sunshine as well as temperature (Berg, 2007). Solarization is recommended for 30–45 days during the months when the soil temperatures exceed 50°C, although when combined with biofumigation, the soil temperature could be as low as 40°C (Medina-Mínguez, 2002). Solarized soils are often more suppressive to certain soilborne pathogens than non-solarized soils. Changes in the microflora following solarization may account for enhanced plant growth and induced suppressiveness to plant pathogens. *Trichoderma* spp. are among the microorganisms that can survive the solar heating process (Pinkerton *et al.*, 2002).

In Florida, soil solarization was initially conducted by covering entire fields (broadcast solarization) with clear or photoselective polyethylene plastic. Subsequently, this method was adapted to local production systems by using clear plastic to solarize raised beds (strip solarization) and then painting the plastic white, so it could be used as a horticultural mulch. Solarization on raised beds achieved higher soil temperatures than broadcast applications, thus improving its efficacy and eliminating any border effects (Rosskopf *et al.*, 2005).

Soil solarization provides control of weeds, but strawberry yields were reduced due to a lack of control of nematode and soil fungi (Locascio *et al.*, 1999b). In field studies begun in the autumn of 1998 in Florida, several soil fumigants and soil solarization treatments were evaluated as alternatives to MeBr for polyethylene-mulched strawberry. Solarization alone generally increased yields significantly over that provided with no fumigant but was not consistent in providing maximum yield (Locascio *et al.*, 1999a).

Zveibil *et al.* (2012) observed a rapid decline in *M. phaseolina* sclerotia viability in soils covered with impermeable polyethylene plastic sheets, which may be attributed to elevated soil temperatures under the polyethylene mulch, even though the mulches remained on the soil surface for only 10 days. Soil solarization alone was not effective in the control of *M. phaseolina* (Lodha *et al.*, 1997). For instance, whereas *F. oxysporum* was eliminated from solar-heated soil down to a 10 cm depth, was reduced at a 25 cm depth and survived at 30 and 40 cm depths, *M. phaseolina* survived at all depths, even when, at the 10 cm depth, the temperature reached 56.2°C (McCain *et al.*, 1982). In strawberry fields, solarization did not reduce the viability of *M. phaseolina* at soil depths of 10 or 20 cm; however, a significant reduction (66%) in survival was determined at a soil depth of 5 cm. This study demonstrated that continuous heat treatment and solarization alone do not provide sufficient control, especially in the deeper layers of the soil (Yildiz *et al.*, 2010). In solarized soils, *Fusarium* wilt severity was lower than that in soils fumigated by either chloropicrin or MeBr, suggesting that *Fusarium* wilt of strawberry could be controlled by soil solarization combined with solar-heated irrigation (Sugimura *et al.*, 2001). Furthermore, in strawberry field trials conducted to evaluate the effectiveness of soil solarization with compost and calcium cyanamide, the soil population of *F. oxysporum* f. sp. *fragariae* was reduced at 100% relative to the untreated control and no incidence of *Fusarium* wilt was observed (Nam *et al.*, 2011b).

Biofumigation uses volatile substances and other products from organic amendments and residue biodegradation as fumigants against soilborne pathogens (Bello *et al.*, 1999). Nitrogen compounds (e.g. ammonium and nitrates), organic acids and a large number of other substances are responsible for this biocidal activity (Mian *et al.*, 1982). Biofumigation is a natural approach using plants containing specific biologically active compounds for the control of soilborne pests and diseases in agricultural crops. It provides a promising sustainable alternative to the use of noxious chemicals, and at the same time will deliver soil and environmental benefits to the farm. The principle is based on the use of plants with high levels of glucosinolates (GCs), which are naturally occurring plant sulphur compounds providing plants with protection. When these plants are incorporated into the soil, the GCs are hydrolysed by the plant enzyme myrosinase, at neutral pH, producing active volatile chemicals called isothiocyanates (ITCs). The most effective plant species have been found to be *Brassica* spp., including oil radish and white mustard. ITCs are related to the active ingredient in the commercial fumigants metam sodium and dazomet, and are highly toxic to pests and pathogens. GCs are sulphur-containing chemicals (thioglucosides) that are produced as secondary metabolites by *Brassica* spp., and their role is thought to be to provide resistance against pests and pathogens (http://www.epa.gov/ozone/air/2000; Mazzola, 2011a). GC profiles and the subsequent ITCs produced vary between *Brassica* species. Researchers have identified over 100 ITCs, 20 of which are commonly produced by *Brassica* spp. and are known to have a biocidal effect. The difference in structure of individual GCs and ITCs depends on their organic side-chain (aliphatic, aromatic or indole), which also influences their biocidal activity (http://www.epa.gov/ozone/air/2000; Mazzola, 2011a). *Brassica* crops can be grown for biomass to be used as a green manure. Another strategy uses *Brassica* products as soil amendments. In this method, *Brassica* seed meal, a by-product of the commercial food and industrial oil extraction process, is incorporated into the soil pre-planting (Mazzola and Mullinix, 2005; Mazzola, 2011b). In addition to the effect of biofumigation, depending on the target organism, pest control attained in response to *Brassica* residue amendments has been attributed to altered soil biology, at times acting through the modulation of host defence responses (Mazzola, 2011a). Mattner *et al.* (2008) showed that volatile compounds from a *Brassica rapa*/*Brassica napus* biofumigant crop suppressed the growth of six different soilborne pathogens of strawberry, among them *F. oxysporum*, but in the field, rotary incorporation of the biofumigant crop did not produce detectable levels of ITCs in the soil, reduce the survival of the strawberry pathogens or affect populations of culturable soil microflora. However, it suppressed the growth of emerging weeds by 40% and the growth of pathogens such as *Phytophthora cactorum* by 20%. They concluded that allelochemicals other than ITCs, such as nitriles, or other biological mechanisms might also play a role in biofumigation. Although, growth of *M. phaseolina* was suppressed when exposed to allyl isothiocyanate produced by *Brassica juncea*, this biologically active volatile was fungistatic towards *M. phaseolina* and was not fungicidal (Mazzola, 2011a). Suppression of *M. phaseolina* was obtained with seed meal sourced from several Brassicaceae species (*Brassica napus*, *Brassica juncea* and *Sinapis alba*), but was not associated with the production of a biologically active chemical (Mazzola, 2011b). Pathogen suppression observed in natural soil was abolished when seed meal-amended soils were pasteurized prior to infestation with *M. phaseolina*. These findings indicated that a biological mechanism contributed significantly to the disease control (Mazzola, 2011a). Brassicaceae seed meal amendments suppressed the proliferation of *M. phaseolina* in soil systems, but optimal seed meal-induced suppression required a functional soil biology (Mazzola, 2011b).

A key to improve the efficacy of biofumigation in the field seems to lie in the development of technological applications that can macerate and incorporate biofumigants evenly in the soil, in addition to incorporating biofumigants under optimal edaphic conditions for the release of ITCs (Mattner *et al.*, 2008). Mazzola (2011a) affirmed that increased severity of *Brassica* tissue segmentation enhanced overall ITC yield and the resulting level of pest suppression.

As well as Brasicaceae residues, other organic materials have been assayed as producers of biofumigators for soil disinfestation, such as green manures, animal manures (chicken manure) and agro-industrial residues, such as rice husks and citrus peelings. An effective integration of solarization treatment with a variety of organic amendments, including composts, crop residues, green manures and animal manures, has been reported for the control of soilborne pathogens; this approach of combining solarization and organic matter application is defined as biological solarization or biosolarization (Ros *et al.*, 2008). This soil treatment process in various forms has also been called 'anaerobic soil disinfestation' (ASD), 'biological soil disinfestation', 'soil reductive sterilization' and 'reductive soil disinfestation' (Butler *et al.*, 2012). Soil treatment by ASD utilizes the principles behind solarization and flooding to create elevated soil temperatures and anaerobic soil conditions through saturation of the topsoil, covering with an oxygen-impermeable plastic tarpaulin and the addition of a labile carbon source to stimulate microbial activity (Butler *et al.*, 2009). ASD can be used where solarization or flooding is not feasible (Shennan *et al.*, 2007). ASD utilizes methods of pre-plant soil disinfestation developed in Japan and the Netherlands (Butler *et al.*, 2012). ASD follows three steps: (i) incorporating organic material (providing a carbon source for soil microorganisms); (ii) irrigation (saturating the soil pores with water); and (iii) covering with an oxygen-impermeable plastic mulch (to limit the oxygen supply, and maintain the soil water conditions above the field capacity for anaerobic decomposition of the soil) (Shennan *et al.*, 2007). Strictly speaking, soil solarization and biosolarization are based on soil irrigation at the field capacity of water content (Katan, 1981).

Lodha *et al.* (1997) demonstrated the efficacy of summer irrigation and soil solarization combined with cruciferous residues in the control of *M. phaseolina*. In irrigated amended soil, polyethylene mulching increased the soil temperature to 57 and 50°C at depths of 0–15 and 16–30 cm, respectively. As a result, population of *M. phaseolina* was almost eradicated (93–99%) at both soil depths. Soil heating and irrigation increased soil temperatures from 50 to 57°C at 0 to 30 cm depths, resulting in an approximately 70% reduction in sclerotia viability. Furthermore, amendment with residues alone or in conjunction with soil solarization increased the population of lytic bacteria active against *M. phaseolina*. However, the results of a test in a strawberry field carried out by Lazzeri *et al.* (2003) showed that biocidal plant green manure treatments (*Brassica juncea* L. sel. ISCI20 and *Eruca sativa* Mill. 'Nemat', containing glucosinolate/myrosinase systems) led to a fruit yield that was lower than with MeBr but higher than with conventional green manure or untreated soil. The marketable yield of strawberries grown in soil treated with biosolarization with *Brassica* pellets (BioFence®, Triumph Italia of Cerealtoscana Group) at 2000 kg ha^{-1} was similar to the untreated control, and significantly lower than the standard MeBr:chloropicrin (50:50) treatment. The percentage of dead plants (caused by both *F. oxysporum* and *M. phaseolina*) was similar for untreated controls (41.8%), biosolarization treatment with *Brassica* pellets (49.9%) and other fumigant treatments such as metam sodium (51.5%), Enzone® (sodium tetrathiocarbonate) (38.8%) and methyl iodide:chloropicrin (98:2) (34.7%), whereas mortality with the standard MeBr:chloropicrin treatment did not reach 4% (Medina-Mínguez *et al.*, 2012). In trials carried out in an experimental farm located in south-western Spain, in soils never fumigated with chemicals, the fruit yield was significantly higher following biosolarization with *Brassica* pellet treatment at 2000 kg ha^{-1} (yield: 60,995 kg ha^{-1}) than in untreated plots (yield: 42,156 kg ha^{-1}). During the first season (2009/10), mortality (caused by *M. phaseolina*) reached 14% in untreated plots in May. Charcoal rot incidence was significantly lower in treated soil. During 2010/11, mortality reached 43% in untreated plots in May and the biosolarization treatment was highly effective in controlling charcoal rot (de los Santos *et al.*, 2014). Similar results were obtained by Daugovish and Fennimore (2011) when they evaluated three non-fumigant combinations: mustard seed meal at 2200 kg ha^{-1} incorporated into beds and followed by solarization via standard clear polyethylene mulch, mustard meal supplemented by steaming for 10 min at 60°C to 25 cm depth, and steaming followed by solarization for 2 months. Early fruit yields were similar for all combinations and 73% greater than in untreated controls. Only combinations that included steam reduced the *M. phaseolina* population in soil compared with the untreated control: 43% at 0–15 cm depth and 88% at 16–30 cm. Daugovish *et al.* (2011b), using the same treatments, reported that the combination of steam with soil solarization was the most effective in suppressing *F. oxysporum* in soil. Declining plants were diagnosed primarily with *M. phaseolina*, infection, which typically causes late-season dieback regardless of pre-plant treatments. The dieback occurred earlier and was greater in treatments with solarization with clear mulch. The increased soil temperatures in spring under clear mulch also promoted infection with *M. phaseolina*, which is known to cause the rapid decline of plants stressed with heavy fruit loads under high temperatures. The results suggested that these non-fumigant treatments may not eliminate pathogens from strawberry rooting depths but may provide substantial yield improvements in non-fumigated buffer zones. However, the non-fumigant combination of mustard pod residues and soil solarization almost eliminated viable propagules of *M. phaseolina* and *F. oxysporum* f. sp. *cumini* at 0–30 cm soil depth (Israel *et al.*, 2005). Biosolarization with *Brassica* pellets and chemical treatments reduced *Fusarium* spp. populations in soil compared with untreated controls (de los Santos *et al.*, unpublished data).

As reported by Pratt (2006), one possible use for animal wastes is as biocontrol materials for soilborne plant diseases. Limited evidence to date indicates that animal manure may reduce or eliminate resting structures of diverse pathogens or diseases following their incorporation into soil as amendments. Animal wastes are major liabilities that result from year-around production of cattle, swine and poultry in confined facilities. Most animal wastes are disposed of by repeated applications to agricultural lands in close proximity to production sites, but this practice may result in chemical and biological pollution of soil, water and air. Pratt (2006) evaluated the potential for commercial poultry litter to provide biocontrol of sclerotia of *M. phaseolina* in soil. Incorporation of litter at 5% by weight resulted in an approximate 75% decrease in survival of sclerotia in comparison with unamended soil. Increasing the poultry litter content to 10% resulted in a further 25% decrease that nearly eliminated survival of sclerotia, although the sclerotia retrieved from soils amended with poultry litter were physically intact, of normal colour and did not differ consistently in appearance from sclerotia retrieved from unamended soil. During germination assays, hyphae of *Fusarium* and other fungi developed much more profusely from and around sclerotia retrieved from litter-amended soil than from unamended soil. He concluded that poultry litter is efficacious as a biocontrol material against sclerotia of *M. phaseolina* in soil. Poultry litter has been used as a carbon source in ASD. In strawberry field trials, the following treatments were applied: untreated control, 1,3-D:chloropicrin (application performed using a standard drip application system) and ASD (poultry litter application in beds). *F. oxysporum* and *M. phaseolina* inoculum survival was decreased significantly by both treatments when compared with the untreated control. ASD was the most effective treatment at eliminating *M. phaseolina*. In the late season, the *Trichoderma* spp. population increased in ASD-treated plots. Mortality and stunting in ASD-treated plots was similar to that in 1,3-D: chloropicrin-treated plots and was significantly lower than untreated control plots. The total weight of fruit was lowest in the untreated control and ASD-treated plots; however there were no statistically significant differences among any of the treatments (Rosskopf *et al.*, 2011).

The marketable yield of strawberry plants grown in soils treated with biosolarization with fresh chicken manure (yield: 835 g per plant) was slightly lower than with the standard MeBr:chloropicrin (50:50) treatment (yield: 878 g per plant) (Medina-Mínguez *et al.*, 2012). Biosolarization with chicken manure was as effective as MeBr in the control of *M. phaseolina* and *F. oxysporum* populations in soil (Miranda *et al.*, 2012; de los Santos *et al.*, unpublished data). Both treatments significantly reduced plant mortality, primarily due to *M. phaseolina* infections, compared with the untreated control. Similar results were obtained when biosolarization with chicken manure was compared with chloropicrin, 1,3-D:chloropicrin or DMDS: chloropicrin treatment. Biofumigation treatments were broadcast applied, whereas chemicals were shank injected with two chisels per bed (Miranda *et al.*, 2012). Analysing *M. phaseolina* and *Fusarium* spp. soil population evolution from 2009 to 2011 in soil samples taken before the application of treatments each year, we observed that, in biofumigated soils, the populations of both pathogens tended to decrease, while in soils treated with bed fumigation, the populations increased (de los Santos *et al.*, unpublished-data). Thus, it is possible that there is a strong correlation between a higher incidence of charcoal rot and *Fusarium* wilt in strawberry plants and bed fumigation with alternative chemicals, as reported by Bolda and Koike (2012). In trials carried out on an experimental farm using non-chemical treatments, biosolarization with chicken manure (at rate of 25,000 kg ha^{-1} and 12,500 kg ha^{-1}) and sugar beet vinasse (15,000 kg ha^{-1}) protected strawberry plants equally well against charcoal rot (<2% mortality). During the second season, mortality in untreated plots reached 43%, whereas mortality in plots biofumigated with chicken manure at 25,000 kg ha^{-1} or 12,500 kg ha^{-1} reached 2 and 15% mortality, respectively, and biofumigation with sugar beet vinasse reached 1% mortality (de los Santos *et al.*, 2014). The main steps of biosolarization with chicken manure are as follow: (i) copious sprinkler irrigation to reach the field capacity content of the soil; (ii) application and immediate soil incorporation (with a disc harrow or rotavator) of 25,000 kg ha^{-1} of fresh chicken manure (rich with the husk of rice); (iii) short sprinkler irrigation to restore water losses due to soil evaporation, just before transparent film application; (iv) application of low-density polyethylene transparent film sheets (50 µm) on strips 3.30 m wide with 0.35–0.40 m of separation among strips; (v) biosolarization for 5–6 weeks; and (vi) removal of the plastic and raking with a disc harrow or cultivator to homogenize the solarized soil (Medina-Mínguez *et al.*, 2004) (Fig. 12.6).

Soil quality after biosolarization with sheep and chicken manure was assessed by measuring physico-chemical, chemical, biological and biochemical parameters. Biosolarization was not toxic, as determined by the luminescence of *Vibrio fischeri*

Fig. 12.6. Biosolarization with chicken manure. (a) Application and immediate soil incorporation (with a disc harrow or rotovator) of fresh chicken manure. (b) Application of low-density polyethylene transparent film sheets.

(a biosensor used to identify qualitatively and quantitatively the presence of different compounds or conditions, including toxicity, in complex environments such as soil), whereas MeBr significantly reduced luminescence by *V. fisheri*. No significant increases in heavy metal content of the soil were observed, although heavy metal accumulation after repetitive applications of manure is one of the commonly mentioned reasons for not recommending biosolarization (Ros *et al.*, 2008). However, Albregts *et al.* (1996) observed that a high rate of chicken manure reduced yield throughout the season, because an average of 36% plants per plot were lost. The authors suggested that the loss of plants in this treatment was due to the high salinity levels in areas of the bed receiving excess manure.

ASD using rice bran as the carbon source was compared with the use of chloropicrin, mustard application and steam + solarization with clear mulch in different combinations. Nine tons per acre of rice bran was pre-plant incorporated and 3–4 acre-inches of irrigation was applied in sandy-loam to clay-loam soils. Only chloropicrin and combinations that included ASD reduced *M. phaseolina* populations in soil compared with the untreated control at a depth of 0–15 cm. At 16–30 cm depth, all treatments, except for ASD, reduced *M. phaseolina* in soil compared with the untreated control. ASD and ASD + mustard did not significantly reduce the *F. oxysporum* population compared with the untreated control. Marketable fruit yield was higher in all treatments than in the untreated control. Mortality reached 17% in untreated plots in June, and was higher in plots treated with: ASD + mustard (23%), solarization + mustard (38%), ASD (42%) and ASD + solarization + mustard (46%). The authors concluded that only steam reduced pathogen levels (http://ceventura.ucanr.edu/files/152781.pdf).

12.5. Conclusions and Future Perspectives

In recent years, novel strawberry plant collapse problems have been detected. Plant diagnostic analyses have confirmed that these problems are associated with two soilborne fungi: *M. phaseolina* (the causal agent of charcoal rot) and *F. oxysporum* (the causal agent of *Fusarium* wilt). Charcoal rot has been reported from Argentina, Australia, Egypt, France, India, Israel, Spain and the USA (California, Florida and Illinois), while *F. oxysporum* has been detected in Argentina, Australia, China, South Korea, Spain, Japan and the USA (California and South Carolina). Both pathogens caused plant collapse, poor growth and reduced yields. Symptoms consist of wilting of foliage, plant stunting, and drying and death of older leaves, although the central youngest leaves often remain green and alive. Plants may eventually collapse and die. Initial problems usually consist of multiple small patches of diseased plants, but in locations where the disease has occurred for more than one season, the patches can be quite large and appear to have spread from the initial problem area. The increase in prevalence of crown and root rot in strawberry crops may be related to the phase-out of MeBr. Some authors have associated the changes in the fumigation products used and methods of applying them prior to planting strawberries with the establishment of these two soilborne pathogens, especially with the use of pre-plant alternatives, applied to the beds, compared with the MeBr or chloropicrin fumigation standard.

Fusarium wilt and charcoal rot can be a threat for strawberry crops, at least in moderately warm and warm climatic areas. Although the outbreak of these diseases has been related to MeBr phase-out and its replacement with other methods, as discussed in this chapter, it is possible that other factors could be involved, as several authors have pointed out the detection of these strawberry diseases in MeBr-fumigated soils. One of the possibilities is climate change, as we have observed a relationship between the increase in temperature and the emergence of disease, in temperate areas. Since the mid-1990s, there has been a large increase in our knowledge and use of products and methods for soil disinfestation for strawberry production. Researchers around the world have tested different techniques for chemical disinfestation in order to minimize their emissions into the atmosphere and to reduce the dosages. The most widespread chemical solutions are mixtures of 1,3-D:chloropicrin, chloropicrin alone, metam sodium and dazomet. In developed countries, the majority of useful chemical alternatives have no future because of regulatory restrictions. To date, only dazomet is in use, while DMDS is not yet registered in Europe but could be considered as a chemical alternative for the future. A list of non-chemical alternatives to MeBr for strawberry fruit production includes: the use of resistant varieties, although the results obtained are not conclusive; crop rotation, which is not always possible; steam or soil-less culture, both of which are expensive; biocontrol practices, which are effective only when applied with other control measures; soil solarization, alone or implemented with chemical compounds, although solarization alone can be applied in limited climatic areas; biofumigation practices and its variants, such as incorporation of plant residues, compost extracts, green manure crops, as dry or fresh biofumigant manure; and ASD techniques, which could be the most promising. As reported by López-Aranda (2014), in the case of non-chemical alternatives, much work is needed in the technology transfer process, and farmers are showing a growing interest in these techniques.

References

Agostini, A. (2011) Effects of Brassicaceae seed meal incorporation on pre-plant weed emergence and strawberry productivity in *Macrophomina phaseolina* infested and non-infested soil. MS thesis, Sonoma State University, California.

Alabouvette, C., Schippers, B., Lemanceau, P. and Bakker, P.A.H.M. (1998) Biological control of *Fusarium* wilts. In: Boland, G.C. and Kuykendall, L.D. (eds) *Plant–Microbe Interactions and Biological Control.* Marcel Dekker, New York, pp. 15–36.

Albregts, E.E., Gilreath, J.P. and Chandler, C.K. (1996) Soil solarization and fumigant alternatives to methyl bromide for strawberry fruit production. *Soil and Crop Science Society of Florida* 55, 16–20.

Almeida, A.M.R., Abdelnoor, R.V., Arrabal Arias C.A., Carvalho, V.P., Jacoud Filho, D.S., Marin, S.R.R., Benato, L.C., Pinto, M.C. and Carvalho, C.G.P. (2003) Genotypic diversity among Brazilian isolates of *Macrophomina phaseolina* revealed by RAPD. *Fitopatología Brasileira* 28, 279–285.

Alves-Santos, F.M., Ramos, B., García-Sánchez, M.A., Eslava, A.P. and Díaz-Mínguez, J.M. (2002) A DNA-based procedure for in planta detection of *Fusarium oxysporum* f. sp. *phaseoli*. *Phytopathology* 92, 237–244.
Angelini, R. and nad Faedi, W. (2010) Malattie e fisiopatie. In: ART Servizi (ed.) *La Fragola*. Bologna, Italy, pp. 228–246.
Arroyo, F.T., Llergo, Y., Aguado, A. and Romero, F. (2009) First report of *Fusarium* wilt caused by *Fusarium oxysporum* on strawberry in Spain. *Plant Disease* 93, 323.
Avilés, M., Castillo, S., Bascón, J., Zea-Bonilla, T., Martín-Sánchez, P.M. and Pérez-Jiménez, R.M. (2008) First report of *Macrophomina phaseolina* causing crown and root rot of strawberry in Spain. *Plant Pathology* 57, 382.
Avilés, M., Castillo, S., Borrero, C., Castillo, M.L., Zea-Bonilla, T. and Pérez-Jiménez, R.M. (2009) Response of strawberry cultivars: 'Camarosa', 'Candonga' and 'Ventana' to inoculation with isolates of *Macrophomina phaseolina*. *Acta Horticulturae* 842, 291–294.
Avilés, M., Castillo, S., Borrero, C. and Refoyo, A. (2012) Strawberry cultivar susceptibility to charcoal rot caused by *Macrophomina phaseolina*. In: *VII International Strawberry Symposium*. International Society for Horticultural Science, Beijing, China, p. 224.
Baayen, R.P., O'Donnell, K., Bonants, P.J.M., Cigelnik, E., Kroon, L.P.N.M., Roebroeck, E.J.A. and Waalwijk, C. (2000) Gene genealogies and AFLP analyses in the *Fusarium oxysporum* complex identify monophyletic and nonmonophyletic formae speciales causing wilt and rot disease. *Phytopathology* 90, 891–900.
Babu, B.K., Saxena, A.K., Srivastava, A.K. and Arora, D.K. (2007) Identification and detection of *Macrophomina phaseolina* by using species-specific oligonucleotide primers and probe. *Mycologia* 99, 797–803.
Babu, B.K., Saikia, R. and Arora, D.K. (2010) Molecular characterization and diagnosis of *Macrophomina phaseolina*: a charcoal rot fungus. In: Gherbawy, Y. and Voigt, K. (eds) *Molecular Identification of Fungi*. Springer, Berlin/Heidelberg, Germany, pp. 179–193.
Babu, B.K., Mesapogu, S., Sharma, A., Somasani, S.R. and Arora, D.K. (2011) Quantitative real-time PCR assay for rapid detection of plant and human pathogenic *Macrophomina phaseolina* from field and environmental samples. *Mycologia* 103, 466–473.
Baino, O., Conci, V.C. and Ramallo, J.C. (2010) Enfermedades de *Fragaria* × *ananassa* Duch. *Atlas Fitopatológico Argentino* 3, 739.
Baino, O.M., Salazar, S.M., Ramallo, A.C. and Kirschbaum, D.S. (2011) First report of *Macrophomina phaseolina* causing strawberry crown and root rot in northwestern Argentina. *Plant Disease* 95, 1477–1477.
Baird, R.E., Watson, C.E. and Scruggs, M. (2003) Relative longevity of *Macrophomina phaseolina* and associated mycobiota on residual soybean roots in soil. *Plant Disease* 87, 563–566.
Baird, R.E., Wadl, P.A., Allen, T., McNeill, D., Wang, X., Moulton, J., Rinehart, T.A., Abbas, H.K., Shier, T. and Trigiano, R.N. (2010) Variability of United States isolates of *Macrophomina phaseolina* based on simple sequence repeats and cross genus transferability to related genera within *Botryosphaeriaceae*. *Mycopathologia* 170, 169–180.
Balajee, S.A., Borman, A.M., Brandt, M.E., Cano, J., Cuenca-Estrella, M., Dannaoui, E., Guarro, J., Haase, G., Kibbler, C.C., Meyer, W., O'Donnell, K., Petti, C.A., Rodriguez-Tudela, J.L., Sutton, D., Velegraki, A. and Wickes, B.L. (2009) Sequence-based identification of *Aspergillus*, *Fusarium*, and *Mucorales* species in the clinical mycology laboratory: where are we and where should we go from here? *Journal of Clinical Microbiology* 47, 877–884.
Baudry, A. and Morzieres, J.P. (1993) First report of charcoal rot of strawberry in France. *Acta Horticulturae* 348, 485–488.
Beas-Fernández, R., Santiago, A.D.S.-D., Hernández-Delgado, S. and Mayék-Pérez, N. (2006) Characterization of Mexican and non-Mexican isolates of *Macrophomina phaseolina* based on morphological characteristics, pathogenicity on bean seeds and endoglucanase genes. *Journal of Plant Pathology* 88, 53–60.
Bello, A., López-Pérez, J.A., Díaz-Viruliche, L., Sanz, R. and Arias, M. (1999) Bio-fumigation and local resources as methyl bromide alternatives. In: *3rd International Workshop Alternatives to Methyl Bromide for the Southern European Countries*. Crete, Greece, p. 17.
Benlioğlu, S., Yildiz, A. and Döken, T. (2004) Studies to determine the causal agents of soil-borne fungal diseases of strawberries in Aydin and to control them by soil disinfestation. *Journal of Phytopathology* 152, 509–513.
Berg, G. (2007) Biological control of fungal soilborne pathogens in strawberries. In: Chincholkar, S.B. and Mujerki, K.G. (eds) *Biological Control of Plant Diseases*. Haworth Press, Binghamton, New York, pp. 1–17.

Berg, G., Roskot, N., Steidle, A., Eberl, L., Zock, A. and Smalla, K. (2002) Plant-dependent genotypic and phenotypic diversity of antagonistic rhizobacteria isolated from different *Verticillium* host plants. *Applied and Environmental Microbiology* 68, 3328–3338.

Bolda, M. and Koike, S. (2012) Third *Macrophomina* find in Watsonville-Salinas production district. Available at: http://ucanr.edu/blogs/blogcore/postdetail.cfm?postnum=8259 (accessed 31 August 2012).

Bowers, G.R. and Russin, J.S. (1999) Soybean disease management. In: Heatherly, L.G. and Hodges, H.F. (eds) *Soybean Production in the Midsouth.* CRC Press, Cleveland, Ohio, pp. 231–271.

Bruns, T.D., White, T.J. and Taylor, J.W. (1991) Fungal molecular systematics. *Annual Review of Ecology and Systematics* 22, 525–564.

Bruton, B.D., Jeger, M.J. and Reuveni, R. (1987) *Macrophomina phaseolina* infection and vine decline in cantaloupe in relation to planting date, soil environment, and plant maturation. *Plant Disease* 71, 259–263.

Bull, C.T., Muramoto, J., Koike, S.T., Leap, J., Shennan, C. and Goldman, P. (2005) Strawberry cultivars and mycorrhizal inoculants evaluated in California organic production fields. *Crop Management* 4, doi: 10.1094/CM-2005-0527-02-RS.

Butler, D.M., Rosskopf, E.N., Burelle, N.K., Muramoto, J. and Shennan, C. (2009) Field evaluation of anaerobic soil disinfestation in a bell pepper-eggplant double crop. In: *Proceedings of the Annual International Research Conference on Methyl Bromide Alternatives and Emissions Reductions.* Methyl Bromide Alternatives Outreach, San Diego, California, pp. 43/1–43/4.

Butler, D.M, Rosskopf, E., Kokalis-Burelle, N., Albano, J., Muramoto, J. and Shennan, C. (2012) Exploring warm-season cover crops as carbon sources for anaerobic soil disinfestation (ASD). *Plant and Soil* 355, 149–165.

Cabrera, J.A., Hanson, B.D., Abit, M.J.M., Gerik, J.S. and Gao, S. (2011) Efficacy of 1,3-dichloropropene plus chloropicrin reduced rates under two different tarps against nematodes, pathogens and weeds. In: *Proceedings of the Annual International Research Conference on Methyl Bromide Alternatives and Emissions Reductions.* Methyl Bromide Alternatives Outreach, San Diego, California, pp. 75/1–75/3.

Chamorro, M., Romero, F. and de los Santos, B. (2012) Evaluación de la resistencia de cultivares de fresa a *Macrophomina phaseolina*, agente causal de podredumbre carbonosa. In: *Libro de Resúmenes XVI Congreso de la Sociedad Española de Fitopatología.* SEF, Málaga, Spain, p. 217.

Chamorro, M., Miranda, L., Domínguez, P., Medina, J.J., Soria, C., Romero, F., López-Aranda, J.M. and De los Santos, B. (2015) Evaluation of biosolarization for the control of charcoal rot disease (*Macrophomina phaseolina*) in strawberry. *Crop Protection* 67, 279–286.

Cho, C.T. and Moon, B.J. (1984) Studies on the wilt of strawberry caused by *Fusarium oxysporum* f. sp. *fragariae* in Korea. *Korean Journal of Plant Protection* 23, 74–81.

Cipriani, M.G., Stea, G., Moretti, A., Altomare, C. and Mulè, G. (2009) Development of a PCR-based assay for the detection of *Fusarium oxysporum* strain FT2, a potential mycoherbicide of *Orobanche ramosa*. *Biological Control* 50, 78–84.

Colla, P., Gilardi, G. and Gullino, M.L. (2012) A review and critical analysis of the European situation of soilborne disease management in the vegetable sector. *Phytoparasitica* 40, 515–523.

Crous, P.W., Slippers, B., Wingfield, M.J., Rheeder, J., Marasas, W.F.O., Philips, A.J.L., Alves, A., Burgess, T., Barber, P. and Groenewald, J.Z. (2006) Phylogenetic lineages in the *Botryosphaeriaceae*. *Studies in Mycology* 55, 235–253.

Daugovish, O. and Fennimore, S.A. (2011) Non-fumigant alternatives for managing *Macrophomina phaseolina* and *Fusarium oxysporum* in California strawberry. *HortScience* 46, S175.

Daugovish, O., Faber, B., Mochizuki, M.J. and Styles, S. (2011a) Strawberry performance during establishment with reduced irrigation. *HortScience* 46, S117.

Daugovish, O., Fennimore, S.A., Gordon, T., Koike, S. and Subbarao, K. (2011b) Non-fumigant combinations for management of San Andreas strawberry in a buffer zone infested with *Fusarium oxysporum* and *Macrophomina phaseolina.* In: *Proceedings of the Annual International Research Conference on Methyl Bromide Alternatives and Emissions Reductions.* Methyl Bromide Alternatives Outreach, San Diego, California, pp. 3/1–3/3.

Daugovish, O., Koike, S., Gordon, T. and Ajwa, H. (2011c) Fumigant and strawberry variety evaluations in *Macrophomina phaseolina* and *Fusarium oxysporum* infested fields. In: *Proceedings of the Annual International Research Conference on Methyl Bromide Alternatives and Emissions Reductions.* Methyl Bromide Alternatives Outreach, San Diego, California, pp. 10/1–10/4.

De Cal, A., Martinez-Treceño, A., Salto, T., López-Aranda, J.M. and Melgarejo, P. (2005) Effect of chemical fumigation on soil fungal communities in Spanish strawberry nurseries. *Applied Soil Ecology* 28, 47–56.

de Haan, L.A.M., Numansen, A., Roebroeck E.J.A. and van Doorn, J. (2000) PCR detection of *Fusarium oxysporum* f.sp. *gladioli* race 1, causal agent of *Gladiolus* yellows disease, from infected corms. *Plant Pathology* 49, 89–100.
de los Santos, B. (2010) Optimización y nuevos desarrollos sostenibles en las alternativas al bromuro de metilo en el cultivo de la fresa en Huelva. Resultados 2009–2010. Available at: http://www.juntadeandalucia.es/agriculturaypesca/ifapa/servlet/controller?action=SwitchRecord&table=555&element=36900&ec=-default (accessed 26 September 2012).
de los Santos, B., Barrau, C., Blanco, C., Arroyo, F., Porras, M., Medina, J.J. and Romero, F. (2003) Relationship between *Trichoderma* soil populations and strawberry fruit production in previously fumigated soil. *HortScience* 38, 1400–1402.
de los Santos, B., Chamorro, M., Domínguez, P., Medina-Mínguez, J.J., Miranda, L., Soria, C., Talavera, M., Velasco, L., López-Aranda, J.M. and Romero, F. (2014) Incidence of charcoal rot, causal agent *Macrophomina phaseolina*, in biosolarized soil. *Acta Horticultura* 1049, 919–922.
D'Ercole, N., Nipoti, P. and Manzali, D. 1989. Research on the root rot complex of strawberry plants. *Acta Horticulturae* 265, 497–506.
Dhingra, O.D. and Sinclair, J.B. (1975) Survival of *Macrophomina phaseolina* sclerotia in soil: effects of soil moisture, carbon: nitrogen ratios, carbon sources, and nitrogen concentrations. *Phytopathology* 65, 236–240.
Duniway, J.M. (2002) Status of chemical alternatives to methyl bromide for pre-plant fumigation of soil. *Phytopathology* 92, 1337–1343.
Edel, V., Steinberg, C., Gautheron, N. and Alabouvette, C. (2000) Ribosomal DNA-targeted oligonucleotide probe and PCR assay specific for *Fusarium oxysporum*. *Mycological Research* 104, 518–526.
Edel-Hermann, V., Aime, S., Cordier, C., Olivain, C., Steinberg, C. and Alabouvette, C. (2011) Development of a strain-specific real-time PCR assay for the detection and quantification of the biological control agent Fo47 in root tissues. *FEMS Microbiology Letters* 322, 34–40.
Elad, Y., Zvieli, Y. and Chet, I. (1986) Biological control of *Macrophomina phaseolina* (Tassi) Goid by *Trichoderma harzianum*. *Crop Protection* 5, 288–292.
Fang, X., Phillips, D., Li, H., Sivasithamparam, K. and Barbetti, M.J. (2011a) Comparisons of virulence of pathogens associated with crown and root diseases of strawberry in Western Australia with special reference to the effect of temperature. *Scientia Horticulturae* 131, 39–48.
Fang, X.L., Phillips, D., Li, H., Sivasithamparam, K. and Barbetti, M.J. (2011b) Severity of crown and root diseases of strawberry and associated fungal and oomycete pathogens in Western Australia. *Australasian Plant Pathology* 40, 109–119.
Fang, X., Kuo, J., You, M.P., Finnegan, P.M. and Barbetti, M.J. (2012a) Comparative root colonisation of strawberry cultivars Camarosa and Festival by *Fusarium oxysporum* f. sp. *fragariae*. *Plant and Soil* 358, 75–89.
Fang, X., Phillips, D., Verheyen, G., Li, H., Sivasithamparam, K. and Barbetti, M.J. (2012b) Yields and resistance of strawberry cultivars to crown and root diseases in the field, and cultivar responses to pathogens under controlled environment conditions. *Phytopathologia Mediterranea* 51, 69–84.
Fennimore, S., Subbarao, K., Daugovish, O., Martin, F., Browne, G., Samtani, J., Serohijos, R., Miller, T., Ajwa, H., Sjulin, T., Thomas, H., Legard, D., Shennan, C., Murimoto, J. and Dara, S. (2012) Facilitating adoption of alternatives to methyl bromide in California strawberries. In: *Proceedings of the Annual International Research Conference on Methyl Bromide Alternatives and Emissions Reductions*. Methyl Bromide Alternatives Outreach, Maitland, Florida, pp. 20/1–20/2.
Fernandez, D., Ouinten, M., Tantaoui, A., Geiger, J.-P., Daboussi, M.-J. and Langin, T. (1998) *Fot* 1 insertions in the *Fusarium oxysporum* f. sp. *albedinis* genome provide diagnostic PCR targets for detection of the date palm pathogen. *Applied and Environmental Microbiology* 64, 633–636.
Fort, S.B., Shaw, D.V. and Larson, K.D. (1996) Performance responses of strawberry seedlings to the sublethal effects of nonfumigated soils. *Journal of the American Society for Horticultural Science* 121, 367–370.
Fravel, D.R., Olivain, C. and Alabouvette, C. (2003) *Fusarium oxysporum* and its biocontrol. *New Phytologist* 157, 493–502.
Gao, S. and Wilhoit, L. (2011) Fumigant use and transition from methyl bromide to alternatives in California. In: *Proceedings of the Annual International Research Conference on Methyl Bromide Alternatives and Emissions Reductions*. Methyl Bromide Alternatives Outreach, San Diego, California, pp. 74/1–74/4.
Gliessman, S.R. and Muramoto, J. (2010) California (U.S.): The conversion of strawberry production. In: Gliessman, S.R. and Rosemeyer, M. (eds) *The Conversion to Sustainable Agriculture. Principles, Processes, and Practices*. CRC Press, Boca Raton, Florida, pp. 117–132.
Golzar H., Phillips, D. and Mack, S. (2007) Occurrence of strawberry root and crown rot in Western Australia. *Australasian Plant Disease Note* 2, 145–147.

González, F.J., Walls, S. and Mancilla, M. (2005) Detección mediante PCR-RFLP de *Fusarium oxysporum* f. sp. *fragariae* agente causal de fusariosis en frutilla. *Boletín Micológico* 20, 63–72.

Goudarzi, A., Banihashemi, Z. and Maftoun, M. (2008) Effect of water potential on sclerotial germination and mycelial growth of *Macrophomina phaseolina*. *Phytopathologia Mediterranea* 47, 107–114.

Gullino, M.L, Savigliano, R., Gasparrini, G. and Clini, C. (2007) Critical use exemption for methyl bromide for soil disinfestation: Italy's experience with the European Union process. *Phytoparasitica* 35,

Hancock, J.F. (1999) *Strawberries, Crop Production Science in Horticulture*. CABI Publishing, Wallingford, UK.

Hirano, Y. and Arie, T. (2006) PCR-based differentiation of *Fusarium oxysporum* ff. sp. *lycopersici* and *radicis-lycopersici* and races of *F. oxysporum* f. sp. *lycopersici*. *Journal of General Plant Pathology* 72, 273–283.

Holliday, P. and Punithalingam, E. (1988) *Macrophomina phaseolina*. *CMI Descriptions of Pathogenic Fungi and Bacteria*, vol. 275. CABI Publishing, Wallingford, UK.

Hutton, D.G. (2010) Alternatives to methyl bromide for strawberry production in Queensland. In: *Final Report Project BSO7017*. Horticulture Australia, Sydney, Australia, pp. 11–12.

Hutton, D.G., Gomez, A.O. and Mattner, S.W. (2013) *Macrophomina phaseolina* and its association with strawberry crown rot in Australia. *International Journal of Fruit Science* 13, 149–155.

Hyun, J.W., Kim, S.O. and Park, W.M. (1996) Vegetative compatibility, isozyme polymorphisms and pathogenicity of isolates of *Fusarium oxysporum* f. sp. *fragariae*. *Korean Journal of Plant Pathology* 12, 33–40.

Inami, K., Yoshioka, C., Hirano, Y., Kawabe, M., Tsushima, S., Teraoka, T., and Arie, T. (2010) Real-time PCR for differential determination of the tomato wilt fungus: *Fusarium oxysporum* f. sp. *lycopersici* and its races. *Journal of General Plant Pathology* 76, 116–121.

Islas, C.M., Gordon, T.R., Daugovish, O. and Koike, S. (2011) *Fusarium* wilt of strawberry, caused by *Fusarium oxysporum* f. sp. *fragariae*, a new disease in California. *Phytopathology* 101, S79.

Israel, S., Mawar, R. and Lodha, S. (2005) Soil solarisation, amendments and bio-control agents for the control of *Macrophomina phaseolina* and *Fusarium oxysporum* f.sp. *cumini* in aridisols. *Annals of Applied Biology* 146, 481–491.

Jana, T., Sharma, T.R., Prasad, R.D. and Arora, D.K. (2003) Molecular characterization of *Macrophomina phaseolina* and *Fusarium* species by a single primer RAPD technique. *Microbiological Research* 158, 249–257.

Jana, T., Sharma, T.R. and Singh, N.K. (2005) SSR-based detection of genetic variability in the charcoal root rot pathogen *Macrophomina phaseolina*. *Mycological Research* 109, 81–86.

Jiménez-Fernández, D., Montes-Borrego, M., Navas-Cortés, J.A., Jiménez-Díaz, R.M. and Landa, B.B. (2010) Identification and quantification of *Fusarium oxysporum* in planta and soil by means of an improved specific and quantitative PCR assay. *Applied Soil Ecology* 46, 372–382.

Jiménez-Gasco, M.M. and Jiménez-Díaz, R.M. (2003) Development of a specific polymerase chain reaction-based assay for the identification of *Fusarium oxysporum* f. sp. *ciceris* and its pathogenic races 0, 1A, 5, and 6. *Phytopathology* 93, 200–209.

Kang, S., Mansfield, M.A., Park, B., Geiser, D.M., Ivors, K.L., Coffey, M.D., Grünwald, N.J., Martin, F.N., Lévesque, C.A. and Blair, J.E. (2010) The promise and pitfalls of sequence-based identification of plant-pathogenic fungi and oomycetes. *Phytopathology* 100, 732–737.

Katan, J. (1981) Solar heating (solarization) of soil for control of soilborne pests. *Annual Review of Phytopathology* 19, 211–236.

Kendig, S.R., Rupe, J.C. and Scott, H.D. (2000) Effect of irrigation and soil water stress on densities of *Macrophomina phaseolina* in soil and roots of two soybean cultivars. *Plant Disease* 84, 895–900.

Khan, S.N. (2007) *Macrophomina phaseolina* as causal agent for charcoal rot of sunflower. *Mycopath* 5, 111–118.

Koike, S.T. (2008a) *Macrophomina* crown rot – possible new production issue for California strawberries. Available at: http://westernfarmpress.com/vegetables/macrophomina-crown-rot-possible-new-production-issue-california-strawberries (accessed 1 October 2012).

Koike, S.T. (2008b) Crown rot of strawberry caused by *Macrophomina phaseolina* in California. *Plant Disease* 92, 1253.

Koike, S.T. (2011) Strawberry plant collapse diseases reach California Central Coast. *2011 Annual Meeting* Available at: http://www.cesantacruz.ucdavis.edu/ files/73969.pdf (accessed 1 October 2012).

Koike, S.T. (2012) Soil-borne diseases of strawberry: a review. *Monterey Crop Notes* 5, 7–10.

Koike, S.T., Gordon, T., Ajwa, H., Daugovish, O. and Legard, D. (2009a) Fumigant and strawberry variety evaluations in *Macrophomina* and *Fusarium* fields. *Proceedings Annual International Research Conference on Methyl Bromide Alternatives and Emissions Reductions*. Methyl Bromide Alternatives Outreach, San Diego, California, pp. 13/1–13/4.

Koike, S.T., Kirkpatrick, S.C. and Gordon, T.R. (2009b) *Fusarium* wilt of strawberry caused by *Fusarium oxysporum* in California. *Plant Disease* 93, 1077.

Koike, S.T., Gordon, T.R, Daugovish, O., Ajwa, H., Bolda, M. and Subbarao, K. (2013) Recent developments on strawberry plant collapse problems in California caused by *Fusarium* and *Macrophomina. International Journal of Fruit Science* 13, 76–83.

Kukkonen, S., Vestberg, M.,Tuovinen, T. and Jarvinen, O. (2004) Influence of soil and planting material on the development of strawberry root rot. *Acta Horticulturae* 635, 19–24.

Larralde-Corona, C.P., Santiago-Mena, M.R., Sifuentes-Rincón, A.M., Rodríguez-Luna, I.C., Rodríguez-Pérez, M.A., Shirai, K. and, Narváez-Zapata, J.A. (2008) Biocontrol potential and polyphasic characterization of novel native *Trichoderma* strains against *Macrophomina phaseolina* isolated from sorghum and common bean. *Applied Microbiology and Biotechnology* 80, 167–177.

Lazzeri, L., Baruzzi, G., Malaguti, L. and Antoniacci, L. (2003) Replacing methyl bromide in annual strawberry production with glucosinolate-containing green manure crops. *Pest Management Science* 59, 983–990.

Legard, D. (2011) Developing non-fumigant based strawberry production system for California. In: *Proceedings of the Annual International Research Conference on Methyl Bromide Alternatives and Emissions Reductions.* Methyl Bromide Alternatives Outreach, San Diego, California, pp. 42/1–42/2.

Lele, V.C. and Phatak, H.C. (1965) Leaf blight and dry stalk rot of strawberry caused by *Rhizoctonia bataticola. Indian Phytopathology* 18, 38–42.

Leslie, J.F. and Summerell, B.A. (2006) *The Fusarium Laboratory Manual.* Blackwell Publishing, Ames, Iowa.

Lievens, B., Claes, L., Vakalounakis, D.J., Vanachter, A.C.R.C. and Thomma, B.P.H.J. (2007) A robust identification and detection assay to discriminate the cucumber pathogens *Fusarium oxysporum* f. sp. *cucumerinum* and f. sp. *radicis-cucumerinum. Environmental Microbiology* 9, 2145–2161.

Lin, Y.-H., Chang, J.-Y., Liu, E.-T., Chao, C.-P., Huang, J.-W. and Chang, P.-F. (2009) Development of a molecular marker for specific detection of *Fusarium oxysporum* f. sp. *cubense* race 4. *European Journal of Plant Pathology* 123, 353–365.

Locascio, S.J., Dickson, D.W., Mitchell, D.J., Olson, S.M., Gilreath, J.P., Noling, J.W., Chase, C.A., Sinclair, T.R. and Rosskopf, E.N. (1999a) Alternative treatments to methyl bromide for strawberry. *Proceedings of the Florida State Horticultural Society* 112, 297–302.

Locascio, S.J., Olson, S.M., Chase, C.A., Sinclair, T.R., Dickson, D.W., Mitchell, D.J. and Chellemi, D.O. (1999b) Strawberry production with alternatives to methyl bromide fumigation. In: *Proceedings of the 28th National Agricultural Plastics Congress.* American Society for Plasticulture, Tallahasse, Florida, pp. 148–154.

Lodha, S., Sharma, S.K. and Aggarwal, R.K. (1997) Solarization and natural heating of irrigated soil amended with cruciferous residues for improved control of *Macrophomina phaseolina. Plant Pathology* 46, 186–190. 321–329.

López-Aranda, J.M. (2014) Methyl bromide alternatives (chemical and non-chemical) for strawberry cultivation in conventional areas of production. *Acta Horticulturae* 1049, 77–88.

López-Aranda, J.M., Medina, J.J., Miranda, L. and Domínguez, F. (2000) Three years of short term alternatives to MB on Huelva strawberries. In: *Proceedings of the Annual International Research Conference on Methyl Bromide Alternatives and Emissions Reductions.* Methyl Bromide Alternatives Outreach, Orlando, Florida, pp. 10/1–10/6.

Luongo, L., Vitale, S., Haegi, A. and Belisario, A. (2012) Development of SCAR markers and PCR assay for *Fusarium oxysporum* f. sp. *melonis* race 2-specific detection. *Journal of Plant Pathology* 94, 193–199.

Maas, J.L. (1998) *Compendium of Strawberry Diseases,* 2nd edn. American Phytopathological Society, St Paul, Minnesota.

Maas, J.L. (2004) Strawberry disease management. In: Naqvi, S.A.M.H. (ed.) *Diseases of Fruits and Vegetables*: Vol. II. Springer, The Netherlands, pp. 441–483.

MacKenzie S.J., Legard, D.E., Timmer, L.W., Chandler, C.K. and Peres, N.A. (2006) Resistance of strawberry cultivars to crown rot caused by *Olletotrichum gloeosporioides* isolates from Florida is nonspecific. *Plant Disease* 90, 1091–1097.

Madkour, M.A. and Aly, M.H. (1981) Cell wall degrading enzymes produced during pathogenesis of *Macrophomina phaseolina* on strawberry plants. *Journal of Phytopathology* 100, 36–43.

Martin, F.N. and Bull, C.T. (2002) Biological approaches for control of root pathogens of strawberry. *Phytopathology* 92, 1356–1362.

Matsubara, Y., Li, Y., Okada, T. and Maya, M.A. (2012) Factors on induced systemic disease resistance in mycorrhizal strawberry plants. *Acta Horticulturae* 926, 497–502.

Mattner, S.W., Porter, I.J., Gounder, R.K., Shanks, A.L., Wren, D.J. and Allen, D. (2008) Factors that impact on the ability of biofumigants to suppress fungal pathogens and weeds of strawberry. *Crop Protection* 27, 1165–1173.

Mayék-Pérez, N., López-Castañeda, C., González-Chavira, M., Garcia-Espinosa, R., Acosta-Gallegos, J.A., Martínez de la Vega, O. and Simpson, J. (2001) Variability of Mexican isolates of *Macrophomina phaseolina*, based on the pathogenesis and AFLP genotype. *Physiological and Molecular Plant Pathology* 59, 257–263.

Mazzola, M. (2011a) Potential of biofumigation for soilborne pest control in strawberry. In: *Proceedings of the Annual International Research Conference on Methyl Bromide Alternatives and Emissions Reductions*. Methyl Bromide Alternatives Outreach, San Diego, California, pp. 47/1–47/2.

Mazzola, M. (2011b) Active manipulation of resident biology to suppress *Macrophomina phaseolina* in strawberry. *Phytopathology* 101, S117.

Mazzola, M. and Mullinix, K. (2005) Comparative field performance of measures containing *Brassica napus* seed meal or green manure for the management of apple replant disease. *Plant Disease* 89, 1207–1213.

McCain, A.H., Bega, R.V. and Jenkinson, J.L. (1982) Solar heating fails to control Macrophomina phaseolina. *Phytopathology* 72, 985.

McDonald, B.A. (1997) The population genetics of fungi: tools and techniques. *Phytopathology* 87, 448–453.

Medina-Mínguez, J.J. (2002) Soil solarization and biofumigation in strawberries in Spain. In: *Proceedings of International Conference on Alternatives to Methyl Bromide*. Methyl Bromide Alternatives Outreach, Orlando, Florida, pp. 108–110.

Medina-Mínguez, J.J., Miranda, L., Romero, F., De Los Santos, B., Montes, F., Vega, J.M., Páez, J.I., Bascón, J., Soria, C. and López-Aranda, J.M. (2004) The use of biofumigation with new types of solarization film for strawberry production in Spain. In: Batchelor, T. and Alfarroba, A. (eds) *Proceedings of the Fifth International Conference on Alternatives to Methyl Bromide*. Lisboa, Portugal, pp. 61–65.

Medina-Mínguez, J.J., Miranda, L., Domínguez, P., Soria, C., Pérez-Jiménez, R.M., Zea, T., Talavera, M., Velasco, L., Romero, F., De Los Santos, B. and López-Aranda, J.M. (2012) Chemical and non-chemical alternatives to methyl bromide on strawberry in Huelva (Spain): 2008–2010 results. *Acta Horticulturae* 926, 637–644.

Mena, A.J., Palacios de Garcia, M.E. and Gonzalez, M.A. (1975) Enfermedades radiculares de la frutilla causadas por *Fusarium oxysporum* Schl. f. sp. *fragariae* Winks y Will y *Rhizoctonia fragariae* Hussain y McKeen. *Revista Agronómica del Noroeste Argentino* 12, 299–307.

Mertely, J., Seijo, T. and Peres, N. (2005) First report of *Macrophomina phaseolina* causing a crown rot of strawberry in Florida. *Plant Disease* 89, 434.

Mian, I., Godoy, G., Shelby, R.A., Rodriguez-Kabana, R. and Morgan-Jones, G. (1982) Chitin amendments for control of *Meloidogyne arenaria* in infested soil. *Nematropica* 12, 71–84.

Mihail, J.D. (1989) *Macrophomina phaseolina*: spatio-temporal dynamics of inoculum and of disease in a highly susceptible crop. *Phytopathology* 79, 848–855.

Mihail, J.D. and Alcorn, S.M. (1987) *Macrophomina phaseolina*: spatial patterns in a cultivated soil and sampling strategies. *Phytopathology* 77, 1126–1131.

Mihail, J.D. and Taylor, S.J. (1995) Interpreting variability among isolates of *Macrophomina phaseolina* in pathogenicity, pycnidium production, and chlorate utilization. *Canadian Journal of Botany* 73, 1596–1603.

Miller, P.W. (1948) Studies on the cause of strawberry root rot in Oregon: second report of progress. *Plant Disease Reporter* 32, 35–38.

Miranda, L., Domínguez, P., Soria, C., Talavera, M., Chamorro, M., Romero, F., De los Santos, B., Medina-Mínguez, J.J. and López-Aranda, J.M. (2012) Nuevos resultados en la evaluación de desinfestantes de suelos para el cultivo de la fresa. *Agrícola Vergel: Fruticultura, Horticultura, Floricultura* 353, 7–13.

Mirmajlessi, S.M., Destefanis, M., Gottsberger, R.A., Mänd, M. and Loit, E. (2015) PCR-based specific techniques used for detecting the most important pathogens on strawberry: a systematic review. *Systematic Reviews* 4, 9.

Mishra, P.K., Fox, R.T.V. and Culham, A. (2003) Development of a PCR-based assay for rapid and reliable identification of pathogenic *Fusaria*. *FEMS Microbiology Letters* 218, 329–332.

Moon, B.J., Park, H. and Chung, H. (1995) Studies on antagonism of *Trichoderma* species to *Fusarium oxysporum* f. sp. *fragariae*. V. Biological control of *Fusarium* wilt of strawberry by a mycoparasite, *Trichoderma harzianum*. *Korean Journal of Plant Pathology* 11, 298–303.

Mulè, G., Susca, A., Stea, G. and Moretti, A. (2004) Specific detection of the toxigenic species *Fusarium proliferatum* and *F. oxysporum* from asparagus plants using primers based on calmodulin gene sequences. *FEMS Microbiology Letters* 230, 235–240.

Muñoz-Cabañas, R.M., Hernández-Delgado, S. and Mayék-Pérez, N. (2005) Análisis patogénico de *Macrophomina phaseolina* (Tassi) Goid. en diferentes hospedadores. *Revista Mexicana de Fitopatología* 23, 11–18.

Nag Raj, T.R. (1993) *Coelomycetous Anamorphs with Appendage-bearing Conidia*. Mycologue Publications, Waterloo, Ontario.

Nagarajan, G., Nam, M.-H., Song, J.-Y., Yoo, S.J. and Kim, H.G. (2004) Genetic variation in *Fusarium oxysporum* f. sp. *fragariae* populations based on RAPD and rDNA RFLP analyses. *Plant Pathology Journal* 20, 264–270.

Nagarajan, G., Kang, S.-W., Nam, M.-H., Song, J.-Y., Yoo, S.-J. and Kim, H.-G. (2006) Characterization of *Fusarium oxysporum* f. sp. *fragariae* based on vegetative compatibility group, random amplified polymorphic DNA and pathogenicity. *Plant Pathology Journal* 22, 222–229.

Nam, M.H., Park, M.S., Kim, H.G. and Yoo, S.J. (2009) Biological control of strawberry *Fusarium* wilt caused by *Fusarium oxysporum* f. sp. *fragariae* using *Bacillus velezensis* BS87 and RK1 formulation. *Journal of Microbiology and Biotechnology* 19, 520–524.

Nam, M.-H., Kang, Y.-J., Lee, I.-H., Kim, H.-G. and Chun, C.-H. (2011a) Infection of daughter plants by *Fusarium oxysporum* f. sp. *fragariae* through runner propagation of strawberry. *Korean Journal of Horticultural Science and Technology* 29, 273–277.

Nam, M.-H., Kim, H.-S. and Kim, H.-G. (2011b) Control of *Fusarium* wilt of the strawberry caused by *Fusarium oxysporum* f. sp. *fragariae* of solarization with compost and calcium cyanamide application. *Research in Plant Disease* 17, 32–37.

Ndiaye, M. (2007) Ecology and management of charcoal rot (*Macrophomina phaseolina*) on cowpea in the Sahel. PhD thesis, Wageningen University, Wageningen, The Netherlands.

Nelson, P.E., Toussoun, T.A. and Marasas, W.F.O. (1983) *Fusarium Species: an Illustrated Manual for Identification*. Pennsylvania State University Press, Filadelfia, Pennsylvania.

Nelson, S.D., Riegel, C., Allen, L.H., Dickson, D.W., Gan, J., Locascio, S.J. and Mitchell, D.J. (2001) Volatilization of 1,3-dichloropropene in Florida plasticulture and effects on fall squash production. *Journal of the American Society for Horticultural Science* 126, 496–502.

Noling, J.W. (2002) The practical realities of alternatives to methyl bromide: concluding remarks. *Phytopathology* 92, 1373.

O'Donnell, K., Kistler, H.C., Cigelnik, E. and Ploetz, R.C. (1998) Multiple evolutionary origins of the fungus causing Panama disease of banana: concordant evidence from nuclear and mitochondrial gene genealogies. *Proceedings of the National Academy of Sciences USA* 95, 2044–2049.

O'Donnell, K., Gueidan, C., Sink, S., Johnston, P.R., Crous, P.W., Glenn, A., Riley, R., Zitomer, N.C., Colyer, P., Waalwijk, C., Lee, T.V., Moretti, A., Kang, S., Kim, H.S., Geiser, D.M., Juba, J.H., Baayen, R.P., Cromey, M.G., Bithell, S., Sutton, D.A., Skovgaard, K., Ploetz, R., Corby Kistler, H., Elliott, M., Davis, M. and Sarver, B.A. (2009) A two-locus DNA sequence database for typing plant and human pathogens within the *Fusarium oxysporum* species complex. *Fungal Genetics and Biology* 46, 936-948.

Odvody, G. and Dunkle, L. (1979) Charcoal stalk rot of sorghum: effect of environment on host–parasite relations. *Phytopathology* 69, 250–254.

Okayama, K. (1991) Selection and effect of antagonistics on *Fusarium* wilt of strawberries. *Bulletin of the Nara-ken Agricultural Experiment Station* 22, 17–22.

Olaya, G. and Abawi, G.S. (1996) Effect of water potential on mycelial growth and on production and germination of sclerotia of *Macrophomina phaseolina*. *Plant Disease* 80, 1347–1350.

Othman, M., Hunzie, J., Koike, S., Gordon, T. and Ajwa, H. (2011) Paladin (DMDS), Midas, and Pic-Clor 60 for strawberry production in California. In: *Proceedings of the Annual International Research Conference on Methyl Bromide Alternatives and Emissions Reductions*. Methyl Bromide Alternatives Outreach, San Diego, California, pp. 36/1–36/3.

Ozbay, N. and Newman, S.E. (2004) Biological control with *Trichoderma* spp. with emphasis on *T. harzianum*. *Pakistan Journal of Biological Sciences* 7, 478–484.

Papavizas, G.C. (1977) Some factors affecting survival of sclerotia of *Macrophomina phaseolina* in soil. *Soil Biology and Biochemistry* 9, 337–341.

Park, B., Park, J., Cheong, K.-C., Choi, J., Jung, K., Kim, D., Lee, Y.-H., Ward, T.J., O'Donnell, K., Geiser, D.M. and Kang, S. (2010) Cyber infrastructure for *Fusarium*: three integrated platforms supporting strain identification, phylogenetics, comparative genomics and knowledge sharing. *Nucleic Acids Research* 39, D640–D646.

Particka, C.A. and Hancock, J.F. (2005) Field evaluation of strawberry genotypes for tolerance to black root rot on fumigated and nonfumigated soil. *Journal of the American Society for Horticultural Science* 130, 688–693.

Particka, C.A. and Hancock, J.F. (2008) Breeding for increased tolerance to black root rot in strawberry. *HortScience* 43, 1698–1702.

Pasquali, M., Marena, L., Fiora, E., Piatti, P., Gullino, M.L. and Garibaldi, A. (2004) Real-time polymerase chain reaction for identification of a highly pathogenic group of *Fusarium oxysporum* f. sp. *chrysanthemi* on *Argyranthemum frutescens* L. *Journal of Plant Pathology* 86, 53–59.

Pasquali, M., Piatti, P., Gullino, M.L. and Garibaldi, A. (2006) Development of a real-time polymerase chain reaction for the detection of *Fusarium oxysporum* f. sp. *basilici* from basil seed and roots. *Journal of Phytopathology* 154, 632–636.

Pasquali, M., Dematheis, F., Gullino, M.L. and Garibaldi, A. (2007) Identification of race 1 of *Fusarium oxysporum* f. sp. *lactucae* on lettuce by inter-retrotransposon sequence-characterized amplified region technique. *Phytopathology* 97, 987–996.

Pastrana, A.M., Capote, N., De los Santos, B., Romero, F. and Basallote-Ureba, M.J. (2013) First report of *Fusarium solani* causing crown and root rot on strawberry crops in southwestern Spain. *Plant Disease* 98, 161.

Perveen, K. and Bokhari, N.A. (2012) Antagonistic activity of *Trichoderma harzianum* and *Trichoderma viride* isolated from soil of date palm field against *Fusarium oxysporum*. *African Journal of Microbiology Research* 6, 3348–3353.

Phillips, D. and Golzar, H. (2008) Strawberry root and crown rot survey. 2005 and 2006 seasons. *Government of Western Australia Bulletin* 4747, 1–12.

Pineda, J.B. and Gonnella, E.R. (1988) Evaluación del control biológico de *Macrophomina phaseolina* en ajonjoli (*Sesamum indicum* L.). *Agronomía Tropical* 38, 43–48.

Pinkerton, J.N., Ivors, K.L., Reeser, P.W., Bristow, P.R. and Windom, G.E. (2002) The use of soil solarization for the management of soilborne plant pathogens in strawberry and red raspberry production. *Plant Disease* 86, 645–651.

Pinzon, M. (2011) Overview of MeBr Projects funded by the multilateral fund of the Montreal Protocol. In: *Proceedings of the Annual International Research Conference on Methyl Bromide Alternatives and Emissions Reduction*. Methyl Bromide Alternatives Outreach, San Diego, California. Available at: http://www.mbao.org/ 2011/ Pinzon.pdf (accessed 26 November 2012).

Pratt, R.G. (2006) A direct observation technique for evaluating sclerotium germination by *Macrophomina phaseolina* and effects of biocontrol materials on survival of sclerotia in soil. *Mycopathologia* 162, 121–131.

Qin, R., Gao, S. and Ajwa, H. (2011) Fumigant degradation as affected by different application rate in five soils. *Proceedings of the Annual International Research Conference on Methyl Bromide Alternatives and Emissions Reductions*. Methyl Bromide Alternatives Outreach, San Diego, California, pp. 79/1–79/3.

Ramezani, H. (2008) Biological control of root-rot of eggplant caused by *Macrophomina phaseolina*. *American–Eurasian Journal of Agricultural and Environmental Science* 4, 218–220.

Rayatpanah, S., Nanagulyan, S.G., Alav, S.V., Razavi, M. and Ghanbari-Malidarreh, A. (2012) Pathogenic and genetic diversity among Iranian isolates of *Macrophomina phaseolina*. *Chilean Journal of Agricultural Research* 72, 40–44.

Redondo, C., Villarino, M., Larenas, J., Melgarejo, P. and De Cal, A. (2012) Enfermedades emergentes en los viveros de altura. In: *Libro de Resúmenes XVI Congreso Nacional de la Sociedad Española de Fitopatología*. SEF, Málaga, Spain, p. 321.

Ros, M., Garcia, C., Hernandez, M., Lacasa, A., Fernandez, P. and Pascual, J. (2008) Effects of biosolarization as methyl bromide alternative for *Meloidogyne incognita* control on quality of soil under pepper. *Biology and Fertility of Soils* 45, 37–44.

Rosskopf, E.N., Mitchell, D.J., Kannwhiser-Mitchell, M.E., Gilreath, J.P., Locascio, S.J. and Olson, S.M. (2000) Effect of transplant type and soil disinfestation treatment on fungal root colonization of strawberry. *Phytopathology* 90, S67.

Rosskopf, E.N., Chellemi, D.O., Kokalis-Burelle, N. and Church, G.T. (2005) *Alternatives to Methyl Bromide: a Florida Perspective*. Phytopathological Society, Saint Paul, Minnesota.

Rosskopf, E.N., Kokalis-Burelle, N., Butler, D.M., Muramoto, J., Shenan, C., Noling, J., He, Z., Booker, B. and Campbell, T. (2011) Field evaluation of non-fumigant pest control for Florida strawberry production. In: *Proceedings of the Annual International Research Conference on Methyl Bromide Alternatives and Emissions Reductions*. Methyl Bromide Alternatives Outreach, San Diego, California, pp. 45/1–45/3.

Rossman, A.Y. and Palm-Hernández, M.E. (2008) Systematics of plant pathogenic fungi: why it matters. *Plant Disease* 92, 1376–1386.

Sanchez, S., Grez, J. and Gambardella, M. (2012) Importance of the disease caused by *Macrophomina phaseolina* in Chilean strawberry. In: *VII International Strawberry Symposium*. ISHS, Beijing, China, p. 379.
Sandoval, I. and López, M.O. (2000) Antagonismo de *Trichoderma harzianum* A34 hacia *Macrophomina phaseolina* y otros patógenos fúngicos del frijol. *Fitosanidad* 4, 69–72.
Santos, B.M. (2007) *Soil Fumigation after Methyl Bromide: Research on 1, 3-Dichloropropene plus Chloropicrin in Florida*. HS1119, one of a series of the Horticultural Sciences Department, UF/IFAS Extension, University of Florida, Florida, pp. 1–6.
Santos, B.M., Gilreath, J.P., López-Aranda, J.M., Miranda, L., Soria, C. and Medina, J.J. (2007) Comparing methyl bromide alternatives for strawberry in Florida and Spain. *Journal of Agronomy* 6, 225–227.
Sharifi, K. and Mahdavi, M. (2012) First report of strawberry crown and root rot caused by *Macrophomina phaseolina* in Iran. *Iranian Journal of Plant Pathology* 47, 161.
Shennan, C., Muramoto, J., Bolda, M., Koike, S.T., Daugovish, O., Rosskopf, E., Burelle, N.K. and Klonsky, K. (2007) Optimizing anaerobic soil disinfestation: an alternative to soil fumigations? In: *Proceedings of the Annual International Research Conference on Methyl Bromide Alternatives and Emissions Reductions*. Methyl Bromide Alternatives Outreach, San Diego, California, pp. 40/1–40/4.
Shokes, F.M., Lyda, S.D. and Jordan, W.R. (1977) Effect of water potential on the growth and survival of *Macrophomina phaseolina*. *Phytopathology* 67, 239–241.
Short, G.E., Wyllie, T.D. and Ammon, V.D. (1978) Quantitative enumeration of *Macrophomina phaseolina* in soybean tissues. *Phytopathology* 68, 736–741.
Short, G.E., Wyllie, T.D. and Bristow, P.R. (1980) Survival of *Macrophomina phaseolina* in soil and in residue of soybean. *Phytopathology* 70, 13–17.
Sivan, A. and Chet, I. (1989) The possible role of competition between *Trichoderma harzianum* and *Fusarium oxysporum* on rhizosphere colonization. *Phytopathology* 79, 198–203.
Smalla, K., Wieland, G., Buchner, A., Zock, A., Parzy, J., Roskot, N., Heuer, H. and Berg, G. (2001) Bacterial bulk and rhizosphere communities studied by denaturing gradient gel electrophoresis of PCR-amplified fragments of 16S rRNA genes: plant-dependent enrichment and seasonal shifts revealed. *Applied and Environmental Microbiology* 67, 4742–4751.
Snyder, W.C. and Hansen, H.N. (1940) The species concept in *Fusarium*. *American Journal of Botany* 27, 64–67.
Sreedevi, B., Charitha Devi, M. and Saigopal, D.V.R. (2011) Isolation and screening of effective *Trichoderma* spp. against the root rot pathogen *Macrophomina phaseolina*. *Journal of Agricultural Technology* 7, 623–635.
Su, G., Suh, S.O., Schneider, R.W. and Russin, J.S. (2001) Host specialization in the charcoal rot fungus, *Macrophomina phaseolina*. *Phytopathology* 91, 120–126.
Suga, H., Hirayama, Y., Suzuki, T., Kageyama, K. and Hyakumachi, M. (2013) Development of PCR primers to identify *Fusarium oxysporum* f. sp. *fragariae*. *Plant Disease* 97, 619–625.
Sugimura, T., Nishizaki, M. and Horimoto, K. (2001) Control of *Fusarium* wilt of strawberry by soil solarization using mulching and tunnel-covering combining with irrigation of solar-heated water. *Bulletin of the Nara Prefectural Agricultural Experiment Station* 32, 1–7.
Summerbell, R.C., Lévesque, C.A., Seifert, K.A., Bovers, M., Fell, J.W., Diaz, M.R., Boekhout, T., De Hoog, G.S., Stalpers, J. and Crous, P.W. (2005) Microcoding: the second step in DNA barcoding. *Philosophical Transactions of the Royal Society B: Biological Sciences* 360, 1897–1903.
Takahashi, H., Yoshida, Y., Kanda, H., Furuya, H. and Matsmoto, T. (2003) Breeding of *Fusarium* wilt-resistant strawberry cultivar suitable for field culture in Northern Japan. *Acta Horticulturae* 626, 113–118.
Tezuka, N. and Makino, T. (1991) Biological control of *Fusarium* wilt of strawberry by nonpathogenic *Fusarium oxysporum* isolated from strawberry. *Annals of the Phytopathological Society of Japan* 57, 506–511.
Tjamos, E., Antoniou, P., Skourtaniotis, A., Kikrilis, E. and Tjamos, S. (2006) Impermeable plastics and methyl bromide alternatives in controlling soilborne fungal pathogens of strawberries in Greece. In: *Proceedings of the 12th Congress of the Mediterranean Phytopathological Union*. Mediterranean Phytopathological Union, Rhodes, Greece, pp. 255–257.
Toyoda, H., Horikoshi, K., Inaba, K. and Ouchi, S. (1990) Plant regeneration of callus tissues induced from leaf explants of strawberry. *Plant Tissue Culture Letters* 7, 38–41.
Toyoda, H., Horikoshi, K., Yamano, Y. and Ouchi, S. (1991) Selection for *Fusarium* wilt disease resistance from regenerants derived from leaf callus of strawberry. *Plant Cell Reports* 10, 167–170.
Tweedy, B. and Powell, D. (1958) Charcoal rot on strawberry in Illinois. *Plant Disease Report* 42, 107.
Vandemark, G., Martinez, O., Pecina, V. and Alvarado, M.J. (2000) Assessment of genetic relationships among isolates of *Macrophomina phaseolina*, using a simplified AFLP technique and two different methods of analyses. *Mycology* 92, 656–664.

von Arx, J.A. (1981) *The Genera of Fungi Sporulating in Pure Culture*, 3rd edn. J. Cramer, Vaduz, Liechtenstein.
von Arx, J.A. (1987) *Plant Pathogenic Fungi. Beiheft zur Nova Hedwigia*, Vol 87. J. Cramer, Berlin, Germany.
Walker, J. (1980) The Coelomycetes: fungi imperfecto with pycnidia, acervuli and stromata. *Australasian Plant Pathology* 9, 120–121.
Wheeler, H. (1975) *Plant Pathogenesis*, Springer-Verlag, Berlin, Germany.
Williamson, M., Fernández-Ortuño, D. and Schnabel, G. (2012) First report of *Fusarium* wilt of strawberry caused by *Fusarium oxysporum* in South Carolina. *Plant Disease* 96, 911.
Winks, B.L. and Williams, Y.N. (1965) A wilt of strawberry caused by a new form of *Fusarium oxysporum*. *Queensland Journal of Agricultural and Animal Sciences* 22, 475–479.
Wyllie, T.D. (1988) Charcoal rot of soybean-current status. In: Wyllie, T.D. and Scott, D.H. (eds) *Soybean Diseases of the North Central Region*. APS Press, Saint Paul, Minnesota, pp. 106–113.
Yates, S.R., Gan, J., Papiernik, S.K., Dungan, R. and Wang, D. (2002) Reducing fumigant emissions after soil application. *Phytopathology* 92, 1344–1348.
Yildiz, A., Benlioğlu, S., Boz, Ö. and Benlioğlu, K. (2010) Use of different plastics for soil solarization in strawberry growth and time–temperature relationships for the control of *Macrophomina phaseolina* and weeds. *Phytoparasitica* 38, 463–473.
Yücel, S., Elekçioglu, I.H., Uludag, A., Can, C., Sögüt, M.A., Özarslandan, A. and Aksoy, E. (2002) The second year results of methyl bromide alternatives in the eastern Mediterranean. In: *Proceedings of the Annual International Research Conference on Methyl Bromide Alternatives and Emissions Reductions*. Methyl Bromide Alternatives Outreach, Orlando, Florida, p. 10.
Zambounis, A.G., Paplomatas, E. and Tsaftaris, A.S. (2007) Intergenic spacer-RFLP analysis and direct quantification of Australian *Fusarium oxysporum* f. sp. *vasinfectum* isolates from soil and infected cotton tissues. *Plant Disease* 91, 1564–1573.
Zhang, Y., Fan, T., Jia, W., Zhang, W., Liu, Q., Li, B. and Zhang, L. (2012) Identification and characterization of a *Bacillus subtilis* strain TS06 as bio-control agent of strawberry replant disease (*Fusarium* and *Verticilium* wilts). *African Journal of Biotechnology* 11, 570–580.
Zhang, Z., Zhang, J., Wang, Y. and Zheng, X. (2005) Molecular detection of *Fusarium oxysporum* f. sp. *niveum* and *Mycosphaerella melonis* in infected plant tissues and soil. *FEMS Microbiology Letters* 249, 39–47.
Zveibil, A. and Freeman, S. (2005) First report of crown and root rot in strawberry caused by *Macrophomina phaseolina* in Israel. *Plant Disease* 89, 1014.
Zveibil, A., Mor, N., Gnayem, N. and Freeman, S. (2012) Survival, host–pathogen interaction, and management of *Macrophomina phaseolina* on strawberry in Israel. *Plant Disease* 96, 265–272.

13 Induced Systemic Resistance to *Fusarium* Wilt and Antioxidative Ability in Mycorrhizal Strawberry Plants

Yoh-ichi Matsubara[1]* and Amjad M. Husaini[2]
[1]*Faculty of Applied Biological Sciences, Gifu University, Gifu, Japan;* [2]*Centre for Plant Biotechnology, Sher-e-Kashmir University of Agricultural Sciences and Technology of Kashmir, Shalimar, Srinagar, Jammu and Kashmir, India*

13.1. Biological Control of *Fusarium* Wilt

Fusarium wilt of strawberry (*Fragaria* × *ananassa* Duch.), caused by *Fusarium oxysporum* f. sp. *fragariae*, is one of the common and difficult-to-control diseases in strawberry worldwide (Koike and Gordon, 2015). In Japan, *Fusarium* wilt and anthracnose are serious diseases in the major strawberry-producing regions and cause heavy losses during the nursery and fruit production period (Tezuka and Makino, 1991; Okayama, 1993; Mori and Kitamura, 2003; Koike and Gordon, 2015). Capillary watering as a cultural method for control of these diseases was introduced into strawberry cultivation, but they are still difficult to control (Okayama, 1993; Akita, 2001). For the management of *Fusarium* wilt, chemical control, biological control, crop rotation and use of resistant cultivars are the most popular methods (Koike and Gordon, 2015). However, it has been difficult to develop resistant cultivars for strawberry, because many traits, such as fruit productivity and quality of fruits, have to be considered in the selection of successful resistant strawberry cultivars (Schaart *et al.*, 2011). Moreover, octoploidy ($2n = 8\times = 56$) and heterozygosity impose

* ymatsu@gifu-u.ac.jp

difficulties when this species is bred according to conventional strategies, and so manipulation on a gene level through transgenic technology is generally also carried out (Husaini, 2010).

For biological control of *Fuarium* wilt, an attempt was made to use non-pathogenic isolates of *Fusarium oxysporum* in strawberry (Tezuka and Makino, 1991). However, the non-pathogenic isolates had no growth-promoting effect and the method was not successful as an independent control measure. *Trichoderma* and *Bacillus* spp. were also used to try to control *Fusarium* wilt in strawberry (Moon *et al.*, 1995; Nam *et al.*, 2009; Zhang *et al.*, 2012), but with very limited success. Arbuscular mycorrhizal fungi (AMF) are considered one of the successful biocontrol agents (Baum *et al.*, 2015). AMF are ubiquitous soil inhabitants and form a symbiotic relationship with the roots of most terrestrial plants. AMF promote host plant growth by enhancing phosphorus uptake through symbiosis (Marschner and Dell, 1994) and hence are an alternative to using high inputs of fertilizers and pesticides in sustainable crop production systems (Baum *et al.*, 2015). For strawberry, growth enhancement through AMF inoculation has been reported in several combinations of fungal species and strawberry cultivars (Robertson *et al.*, 1988; Chávez and Ferrera-Cerrato, 1990; Niemi and Vestberg, 1992; Williams *et al.*, 1992; Varma and Schüepp, 1994; Husaini *et al.*, 2012). *Piriformospora indica* and *Sebacina vermifera*, living as endophytes in a variety of hosts, have gained increased attention recently. *Piriformospora indica* is a newly described, axenically cultivable, phytopromotional endosymbiont, which mimics the capabilities of AMF and shows pronounced growth-promotional effects in strawberry, as well as in many other host species. *In vitro* and *ex vitro* co-culture studies have verified that both *Piriformospora indica* and *S. vermifera* possess pronounced plant growth-stimulating and stress amelioration capacity, most likely by inducing secondary metabolite accumulation in strawberry plants (Husaini *et al.*, 2012). *Piriformospora indica* promotes the synthesis and expression of defence-related proteins and enzymes (isoflavonoid phytoalexins, isoflavonoid glyceollin coumestrol and coumestin isosojagol), which provide protection and result in higher plant survival (Sahay and Varma, 1999).

In addition, the inoculation of strawberry plants with AMF results in reduced disease when challenged with *Phytophthora fragariae*, this effect being dependent on differences in host cultivar and AMF species (Baath and Hayman, 1984; Mark and Cassells, 1996; Norman *et al.*, 1996). Matsubara *et al.* (2004) were the first to report tolerance to *Fusarium* wilt in strawberry (*Fragaria* × *ananassa* Duch. 'Nohime') plants inoculated with five species of AMF (*Gigaspora margarita*, *Glomus fasciculatum*, *Glomus mosseae*, *Glomus* sp. R10 and *Glomus aggregatum*) in greenhouse experiments, with the effect mostly appearing in *Glomus mosseae*-inoculated plants (Matsubara *et al.*, 2004). Later, Li *et al.* (2010) reported tolerance to *Fusarium* wilt and anthracnose in mycorrhizal strawberry plants, but the mechanism of disease tolerance in mycorrhizal plants remains unclear.

13.2. Induced Systemic Resistance and Antioxidative Ability

Induced systemic resistance (ISR) in plants is explained as plant defences preconditioned by prior infection or treatment of fungi, for example, leading to the development of resistance in the whole plant against any subsequent challenge by the pathogen or parasite (Koike *et al.*, 2001; Choudhary *et al.*, 2007). He *et al.* (2002) reported that ISR to *Fusarium* root rot appeared in asparagus plants with a split root system inoculated with non-pathogenic isolates of *F. oxysporum*, and lignin-related enzymes and lignification was activated in these plants. In strawberry, Li *et al.* (2010) described ISR to anthracnose and *Fusarium* wilt in mycorrhizal strawberry plants with a split root system. However, the mechanisms of ISR in mycorrhizal plants are still unclear.

Generally, when plants suffer environmental stresses, reactive oxygen species (ROS) are generated in cells, and the ensuing oxidative stress to intercellular structures causes major damage in the plant (Wahid *et al.*, 2007). In order to cope with these toxic ROS, plants produce antioxidants as detoxification factors (Kuźniak and Skłodowska, 2004; Wu *et al.*, 2006). Antioxidative defence mechanisms consist of antioxidative enzymes such as superoxide dismutase (SOD), ascorbate peroxidase (APX) and catalase (CAT), and non-enzymatic antioxidants such as polyphenols and ascorbic acid (Asada, 1999; Husaini, 2010). In pathogen stress conditions, the production of higher concentrations of ROS such as H_2O_2, $O_2^{\bullet -}$ and ·OH have been shown to create cytotoxic conditions (Sahoo *et al.*, 2007). To overcome this negative consequence of ROS, plants have evolved various protective mechanisms either to reduce or completely eliminate such ROS through antioxidative abilities of antioxidative enzymes and substances (Ehsani-Moghaddam *et al.*, 2006; Sahoo *et al.*, 2007). For mycorrhizal plants, there is a correlation between disease tolerance and an increase in SOD activity in tomato (Pozo *et al.*, 2002) and pepper (Garmendia *et al.*, 2006). Additionally, Li *et al.* (2008) demonstrated that there seems to be a correlation between increased antioxidative enzyme production in mycorrhizal strawberry plants and their tolerance to high-temperature stress. Later, Li *et al.* (2010) reported tolerance to *Fusarium* wilt and anthracnose, with an increase in the antioxidative ability of mycorrizal strawberry plants. In addition, Matsubara (2011) showed that free amino acids and antioxidative ability were associated with tolerance to *Fusarium* wilt in mycorrhizal (*Glomus mosseae*) strawberry plants. However, in the reports by Li *et al.* (2010) and Matsubara (2011), analysis of antioxidative ability was carried out only twice (before and after pathogen inoculation), and it remains unclear how AMF colonization per se affected the antioxidative ability, and how long this increase in antioxidative ability would continue under the pathogen-inoculated conditions. In this chapter, we discuss this issue in light of a more comprehensive study about the influence of AMF colonization on tolerance to *Fusarium* wilt including ISR and antioxidative ability in strawberry plants, in order to develop a deeper understanding of the mechanisms of their disease tolerance.

13.3. Estimation of Induced Systemic Resistance to *Fusarium* Wilt in Mycorrhizal Strawberry Plants

Strawberry (*Fragaria* × *ananassa* Duch. 'Nohime', a commercial cultivar in Japan) runner plants were treated with a split-root system method. The plant root system was separated into halves and each was placed in a different pot (10.5 cm in diameter): one with disinfested bedding soil (autoclaved at 1.2 kg cm^{-2} at 121°C for 1 h) and one with the same soil inoculated with *Glomus mosseae* (supplied by Idemitsukosan Co.), according to Matsubara *et al.* (2004). Two weeks after AMF inoculation, the plants were administered mixed fertilizer (N:P:K = 13:11:13, 1 g per pot). Twenty plants with split root systems were raised per treatment with two replications, and were irrigated by the capillary watering method (Matsubara *et al.*, 2004) and grown in a greenhouse at 28 ± 3°C with natural light and day length.

F. oxysporum f. sp. *fragariae* (strain 2S; supplied by the National Agricultural Research Center for Kyusyu Okinawa Region in Japan) was grown on potato dextrose agar medium. The conidia were harvested in potato sucrose liquid medium (Nissui Pharmaceutical Co.) and incubated at 25°C in the dark for 7 days. The conidial suspension of *F. oxysporum* f. sp. *fragariae* was sieved (45 µm), and the concentration was adjusted to 10^6 conidia ml^{-1}. Six weeks after AMF inoculation, all plants were inoculated with 10 ml of the *F. oxysporum* f. sp. *fragariae* conidial suspension onto all the split roots. The plants were grown in a growth chamber with capillary watering at 28 ± 3°C, with relative humidity of 60 ± 5% under natural light and day-length conditions. Symptoms of *Fusarium* wilt were investigated 14 and 42 days after inoculation. The severity of *Fusarium* wilt was categorized into five classes on the basis of the percentage of diseased petioles in a plant: <20%, 20–40%, 40–60%, 60–80%, >80–100%. Most of the values were expressed as the mean of three measurements for each treatment. Mean values were separated using a *t*-test at $P < 0.05$ and Tukey's multiple range test at $P < 0.05$ via analysis of variance.

Plants were sampled after AMF inoculation (41 and 52 days) and after *F. oxysporum* f. sp. *fragariae* inoculation (14 and 42 days) to investigate the continuity of the treatment on the resultant effects. Plant samples were separated into shoots and roots, and dry matter was weighed after drying at 100°C for 2 days. AMF plants had a higher dry weight of shoots and roots than non-AMF-inoculated (NAM) plants in all the investigated periods before and after *F. oxysporum* f. sp. *fragariae* inoculation (Fig. 13.1a), suggesting that the growth promotion effect appeared in all the mycorrhizal strawberry plants.

The roots of AMF plants were harvested at the same time intervals. The root samples were stained according to Phillips and Hayman (1970), and the rate of AMF colonization in 1 cm segments of lateral roots (RFCSL) was calculated. Hence, RFCSL expresses the percentage of 1 cm AMF-colonized segments out of the total number of 1 cm segments of all the lateral roots; the number of total segments was approximately 50 per plant. The mean colonization value was calculated from the values of three plants. AMF colonization occurred in all the inoculated plants, and no major difference in colonization level appeared over the duration of the experiment (Fig. 13.2).

The incidence of *Fusarium* wilt in shoots reached 100% in NAM and AMF-inoculated plants 42 days after *F. oxysporum* f. sp. *fragariae* inoculation; however, the severity of symptoms showed a significant decrease in AMF-inoculated plants (Fig. 13.3). In addition, the severity of symptoms in both AMF-inoculated roots (AMF$^+$) and non-inoculated roots (AMF$^-$) in AMF-inoculated plants was lesser than in NAM plants; thus, AMF roots showed induced systemic resistance. These findings illustrated that tolerance to *Fusarium* wilt can be as a result of ISR in mycorrhizal strawberry plants with a split root system. Davis and Menge (1980) showed that, in citrus seedlings with a split root system inoculated with *Phytophthora parasitica*,

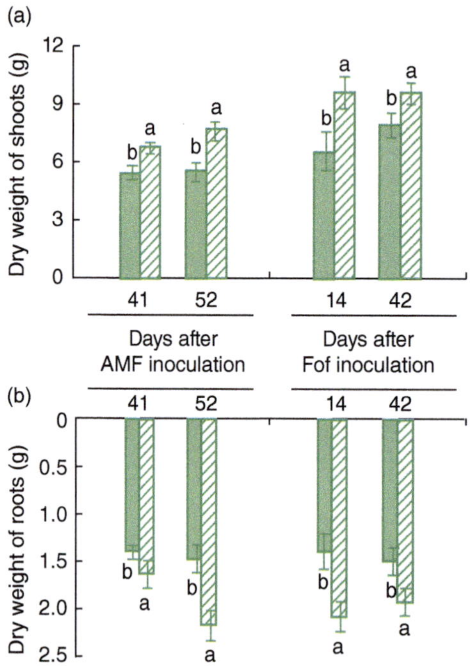

Fig. 13.1. Dry weight of strawberry plant shoots (a) and roots (b). Filled columns, non-arbuscular mycorrhizal fungi (AMF)-inoculated plants; hatched columns, AMF-inoculated plants. Bars represent mean±standard error. Different lower-case letters indicate a significant difference according to a *t*-test ($P<0.05$). Fof, *Fusarium oxysporum* f. sp. *fragariae*.

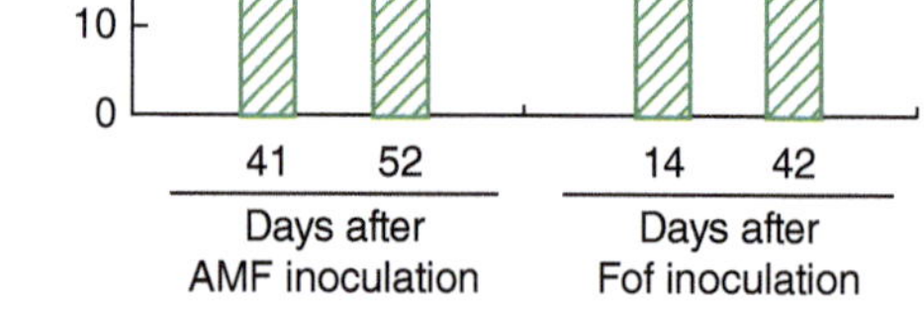

Fig. 13.2. Arbuscular mycorrhizal fungi (AMF) colonization levels, measured as the rate of AMF colonization in 1 cm segments of lateral roots (RFCSL) in mycorrhizal strawberry plants. Bars represent mean±standard error. Different lower-case letters indicate a significant difference according to a *t*-test ($P<0.05$). Fof, *Fusarium oxysporum* f. sp. *fragariae*.

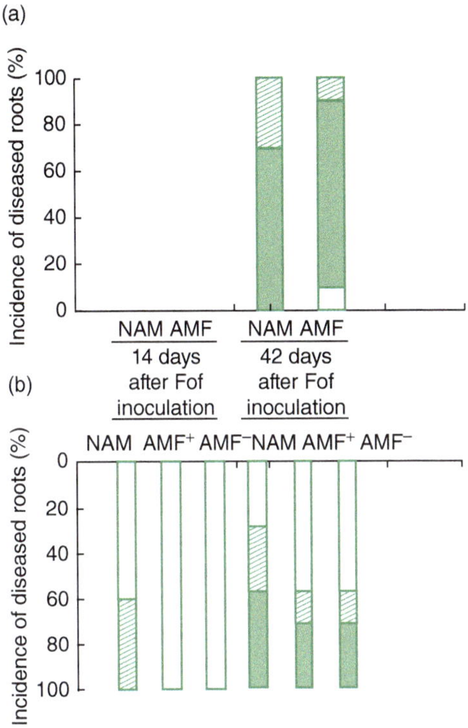

Fig. 13.3. Incidence of *Fusarium* wilt in strawberry plants with a split-root system. (a) Percentage of diseased shoots in the plant: open columns, 0–20%; filled columns, 20–40%; hatched columns, 40–60%. (b) Percentage of diseased roots in the root system: open columns, part diseased; hatched, half diseased; filled columns, all diseased. Fof, *Fusarium oxysporum* f. sp. *fragariae*; NAM, non-arbuscular mycorrhizal fungi (AMF)-inoculated plants; AMF, AMF-inoculated plants; AMF⁻, AMF-non-inoculated roots; AMF⁺, AMF-inoculated roots.

tolerance appeared in non-mycorrhizal roots, even when these roots were separated from mycorrhizal (*Glomus fasciculatum*) roots. Similarly, Pozo *et al.* (2002) reports that, in tomato plants with a split root system, tolerance to *Phytophthora parasitica* appeared in both NAM roots and AMF-inoculated roots in plants, implicating a role of ISR. Li *et al.* (2010) reported that the disease symptoms of *Fusarium* wilt were suppressed in AMF-inoculated and NAM root parts of strawberry plants. These studies confirmed the role of ISR in mycorrhizal plants with a split root system.

13.4. Effect of Arbuscular Mycorrhizal Fungi Inoculation on Antioxidative Ability

In mycorrhizal plants, disease tolerance and an increase in SOD activity showed a correlation in tomato (Pozo *et al.*, 2002) and pepper (Garmendia *et al.*, 2006), and drought tolerance and a concomitant increase in SOD take place in lettuce (Ruiz-Lozano *et al.*, 1996) and shrubs (Roldán *et al.*, 2008), while tolerance to sulphur dioxide concomitant with a SOD increase occur in *Avena nuda* (Huang *et al.*, 2008). In strawberry plants, tolerance to *Fusarium* wilt and anthracnose, and a concomitant increase in the antioxidative ability of mycorrhizal strawberry plants was illustrated by Li *et al.* (2010) and Matsubara (2011), although the antioxidative ability was measured only twice (before and after pathogen inoculation), so the effect of AMF colonization on antioxidative ability and whether the increase in antioxidative ability in the pathogen-inoculated condition would be maintained was unclear. Below, we discuss additional detailed experiments that were carried out to clarify these points. After AMF inoculation (at 41 and 52 days) and *F. oxysporum* f. sp. *fragariae* inoculation (at 14 and 42 days), the strawberry plants were sampled and partitioned into petioles, crowns and main roots with no colonization and were frozen in liquid nitrogen. Analyses of antioxidative abilities were carried out and the results are discussed in the following sections.

13.4.1. SOD activity

SOD plays a primary role among the antioxidative enzymes in defence reactions and detoxifying superoxide ($O_2^{\bullet-}$); thus, SOD activity is considered the most important key enzyme in antioxidative abilities in plants (Fridovich, 1986). Garmendia *et al.* (2006) reported that tolerance to *Verticillium dahliae* and an increase in SOD activity occurred in mycorrhizal pepper plants. Ehsani-Moghaddam *et al.* (2006) and Sahoo *et al.* (2007) showed that a cultivar resistant to *Mycosphaerella fragariae* and *Phytophthora colocasiae* showed higher levels of SOD activity than a susceptible cultivar of strawberry and taro, respectively. In the investigation discussed here, 1 g of sample was ground in 4 ml of 0.1 M phosphate/borate buffer (pH 7.8) with 1 mM EDTA, 3 mM dichlorodiphenyltrichloroethane and 4% (w/v) polyvinylpyrrolidone. The filtrate was centrifuged (EF-1300; Tomy Co.) at 13,000 r.p.m. for 20 min and the supernatant used as a crude enzyme extract. The enzyme activity was determined using the nitro blue tetrazolium reduction method (Beauchamp and Fridovich, 1971). One unit of SOD activity was defined as the amount of enzyme required to cause 50% inhibition of the rate of nitro blue tetrazolium reduction measured at 560 nm using a spectrophotometer (U-1900: Hitachi). The results indicated that the increase in SOD activity was related to disease tolerance. In this study, before *F. oxysporum* f. sp. *fragariae* inoculation, AMF plants showed no increase in SOD activity compared with NAM plants, but after pathogen inoculation, SOD activity was higher in AMF plants than in NAM plants (Fig. 13.4a). Previously, Pozo *et al.* (2002) demonstrated a relationship between the tolerance to *Phytophthora parasitica* of mycorrhizal tomato plants and SOD activity in both non-AMF-inoculated and AMF-inoculated roots in split root systems. Li *et al.* (2008) reported that tolerance to high temperature stress and the increase in activities of SOD and APX under the stress condition occurred in strawberry (cultivar 'Nohime') plants inoculated with *Glomus mosseae*, though the antioxidative enzyme activities differed little between AMF and NAM plants before the high temperature stress condition. Hence, the increase in antioxidative ability might be induced especially under stress conditions in mycorrhizal strawberry plants, resulting in tolerance to *Fusarium* wilt.

13.4.2. 1,1-Diphenyl-2-picrylhydrazyl radical-scavenging activity

1,1-Diphenyl-2-picrylhydrazyl (DPPH) is a common organic chemical compound and is a well-known radical scavenger. Therefore, a rate reduction in a chemical reaction on addition of DPPH is used as an indicator of the radical nature of the reaction. The DPPH radical-scavenging activity was measured according to the method of Burits and Bucar (2000). One gram of each sample was ground in 40 ml of 90% methanol. Then, 10 µl of the extract was added to a test tube, and 4 ml of distilled water and 1 ml of 250 µM DPPH solution was added. The tubes were mixed and allowed to stand for 30 min in the dark. Absorbance was read against a blank at 517 nm using a spectrophotometer (U-1900). DPPH radical-scavenging activity was calculated as the percentage inhibition relative to the control (Fig. 13.4b).

DPPH radical-scavenging activity showed higher levels in shoots of AMF plants except at 41 days after AMF inoculation, and also had higher levels in the main roots (AMF^+ and AMF^-) 14 days after *F. oxysporum* f. sp. *fragariae* inoculation. In mycorrhizal asparagus plants, tolerance to *Fusarium* root rot was apparent and the increase in DPPH radical-scavenging activity occurred before and after *Fusarium* inoculation in the mycorrhizal plants (Mohammad Nahiyan and Matsubara, 2012). Our results resembled those of that study.

13.4.3. Polyphenol content

Polyphenols are a structural class of mainly natural but also some synthetic organic chemicals characterized by the presence of large multiples of phenol structural units. The number and characteristics of these phenol structures underlie the unique biological properties of these compounds. Polyphenol content was determined with the Folin and Denis (1915) method. One gram of sample was ground in 10 ml of 90% methanol. Then, 400 µl of the extract was placed in a test tube, and 3 ml of distilled water and 200 µl of Folin–Denis reagent (10% $Na_2WO_4.2H_2O$, 2% $H_3(PMo_{12}O_{40}).nH_2O$, 5% phosphoric acid). After 3 min, 0.4 ml of 10% Na_2CO_3 was added and the mixture was allowed to stand for 30 min in the dark. Absorbance was measured at 700 nm using a spectrophotometer (U-1900).

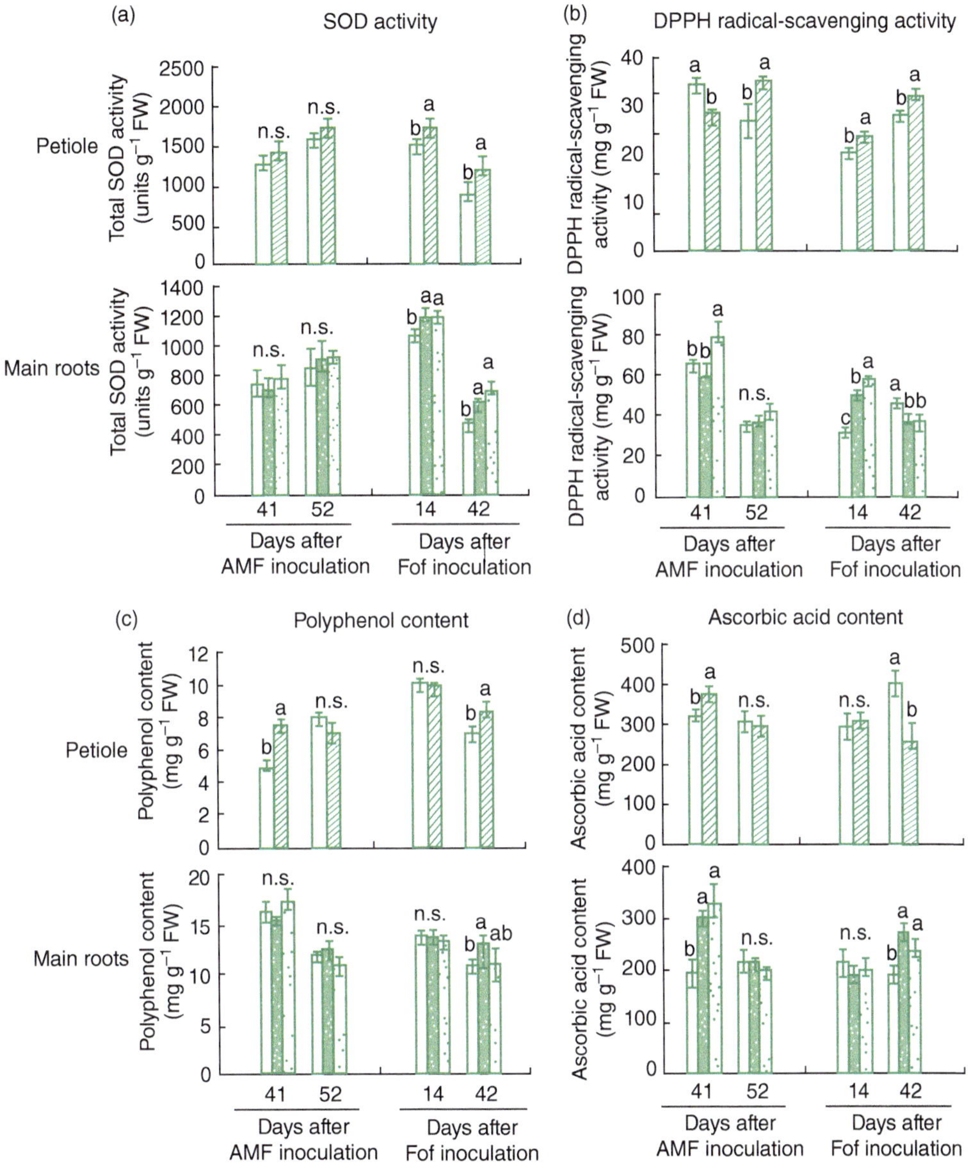

Fig. 13.4. Antioxidative ability in mycorrhizal strawberry plants. The petioles and roots were tested for SOD activity (a), DPPH radical-scavenging activity (b), polyphenol content (c) and ascorbic acid content (d). Open columns, non-AMF-inoculated plants; hatched columns, AMF-inoculated plants; filled columns with white dots, AMF-inoculated roots; open columns with green dots, AMF⁻, non-AMF-inoculated roots. Bars represent mean±standard error. Different lower-case letters indicate a significant difference according to a *t*-test (*P*=0.05). DPPH, 1,1-diphenyl-2-picrylhydrazyl; FoF, *Fusarium oxysporum* f. sp. *fragariae*; FW, fresh weight; n.s., not significant; SOD, superoxide dismutase.

Polyphenol content was determined as quercetin equivalent using an equation obtained from a standard quercetin graph. The results (Fig. 13.4c) demonstrated that polyphenol contents were higher in the petiole and main roots (AMF⁺) 42 days after *F. oxysporum* f. sp. *fragariae* inoculation. With respect to antioxidative substances, polyphenols have a lower electron reduction potential than that of oxygen radicals; as a result, polyphenols directly scavenge reactive oxygen intermediates without promoting further oxidative reactions (Ainsworth and Gillespie, 2007). Vanitha *et al.* (2009) found that total phenol content increased in bacterial wilt in tomato following pathogen inoculation. Our results showed a similar pattern with respect to the increase in polyphenol contents in AMF plants.

13.4.4. Ascorbic acid content

The ascorbate ion is a mild reducing agent and antioxidant, and is the predominant species at typical biological pH values. It is oxidized with the loss of one electron to form a radical cation and then with the loss of a second electron to form dehydroascorbic acid. It typically reacts with oxidants of ROS, such as the hydroxyl radical. An assay to determine ascorbic acid content was performed using the 2,4-dinitrophenylhydrazine method (Roe *et al.*, 1948). One gram of sample was ground in 20 ml of 5% metaphosphate. Then, 2 ml of the extract was placed in a test tube, and 1 ml of 2,6-dichloroindophenol sodium (0.03%) was added. Next, 1 ml of 2% 2,4-dinitrophenylhydrazine was added and the test tube was incubated for 3 h. Absorbance was measured at 520 nm using a spectrophotometer (U-1900). Ascorbic acid content was determined using an equation obtained from a standard l-ascorbic acid graph. The results showed that ascorbic acid content increased, especially in the main roots (AMF^+ and AMF^-) of AMF plants 41 days after AMF inoculation and 42 days after *F. oxysporum* f. sp. *fragariae* inoculation (Fig. 13.4d).

In mycorrhizal St John's wort plants, an increase in ascorbic acid content and disease tolerance was seen under stress conditions induced by the pathogen *Colletotrichum gloeosporioides* (Richter *et al.*, 2011). Our results showed similar patterns. The changes in antioxidative abilities in mycorrhizal strawberry plants varied depending on the plant part, as well as before and after pathogen inoculation.

In the horticultural industry including strawberry, the main focus traditionally has been on yield. However, recently, improving product quality through AMF has been also attempted (Baum *et al.*, 2015). Numerous research results have confirmed the positive effects of AMF not only on plant growth but also on plant quality, for example in tomato (Ulrichs *et al.*, 2008), asparagus (Okada and Matsubara, 2012) and strawberry (Castellanos-Morales *et al.*, 2010). The observations regarding accumulation of ascorbic acid in mycorrhizal strawberry plants are indicative of the expected improvement in strawberry fruit quality.

13.5. Conclusions

Fusarium wilt is a serious disease in the major strawberry producing regions, and the disease causes heavy losses during nursery and fruit production. AMF are ubiquitous soil inhabitants and form a symbiotic relationship with the roots of most terrestrial plants. AMF promote host plant growth by enhancing phosphorus uptake through symbiosis and hence offer an alternative to high inputs of fertilizers and pesticides in sustainable crop production systems. Mycorrhizal strawberry plants accumulate a higher dry weight of shoots and roots than non-mycorrhizal plants in most cases after AMF inoculation. Disease incidence and the severity of symptoms are lower in the shoots and roots of mycorrhizal strawberry plants. The disease symptoms are suppressed in both AMF-inoculated and non-inoculated root parts with a split-root system, indicating the development of ISR in mycorrhizal strawberry plants. Davis and Menge (1980) found that, in citrus seedlings with a split-root system inoculated with *Phytophthora parasitica*, tolerance appeared in non-mycorrhizal roots when these roots were separated from the mycorrhizal (*Glomus fasciculatus*) roots; this effect was thought to be caused by a phosphorus increase through symbiosis. Pozo *et al.* (2002) demonstrated that ISR against *Phytophthora parasitica* appears in mycorrhizal (*Glomus mosseae*) tomato plants with a split-root system. In their report, the systemically induced resistance in non-mycorrhizal root parts was characterized by elicitation of host wall thickenings containing non-esterified pectins and PR-1a proteins in reaction to the intercellular hyphae of the pathogen, and by formation of callose-rich encasement material around the *Phytophthora parasitica* hyphae that were penetrating the root cells. Mycorrhizal strawberry plants show higher SOD activity, DPPH radical-scavenging activity, and total polyphenol and ascorbic acid contents after pathogen inoculation, although with variations between the different plant parts. Notably, no major difference in antioxidative abilities takes place before pathogen inoculation. Some reports have indicated that mycorrhizal colonization induces temporal increases in antioxidative enzymes (APX and CAT), H_2O_2 and flavonoid content, suggesting that colonization acts as a temporal stress for host plants (Volpin *et al.*, 1995; Salzer *et al.*, 1999; Blilou *et al.*, 2000). However, in our opinion, the antioxidative response is higher in mycorrhizal plants after pathogen inoculation, rather than before it. Li *et al.* (2010) also mentioned this tendency; however, it remained unclear as to when the effects appear and how long they continue. From the data discussed in this chapter, it is clear that AMF colonization itself does not stimulate host antioxidative ability strongly before pathogen inoculation; rather, the mycorrhizal strawberry plants acquire the ability to quickly increase their antioxidative potential and maintain these higher levels after pathogen (*Fusarium*) inoculation for a long time.

References

Ainsworth, E.A. and Gillespie, K.M. (2007) Estimation of total phenolic content and other oxidation substrates in plant tissues using Folin–Ciocalteu reagent. *Nature Protocols* 2, 875–877.

Akita, S. (2001) Developmental distribution and cultural control of anthracnose in strawberry. *Agriculture and Horticulture* 56, 76–79.

Asada, K. (1999) The water-water cycle in chloroplasts: scavenging of active oxygens and dissipation of excess photons. *Annual Review of Plant Physiology and Plant Molecular Biology* 50, 601–639.

Baath, E. and Hayman, D.S. (1984) No effect of VA mycorrhiza on red core disease of strawberry. *Transactions of the British Mycological Society* 82, 534–536.

Baum, C., El-Tohamy, W. and Gruda, N. (2015) Increasing the productivity and product quality of vegetable crops using arbuscular mycorrhizal fungi: a review. *Scientia Horticulturae* 187, 131–141.

Beauchamp, C. and Fridovich, I. (1971) Superoxide dismutase: Improved assays and an assay applicable to acrylamide gels. *Analytical Biochemistry* 44, 276–287.

Blilou, I., Bueno, P., Ocampo, J.A. and García-Garrido, J.M. (2000) Induction of catalase and ascorbate peroxidase activities in tobacco roots inoculated with the arbuscular mycorrhizal *Glomus mosseae. Mycological Research* 104, 722–725.

Burits, M. and Bucar, F. (2000) Antioxidant activity of *Nigella sativa* essential oil. *Phytotherapy Research* 14, 323–328.

Castellanos-Morales, V., Villegas, J., Wendelin, S., Vierheilig, H., Eder, R. and Cárdenas-Navarro, R. (2010) Root colonisation by the arbuscular mycorrhizal fungus *Glomus intraradices* alters the quality of strawberry fruits (*Fragaria* × *ananassa* Duch.) at different nitrogen levels. *Journal of the Science of Food and Agriculture* 90, 1774–1782.

Chávez, M.G. and Ferrera-Cerrato, R. (1990) Effect of vesicular-arbuscular mycorrhizae on tissue culture-derived plantlets of strawberry. *HortScience* 25, 903–905.

Choudhary, D.K., Prakash, A. and Johri, B.N. (2007) Induced systemic resistance in plants: mechanism of action. *Indian Journal of Microbiology* 47, 289–297.

Davis, R.M. and Menge, J.A. (1980) Influence of *Glomus fasciculatus* and soil phosphorus on *Phytophthora* root rot of citrus. *Phytopathology* 70, 447–452.

Ehsani-Moghaddam, B., Charles, M.T., Carisse, O. and Khanizadeh, S. (2006) Superoxide dismutase responses of strawberry cultivars to infection by *Mycosphaerella fragariae*. *Journal of Plant Physiology* 163, 147–153.

Folin, O. and Denis, W. (1915) A colorimetric method for the determination of phenols (and phenol derivatives) in urine. *Journal of Biological Chemistry* 22, 305–308.

Fridovich, I. (1986) Biological effects of the superoxide radical. *Archives of Biochemistry and Biophysics* 247, 1–11.
Garmendia, I., Aguirreolea, J. and Goicoechea, N. (2006) Defence-related enzymes in pepper roots during interactions with arbuscular mycorrhizal fungi and/or *Verticillium dahliae*. *BioControl* 51, 293–310.
He, C.Y., Hsiang, T. and Wolyn, D.J. (2002) Induction of systemic disease resistance and pathogen defence responses in *Asparagus officinalis* inoculated with nonpathogenic strains of *Fusarium oxysporum*. *Plant Pathology* 51, 225–230.
Huang, L.L., Yang, C., Zhao, Y., Xu, X., Xu, Q., Li, G.Z., Cao, J., Herbert, S.J. and Hao, L. (2008) Antioxidant defenses of mycorrhizal fungus infection against SO_2-induced oxidative stress in *Avena nuda* seedlings. *Bulletin of Environmental Contamination and Toxicology* 81, 440–444.
Husaini, A.M. (2010) Pre- and post-agroinfection strategies for efficient leaf disk transformation and regeneration of transgenic strawberry plants. *Plant Cell Reports* 29, 97–110.
Husaini, A.M., Abdin, M.Z., Khan, S., Xu, Y.W., Aquil, S. and Anis, M. (2012) Modifying strawberry for better adaptability to adverse impact of climate change. *Current Science* 102, 1660–1673.
Koike, N., Hyakumachi, M., Kageyama, K., Tsuyumu, S. and Doke, N. (2001) Induction of systemic resistance in cucumber against several diseases by plant growth-promoting fungi: lignification and superoxide generation. *European Journal of Plant Pathology* 107, 523–533.
Koike, S.T. and Gordon, T.R. (2015) Management of *Fusarium* wilt of strawberry. *Crop Protection* 73, 67–72.
Kuźniak, E. and Skłodowska, M. (2004) The effect of *Botrytis cinerea* infection on the antioxidant profile of mitochondria from tomato leaves. *Journal of Experimental Botany* 55, 605–612.
Li, Y., Matsubara, Y., Miyawaki, C., Liu, Y. and Koshikawa, K. (2008) Temperature stress tolerance and increase in antioxidative enzyme activities in mycorrhizal strawberry plants. *Acta Horticulturae* 774, 391–395.
Li, Y., Yanagi, A., Miyawaki, Y., Okada, T. and Matsubara, Y. (2010) Disease tolerance and changes in antioxidative abilities in mycorrhizal strawberry plants. *Journal of the Japanese Society for Horticultural Science* 79, 174–178.
Mark, G.L. and Cassells, A.C. (1996) Genotype-dependence in the interaction between *Glomus fistulosum*, *Phytophthora fragariae* and the wild strawberry (*Fragaria vesca*). *Plant and Soil* 185, 233–239.
Marschner, H. and Dell, B. (1994) Nutrient uptake in mycorrhizal symbiosis. *Plant and Soil* 159, 89–102.
Matsubara, Y. (2011) Tolerance to *Fusarium* wilt and changes in antioxidative ability and free amino acid content in mycorrhizal strawberry plants. In: *Genomics, Transgenics, Molecular Breeding and Biotechnology of Strawberry*. Husaini, A.M. and Mercado, J.A. (eds) Global Science Books, UK, pp. 126–131.
Matsubara, Y., Hirano, I., Sassa, D. and Koshikawa, K. (2004) Increased tolerance to *Fusarium* wilt in mycorrhizal strawberry plants raised by capillary watering methods. *Environmental Control in Biology* 42, 185–191.
Mohammad Nahiyan, A.S. and Matsubara, Y.-I. (2012) Tolerance to *Fusarium* root rot and changes in antioxidative ability in mycorrhizal asparagus plants. *HortScience* 47, 356–360.
Moon, B.J., Park, H. and Chung, H. (1995) Studies on antagonism of *Trichoderma* species to *Fusarium oxysporum* f. sp. *fragariae*. V. Biological control of *Fusarium* wilt of strawberry by a mycoparasite, *Trichoderma harzianum*. *Korean Journal of Plant Pathology* 11, 298–303.
Mori, T. and Kitamura, H. (2003) Comparison of fruit quality and earliness of strawberry between inoculated and non-inoculated progenies for resistance to anthracnose. *Journal of the Japanese Society for Horticultural Science* 72, 64–68.
Nam, M.H., Park, M.S., Kim, H.G. and Yoo, S.J. (2009) Biological control of strawberry *Fusarium* wilt caused by *Fusarium oxysporum* f. sp. *fragariae* using *Bacillus velezensis* BS87 and RK1 formulation. *Journal of Microbiology and Biotechnology* 19, 520–524.
Niemi, M. and Vestberg, M. (1992) Inoculation of commercially grown strawberry with VA mycorrhizal fungi. *Plant and Soil* 144, 133–142.
Norman, J.R., Atkinson, D. and Hooker, J.E. (1996) Arbuscular mycorrhizal fungal-induced alteration to root architecture in strawberry and induced resistance to the root pathogen *Phytophthora fragariae*. *Plant and Soil* 185, 191–198.
Okada, T. and Matsubara, Y.-I. (2012) Tolerance to *Fusarium* root rot and the changes in free amino acid contents in mycorrhizal asparagus plants. *HortScience* 47, 751–754.
Okayama, K. (1993) Effects of rain shelter and capillary watering on disease development of symptomless strawberry plants infected with *Glomerella cingulata* (*Colletotrichum gloeosporioides*). *Annals of the Phytopathological Society of Japan* 59, 514–519.
Phillips, J.M. and Hayman, D.S. (1970) Improved procedures for clearing roots and staining parasitic and vesicular-arbuscular mycorrhizal fungi for rapid assessment of infection. *Transactions of the British Mycological Society* 55, 158–161.

Pozo, M.J., Cordier, C., Dumas-Gaudot, E., Gianinazzi, S., Barea, J.M. and Azcón-Aguilar, C. (2002) Localized versus systemic effect of arbuscular mycorrhizal fungi on defence responses to *Phytophthora* infection in tomato plants. *Journal of Experimental Botany* 53, 525–534.
Richter, J., Baltruschat, H., Kabrodt, K. and Schellenberg, I. (2011) Impact of arbuscular mycorrhiza on the St John's wort (*Hypericum perforation*) wilt disease induced by *Colletotrichum* cf. *gloeosporioides*. *Journal of Plant Diseases and Protection* 118, 109–118.
Robertson, W., Boyle, C. and Brown, H. (1988) Endomycorrhizal status of certified strawberry nursery stock. *Journal of the American Society for Horticultural Science* 113, 525–529.
Roe, J.H., Mary, B.M., Oesterking, M.J. and Charlotle, M.D. (1948) The determination of diketo-L-gluconic acid, dehydro-L-ascorbic acid, and L-ascorbic acid in the same tissues by 2,4-dinitrophenyl hydrazine method. *Journal of Biological Chemistry* 174, 201–208.
Roldán, A., Díaz-Vivancos, P., Hernández, J.A., Carrasco, L. and Caravaca, F. (2008) Superoxide dismutase and total peroxidase activities in relation to drought recovery performance of mycorrhizal shrub seedlings grown in an amended semiarid soil. *Journal of Plant Physiology* 165, 715–722.
Ruiz-Lozano, J.M., Azcón, R. and Palma, J.M. (1996) Superoxide dismutase activity in arbuscular mycorrhizal *Lactuca sativa* plants subjected to drought stress. *New Phytologist* 134, 327–333.
Sahay, N.S. and Varma, A. (1999) *Piriformospora indica*: a new biological hardening tool for micropropagated plants. *FEMS Microbiology Letters* 181, 297–302.
Sahoo, M., DasGupta, M., Kole, P., Bhat, J. and Mukherjee, A. (2007) Antioxidative enzymes and isozymes analysis of taro genotypes and their implications in *Phytophthora* blight disease resistance. *Mycopathologia* 163, 241–248.
Salzer, P., Corbière, H. and Boller, T. (1999) Hydrogen peroxide accumulation in *Medicago truncatula* roots colonized by the arbuscular mycorrhiza-forming fungus *Glomus intraradices*. *Planta* 208, 319–325.
Schaart, J.G., Kjellsen, T.D., Mehli, L., Heggem, R., Iversen, T.-H., Schouten, H.J. and Krens, F.A. (2011) Towards the production of genetically modified strawberries which are acceptable to consumers. In: Husaini, A.M. and Mercado, J.A. (eds) *Genomics, Transgenics, Molecular Breeding and Biotechnology of Strawberry. Global Science Books*, UK, pp. 102–107.
Tezuka, N. and Makino, T. (1991) Biological control of *Fusarium* wilt of strawberry by nonpathogenic *Fusarium oxysporum* isolated from strawberry. *Annals of the Phytopathological Society of Japan* 57, 506–511.
Ulrichs, C., Fischer, G., Büttner, C. and Mewis, I. (2008) Comparison of lycopene, β-carotene and phenolic contents of tomato using conventional and ecological horticultural practices, and arbuscular mycorrhizal fungi (AMF). *Agronomia Colombiana* 26, 40–46.
Vanitha, S.C., Niranjana, S.R. and Umesha, S. (2009) Role of phenylalanine ammonial yase and polyphenol oxidase in host resistance to bacterial wilt of tomato. *Journal of Phytopathology* 157, 552–557.
Varma, A. and Schüepp, H. (1994) Infectivity and effectiveness of *Glomus intraradices* on micropropagated plants. *Mycorrhiza* 5, 29–37.
Volpin, H., Phillips, D.A., Okon, Y. and Kapulnik, Y. (1995) Suppression of an isoflavonoid phytoalexin defense response in mycorrhizal alfalfa roots. *Plant Physiology* 108, 1449–1454.
Wahid, A., Gelani, S., Ashraf, M. and Foolad, M.R. (2007) Heat tolerance in plants: an overview. *Environmental and Experimental Botany* 61, 199–223.
Williams, S.C.K., Vestberg, M., Uosukainen, M., Dodd, J.C. and Jeffries, P. (1992) Effects of fertilizers and arbuscular mycorrhizal fungi on the post-vitro growth of micropropagated strawberry. *Agronomie* 12, 851–857.
Wu, Q.S., Zou, Y.N. and Xia, R.X. (2006) Effects of water stress and arbuscular mycorrhizal fungi on reactive oxygen metabolism and antioxidant production by citrus (*Citrus tangerine*) roots. *European Journal of Soil Biology* 42, 166–172.
Zhang, Y., Fan, T., Jia, W., Zhang, W., Liu, Q., Li, B. and Zhang, L. (2012) Identification and characterization of a *Bacillus subtilis* strain TS06 as bio-control agent of strawberry replant disease (*Fusarium* and *Verticilium* wilts). *African Journal of Biotechnology* 11, 570–580.

14 Challenges of Climate Change to Strawberry Cultivation: Uncertainty and Beyond

Amjad M. Husaini[1]* and Y. Wen Xu[2]

[1]*Centre for Plant Biotechnology, Sher-e-Kashmir University of Agricultural Sciences and Technology of Kashmir, Shalimar, Jammu and Kashmir, India;* [2]*Pharmaceutical Botany, Sichuan Agricultural University, Ya'an, PR China*

14.1. Climate Change and Forecasting the Impact on Strawberry

The United Nations (UN) Framework Convention on Climate Change, which was opened for signature during the 1992 UN Conference on the Environment and Development in Rio, was the first attempt to deal with the threat of global climate change. In June 2008, the UN Food and Agriculture Organization (FAO), together with International Fund for Agricultural Development (IFAD), UN World Food Programme (WFP) and Consultative Group on International Agricultural Research (CGIAR) system, convened the High-Level Conference on World

* amjadhusaini@skuastkashmir.ac.in

Food Security: the Challenges of Climate Change and Bioenergy, and 181 countries adopted a declaration that 'It is essential to address the question of how to increase the resilience of present food production systems to challenges posed by climate change' (http://www.fao.org/foodclimate). Recently, it was again reiterated in the UN Sustainable Development Goals: 'by 2030, ensure sustainable food production systems and implement resilient agricultural practices that increase productivity and production, that help maintain ecosystems, that strengthen capacity for adaptation to climate change, extreme weather, drought, flooding and other disasters and that progressively improve land and soil quality' (http://www.un.org/sustainabledevelopment/climate-change-2/).

What actually constitutes climate? Climate is an extraordinarily complex system with many delicately interrelated components that have significant roles in forming climate through their interactions with the atmosphere (FAO, 2008) (Fig. 14.1). Climate is the statistical aggregate of weather conditions such as temperature, precipitation, wind, cloudiness and storms over a period of time (Glover *et al.*, 2008). Five primary environmental factors are critical for plants to germinate, grow and reproduce, and strawberry is no exception. These factors are carbon dioxide, sunlight, water, an optimal temperature range and nutrients. Of these, the first four are directly related to climate and vary spatially, diurnally and seasonally. In the context of global climate change, these factors behave erratically and therefore need to be considered foremost when designing strawberry plants adapted to changing climatic conditions.

Strawberry is cosmopolitan in distribution, and therefore predicting the precise impact of climate change on strawberry across all current areas of distribution is very complex and perhaps not possible. The cultivated strawberry, *Fragaria* × *ananassa* Duch., is a natural hybrid of *Fragaria chiloensis* (L.) P. Mill. and *Fragaria virginiana* (Duch.) and is octoploid ($2n = 8x = 56$). There are about 34 other species of *Fragaria* found in Asia, North and South America, and Europe, of which two are cultivated commercially for their fruit: *Fragaria moschata*, the musky or hautboy strawberry, and *Fragaria vesca*, the wood or alpine strawberry. These species have been cultivated for centuries, but there is little production of them today, due to the success of *Fragaria* × *ananassa*. About a dozen diploid strawberry species are distributed in Eurasia, although one of them, *F. vesca*, has a wider distribution and is also found in America. Tetraploid species, namely *Fragaria corymbosa*, *Fragaria orientalis*, and *Fragaria moupinensis*, are mostly restricted

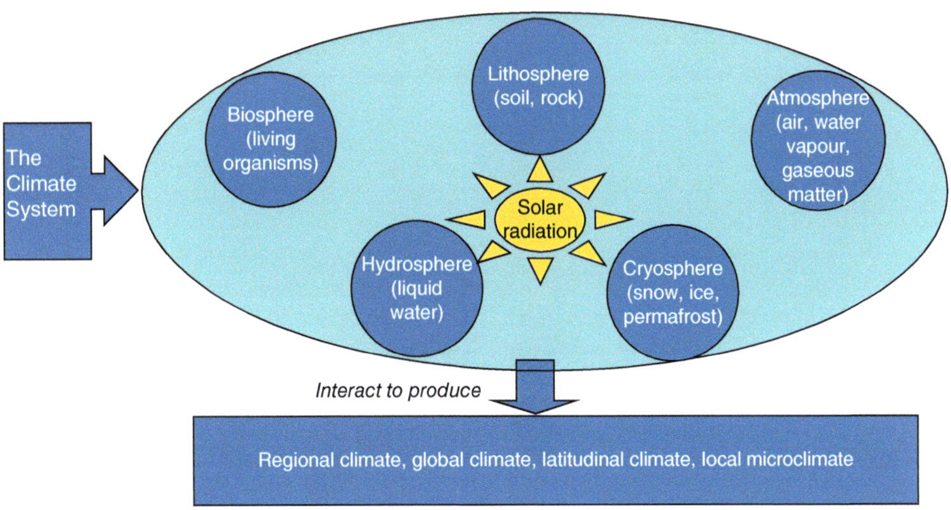

Fig. 14.1. The climate system.

to East Asia, while *F. moschata*, the only hexaploid species, is found in Europe. The octoploid species *F. virginiana* L. is native to North America, north of Mexico, while *F. chiloensis* Duch. is native to west coastal North America and the west coast and Andes Mountains of South America. *Fragaria iturupensis* is the only Asian octoploid species, as described by Staudt (1973). Given the lack of sufficient detail regarding climatic variations and their interactions, as well as the precise climate changes for each of these regions/localities, the impact of climate change on strawberry can be predicted only in general terms. Broadly speaking, global climate change will alter many elements of future strawberry production, primarily because atmospheric carbon dioxide concentration, average temperature and tropospheric ozone concentration will be higher, droughts will be more frequent and severe, more intense precipitation events will lead to increased flooding, some soils will degrade and weather extremes will be more frequent. Climate change will also affect host–pathogen interactions, in particular by: (i) increasing pathogen development rates (the number of generations per year); (ii) relaxing overwintering restrictions on pathogen life cycles due to milder winters; and (iii) modifying host susceptibility to infection (Harvell *et al.*, 2002). Changes in temperature, precipitation and humidity influence the growth, spread and survival of plant diseases, as climate and weather have a major role in influencing disease epidemiology (Rosenzweig *et al.*, 2001). This means that knowledge of how climate change will bring about specific changes in regional weather patterns is crucial in determining how plant disease outbreaks will change in the future. Many disease prediction models have been published in the last decade (Booth *et al.*, 2000; Bergot *et al.*, 2004; Francesca *et al.*, 2006; Evans *et al.*, 2008). By combining these models with climate change models, it becomes possible to study the change in likelihood of the diseases occurring and assess the probability of disease outbreaks. However, most of these studies do not take into account factors such as changes in agricultural and environmental policy, mitigation activity and consumer preferences. Furthermore, being simplistic in form, these models assume disease inoculum to be present equally throughout the region and everything else to be constant except the weather. In a real case scenario, even though the predictions of disease potential might be valid, the likelihood of actual outbreaks occurring would also depend on the availability of inoculum. In an interesting study about the effect of climate change on strawberry production in the UK, Calleja (2011) predicted that the incidence of powdery mildew in strawberries will increase under climate change projections across most of the UK, with disease incidence increasing more in Scotland, western Wales and the north of England. Similarly, grey mould incidence was predicted to change across most of the UK, with increases north of the Midlands and slight decreases in the south. The incidence of strawberry black spot was predicted to increase drastically across most of the UK, with potential disease incidence increasing most in Scotland and on the west coast of the UK. However, Calleja (2011) cautioned on the usefulness of these disease prediction models, because of their simplicity and assumptions that inocula are present throughout the area being studied and that the same production and disease management practices are used throughout. Thus, while being indicative of potential changes in disease incidence in the UK for these three diseases, more work needs to be done to build more detailed disease models that can provide further information for policymakers and the industry. Additionally, the predictions of these models depend on the appropriateness of the chosen conditions for infection, and might vary slightly if a different temperature range, precipitation level or duration is chosen.

14.1.1. Photoperiod and vernalization

Photoperiod and temperature are major environmental signals affecting flowering in plants. Although molecular pathways mediating these signals have been well characterized in the annual model plant

Arabidopsis, much less information is known in perennials. One of the primary ways of classifying strawberry is on the basis of photoperiod requirements on the seasonal flowering characteristics of the cultivars. Shortening days and lower temperatures normally induce flower development in the crown. However, different genotypes have a wide range of growth habits, from the single fruiting habit of short-day plants to the almost continually blooming and fruiting habit of day-neutral plants. Two main types of cultivar are recognized based on environmental control of flowering:

- Short-day or June-bearing: plants that flower in the spring, produce a single crop and form flower buds in the autumn. These short-day plants require photoperiods under 14 h and/or temperatures under 17°C for floral initiation.
- Ever-bearing: plants that flower and fruit two or more times per season. The ever-bearing habit is governed by long photoperiods. These cultivars fruit throughout the growing season provided temperatures are not too high.

There are two types of ever-bearing plants: (i) Long-day plants, in which lengthening days promote more-or-less continuous flowering during the summer, provided temperatures are not too high; and (ii) Day-neutral plants, in which photoperiod has no effect on flowering; they will also flower several times per year, but do so during short as well as long days.

The flowering habit of the strawberry is directly related to temperature, and strongly interacts with photoperiod. These factors affect induction, initiation and differentiation of flower buds. It has been shown that cool summer temperatures (17°C) allow induction to occur under long photoperiods, even in short-day cultivars. This means that some cultivars considered as short-day types behave as remontants in cool climates. High temperatures generally inhibit flowering under any condition (regarding photoperiod or cultivar), although it has been observed that critical temperature is higher in day-neutral cultivars (Manakasem and Goodwin, 2001; Stewart and Folta, 2010). Thus, cool temperatures promote flowering, while warm temperatures inhibit flowering, and temperature sensitivity is greater in short-day cultivars and least in day-neutral cultivars.

The day-neutral characteristic, introduced by Bringhurst and Voth (1984) using the genotype *F. virginiana* subsp. *glauca*, collected from the Wasatch Mountains in Utah (Hancock and Luby, 1993), has been investigated widely to elucidate its mode of inheritance. Most studies indicate it to be governed by a dominant locus affected by some minor genes (Ahmadi *et al.*, 1990; Shaw and Famula, 2005), although some suggest it to be a quantitative character having polygenic inheritance (Serçe and Hancock, 2005b; Weebadde *et al.*, 2008). In recent decades, some wild accessions of *F. virginiana* have been identified with varying degrees of photoperiod insensitivity or continuous flowering, but more information and crosses are needed to incorporate these new sources into commercial varieties (Hancock *et al.*, 2001a; Serçe and Hancock, 2005a,b), as the diversity of responses to photoperiod and the interaction with other environmental and epigenetic factors make the analysis difficult. Nevertheless, through the availability of *F. vesca* genomic information and through its genetic manipulation, it has been possible to characterize a gene that inhibits photoperiod sensitivity, named *Terminalflower1* (*FvTFL1*), and to develop molecular markers for its assisted selection (Koskela *et al.*, 2012). By using transgenic lines and gene expression analyses, Rantanen *et al.* (2015) recently observed that the temperature-mediated changes in the *FvTFL1* mRNA expression set critical temperature limits for the photoperiodic flowering in strawberry. Although *F.* × *ananassa* and *F. vesca* show similar environmental responses, they seem to differ in the genetic control of the perpetual flowering trait. It has been shown that perpetual flowering is controlled by a single dominant locus (Morishita *et al.*, 2012) in *F.* × *ananassa*. One major quantitative trait locus, namely *FaPFRU*, has been identified in *F.* × *ananassa*, and is not orthologous to the *Seasonal Flowering Locus* (*SFL*) or *Runnering* locus (*R*) of *F. vesca*, and is positively linked to perpetual flowering but negatively linked to runnering in *F.* × *ananassa* (Gaston *et al.*, 2013). In another recent development, Castro *et al.* (2015) found a marker showing association with day-neutral flowering and runnering in a population derived from a cross of cultivars 'Tribute' × 'Honeoye'. The marker ChFaM148-184T was derived from an expressed sequence tag (EST) (GenBank accession no. GU815864) with homology to *GalUR* (Zorrilla-Fontanesi *et al.*, 2011), a gene that encodes an NAPDH-dependent d-galacturonate reductase. This finding supports the hypothesis that photoperiod is mainly a monogenic or oligogenic character.

The chilling requirement (accumulation of chilling hours at 0–7°C) is an important factor affecting reproductive and vegetative growth in strawberry plants. This vernalization period is required to break bud dormancy, and varies with genotypes. It is a mechanism to prevent plants from budding and flourishing early in the season, when chance of spring frost is still high. Therefore, in areas where climate change may cause a drop in temperature during the spring, high-chilling-requirement cultivars may be suitable for cultivation. In Norway, the *F. vesca* cultivar 'Alta', with a much delayed budding and flowering, was discovered. This characteristic was attributed to a high winter chilling requirement, compared with other studied populations of the same species (Heide and Sønsteby, 2007), and therefore it might be possible that such a population will adapt better to regions where climate change has caused prolonging of the winter season.

14.1.2. Disease and pest outbreaks

Climate change will undoubtedly modify pest dynamics. Changes in disease incidence and movement of diseases to areas where they were previously not a problem can lead to substantive regional impacts. Moreover, the impacts of climatic change on crop physiology could affect the interaction between pathogen and host, and potentially also induce changes in the epidemiology of the pathogen (Garrett *et al.*, 2006). The current understanding of these changes is quite limited, despite some past attempts of modelling these changes (Aggarwal *et al.*, 2006; Easterling *et al.*, 2007). The greatest impact of climate change on pathogens is related to the altered condition of the plant host in the changed environment, rather than to any direct impact on the organisms (Chakraborty *et al.*, 2002; Ziska and Runion, 2007). A strawberry plant (host) stressed by the environment could have an altered physiology and chemical composition, which may render it more susceptible to pests or diseases due to altered concentrations of defensive compounds (Ziska and Runion, 2007).

There is a complex interaction between biotic and abiotic factors in disease incidence, development and spread. For example, in recent years, an increase in collapsed and dead strawberry plants due to crown rot has been observed in several fields in central Chile, an important area for strawberry cultivation in this country (Sánchez *et al.*, 2013). According to this study, this disease, caused by *Macrophomina phaseolina*, has also been reported recently in Spain, the USA and Argentina. The microsclerotia of *M. phaseolina* are produced in the host tissue and released into the soil as the infected plant decays. These multicelled structures allow persistence of the fungus under adverse conditions such as low soil nutrient levels and temperatures above 30°C. Germination of the microsclerotia occurs when temperatures are between 28 and 35°C. Soil moisture content is considered a major factor that affects survival of *M. phaseolina* in soil (Pratt, 2006). Zveibil *et al.* (2012) reported that high soil moisture content and flooding caused a decline in sclerotia viability compared with dry soils. Olaya and Abawi (1996) reported that high temperatures and low soil water potentials are important factors in causing this disease, which is important from the perspective of climate change.

Cool climates will favour the development of diseases such as root rot. *Pythium* spp., especially *Pythium ultimum*, are the most widespread strawberry root pathogens and their growth is favoured by cool climates. This species is a major cause of black root rot disease in strawberry and can also attack many other crops. In strawberry, *Pythium* spp. destroy juvenile root tissue, such as feeder rootlets. In contrast, *Fusarium* wilt development in strawberry is favoured by high temperatures, which cause the infected leaves to wilt and die rapidly (Maas, 1998). With an increase of temperature from 17 to 27°C, the disease severity of plants infected with *Fusarium oxysporum* increases significantly, causing severe disease symptoms in the crown, root and vascular tissues, along with poor development of roots, eventually resulting in death of the plants within 4 weeks (Fang *et al.*, 2011).

Warmer winters will also have a large influence on the survival of insect pests. The main effect of temperature in temperate regions is on the winter survival of pests. With warmer winters, the natural shield that protects the strawberry industry in cooler areas of Europe will be eliminated, subjecting plants to the invasion of insect pests. In the absence of improved varieties, growers will need to resort to other effective methods of controlling disease, such as increased dependence on pesticides. This will not be a sustainable solution, as access to chemical pesticides is likely to be restricted in the future. As a result of the enactment of legislations such as EC Regulation No. 1107/2009 (this regulation repealed Council Directives 79/117/EEC and 91/414/EEC in 21 October 2009 and applied the precautionary principle to 'ensure that industry demonstrates that substances or products produced or placed on the market do not have any harmful effect on human or animal health or any unacceptable effects on the environment' (http://eur-lex.europa.eu/legal-content/EN/TXT/?uri=celex:32009R1107), the availability of pesticides for the strawberry sector has been reduced, making this crop more vulnerable to the impacts of plant disease and climate change.

Calleja (2011) found evidence that climate change is already affecting the strawberry sector in the UK. Many farms have already been adversely affected and have suffered financial loss. However, this has driven innovation within the sector, sometimes as a direct result of companies involved in strawberry production funding research and development to develop new products, either by the company themselves or through collaboration with universities by funding their research. The need to minimize the use of chemical pesticides for environment-friendly cultivation will require the development of improved cultivars containing pest- and disease-resistant genes. Furthermore, the need to maintain production in different areas and to face changing climates will also require the presence of abiotic stress-resistance genes.

14.2. Major Environmental Factors of Critical Importance

It is evident from the above discussion that one of the major environmental factors affecting plant growth and productivity is temperature, with optimum growth temperatures between 10 and 26°C for strawberry (Ledesma *et al.*, 2004). Field-grown strawberry plants are often exposed to fluctuating temperatures and soil moisture, which have profound effects on plant metabolism. These factors are discussed further in the following sections.

14.2.1. Temperature rise

The increase in temperature above a threshold level for a period of time sufficient to cause irreversible damage to plant growth and development is called 'heat stress'. Heat stress is a complex function of heat intensity (temperature in degrees), duration (seasonal compared with daily temperature extremes) and rate of increase in temperature (Wahid *et al.*, 2007). According to Bray *et al.* (2000), a heat-stress response is observed in plants exposed to temperatures at least 5°C above optimal conditions. Studies on the mechanism of stress tolerance in each species or cultivar are very important for the development of heat-tolerant plants. However, knowledge about the mechanisms of heat tolerance in common strawberry cultivars is limited.

Heat acclimation studies in strawberry plants by Gulen and Eris (2004) revealed that gradually heat-stressed plants showed less membrane injury and exhibited greater heat-stress tolerance than shock heat-stressed plants. Thus, gradual high-temperature stress with long exposure increases heat tolerance in strawberry leaves. Increased heat tolerance was correlated with the accumulation of several heat-stable proteins in the gradually heat-stressed plants. In addition, significant differences between gradual and shock heat-stressed plants were reported in leaf relative water content, loss of turgidity, and total protein and DNA content. Surprisingly, heat stress increased the chlorophyll content in strawberry plants (Gulen and Eris, 2003; Kesici *et al.*, 2013). Moreover, high temperatures increased the activities of antioxidant enzymes ascorbate peroxidase (APX), peroxidase (PRX) and catalase (CAT) (Ergin, 2012). While no effect on glutathione reductase (GR) activity was reported under heat stress, the activity of PRX enzyme was closely related to cell membrane stability and lignification, thus, qualifying as a potential enzyme marker (Gulen and Eris, 2004; Ergin *et al.*, 2012). Heat-shock proteins (HSPs) under the control of heat-stress transcription factors play a central role in the heat-stress response. These were first identified as proteins that are strongly induced by heat stress and subsequently were shown also to be essential for normal growth and metabolism. The role of HSPs in heat tolerance of strawberry is described in a limited number of studies. Ledesma *et al.* (2004) reported 19–29 kDa HSPs in leaves and 16–26 kDa HSPs in flowers of the strawberry cultivars 'Nyoho' and 'Toyonaka' at high temperatures. Ergin (2012) detected a 23 kDa HSP in leaf tissues during heat acclimation of strawberry cultivars. Accumulation of this 23 kDa HSP was found to be higher in heat-tolerant cultivars than in heat-sensitive cultivars, and thus correlated with heat tolerance in strawberry.

Recently, Kesici *et al.* (2013) evaluated the heat tolerance of 15 commonly grown strawberry cultivars, and found that 'Elsanta', 'R. Hope' and 'Camarosa' were relatively heat tolerant, while 'Whitney', 'Fern', 'Festival' and 'CG3' were relatively sensitive. More such studies need to be undertaken to identify cultivars suitable for different temperature regimes.

14.2.2. Temperature drop

Strawberry is a moderately cold-hardy species. Barney *et al.* (1992) report the cold hardiness of strawberry cultivars over a wide range (from –3.8 to –45.0°C). It is one of the most important small fruit crops in several northern parts of the world (Ndong *et al.*, 1997). Genotypes collected in cold environments usually possess greater hardiness and are usually more tolerant to spring frost damage during flowering. In a study on germplasm collections of *F. chiloensis* and *F. virginiana*, a high degree of cold resistance was found in clone PI 552091 of *F. chiloensis*, collected in Chile at an elevation of 1900 m (Hancock *et al.*, 2001a). However, in a study that compared accessions of native American octoploid genotypes, it was determined that those of *F. virginiana*, regardless of their origin, had a greater resistance to cold weather than those of *F. chiloensis*. Within the latter, clones collected in North America were more resistant than those collected in South America (Hancock *et al.*, 2001b).

Chilling injury of strawberry is a serious problem in particularly subtropical cultivated areas (Zhang *et al.*, 2008). Freezing tolerance is evaluated by an ion-leakage test at the cell membrane (Arora *et al.*, 1992). The level of LT_{50} shows the cold-hardiness level of a cultivar and is defined as the temperature killing 50% of the whole plant or tissues. For strawberry cultivars, the LT_{50} of 'Chamly', 'Red Coat' and 'Oka' was reported as –12°C, while the LT_{50} of the freezing-sensitive 'Chandler' was –2°C (Ndong *et al.*, 1997). Carbohydrate metabolism is also very important in determining the cold hardiness of plants. Sugars increase the intracellular osmotic potential and reduce the freezing point (Levitt, 1980; Paquin *et al.*, 1989). Seasonal alteration of apoplastic and symplastic sugar metabolism occurs in plants. Symplastic sugar metabolism was found to be more effective in cold tolerance, as the total soluble sugar, reducing sugar and sucrose contents of the symplast were higher in the cold-acclimated stage than in the non-acclimated stage (Turhan, 2012). Turhan *et al.* (2012a) evaluated the cold hardiness of eight strawberry cultivars and reported that 'Ventane' was the most cold hardy, while 'Kabarla' was the least cold hardy among the eight cultivars.

Strawberries differ considerably in their 'ability' to develop cold or freezing tolerance through a process known as cold acclimation. Thus, understanding the mechanisms involved in cold acclimation has great importance in enhancing the cold tolerance of species, and would be of considerable interest for preventing cold damage. Biochemical changes associated with cold acclimation in various plant species include alterations in lipid composition, increased sugar and soluble protein content, expression of specific proteins, the appearance of new isozymes and alterations in the activity of antioxidative enzymes (Thomashow, 1999; Sarnighausen *et al.*, 2004; Eris *et al.*, 2007; Gulen *et al.*, 2008; Cansev *et al.*, 2009). Zhang *et al.* (2008) reported that chilling acclimation at 0°C resulted in enhancement of superoxide dismutase, CAT, PRX, dehydroascorbate reductase, monodehydroascorbate reductase, APX and GR activities and an increase in reduced ascorbate, dehydroascorbate and reduced glutathione (GSH) contents in the strawberry cultivar 'Toyonaka'. They concluded that the ascorbate–GSH cycle plays an important role in the detoxification of H_2O_2, thereby enhancing chilling resistance in strawberry. As ice formation occurs primarily in the intercellular spaces during freezing injury, Turhan *et al.* (2012b) studied apoplastic antioxidant enzymes in the leaves of two strawberry cultivars, 'Aromas' and 'Diamante'. The activities of apoplastic CAT, PRX and APX varied significantly depending on the cold-hardiness level of the cultivar. Considerable increases in APX and GR enzyme activity in the cold-acclimated stage might serve as markers of cold hardiness in strawberry plants (Turhan *et al.*, 2012a). All these biochemical processes are regulated at the gene expression level and many cold-induced genes/proteins have been characterized in many species. These proteins accumulate at a high level under cold stress. Recently, Koehler *et al.* (2012) reported stress-responsive proteins in crown tissue of strawberry plants exposed to cold treatments. Ndong *et al.* (1997) conducted a study to identify genes associated with cold acclimation and reported that the expression of *Fcor1* (*Fragaria Cold-Regulated1*) correlated with the freezing tolerance of the strawberry cultivars 'Champly', 'Red Coat', 'Oka' and 'Chandler' and wild strawberry (*F. virginiana*), suggesting that *Fcor1* might serve as a molecular marker for cold tolerance.

14.2.3. Water scarcity/surplus

Another factor that determines the degree of a plant's hardiness towards climate change is its ability to adapt to extreme drought or flooding conditions. Strawberry plants are known for their high demand for water because of their shallow root system (almost all of the roots are in the 0–15 cm depth zone), large leaf area and high water content of the fruit (Chandler and Ferree, 1990; Klamkowski and Treder, 2006). According to Li *et al.* (2010), strawberry plant water stress occurs on the slope and shoulder areas, where there is high irradiance and temperature and low soil water content, resulting in significantly lower fruit yield. Drought stress has a negative effect on strawberry plant growth and yield (Chandler and Ferree, 1990; Husaini and Tuteja, 2012). While drought stress decreases strawberry leaf water potential, transpiration rate and water channel activity, flooding does not cause any change in leaf water potentials and water channel activity in strawberry plants (Blanke and Cooke, 2006). Stomata close more rapidly in flooded strawberry leaves than in drought-stressed ones. The evidence suggests that strawberry plants are generally more adversely affected by drought than by flooding.

In order to develop drought-tolerant strawberry varieties, knowledge about species-level and cultivar-level tolerance is imperative for a successful breeding programme. However, there are limited studies on this aspect in strawberry. The most drought-tolerant species within the genus *Fragaria* is reported to be *F. chiloensis* (Zhang and Archbold, 1993). Hancock *et al.* (2001a) selected plants of *F. chiloensis* from extremely arid zones, in coastal regions of Chile or the USA (PI 602567, PI 612317, PI 551728 and PI 612490), and used them as donors for tolerance to water/salt stress for modern cultivars. Grant *et al.* (2010) compared the performance of ten strawberry cultivars under water deficit and reported that osmotic adjustment, small leaf area and transpiration rate were some of the drought-tolerant characteristics useful in breeding programmes. Strawberry plants exposed to water stress show a tolerance mechanism based on high elasticity of tissues as a morphological adaptation. The changes in whole-plant morphology and canopy architecture, from monolayer to polylayer leaf distribution, and leaf orientation from south to north, have been reported as an indication of stress-avoidance mechanisms. Based on such morphological and physiological measurements, Klamkowski and Treder (2008) report that the cultivar 'Elsanta' was the most drought tolerant among the three strawberry cultivars evaluated in their study.

Annual average temperatures are showing an upward trend, and evapotranspiration requirements for water are growing. Moreover, water deficit/drought usually causes inhibition of transpiration, because of which the plants fail to dissipate heat, and thus heat stress can also ensue (Buchanan *et al.*, 2000). Studies on the combined effects of heat and water stress, using physiological and molecular tools, have indicated a direct relationship between heat and drought tolerance of strawberry cultivars (Çetinkaya, 2013). In this study, the heat-tolerant cultivars 'Redlands Hope' and 'Camarosa' also showed drought tolerance, while the heat-sensitive 'Festival' and 'CG3' were also drought sensitive.

14.3. Engineering Strawberry with 'Adaptive' Advantages

Plants respond to extreme levels of environmental conditions in a wide variety of ways depending on the species, cultivar, age and physiological stage of the plant. While it is important to continue striving to reduce greenhouse gas emissions, developing resilient plants that can adapt to climate change is equally important. Abiotic stress conditions can make a difference between having a crop to harvest and crop failure. Tolerance to heat, drought, water-logging and frost, resistance to pests and diseases, and water-use and nitrogen-use efficiency are the most important traits for adaptation to climate change. One way could be through a steady search for new genes among wild strawberry populations, which could then be employed through classical breeding or recombinant DNA technology to develop plants with better built-in resistance to adverse conditions.

14.3.1. Classical breeding

It is not always possible to find the desired characteristics in commercial varieties of the hybrid *F.* × *ananassa*; therefore, searching for and characterization of wild genotypes is required. The species mostly used as sources of genetic variability are the octoploid species *F. chiloensis* and *F. virginiana*, as these species are of the same ploidy level and easily can be crossed with *F.* × *ananassa*. These species act as a source of biotic and abiotic stress-resistance genes because of their geographical distribution over a wide area and having been subjected to selection pressure in extreme environments. These genotypes therefore have a long history of involvement in strawberry improvement programmes. Albert Etter, in the early 20th century (1903–1920) in California, obtained more than 50 strawberry cultivars by crossing plants of *F.* × *ananassa* with wild accessions of *F. chiloensis*, while C.L. Powers and A.C. Hildreth, from the US Department of Agriculture, Beltsville, Maryland, used *F. virginiana* subsp. *glauca* in their programmes (1930s and 1940s). R.S. Bringhurst and V. Voth, from the University of California, used *F. virginiana* to produce day-neutral varieties, and used *F. chiloensis* to increase fruit size in short-day cultivars (Hancock and Luby, 1993). Thus, strawberry cultivars were developed by plant breeders to fit particular environmental or marketing niches, and generally, no single cultivar is grown worldwide or even nationwide.

According to Sjulin and Dale (1987), *Fragaria* spp. that have shown cold tolerance are *F. viridis* (Europe, Asia), *F. daltoniana* J. Gay (Himalayas), *F. nipponica* Makino (Japan), *F. mandshurica* Staudt (Manchuria), *F. orientalis* Losinsk (West Asia), *F. moupinensis* (Fr.) *Card.* (South-west China) and *F. moschata* (North Europe). *F. virginiana* (North America) has been shown to possess disease resistance (red stele, *Verticillium* wilt, powdery mildew, leaf spot, scorch, leaf blight), nematode resistance (root-knot nematode, root-lesion nematode), stress tolerance (waterlogged soil, heat, drought, high soil pH, cold, frost) and photosynthetic efficiency at high temperatures. Similarly, *F. chiloensis* (North and South America) has been shown to possess disease resistance (red stele, *Verticillium* wilt, powdery mildew, leaf blight, viruses), nematode resistance (root-lesion nematode, pest resistance (two-spotted mite, strawberry weevil, black vine weevil) and stress tolerance (drought, high soil pH, salinity, cold). In relation to pest and disease resistance, *F. vesca* could be a natural genetic source of resistance against important diseases, such as powdery mildew, *Verticillium* wilt, and root and crown rot (Gooding *et al.*, 1981; Hancock and Luby, 1993; Korbin, 2011). Powdery mildew-resistant *F. moschata* and red stele-, powdery mildew- and leaf spot-resistant clones of *F. chiloensis* could act as donors for these traits (Hancock *et al.*, 1989).

The US National Clonal Germplasm Repository in Corvallis, Oregon, houses a collection of native germplasms from *F. virginiana* and *F. chiloensis*, which have been characterized for resistance against several foliar diseases, black root rot, root-knot nematode (*Meloidogyne hapla*) and root-lesion nematode (*Pratylenchus penetrans*). Studies have been able to identify various genotypes resistant to a number of important pathogens (Hancock *et al.*, 2003). Similarly, extensive studies carried out at the US Department of Agriculture helped to identify various sources of resistance to different pathogens. For example, they found resistance to *Xanthomona fragariae* in a clone of *F. virginiana* from Minnesota and in a hybrid between *F. virginiana* and *F.* × *ananassa* (Maas *et al.*, 2000), and identified germplasm tolerant to red stele (*Phytophthora fragariae*), *Verticillium* wilt, leaf spot (*Mycosphaerella fragariae*), leaf scorch (*Diplocarpon earlianum*), leaf blight (*Phomopsis obscurans*), powdery mildew (*Sphaerotheca macularis* f. sp. *fragariae*), fruit rot or grey mould (*Botrytis cinerea*) and crown rot (Galletta *et al.*, 1997). For successful use of these wild germplasms to be implemented, much effort will be needed in terms of collection, maintenance and particularly morphological, physiological and biochemical characterization of the different accessions collected in the areas of origin. Detailed information needs to be generated regarding genotypic and phenotypic characterization of collections in germplasm banks and the study of hereditary mechanisms involved in each characteristic. This wild material has great potential to be used as source of genes for developing varieties with enhanced adaptive abilities.

14.3.2. Advanced genetics

Recombinant DNA technology can play a significant role in enabling strawberry farmers to develop plants that cope with the challenges of climate change. Genetic modification techniques are becoming increasingly feasible due to improved techniques and demystification of the various side effects of transgenic technology (Husaini, 2010; Husaini *et al.*, 2011b; Husaini and Tuteja, 2012).

Agrobacterium-mediated gene transfer is the most widely used method for developing transgenic strawberry plants (Husaini *et al.*, 2011a). This genetic transformation protocol has been standardized and perfected for higher transformation success (Husaini and Srivastava, 2006a,b; Husaini, 2010). Several traits that confer adaptive advantage to strawberry plants with regard to global climate change have been improved using this genetic transformation technique (Table 14.1). Transgenic strawberry plants

Table 14.1. Genes conferring an adaptive advantage to transgenic strawberry plants against biotic and abiotic stresses.

Gene	Resistance against:	Cultivar	Reference
Fungal resistance			
Chitinase from rice (*rcc2*)	*Sphaeroteca humuli*	'Toyonoka '	Asao *et al.* (1997)
Chitinase from *Lycopersicon chilense* (*pcht28*)	*Verticillium dahliae*	'Joliette '	Chalavi *et al.* (2003)
Thaumatin II from *Thaumatococcus danielli* (*thauII*)	*Botrytis cinerea*	'Firework'	Schestibratov and Dolgov (2005)
Chitinase and glucanase from *Trichoderma harzianum*	*Colletotrichum acutatum*	'Camarosa'	Mercado *et al.* (2005)
Chitinase from *Phaseolus vulgaris* (*ch5B*), Glucanase (*gln2*) and thaumatin-like protein (*ap24*) from *Nicotiana tabacum*	*Botrytis cinerea* and *Colletotrichum acutatum*	'Pajaro'	Porcel *et al.* (2006)
Bgn13.1 from *Trichoderma harzianum*	*Colletotrichum acutatum* and *Rosellinia necatrix*	'Camarosa'	Mercado *et al.* (2015)
AtNPR1 from *Arabidopsis thaliana*	*Anthracnose, Powdery mildew*	*Fragaria vesca*[a]	Silva *et al.* (2015)
Insect resistance			
Cowpea protease inhibitor (*CpTi*)	*Otiorynchus* spp.	'Rapella'	James *et al.* (1992, 1993)
		'Symphony'	Graham *et al.* (1995)
Virus resistance			
Coat protein	Strawberry mild yellow edge virus		Finstad and Martin (1995)
Cold tolerance			
Cold-induced transcription factor (*CBF1*) from *Arabidopsis*		'Honeoye'	Owens *et al.* (2002, 2003)
Acidic dehydrin from wheat (*wcor410a*)		'Chambly'	Houde *et al.* (2004)
Antifreeze protein from Antarctic fish (*AFP*)		'Tioga'	Khammuang *et al.* (2005)
Salt tolerance			
Late embryogenesis abundance (*LEA3*) from barley		'Toyonaka'	Wang *et al.* (2004)
Osmotin from *Nicotiana tabacum*		'Chandler'	Husaini and Abdin (2008)
Drought tolerance			
Osmotin from *Nicotiana tabacum*		'Chandler'	Husaini *et al.* (2012)
P5CS from *Nicotiana tabacum*		'Camaros', 'Kurdistan'	Bahramnejad *et al.* (2015)

[a]*Fragaria vesca* is diploid and is not a cultivar of *F.* × *ananassa*.

overexpressing *pcht28* chitinase from *Lycopersicon chilense* show enhanced resistance to *Verticillum* wilt (*Verticillium dahliae*) (Chalavi *et al.*, 2003), while plants containing a rice chitinase gene (*RCC2*) exhibit improved resistance to powdery mildew (*Sphaerotheca humuli*) (Asao *et al.*, 1997). There are some reports that present the possibility of using pathogenesis-related (*PR*) genes to decrease the severity of fungal diseases (*Rhizoctonia solani*, *Phytopthora fragariae*, *B. cinerae*) in strawberry (Schestibratov and Dolgov, 2005; Vellicce *et al.*, 2006). In a recent study, Mercado *et al.* (2015) expressed the gene *bgn13.1* from *Trichoderma harzianum* CECT 2413 in strawberry plants and reported that the constitutive expression of *bgn13.1* improved strawberry tolerance to crown rot diseases caused by *Colletotrichum acutatum* or *Rosellinia necatrix*. In an interesting recent study, Silva *et al.* (2015) showed that ectopic expression of *AtNPR1* from *Arabidopsis thaliana* in strawberry increased resistance to anthracnose, powdery mildew and angular leaf spot, which are caused by different fungal or bacterial pathogens. Consistently, most of the transgenic lines constitutively expressed the defence gene *FvPR5*, suggesting that the systemic acquired resistance activation mechanisms in strawberry and *Arabidopsis* are different. Nevertheless, these findings indicate that increasing the transcript levels of *AtNPR1* or its strawberry orthologues may be a promising approach for development of strawberry cultivars with enhanced broad-spectrum disease resistance.

Genes conferring tolerance to viruses, such as strawberry mild yellow edge virus coat protein (Finstad and Martin, 1995) and cowpea trypsin inhibitor (*CpTi*) (Graham *et al.*, 1995) have also been transferred successfully into strawberry. Coincidentally, the overexpression of *CpTi* also resulted in significantly higher root mass than in control plants.

Transgenic strawberry with enhanced salinity tolerance has been developed using the late embryogenesis abundance (*LEA3*) gene (Wang *et al.*, 2004) and osmotin gene (Husaini and Abdin, 2008). Houde *et al.* (2004) transferred the acidic dehydrin gene of wheat, *Wcor410a*, into strawberry, and this modification prevented membrane injury and greatly improved frost resistance in the leaves. The transfer of cold-induced transcription factor (*CBF1*) and antifreeze protein genes into strawberries has also been reported (Owens *et al.*, 2002; Khammuang *et al.*, 2005). Transfer of the *Nicotiana* osmotin gene into strawberry has been shown to confer drought tolerance to the transgenic plants of strawberry cultivar 'Chandler' (Husaini *et al.*, 2012). Recently, two economically important cultivars of strawberry ('Camarosa' and 'Kurdistan') were transformed with the *P5CS* gene, which encodes Δ1-pyrroline-5-carboxylate synthetase, the key enzyme in proline biosynthesis (Bahramnejad *et al.*, 2015). Overproduction of P5CS increased proline content, chlorophyll content, shoot length, shoot fresh and dry weight in the transgenic plants under drought-stress conditions.

Under climate change conditions, crop plants will often experience more than one biotic and abiotic stress. An array of genes, available for use in both cisgenic and transgenic approaches, can be used for the development of varieties with better resilience to the vagaries of climate. Transgenes encoding reactive oxygen species scavenger proteins, transcription factors and protein kinases will be the most suitable candidate genes for conferring adaptive advantages to strawberry against multiple stresses. Moreover, a good strategy would be to use major-effect multirole genes to develop climate resilience in strawberry. In our opinion, one such major-effect multirole gene is osmotin (from *Nicotiana tabacum*). A brief description of the possible role of this gene in developing plants with better resilience through a single gene transfer is described below.

Osmotin: a multirole PR-5 protein

Osmotin can be used effectively in designing plants resilient to multiple stresses (Husaini and Rafiqi, 2012). Osmotin is a stress-responsive multifunctional tobacco PR-5c protein isolated from tobacco (*N. tabacum* var. Wisconsin 38; Singh *et al.*, 1985). The tertiary structure of osmotin is similar to thaumatin and other PR proteins, and is composed of three domains (Min *et al.*, 2004). Domain I consists of an 11-strand, flattened β-sandwich (residues 1–53, 82–125 and 175–205), which forms the compact core of the molecule. Domain II consists of several loops extending from domain I and is stabilized by four disulphide bonds (residues 126–174). Domain III consists of a small loop (residues 54–81) with two disulphide bonds. Osmotin has a pronounced cleft formed by domains I and II.

Osmotin acts as compatible osmolyte that enhances the osmotic potential of the cells and upregulates the levels of proline (Barthakur *et al.*, 2001; Sarad *et al.*, 2004; Sokhansanj *et al.*, 2006; Husaini and Abdin, 2008), another compatible osmolyte and scavenger of reactive oxygen species. Identification of an osmotin-like protein from the intercellular space of the halophyte

Mesaembryanthemum crystallinum and the association of osmotin with the tonoplast of tobacco (Singh *et al.*, 1987; Yen *et al.*, 1994) indicate that osmotin plays a role in the intracellular compartmentation of Na^+ ions to both the intercellular space and the vacuole, thereby minimizing the build-up of Na^+ ions in the cytoplasm (Yen *et al.*, 1994).

In addition to providing salt tolerance (Singh *et al.*, 1989; Bol *et al.*, 1990; Zhu *et al.*, 1993), drought tolerance (Parkhi *et al.*, 2009; Husaini *et al.*, 2012) and cryoprotection (D'Angeli and Altamura, 2007), osmotin also provides protection from fungal pathogens (Raghothama *et al.*, 1993; Zuker *et al.*, 2001). The protein has shown antifungal activity *in vitro* against a broad range of fungal pathogens (Yun *et al.*, 1998). Specifically, osmotin and osmotin-like proteins have demonstrated antifungal activity against *Phytophthora infestans* (Woloshuk *et al.*, 1991; Zhu *et al.*, 1993; Liu *et al.*, 1994), *Phomopsis viticola* and *B. cinerea* (Monteiro *et al.*, 2003), *Fusarium solani*, *Neurospora* sp. and *Colletotrichum gloeosporioides* (de Freitas *et al.*, 2011a). The fungal growth inhibition by osmotin and osmotin-like proteins is correlated with plasma membrane permeabilization and dissipation of the plasma membrane potential of sensitive fungi (Abad *et al.*, 1996; de Freitas *et al.*, 2011b) (Fig 14.2).

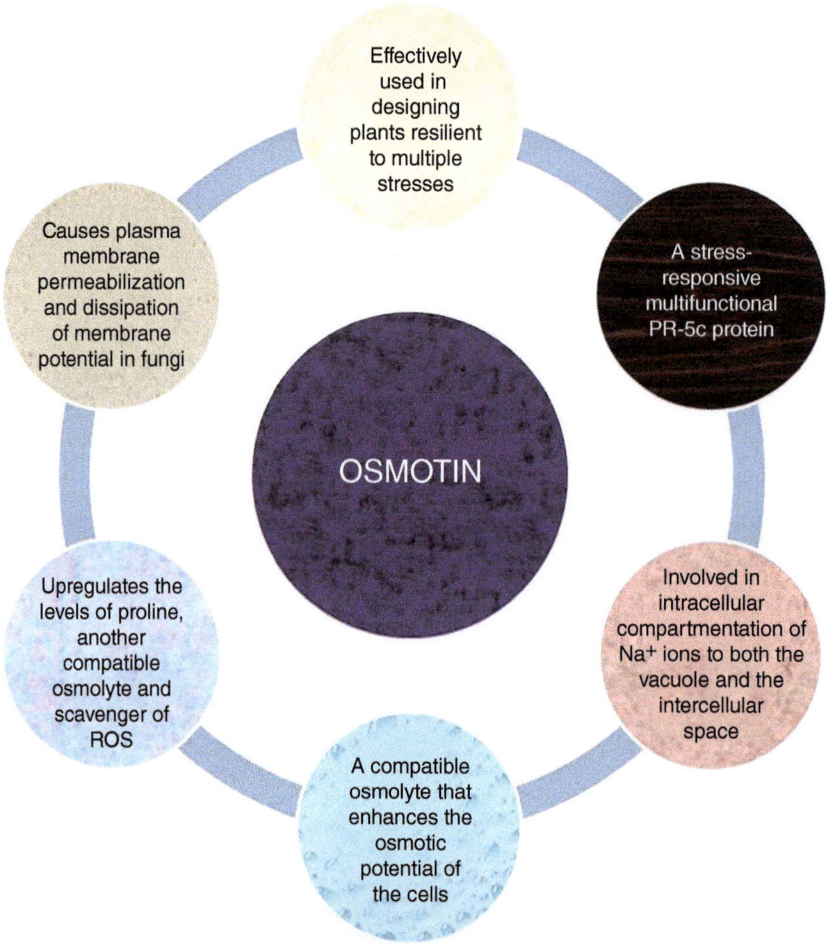

Fig. 14.2. Roles of osmotin.

Clustered regulatory interspaced short palindromic repeats (CRISPRs) and CRISPR-associated (Cas) systems are other emerging techniques that can be employed to introduce useful genome modifications by genome engineering of plants (Sander and Young, 2014). However, no published report about the use of this technology in strawberry is available so far.

14.4. Microorganisms for Enhancing Resilience

The European Commission's working document on *Adapting to Climate Change: the Challenge for European Agriculture and Rural Areas*, published in April 2009, states that high water-stress areas are expected to increase from 19 to 35% by 2070, implying 'significant changes in the quality and availability of water resources' (http://ec.europa.eu/health/ph_threats/climate/docs/com_2009_147_en.pdf). The predicted severe drought conditions will also affect the soil nutrient status. Drought will lead to diminished plant uptake of nutrients such as phosphorus and potassium, causing an additional stress on plants, as both phosphorus and potassium are necessary for high water-use efficiency and stomatal control (Sardans and Peñuelas, 2007). This will influence crop establishment, the crop stand maturation period and total production. Symbiotic soil microorganisms can play a crucial role in buffering crop production against these adverse effects (Husaini, 2014).

Mycorrhizal technology has been used to improve growth of a number of micropropagated horticultural crops, as well as to enable host plants to tolerate or withstand the impairing effects of abiotic and biotic stresses (Borkowska, 2002; Yin *et al.*, 2010). Mycorrhization strongly affects growth and tolerance of plants to water deficiency. Mycorrhizal fungi influence the development of a superior root system, enhance the water-conducting capacity, increase the uptake of macro-, micro- and immobile nutrients, and result in higher photosynthetic rates due to better carbon dioxide assimilation (El-Tohamy *et al.*, 1999; Estrada-Luna *et al.*, 2000; Augé, 2001).

14.4.1. Arbuscular mycorrhizal fungi

About 90% of vascular plants establish a symbiotic relationship with arbuscular mycorrhizal fungi (AMF) (Gai *et al.*, 2006). The association between AMF and a plant makes the host plant more tolerant to drought in terms of plant growth (Ruiz-Lozano, 2003; Wu and Xia, 2006; Bolandnazar *et al.*, 2007). Generally, AMF-colonized plants have higher activities of several antioxidant enzymes compared with non-colonized plants (Wu *et al.*, 2006ab), but this depends on the enzyme activity, plant organ and AMF genotype involved (Lambais *et al.*, 2003; Wu *et al.*, 2006ab). One of the main benefits of AMF symbiosis for strawberry plants is increased phosphorus uptake under both low-phosphorus fertigation regimes (Boyer *et al.*, 2015) and a high-phosphorus soil environment (Stewart *et al.*, 2005). Opinion on the beneficial effects of AMF inoculation on strawberry is not, however, unanimous. Some studies have reported the beneficial effects of mycorrhiza on strawberry plant growth and yield (Hršelová *et al.*, 1989; Niemi and Vestberg, 1992; Borkowska, 2002; Vestberg *et al.*, 2004; Stewart *et al.*, 2005; Castellanos-Morales *et al.*, 2010; Fan *et al.*, 2011). Yin *et al.* (2010) showed that inoculation of AMF in drought-stressed strawberry plants increased the enzymes associated with the plant's protective system against water stress. Borkowska (2002) indicated that application of AMF strongly affected growth and tolerance to drought stress of the strawberry cultivar 'Senga Sengana' under greenhouse conditions. However, other studies have shown either limited (Garland *et al.*, 2011) or no (Vestberg *et al.*, 2004) beneficial effects. Borowicz (2010) examined the effect of AMF inoculation on water stress in wild strawberry (*F. virginiana*) and reported no strong effects of AMF on strawberry tolerance to drought.

In a recent study, it was found that the water-use efficiency of AMF-colonized strawberry plants was greater under a regulated deficit irrigation (water-stressed) regime than under a fully watered regime (Boyer *et al.*, 2015). The most notable consequence of inoculation with AMF (*Funneliformis mosseae* BEG25, *Funneliformis geosporus* BEG11 or a 50:50 mixed inoculation) was a significant increase in the survival time of the plants when water was not provided: mycorrhizal plants survived on average 4.4 days longer than control plants. Inoculation with AMF increased all growth characteristics of the strawberry plants measured under both fully watered and water-stressed conditions, regardless of single- or mixed-species inoculation. The addition of AMF inocula (in any combination) to plants subjected to reduced irrigation of up to 40% restored plant growth back to the same or higher values as the non-mycorrhizal, fully watered plants.

In an interesting study, Matsubara *et al.* (2004) investigated the tolerance to *Fusarium* wilt (*Fusarium oxysporum* f. sp. *fragariae*) in response to inoculation with five AMF species (*Glomus margarita*, *Glomus fasciculatum*, *Glomus mosseae*, *Glomus aggregatum*, *Glomus* sp. R10) in strawberry. At 30 days after pathogen inoculation, *Fusarium* wilt incidence ranged from 22% in *G. mosseae* plots to 100% in non-mycorrhizal plots. The incidence and severity of infected roots and vessels were lower in AMF-inoculated than in non-inoculated plots. Non-inoculated and AMF-inoculated plants had higher shoot and root weights compared with non-mycorrhizal plants. The authors claimed that inoculation with AMF improved *Fusarium* wilt tolerance in strawberry.

14.4.2. Vesicular–arbuscular mycorrhiza

Many studies have shown that the drought resistance of crops can be improved by vesicular–arbuscular mycorrhiza (VAM) (Sylvia and Williams, 1992; Al-Karaki, 1998; Li *et al.*, 1999; Liang *et al.*, 2003; Lu *et al.*, 2003; Zhang and He, 2007). Recently, VAM were reported to improve plant drought resistance abilities in strawberry. VAM fungi inoculation slowed down the reduction of chlorophyll and increased the drought resistance of plants by promoting the defence response of the protective enzyme system in host plants (Yin *et al.*, 2010). The formation of VAM enhanced the activity of the antioxidant enzymes superoxide dismutase, PRX and CAT, and H+-ATPase in host plants, and reduced the content of malondialdehyde and electrical conductivity of the plasma membrane significantly (Wu and Xia, 2003; Wu *et al.*, 2007), but the exact mechanism of enhancing drought resistance in plants is still not clear.

14.4.3. Plant growth-promoting rhizobacteria

Plant growth-promoting rhizobacteria (PGPR) are free-living bacteria present in the rhizosphere or on the root surface, or inhabiting spaces between cortical cells (Gray and Smith, 2005). Several free-living soil bacteria induce drought tolerance in plants (Arkhipova *et al.*, 2007; Sziderics *et al.*, 2007). Timmusk and Wagner (1999) report that inoculation of *Arabidopsis* plants with the PGPR *Paenibacillus polymyxa*, prior to drought stress, caused an elevation in mRNA copies of abscisic acid-related genes (*ERD15* and *RAB18*), and that these plants were more tolerant to drought than non-inoculated ones. The authors hypothesized that inoculation with the bacteria caused a mild biotic stress, preparing the plants to cope with subsequent drought stress. Sziderics *et al.* (2007) reported that, after a mild osmotic stress, pepper plants inoculated with several PGPRs showed lower expression of genes related to abiotic stresses than non-inoculated plants, probably because the inoculated plants suffered less stress and therefore had a lower expression of stress-related genes.

Recently, there was a report discussing the effects of five PGPRs (*Bacillus subtilis* EY2, *Bacillus atrophaeus* EY6, *Bacillus sphaericus* GC subgroup B EY30, *Staphylococcus kloosii* EY37, and *Kocuria erythromyxa* EY43) on the growth, chlorophyll content, nutrient element content and yield of strawberry plants under natural field salinity stress conditions (Karlidag *et al.*, 2009). PGPR inoculation significantly increased the growth, chlorophyll content, nutrient element content and yield of the strawberry plants, while lowering the electrolyte leakage of the plants under saline conditions. The highest efficiency in terms of alleviating salinity stress on the yield and nutrient uptake of strawberry plants was obtained from *K. erythromyxa* EY43 (228 g per plant) and *Staphylococcus kloosii* EY37 (225 g per plant) treatment, and the yield increasing ratio of plants was 48% for *K. erythromyxa* EY43 and 46% for *Staphylococcus kloosii* EY37 compared with the control treatment (154 g per plant).

14.4.4. *Piriformospora indica*

Piriformospora indica (Basidiomycota) is an axenically cultivable phytopromotional endosymbiont, which mimics the capabilities of AMF and shows pronounced growth-promotional effects (Varma *et al.*, 1999, 2001). *Piriformospora indica* promotes the synthesis and expression of defence-related proteins and enzymes (isoflavonoid phytoalexins, isoflavonoid glyceollin coumestrol, coumestin isosojagol) that provide protection and result in higher plant survival (Sahay and Varma, 1999, 2000). In addition, *Piriformospora indica* produces significant amounts of acid phosphatases for the mobilization of a broad range of insoluble, condensed or complex forms of phosphate, enabling the host plant the accessibility to adequate phosphorus from immobilized reserves in the soil (Sahay and Varma, 2000).

Piriformospora indica is a root-colonizing endophytic fungus with a wide host range and helps plants to grow under extreme physical and nutrient stress conditions. This symbiotrophic fungal endophyte was discovered in the Great Indian Desert of Western Rajasthan in India (Varma *et al.*, 1999). The fungus colonizes the roots of the desert plants growing under water-scarcity conditions. It has been established that the fungus promotes the acquisition of drought tolerance in plants (Waller *et al.*, 2005; Sherameti *et al.*, 2008). Strawberry plants inoculated with *Piriformospora indica* or *Sebacina vermifera* show a higher percentage of survival under drought conditions with increased chlorophyll contents, total soluble proteins and relative water content, and higher biomass. *Piriformospora indica*-inoculated plants showed a markedly higher tolerance level than *Sebacina vermifera*-inoculated plants (Husaini *et al.*, 2012). The use of *in vitro* and *ex vitro* co-culture systems verified that *Piriformospora indica* possesses pronounced plant growth-stimulating capacity and induces plant secondary metabolite accumulation (Rai *et al.*, 2001; Baldi *et al.*, 2008).

Piriformospora indica has a wide geographical distribution and is reported from Asia, Australia and South America. The fungus seems to have promising potential under changing climate scenarios because it functions as a biofertilizer in nutrient-deficient soils, as a bioprotector against biotic and abiotic stresses, as a bioregulator for plant growth development and enhanced seed production, and as a bioagent for the hardening of plants (Oelmüller *et al.*, 2009).

14.5. Extreme Events, Policy Response and Social Action

Although the climate is warming, the incidence of inter-annual climate variability (extreme events) accompanying the mean climate changes is actually increasing. It has been widely accepted by the Intergovernmental Panel on Climate Change that increases in the frequency of climate extremes may lower crop yields beyond the impacts of 'mean' climate change (Easterling and Apps, 2005; Easterling *et al.*, 2007). According to the World Development Report on *Development and Climate Change* commissioned by the World Bank in 2009, the public in the 15 countries polled (both developed and developing) saw climate change as a serious problem: either as very serious or somewhat serious (WorldPublicOpinion.org, 2010).

From a social and economic point of view, climate impact studies should take into consideration the geographical dimensions of the area of study, as different climatic patterns affect regions differently, as well as local policies and legislations. For example, the introduction of polytunnels (in the mid-1990s) was one of the most important factors affecting the expansion of strawberry cultivation in the UK from being a minor crop to becoming an industry with a specialist supply chain. The use of polytunnels decreased the impact of weather on the crop, thus bridging the difference in yields obtained between Scottish and English farms. As a result, growers in areas previously disadvantaged by the weather could now grow strawberries competitively. This brought a shift in the geography of strawberry cultivation in the UK. However, damage to polytunnels by extreme weather events such as floods and wind could have a worse impact on strawberry growers in the absence of crop insurance; thus, any damage to the structures would result in a year's worth of investment being lost in one event, the cost of which would be borne by the grower. Crop insurance cover may be a viable solution for strawberry growers, which they may opt for voluntarily. However, policy support or legislation can often act as a driver for such changes. For instance, according to Calleja (2011), in Kent, UK, the county council puts water bans on agricultural enterprises during periods of drought, due to them being close to a large metropolitan area where a constant water supply is needed for human consumption. In response to this, growers in this area have been building water-storage facilities to eliminate their dependency on the council's supply of water. This is an interesting case, whereby the growers have not built water-storage facilities as a conscious response to climate change, but are responding to pressures caused by man-made policies in response to water shortages during periods of drought.

Keeping this in mind, there can be positive ways of using the increased soil temperature in strawberry cultivation. With increasing temperature, soil solarization may become a more feasible option for pest control. The drawback of solarization is that it is restricted in some areas by the length and intensity of sunshine, as well as by the temperature (Berg, 2007). Solarization is recommended for 30–45 days during the months when the soil temperatures exceed 50°C, although when combined with biofumigation, the soil temperature could be as low as 40°C (Medina-Mínguez, 2002). Solarized soils are often more suppressive to certain soilborne pathogens than non-solarized soils, while *Trichoderma* spp., which act as biocontrol agents against several diseases, are among the microorganisms that can survive the solarization process (Pinkerton *et al.*, 2002).

14.6. Future Perspective

The projections of global climate changes on the earth predict a rise in the concentration of greenhouse gases, an increase in temperature, increased frequency of extreme weather events and aridization of the environment. The implications of these changes on horticulture in general, and on the ability of plants such as strawberry in particular, to adapt to such changes need rigorous analysis. With a view to ensuring the sustainability of strawberry cultivation under these changing climatic conditions, the possible roles of different technological interventions need to be evaluated. The diploid strawberry *F. vesca* is an attractive model plant for genomic analysis because of its small genome. The *F. vesca* EST collection and recent data from proteomic studies in abiotic stresses could also be very useful for further investigations. Unlike in the past, multidisciplinary studies focusing on the effect of multiple stress factors are necessary to understand the impact of stressful conditions on plant growth and development in the open field. Scientists with expertise in climatology, biochemistry, microbiology, pathology, entomology, agricultural physics, plant breeding, biotechnology and systems biology need to collaborate and develop multipronged strategies to address these challenges. Such research studies should not only be based on the current climate conditions but also planned according to the potential future challenges.

Using genetic engineering approaches, transgenic strawberry plants with tolerance against certain stress factors have already been developed. However, these studies are mostly at the laboratory level, and there is an urgent need to test and evaluate these plants under field conditions. A detailed understanding of the complex mechanisms involved the genome, epigenome, transcriptome and metabolome levels will be useful in designing better strategies for improving stress tolerance in plants. Screening studies, investigating the stress tolerance of genotypes for each biotic/abiotic stress factor, should also be undertaken in parallel for successful breeding programmes. Establishment of a symbiotic relationship between microorganisms and plants also shows promising potential in enhancing the tolerance of plants to drought and temperature stresses. The beneficial effect of mycorrhization on plant drought tolerance, resulting from the development of a superior root system, enhanced water-conducting capacity and increased uptake of macro-, micro- and immobile nutrients, will be very helpful in understanding the impact on carbon dioxide assimilation and higher photosynthetic rates. Through these technologies, we can aim to develop strawberry plants that have better water- and nutrient-use efficiencies, and that are resilient to high or low temperatures, and to water scarcity or flooding.

References

Abad, L.R., D'Urzo, M.P., Liu, D., Narasimhan, M.L., Reuveni, M., Zhu, J.K., Niu, X., Singh, N.K., Hasegawa, P.M. and Bressan, R.A. (1996) Antifungal activity of tobacco osmotin has specificity and involves plasma membrane permeabilization. *Plant Science* 118, 11–23.

Aggarwal, P.K., Banerjee, B., Daryaei, M.G., Bhatia, A., Bala, A., Rani, S., Chander, S., Pathak, H. and Kalra, N. (2006) InfoCrop: a dynamic simulation model for the assessment of crop yields, losses due to pests, and environmental impact of agro-ecosystems in tropical environments. II. Performance of the model. *Agricultural Systems* 89, 47–67.

Ahmadi, H., Bringhurst, R.S. and Voth, V. (1990) Modes of inheritance of photoperiodism in *Fragaria*. *Journal of the American Society for Horticultural Science* 115, 146–152.

Al-Karaki, G. (1998) Benefit, cost and water-use efficiency of arbuscular mycorrhizal durum wheat grown under drought stress. *Mycorrhiza* 8, 41–45.

Arkhipova, T.N., Prinsen, E., Veselov, S.U., Martinenko, E.V., Melentiev, A.I. and Kudoyarova, G.R. (2007) Cytokinin producing bacteria enhance plant growth in drying soil. *Plant and Soil* 292, 305–315.

Arora, R., Wisniewski, M.E. and Scorza, R. (1992) Cold acclimation in genetically related (sibling) deciduous and evergreen peach (*Prunus persica* [L.] Batsch): I. seasonal changes in cold hardiness and polypeptides of bark and xylem tissues. *Plant Physiology* 99, 1562–1568.

Asao, H., Nishizawa, Y., Arai, S., Sato, T., Hirai, M., Yoshida, K., Shinmyo, A. and Hibi, T. (1997) Enhanced resistance against a fungal pathogen *Sphaerotheca humuli* in transgenic strawberry expressing a rice chitinase gene. *Plant Biotechnology* 14, 145–150.

Augé, R.M. (2001) Water relations, drought and vesicular-arbuscular mycorrhizal symbiosis. *Mycorrhiza* 11, 3–42.

Bahramnejad, B., Nasri, S., Mozafari, A. and Siosemardeh, A. (2015) *Agrobacterium tumefaciens*-mediated transformation of two economically important strawberry cultivars with *P5CS* gene. *Journal of Plant Physiology and Breeding* 5, 55–66.

Baldi, A., Jain, A., Gupta, N., Srivastava, A. and Bisaria, V. (2008) Co-culture of arbuscular mycorrhiza-like fungi (*Piriformospora indica* and *Sebacina vermifera*) with plant cells of *Linum album* for enhanced production of podophyllotoxins: a first report. *Biotechnology Letters* 30, 1671–1677.

Barney, D.L., Davis, B.B. and Fellman, J.K. (1992) *Strawberry Production: Overview*. College of Agriculture, University of Idaho, Moscow, Idaho.

Barthakur, S., Babu, V. and Bansal, K.C. (2001) Over-expression of osmotin induces proline accumulation and confers tolerance to osmotic stress in transgenic tobacco. *Journal of Plant Biochemistry and Biotechnology* 10, 31–37.

Berg, G. (2007) Biological control of fungal soilborne pathogens in strawberries. In: Chincholkar, S.B. and Mujerki, K.G. (eds) *Biological Control of Plant Diseases*. Haworth Press, Binghamton NY, pp. 1–17.

Bergot, M., Cloppet, E., Pérarnaud, V., Déqué, M., Marçais, B. and Desprez-Loustau, M.L. (2004) Simulation of potential range expansion of oak disease caused by *Phytophthora cinnamomi* under climate change. *Global Change Biology* 10, 1–14.

Blanke, M.M. and Cooke, D.T. (2006) Water channels in strawberry, and their role in the plants response to water stress. *Acta Horticulturae* 708, 65–68.

Bol, J.F., Linthorst, H.J.M. and Cornelissen, B.J.C. (1990) Plant pathogenesis-related proteins induced by virus infection. *Annual Review of Phytopathology* 28, 113–138.

Bolandnazar, S., Aliasgarzad, N., Neishabury, M.R. and Chaparzadeh, N. (2007) Mycorrhizal colonization improves onion (*Allium cepa* L.) yield and water use efficiency under water deficit condition. *Scientia Horticulturae* 114, 11–15.

Booth, T.H., Jovanovic, T., Old, K.M. and Dudzinski, M.J. (2000) Climatic mapping to identify high-risk areas for *Cylindrocladium quinqueseptatum* leaf blight on eucalypts in mainland South East Asia and around the world. *Environmental Pollution* 108, 365–372.

Borkowska, B. (2002) Growth and photosynthetic activity of micropropagated strawberry plants inoculated with endomycorrhizal fungi (AMF) and growing under drought stress. *Acta Physiologiae Plantarum* 24, 365–370.

Borowicz, V.A. (2010) The impact of arbuscular mycorrhizal fungi on strawberry tolerance to root damage and drought stress. *Pedobiologia* 53, 265–270.

Boyer, L.R., Brain, P., Xu, X.M. and Jeffries, P. (2015) Inoculation of drought-stressed strawberry with a mixed inoculum of two arbuscular mycorrhizal fungi: effects on population dynamics of fungal species in roots and consequential plant tolerance to water deficiency. *Mycorrhiza* 25, 215–227.

Bray, E.A., Bailey-Serres, J. and Weretilnyk, E. (2000) Responses to abiotic stresses. In: Gruissem, W., Buchnnan, B. and Jones, R. (eds) *Biochemistry and Molecular Biology of Plants*. American Society of Plant Physiologists, Rockville, Maryland, pp. 1158–1249.

Bringhurst, R.S. and Voth, V. (1984) Breeding octoploid strawberries. *Iowa State Journal of Research* 58, 371–381.

Buchanan, B.B., Gruissem, W. and Jones, R.L. (2000) *Biochemistry and Molecular Biology of Plants*, American Society of Plant Physiologists, Rockville, Maryland.

Calleja, E.J. (2011) The potential impacts of climate change on diseases affecting strawberries and the UK strawberry industry. PhD thesis, University of Warwick, Coventry, UK.

Cansev, A., Gulen, H. and Eris, A. (2009) Cold-hardiness of olive (*Olea europaea* L.) cultivars in cold-acclimated and non-acclimated stages: seasonal alteration of antioxidative enzymes and dehydrin-like proteins. *Journal of Agricultural Science* 147, 51–61.

Castellanos-Morales, V., Villegas, J., Wendelin, S., Vierheilig, H., Eder, R. and Cárdenas-Navarro, R. (2010) Root colonisation by the arbuscular mycorrhizal fungus *Glomus intraradices* alters the quality of strawberry fruits (*Fragaria* × *ananassa* Duch.) at different nitrogen levels. *Journal of the Science of Food and Agriculture* 90, 1774–1782.

Castro, P., Bushakra, J.M., Stewart, P., Weebadde, C.K., Wang, D., Hancock, J.F., Finn, C.E., Luby, J.J. and Lewers, K.S. (2015) Genetic mapping of day-neutrality in cultivated strawberry. *Molecular Breeding* 35, 1–16.

Çetinkaya, C. (2013) Physiological effects of drought and recovery treatments in some strawberry cultivars. MSc thesis, Uludag University, Bursa, Turkey.

Chakraborty, S., Murray, G. and White, N. (2002) *Impact of Climate Change on Important Plant Diseases in Australia*. Rural Industries Research and Development Corporation, Canberra, Australia.

Chalavi, V., Tabaeizadeh, Z. and Thibodeau, P. (2003) Enhanced resistance to *Verticillium dahliae* in transgenic strawberry plants expressing a *Lycopersicon chilense* chitinase gene. *Journal of the American Society for Horticultural Science* 128, 747–753.

Chandler, C.K. and Ferree, D.C. (1990) Response of 'Raritan' and 'Surecrop' strawberry plants to drought stress. *Fruit Varieties Journal* 44, 183–184.

D'Angeli, S. and Altamura, M. (2007) Osmotin induces cold protection in olive trees by affecting programmed cell death and cytoskeleton organization. *Planta* 225, 1147–1163.

de Freitas, C.D.T., Lopes, J.L.d.S., Beltramini, L.M., de Oliveira, R.S.B., Oliveira, J.T.A. and Ramos, M.V. (2011a) Osmotin from *Calotropis procera* latex: new insights into structure and antifungal properties. *Biochimica et Biophysica Acta – Biomembranes* 1808, 2501–2507.

de Freitas, C.D.T., Sousa Nogueira, F.C., Vasconcelos, I.M., Abreu Oliveira, J.T., Domont, G.B. and Ramos, M.V. (2011b) Osmotin purified from the latex of *Calotropis procera*: Biochemical characterization, biological activity and role in plant defense. *Plant Physiology and Biochemistry* 49, 738–743.

Easterling, W. and Apps, M. (2005) Assessing the consequences of climate change for food and forest resources: a view from the IPCC. In: Salinger, J., Sivakumar, M.V.K. and Motha, R.P. (eds) *Increasing Climate Variability and Change*. Springer, The Netherlands, pp. 165–189.
Easterling, W.E., Aggarwal, P.K., Batima, P., Brander, K.M., Erda, L., Howden, S.M., Kirilenko, A., Morton, J., Soussana, J.F., Schmidhuber, J. and Tubiello, F.N. (2007) Food, fibre, and forest products. In: Parry, M.L., Canziani, O.F., Palutikof, J.P., van der Linden, P.J. and Hanson, C.E. (eds) *Climate Change 2007: Impacts, Adaptation and Vulnerability: Contribution of Working Group II to the Fourth Assessment Report of the Intergovernmental Panel on Climate Change*. Cambridge University Press, Cambridge, pp. 273–313.
El-Tohamy, W., Schnitzler, W.H., El-Behairy, U. and El-Beltagy, M.S. (1999) Effect of VA mycorrhiza on improving drought and chilling tolerance of bean plants (*Phaseolus vulgaris* L.). *Angewandte Botanik* 73, 178–183.
Ergin, S. (2012) Effects of high temperature stress on enzymatic and non-enzymatic antioxidants and protein metabolism in strawberry plants. PhD Thesis, Uludag University, Bursa, Turkey.
Ergin, S., Kesici, M. and Gulen, H. (2012) Changes in H_2O_2 and peroxidase activities in strawberry plants under heat stress. *Harran University Journal of the Faculty of Agriculture* 16, 25–35.
Eris, A., Gulen, H., Barut, E. and Cansev, A. (2007) Annual patterns of total soluble sugars and proteins related to cold-hardiness in olive (*Olea europaea* L. 'Gemlik'). *Journal of Horticultural Science and Biotechnology* 82, 597–604.
Estrada-Luna, A.A., Davies F.T. Jr and Egilla, J.N. (2000) Mycorrhizal fungi enhancement of growth and gas exchange of micropropagated guava plantlets (*Psidium guajava* L.) during *ex vitro* acclimatization and plant establishment. *Mycorrhiza* 10, 1–8.
Evans, N., Baierl, A., Semenov, M.A., Gladders, P. and Fitt, B.D.L. (2008) Range and severity of a plant disease increased by global warming. *Journal of the Royal Society Interface* 5, 525–531.
Fan, L., Dalpé, Y., Fang, C., Dubé, C. and Khanizadeh, S. (2011) Influence of arbuscular mycorrhizae on biomass and root morphology of selected strawberry cultivars under salt stress. *Botany* 89, 397–403.
Fang, X., Phillips, D., Li, H., Sivasithamparam, K. and Barbetti, M.J. (2011) Comparisons of virulence of pathogens associated with crown and root diseases of strawberry in Western Australia with special reference to the effect of temperature. *Scientia Horticulturae* 131, 39–48.
FAO (2008) *Climate Change and Food Security: a Framework Document*. Food and Agriculture Organisation, Rome, Italy.
Finstad, K. and Martin, R.R. (1995) Transformation of strawberry for virus resistance. *Acta Horticulturae* 385, 86–90.
Francesca, S., Simona, G., Francesco Nicola, T., Andrea, R., Vittorio, R., Federico, S., Cynthia, R. and Maria Lodovica, G. (2006) Downy mildew (*Plasmopara viticola*) epidemics on grapevine under climate change. *Global Change Biology* 12, 1299–1307.
Gai, J.P., Christie, P., Feng, G. and Li, X.L. (2006) Twenty years of research on community composition and species distribution of arbuscular mycorrhizal fungi in China: a review. *Mycorrhiza* 16, 229–239.
Galletta, G.J., Maas, J.L., Finn, C.E., Smith, B.J. and Gupton, C.L. (1997) The United States Department of Agriculture strawberry breeding program. *Fruit Varieties Journal* 51, 204–210.
Garland, B.C., Schroeder-Moreno, M.S., Fernandez, G.E. and Creamer, N.G. (2011) Influence of summer cover crops and mycorrhizal fungi on strawberry production in the southeastern United States. *HortScience* 46, 985–991.
Garrett, K.A., Dendy, S.P., Frank, E.E., Rouse, M.N. and Travers, S.E. (2006) Climate change effects on plant disease: genomes to ecosystems. *Annual Review of Phytopathology* 44, 489–509.
Gaston, A., Perrotte, J., Lerceteau-Köhler, E., Rousseau-Gueutin, M., Petit, A., Hernould, M., Rothan, C. and Denoyes, B. (2013) *PFRU*, a single dominant locus regulates the balance between sexual and asexual plant reproduction in cultivated strawberry. *Journal of Experimental Botany* 64, 1837–1848.
Glover, J., Johnson, H., Lizzio, J., Wesley, V., Hattersley, P. and Knight, C. (2008) *Australia's Crops and Pastures in a Changing Climate – Can Biotechnology Help?* Australian Government Bureau of Rural Sciences, Canberra, Australia.
Gooding, H.J., McNicol, R.J. and MacIntyre, D. (1981) Methods of screening strawberries for resistance to *Sphaerotheca macularis* (Wall ex Frier) and *Phytophthora cactorum* (Leb. and Cohn). *Journal of Horticultural Science* 56, 239–245.
Graham, J., McNicol, R.J. and Greig, K. (1995) Towards genetic based insect resistance in strawberry using the cowpea trypsin inhibitor gene. *Annals of Applied Biology* 127, 163–173.

Grant, O.M., Johnson, A.W., Davies, M.J., James, C.M. and Simpson, D.W. (2010) Physiological and morphological diversity of cultivated strawberry (*Fragaria × ananassa*) in response to water deficit. *Environmental and Experimental Botany* 68, 264–272.

Gray, E.J. and Smith, D.L. (2005) Intracellular and extracellular PGPR: commonalities and distinctions in the plant-bacterium signaling processes. *Soil Biology & Biochemistry* 37, 395–412.

Gulen, H. and Eris, A. (2003) Some physiological changes in strawberry (*Fragaria × ananassa* 'Camarosa') plants under heat stress. *Journal of Horticultural Science and Biotechnology* 78, 894–898.

Gulen, H. and Eris, A. (2004) Effect of heat stress on peroxidase activity and total protein content in strawberry plants. *Plant Science* 166, 739–744.

Gulen, H., Çetinkaya, C., Kadıoğlu, M., Kesici, M., Cansev, A. and Eris, A. (2008) Peroxidase activity and lipid peroxidation in strawberry (*Fragaria × ananassa*) plants under low temperature. *Journal of Biological and Environmental Science* 2, 95–100.

Hancock, J.F. and Luby, J.J. (1993) Genetic resources at our doorstep: the wild strawberries. *Bioscience* 43, 141–147.

Hancock, J.F., Flore, J.A. and Galletta, G.J. (1989) Variation in leaf photosynthetic rates and yield in strawberries. *Journal of Horticultural Science* 64, 449–454.

Hancock, J.F., Callow, P.W., Dale, A., Luby, J.J., Finn, C.E., Hokanson, S.C. and Hummer, K.E. (2001a) From the Andes to the Rockies: native strawberry collection and utilization. *HortScience* 36, 221–225.

Hancock, J.F., Finn, C.A., Hokanson, S.C., Luby, J.J., Goulart, B.L., Demchak, K., Callow, P.W., Serçe, S., Schilder, A.M.C. and Hummer, K.E. (2001b) A multistate comparison of native octoploid strawberries from North and South America. *Journal of the American Society for Horticultural Science* 126, 579–586.

Hancock, J.F., Callow, P.W., Serçe, S. and Son, P.Q. (2003) Variation in the horticultural characteristics of native *Fragaria virginiana* and *F. chiloensis* from North and South America. *Journal of the American Society for Horticultural Science* 128, 201–208.

Harvell, C.D., Mitchell, C.E., Ward, J.R., Altizer, S., Dobson, A.P., Ostfeld, R.S. and Samuel, M.D. (2002) Climate warming and disease risks for terrestrial and marine biota. *Science* 296, 2158–2162.

Heide, O.M. and Sønsteby, A. (2007) Interactions of temperature and photoperiod in the control of flowering of latitudinal and altitudinal populations of wild strawberry (*Fragaria vesca*). *Physiologia Plantarum* 130, 280–289.

Houde, M., Dallaire, S., N'Dong, D. and Sarhan, F. (2004) Overexpression of the acidic dehydrin WCOR410 improves freezing tolerance in transgenic strawberry leaves. *Plant Biotechnology Journal* 2, 381–387.

Hršelová, H., Vejsadová, H., Přrikryl, Z., Váchová, J., Vančura, V. and Vít, A. (1989) Effect of inoculation with vesicular-arbuscular mycorrhizal fungi on growth of strawberries. In: Vlastimil, V. and Františk, K. (eds) *Developments in Soil Science. Elsevier*, Amsterdam, The Netherlands, pp. 109–114.

Husaini, A.M. (2010) Pre- and post-agroinfection strategies for efficient leaf disk transformation and regeneration of transgenic strawberry plants. *Plant Cell Reports* 29, 97–110.

Husaini, A.M. (2014) Challenges of climate change. *GM Crops and Food* 5, 97–105.

Husaini, A.M. and Abdin, M.Z. (2008) Development of transgenic strawberry (*Fragaria × ananassa* Duch.) plants tolerant to salt stress. *Plant Science* 174, 446–455.

Husaini, A.M. and Rafiqi, A.M. (2012) Role of osmotin in strawberry improvement. *Plant Molecular Biology Reporter* 30, 1055–1064.

Husaini, A.M. and Srivastava, D.K. (2006a) Plant regeneration and *Agrobacterium*-mediated gene transfer studies in strawberry tissues (*Fragaria × ananassa*). *Asian Journal of Microbiology Biotechnology and Environmental Sciences* 8, 671.

Husaini, A.M. and Srivastava, D.K. (2006b) Genetic transformation in strawberry – a review. *Asian Journal of Microbiology Biotechnology and Environmental Sciences* 8, 75–81.

Husaini, A.M. and Tuteja, N. (2012) Biotech crops: imperative for achieving the Millenium Development Goals and sustainability of agriculture in the climate change era. *GM Crops and Food* 4, 1–9.

Husaini, A.M., Mercado, J.A., Schaart, J.G. and Teixeira da Silva, J.A. (2011a) Review of factors affecting organogenesis, somatic embryogenesis and *Agrobacterium tumefaciens*-mediated transformation of strawberry. In: Husaini, A.M. and Mercado, J.A. (eds) *Genomics, Transgenics, Molecular Breeding and Biotechnology of Strawberry*. Global Science Books, UK, pp. 1–11.

Husaini, A.M., Rashid, Z., Mir, R.u.R. and Aquil, B. (2011b) Approaches for gene targeting and targeted gene expression in plants. *GM Crops* 2, 150–162.

Husaini, A.M., Abdin, M.Z., Khan, S., Xu, Y.W., Aquil, S. and Anis, M. (2012) Modifying strawberry for better adaptability to adverse impact of climate change. *Current Science* 102, 1660–1673.

James, D.J., Passey, A.J., Eastebrook, M.A., Solomon, M.G. and Barbara, D.J. (1992) Transgenes for pest and disease resistance: progress in the introduction of transgenes for pest and disease resistance into strawberries. *Phytoparasitica* 20, S83–S87.
James, D.J., Uratsu, S., Cheng, J., Negri, P., Viss, P. and Dandekar, A.M. (1993) Acetosyringone and osmoprotectants like betaine or proline synergistically enhance *Agrobacterium*-mediated transformation of apple. *Plant Cell Reports* 12, 559–563.
Karlidag, H., Yildirim, E. and Turan, M. (2009) Salicylic acid ameliorates the adverse effect of salt stress on strawberry. *Scientia Agricola* 66, 180–187.
Kesici, M., Gulen, H., Ergin, S., Turhan, E., Ipek, A. and Koksal, N. (2013) Heat-stress tolerance of some strawberry (*Fragaria* × *ananassa*) cultivars. *Notulae Botanicae Horti Agrobotanici Cluj-Napoca* 41, 238–243.
Khammuang, S., Dheeranupattana, S., Hanmuangjai, P. and Wongroung, S. (2005) *Agrobacterium*-mediated transformation of modified antifreeze protein gene in strawberry. *Songklanakarin Journal of Science and Technology* 27, 693–703.
Klamkowski, K. and Treder, W. (2006) Morphological and physiological responses of strawberry plants to water stress. *Agriculturae Conspectus Scientificus* 71, 159–165.
Klamkowski, K. and Treder, W. (2008) Response to drought stress of three strawberry cultivars grown under greenhouse conditions. *Journal of Fruit and Ornamental Plant Research* 16, 179–188.
Koehler, G., Wilson, R.C., Goodpaster, J.V., Sønsteby, A., Lai, X., Witzmann, F.A., You, J.-S., Rohloff, J., Randall, S.K. and Alsheikh, M. (2012) Proteomic study of low-temperature responses in strawberry cultivars (*Fragaria* × *ananassa*) that differ in cold tolerance. *Plant Physiology* 159, 1787–1805.
Korbin, M. (2011) Molecular approaches to disease resistance in *Fragaria* spp. *Journal of Plant Protection Research* 51, 60–65.
Koskela, E.A., Mouhu, K., Albani, M.C., Kurokura, T., Rantanen, M., Sargent, D.J., Battey, N.H., Coupland, G., Elomaa, P. and Hytönen, T. (2012) Mutation in *TERMINAL FLOWER1* reverses the photoperiodic requirement for flowering in the wild strawberry *Fragaria vesca*. *Plant Physiology* 159, 1043–1054.
Lambais, M.R., Ríos-Ruiz, W.F. and Andrade, R.M. (2003) Antioxidant responses in bean (*Phaseolus vulgaris*) roots colonized by arbuscular mycorrhizal fungi. *New Phytologist* 160, 421–428.
Ledesma, N.A., Kawabata, S. and Sugiyama, N. (2004) Effect of high temperature on protein expression in strawberry plants. *Biologia Plantarum* 48, 73–79.
Levitt, J. (1980) *Responses of Plants to Environmental Stresses*, 2nd edn. Academic Press, New York.
Li, H., Li, T., Gordon, R.J., Asiedu, S.K. and Hu, K. (2010) Strawberry plant fruiting efficiency and its correlation with solar irradiance, temperature and reflectance water index variation. *Environmental and Experimental Botany* 68, 165–174.
Li, M., Jiang, D., Meng, X., Liu, R. and Li, X. (1999) Effects of arbuscular mycorrhizal fungi on growth, yield and quality of *Phaseolus vulgaris* L. in field. *Eco-Agriculture Research* 7, 43–46.
Liang, J., Zhang, Y., Jia, X.Z., Lv, Q. and Zhang, X.Y. (2003) Effects of ectomycorrhizae on growth and resistance of poplar. *Journal of Nanjing Forestry University* 27, 39–43.
Liu, D., Raghothama, K.G., Hasegawa, P.M. and Bressan, R.A. (1994) Osmotin overexpression in potato delays development of disease symptoms. *Proceedings of the National Academy of Sciences USA* 91, 1888–1892.
Lu, J.Y., Mao, Y.M., Shen, L.Y., Peng, S.Q. and Li, X.L. (2003) Effects of VA mycorrhizal fungi inoculated on drought tolerance of wild jujube (*Zizyphus spinosus* Hu) seedlings. *Acta Horticulturae Sinica* 30, 29–33.
Maas, J.L. (1998) *Compendium of Strawberry Diseases*. American Phytopathological Society, St Paul, Minnesota.
Maas, J.L., Hartung, J.S., Gouin-Behe, C. and Hokanson, S.C. (2000) Sources of resistance for two differentially pathogenic strains of *Xanthomonas fragariae* in *Fragaria* genotypes. *HortScience* 35, 128–131.
Manakasem, Y. and Goodwin, P. (2001) Responses of dayneutral and Junebearing strawberries to temperature and daylength. *Journal of Horticultural Science and Biotechnology* 76, 629–635.
Matsubara, Y., Hirano, I., Sassa, D. and Koshikawa, K. (2004) Increased tolerance to *Fusarium* wilt in mycorrhizal strawberry plants raised by capillary watering methods. *Environmental Control in Biology* 42, 185–191.
Medina-Mínguez, J.J. (2002) Soil solarization and biofumigation in strawberries in Spain. In: *Proceedings of International Conference on Alternatives to Methyl Bromide*, 5–8 March, Seville, Spain, pp. 108–110.
Mercado, J.A., Martín-Pizarro, C., Pascual, L., Quesada, M.A., Pliego-Alfaro, F., Santos, B.d.l., Romero, F., Galvez, J., Rey, M., Viña, G.d.l., Llobell, A., Yubero-Serrano, E.-M., Muñoz-Blanco, J. and Caballero, J.L. (2005) Evaluation of tolerance to *Colletotrichum acutatum* in strawberry plants transformed with *Trichoderma*-derived genes. *Acta Horticulturae* 738, 383–388.
Mercado, J.A., Barceló, M., Pliego, C., Rey, M., Caballero, J., Muñoz-Blanco, J., Ruano-Rosa, D., López-Herrera, C., de los Santos, B., Romero-Muñoz, F. and Pliego-Alfaro, F. (2015). Expression of the β-1,3-glucanase gene

bgn13.1 from *Trichoderma harzianum* in strawberry increases tolerance to crown rot diseases but interferes with plant growth. *Transgenic Research* 24, 979–989.

Min, K., Ha, S.C., Hasegawa, P.M., Bressan, R.A., Yun, D.J. and Kim, K.K. (2004) Crystal structure of osmotin, a plant antifungal protein. *Proteins: Structure, Function, and Bioinformatics* 54, 170–173.

Monteiro, S., Barakat, M., Piçarra-Pereira, M.A., Teixeira, A.R. and Ferreira, R.B. (2003) Osmotin and thaumatin from grape: a putative general defense mechanism against pathogenic fungi. *Phytopathology* 93, 1505–1512.

Morishita, M., Honjo, M., Hamano, M., Yamazaki, H. and Yano, T. (2012) Genetic analysis of the everbearing habit in strawberry cultivars under 24-hour daylength condition. *Horticultural Research* 11, 301–307.

Ndong, C., Ouellet, F., Houde, M. and Sarhan, F. (1997) Gene expression during cold acclimation in strawberry. *Plant and Cell Physiology* 38, 863–870.

Niemi, M. and Vestberg, M. (1992) Inoculation of commercially grown strawberry with VA mycorrhizal fungi. *Plant and Soil* 144, 133–142.

Oelmüller, R., Sherameti, I., Tripathi, S. and Varma, A. (2009) *Piriformospora indica,* a cultivable root endophyte with multiple biotechnological applications. *Symbiosis* 49, 1–17.

Olaya, G. and Abawi, G.S. (1996) Effect of water potential on mycelial growth and on production and germination of sclerotia of *Macrophomina phaseolina. Plant Disease* 80, 1347–1350.

Owens, C.L., Thomashow, M.F., Hancock, J.F. and Iezzoni, A.F. (2002) *CBF1* orthologs in sour cherry and strawberry and the heterologous expression of *CBF1* in strawberry. *Journal of the American Society for Horticultural Science* 127, 489–494.

Owens, C.L., Iezzoni, A.F. and Hancock, J.F. (2003) Enhancement of freezing tolerance of strawberry by heterologous expression of *CBF1. Acta Horticulturae* 626, 93–100.

Paquin, R., Bolduc, R., Zizka, J., Pelletier, G. and Lechausser, P. (1989) Frost tolerance, sugar and proline content of strawberry (*Fragaria ananassa* Duch) during winter. *Canadian Journal of Plant Science* 69, 945–954.

Parkhi, V., Kumar, V., Sunilkumar, G., Campbell, L.A.M., Singh, N.K. and Rathore, K.S. (2009) Expression of apoplastically secreted tobacco osmotin in cotton confers drought tolerance. *Molecular Breeding* 23, 625–639.

Pinkerton, J.N., Ivors, K.L., Reeser, P.W., Bristow, P.R. and Windom, G.E. (2002) The use of soil solarization for the management of soilborne plant pathogens in strawberry and red raspberry production. *Plant Disease* 86, 645–651.

Porcel, R., Aroca, R., Azcón, R. and Ruiz-Lozano, J. (2006) PIP aquaporin gene expression in arbuscular mycorrhizal *Glycine max* and *Lactuca sativa* plants in relation to drought stress tolerance. *Plant Molecular Biology* 60, 389–404.

Pratt, R.G. (2006) A direct observation technique for evaluating sclerotium germination by *Macrophomina phaseolina* and effects of biocontrol materials on survival of sclerotia in soil. *Mycopathologia* 162, 121–131.

Raghothama, K.G., Liu, D., Nelson, D.E., Hasegawa, P.M. and Bressan, R.A. (1993) Analysis of an osmotically regulated pathogenesis-related osmotin gene promoter. *Plant Molecular Biology* 23, 1117–1128.

Rai, M., Acharya, D., Singh, A. and Varma, A. (2001) Positive growth responses of the medicinal plants *Spilanthes calva* and *Withania somnifera* to inoculation by *Piriformospora indica* in a field trial. *Mycorrhiza* 11, 123–128.

Rantanen, M., Kurokura, T., Jiang, P., Mouhu, K. and Hytönen, T. (2015) Strawberry homologue of TERMINAL FLOWER1 integrates photoperiod and temperature signals to inhibit flowering. *The Plant Journal* 82, 163–173.

Rosenzweig, C., Iglesias, A., Yang, X.B., Epstein, P. and Chivian, E. (2001) Climate change and extreme weather events; implications for food production, plant diseases, and pests. *Global Change and Human Health* 2, 90–104.

Ruiz-Lozano, J.M. (2003) Arbuscular mycorrhizal symbiosis and alleviation of osmotic stress. New perspectives for molecular studies. *Mycorrhiza* 13, 309–317.

Sahay, N.S. and Varma, A. (1999) *Piriformospora indica*: a new biological hardening tool for micropropagated plants. *FEMS Microbiology Letters* 181, 297–302.

Sahay, N.S. and Varma, A. (2000) Biological approach towards increasing the survival rates of micropropagated plants. *Current Science* 78, 126–129.

Sánchez, S., Gambardella, M., Henríquez, J.L. and Díaz, I. (2013) First report of crown rot of strawberry caused by *Macrophomina phaseolina* in Chile. *Plant Disease* 97, 996–996.

Sander, J.D. and Joung, J.K. (2014) CRISPR-Cas systems for editing, regulating and targeting genomes. *Nature Biotechnology* 32, 347–355.

Sarad, N., Rathore, M., Singh, N.K. and Kumar, N. (2004) Genetically engineered tomatoes: new vista for sustainable agriculture in high altitude aegions. In: *Proceedings of the 4th International Crop Science Congress*, 26 September–1 October, Brisbane, Australia, pp. 234–238.
Sardans, J. and Peñuelas, J. (2007) Drought changes phosphorus and potassium accumulation patterns in an evergreen Mediterranean forest. *Functional Ecology* 21, 191–201.
Sarnighausen, E., Karlson, D.T., Zeng, Y., Goldsbrough, P.B., Raghothama, K.G. and Ashworth, E.N. (2004) Characterization of a novel Y_nSK_n class of dehydrin-like cDNAs from cold acclimated red-osier dogwood (*Cornus sericea* L.) xylem. *Journal of Crop Improvement* 10, 17–35.
Schestibratov, K.A. and Dolgov, S.V. (2005) Transgenic strawberry plants expressing a thaumatin II gene demonstrate enhanced resistance to *Botrytis cinerea*. *Scientia Horticulturae* 106, 177–189.
Serçe, S. and Hancock, J.F. (2005a) The temperature and photoperiod regulation of flowering and runnering in the strawberries, *Fragaria chiloensis*, *F. virginiana*, and *F.* × *ananassa*. *Scientia Horticulturae* 103, 167–177.
Serçe, S. and Hancock, J.F. (2005b) Inheritance of day-neutrality in octoploid species of *Fragaria*. *Journal of the American Society for Horticultural Science* 130, 580–584.
Shaw, D.V. and Famula, T.R. (2005) Complex segregation analysis of day-neutrality in domestic strawberry (*Fragaria* × *ananassa* Duch.). *Euphytica* 145, 331–338.
Sherameti, I., Tripathi, S., Varma, A. and Oelmüller, R. (2008) The root-colonizing endophyte *Pirifomospora indica* confers drought tolerance in *Arabidopsis* by stimulating the expression of drought stress-related genes in leaves. *Molecular Plant–Microbe Interactions* 21, 799–807.
Silva, K.J.P., Brunings, A., Peres, N.A., Mou, Z. and Folta, K.M. (2015) The *Arabidopsis NPR1* gene confers broad-spectrum disease resistance in strawberry. *Transgenic Research* 24, 693–704.
Singh, N.K., Handa, A.K., Hasegawa, P.M. and Bressan, R.A. (1985) Proteins associated with adaptation of cultured tobacco cells to NaCl. *Plant Physiology* 79, 126–137.
Singh, N.K., Bracker, C.A., Hasegawa, P.M., Handa, A.K., Buckel, S., Hermodson, M.A., Pfankoch, E., Regnier, F.E. and Bressan, R.A. (1987) Characterization of osmotin: a thaumatin-like protein associated with osmotic adaptation in plant cells. *Plant Physiology* 85, 529–536.
Singh, N.K., Nelson, D.E., Kuhn, D., Hasegawa, P.M. and Bressan, R.A. (1989) Molecular cloning of osmotin and regulation of its expression by ABA and adaptation to low water potential. *Plant Physiology* 90, 1096–1101.
Sjulin, T.M. and Dale, A. (1987) Genetic diversity of North American strawberry cultivars. *Journal of the American Society for Horticultural Science* 112, 375–385.
Sokhansanj, A., Sadat Noori, S. and Niknam, V. (2006) Comparison of bacterial and plant genes participating in proline biosynthesis with *Osmotin* gene, with respect to enhancing salinity tolerance of transgenic tobacco plants. *Russian Journal of Plant Physiology* 53, 110–115.
Staudt, G. (1973) *Fragaria iturupensis*: a new species of strawberry from East Asia. *Willdenowia* 7, 101–104.
Stewart, L.I., Hamel, C., Hogue, R. and Moutoglis, P. (2005) Response of strawberry to inoculation with arbuscular mycorrhizal fungi under very high soil phosphorus conditions. *Mycorrhiza* 15, 612–619.
Stewart, P.J. and Folta, K.M. (2010) A review of photoperiodic flowering research in strawberry (*Fragaria* spp.). *Critical Reviews in Plant Sciences* 29, 1–13.
Sylvia, D.M. and Williams, S.E. (1992) Vesicular-arbuscular mycorrhizae and environmental stress. *ASA Special Publication* 54, 101–124.
Sziderics, A.H., Rasche, F., Trognitz, F., Sessitsch, A. and Wilhelm, E. (2007) Bacterial endophytes contribute to abiotic stress adaptation in pepper plants (*Capsicum annuum* L.). *Canadian Journal of Microbiology* 53, 1195–1202.
Thomashow, M.F. (1999) Plant cold acclimation: freezing tolerance genes and regulatory mechanisms. *Annual Review of Plant Physiology and Plant Molecular Biology* 50, 571–599.
Timmusk, S. and Wagner, E.G.H. (1999) The plant-growth-promoting rhizobacterium *Paenibacillus polymyxa* induces changes in *Arabidopsis thaliana* gene expression: a possible connection between biotic and abiotic stress responses. *Molecular Plant–Microbe Interactions* 12, 951–959.
Turhan, E. (2012) Seasonal alteration of sugar metabolism in strawberry (*Fragaria* × *ananassa*) plants during cold-acclimated and non-acclimated stages. *African Journal of Biotechnology* 11, 4558–4565.
Turhan, E., Aydogan, C., Akoglu, A., Baykul, A. and Evrenosoğlu, Y. (2012a) Relationship of seasonal changes in antioxidative enzymes and cold-hardiness in strawberry plant. *Journal of Food, Agriculture and Environment* 10, 445–450.
Turhan, E., Aydogan, C., Baykul, A., Akoglu, A., Evrenosoğlu, Y. and Ergin, S. (2012b) Apoplastic antioxidant enzymes in the leaves of two strawberry cultivars and their relationship to cold-hardiness. *Notulae Botanicae Horti Agrobotanici Cluj-Napoca* 40, 114–122.

Varma, A., Verma, S., Sudha, N.S., Bütehorn, B. and Franken, P. (1999) *Piriformospora indica*, a cultivable plant-growth-promoting root endophyte. *Applied and Environmental Microbiology* 65, 2741–2744.
Varma, A., Singh, A., Sudha, Sahay, N.S., Sharma, J., Roy, A., Kumari, M., Rana, D., Thakran, S., Deka, D., Bharti, K., Hurek, T., Blechert, O., Rexer, K.H., Kost, G., Hahn, A., Maier, W., Walter, M., Strack, D. and Kranner, I. (2001) *Piriformospora indica*: an axenically culturable mycorrhiza-like endosymbiotic fungus. In: Hock, B. (ed.) *Fungal Associations*. Springer, Berlin/Heidelberg, pp. 125–150.
Vellicce, G., Ricci, J.D., Hernández, L. and Castagnaro, A. (2006) Enhanced resistance to *Botrytis cinerea* mediated by the transgenic expression of the chitinase gene *ch5B* in strawberry. *Transgenic Research* 15, 57–68.
Vestberg, M., Kukkonen, S., Saari, K., Parikka, P., Huttunen, J., Tainio, L., Devos, N., Weekers, F., Kevers, C., Thonart, P., Lemoine, M.C., Cordier, C., Alabouvette, C. and Gianinazzi, S. (2004) Microbial inoculation for improving the growth and health of micropropagated strawberry. *Applied Soil Ecology* 27, 243–258.
Wahid, A., Gelani, S., Ashraf, M. and Foolad, M.R. (2007) Heat tolerance in plants: an overview. *Environmental and Experimental Botany* 61, 199–223.
Waller, F., Achatz, B., Baltruschat, H., Fodor, J., Becker, K., Fischer, M., Heier, T., Hückelhoven, R., Neumann, C., von Wettstein, D., Franken, P. and Kogel, K.H. (2005) The endophytic fungus *Piriformospora indica* reprograms barley to salt-stress tolerance, disease resistance, and higher yield. *Proceedings of the National Academy of Sciences USA* 102, 13386–13391.
Wang, J.L., Ge, H.B., Peng, S.Q., Zhang, H.M., Chen, P.L. and Xu, J.R. (2004) Transformation of strawberry (*Fragaria ananassa* Duch.) with late embryogenesis abundant protein gene. *Journal of Horticultural Science and Biotechnology* 79, 735–738.
Weebadde, C.K., Wang, D., Finn, C.E., Lewers, K.S., Luby, J.J., Bushakra, J., Sjulin, T.M. and Hancock, J.F. (2008) Using a linkage mapping approach to identify QTL for day-neutrality in the octoploid strawberry. *Plant Breeding* 127, 94–101.
Woloshuk, C.P., Meulenhoff, J.S., Sela-Buurlage, M., van den Elzen, P. and Cornelissen, B. (1991) Pathogen-induced proteins with inhibitory activity toward *Phytophthora infestans*. *Plant Cell* 3, 619–628.
WorldPublicOpinion.org (2010) *Public Attitudes Toward Climate Change: Findings From a Multicountry Poll*. The World Bank, Washington, DC.
Wu, Q.S. and Xia, R.X. (2003) Research and application on vesicular-arbuscular mycorrhiza of fruit trees. *Plant Physiology Communications* 39, 536–540.
Wu, Q.-S. and Xia, R.-X. (2006) Arbuscular mycorrhizal fungi influence growth, osmotic adjustment and photosynthesis of citrus under well-watered and water stress conditions. *Journal of Plant Physiology* 163, 417–425.
Wu, Q.S., Xia, R.X. and Zou, Y.N. (2006a) Reactive oxygen metabolism in mycorrhizal and non-mycorrhizal citrus (*Poncirus trifoliata*) seedlings subjected to water stress. *Journal of Plant Physiology* 163, 1101–1110.
Wu, Q.S., Zou, Y.N. and Xia, R.X. (2006b) Effects of water stress and arbuscular mycorrhizal fungi on reactive oxygen metabolism and antioxidant production by citrus (*Citrus tangerine*) roots. *European Journal of Soil Biology* 42, 166–172.
Wu, Q.S., Zou, Y.N. and Wang, G.Y. (2007) Effect of inoculation with arbuscular mycorrhizal fungal on citrus under water stress conditions. *Journal of Yangtze University (Natural Science Edition)* 4, 18–21.
Yen, H.E., Edwards, G.E. and Grimes, H.D. (1994) Characterization of a salt-responsive 24-kilodalton glycoprotein in *Mesembryanthemum crystallinum*. *Plant Physiology* 105, 1179–1187.
Yin, B., Wang, Y., Liu, P., Hu, J. and Zhen, W. (2010) Effects of vesicular-arbuscular mycorrhiza on the protective system in strawberry leaves under drought stress. *Frontiers of Agriculture in China* 4, 165–169.
Yun, D.-J., Ibeas, J.I., Lee, H., Coca, M.A., Narasimhan, M.L., Uesono, Y., Hasegawa, P.M., Pardo, J.M. and Bressan, R.A. (1998) Osmotin, a plant antifungal protein, subverts signal transduction to enhance fungal cell susceptibility. *Molecular Cell* 1, 807–817.
Zhang, B. and Archbold, D.D. (1993) Solute accumulation in leaves of a *Fragaria chiloensis* and a *F. virginiana* selection responds to water deficit stress. *Journal of the American Society for Horticultural Science* 118, 280–285.
Zhang, H.S. and He, X.L. (2007) Effect of AM fungal on the protective system in leaves of *Artemisia ordosica* under drought stress. *Biotechnology Bulletin* 3, 129–133.
Zhang, Y., Luo, Y., Hou, Y.-X., Jiang, H., Chen, Q. and Tang, H.-R. (2008) Chilling acclimation induced changes in the distribution of H_2O_2 and antioxidant system of strawberry leaves. *Agriculture Journal* 3, 286–291.
Zhu, B., Chen, T.H.H. and Li, P.H. (1993) Expression of an ABA-responsive osmotin-like gene during the induction of freezing tolerance in *Solanum commersonii*. *Plant Molecular Biology* 21, 729–735.

Ziska, L.H. and Runion, G.B. (2007) Future weed, pest and disease problems for plants. In: Newton, P.C.D., Carran, R.A., Edwards, G.R. and Niklaus, P.A. (eds) *Agroecosystems in a Changing Climate*. CRC Press, Boca Raton, Florida, pp. 261–287.

Zorrilla-Fontanesi, Y., Cabeza, A., Domínguez, P., Medina, J., Valpuesta, V., Denoyes-Rothan, B., Sánchez-Sevilla, J. and Amaya, I. (2011) Quantitative trait loci and underlying candidate genes controlling agronomical and fruit quality traits in octoploid strawberry (*Fragaria* × *ananassa*). *Theoretical and Applied Genetics* 123, 755–778.

Zuker, A., Shklarman, E., Scovel, G., Ben-Meir, H., Ovadis, M., Neta-Sharir, I., Ben-Yephet, Y., Weiss, D., Watad, A. and Vainstein, A. (2001) Genetic engineering of agronomic and ornamental traits in carnation. *Acta Horticulturae* 560, 91–94.

Zveibil, A., Mor, N., Gnayem, N. and Freeman, S. (2012) Survival, host–pathogen interaction, and management of *Macrophomina phaseolina* on strawberry in Israel. *Plant Disease* 96, 265–272.

15 Molecular and Physiological Responses of Strawberry Plants to Abiotic Stress

Hatice Gulen[1]*, Ece Turhan[2] and Atilla Eris[1]

[1]Istanbul Bilgi University, Faculty of Engineering and Natural Sciences, Department of Genetics and Bioengineering, Istanbul, Turkey; [2]Eskisehir Osmangazi University, Faculty of Agriculture, Department of Agricultural Biotechnology, Eskisehir, Turkey

15.1. Introduction

Only about 38.5% of land in the world is farmed. However, most of the farmable land is under suboptimal conditions for plant growth and development due to various stress factors (FAO, 2015a). In addition to stress factors that already exist, farmable lands are being lost every year due to climate change-induced global warming. However, the population is increasing and is estimated to reach over 9 billion in 2050, so a considerable increase in world food production will be necessary. Stress factors are the principal cause of crop failure worldwide, reducing average yields for most major crops by more than 50% (Bray *et al.*, 2000). Thus, an understanding of stress mechanisms in plants is very important.

According to Levitt (1980), biological stress is any change in environmental conditions that might reduce or adversely change a plant's growth or development; biological strain is the reduced or changed function. Any change in environmental conditions that results in a plant response that is less than optimal is considered stressful (Salisbury and Ross, 1992). The changes in an organism's function that enable a return to the optimal level when conditions are again optimal are known as elastic biological strain. If the functions do not return to normal, this is defined as plastic biological strain (Levitt, 1980). Plants may possess avoidance or tolerance

* hatice.gulen@bilgi.edu.tr

mechanisms to any stress factor to help them to cope with the stress. Levitt (1980) defined avoidance as a response of the organism that somehow reduces the impact of the stress factor. Tolerance implies that the plant simply tolerates or endures the adverse environment. Salisbury and Ross (1992) identified three main stages that occur when plants are exposed to a stress factor: (i) the alarm stage, in which the function of interest deviates markedly from the norm; (ii) the resistance stage, in which the organism adapts to the stress factor and the function often returns towards its normal level (but may not completely achieve it); and (iii) the exhaustion stage, in which the function may again strongly deviate from the norm if the stress factor increases or continues for a long time (this can eventually lead to death of the plant).

Plants can respond and adapt to any stress factor using a number of physiological, biochemical and molecular responses. One example is the antioxidant system: evidence shows that stress factors can lead to the production of reactive oxygen species (ROS). This increases the level of reactive oxygen injury, which leads to lipid peroxidation, membrane deterioration, protein degradation, nucleic acid damage, chlorophyll bleaching and metabolic function disruption (Tan *et al.*, 2011). The degree of damage depends on the balance between the formation of ROS and their detoxification by the antioxidative scavenging system, specific protein accumulation, hormonal system and other mechanisms. Thus, a high level of protective enzymes and antioxidants such as superoxide dismutase (SOD), peroxidase (PRX), ascorbate peroxidase (APX), catalase (CAT) and glutathione reductase (GR) are essential for maintenance of the concentration of ROS at a relatively low level, which is required for the survival of plants under many types of stress (Scebba *et al.*, 1998; Wang, 1999; Gulen and Eris, 2004; Gulen *et al.*, 2006; Turhan *et al.*, 2008; Zhang *et al.*, 2008a; Ergin, 2012; Ergin *et al.*, 2012; Turhan *et al.*, 2012a,b). In plant systems, tolerance mechanisms are activated by induction of the antioxidant system, and production of chaperone proteins, osmolytes and hormones to stop the cellular homeostasis and to repair the damaged membrane. However, if the stress is too strong, the defence system of plants cannot remove the ROS effectively and the activities of antioxidant enzymes decrease, resulting in severe damage to the plant and even death (Zhang *et al.*, 2008b). Cellular membranes are an important factor in the ability of plants to cope with stressful conditions. Cellular membranes have been implicated as the site of primary lesions in many stress-caused injuries. It has been shown that plants exhibit a capacity for considerable acclimation at the membrane level by increasing the amount of specific membrane components useful in mitigating cellular stresses and, in some instances, by undergoing relatively rapid molecular rearrangements within the membrane when challenged by stress factors (O'Neill, 1983).

The stress factors limiting plant growth and development are classified into two main groups: biotic and abiotic. The biotic group comprises pathogens such as bacteria, fungi, and viruses, as well as herbivores, while temperature, salt, drought, flooding and heavy metal toxicity are examples of abiotic stress factors. All of these stress factors are a menace for plants and prevent them from reaching their full genetic potential, limiting crop productivity worldwide (Mahajan and Tuteja, 2005).

Strawberry is an economically important world berry fruit whose production continues to increase throughout the world. In 2012, strawberry production throughout the world was approximately 4,516,810 t on 241,109 ha (FAO, 2015b). The ploidy levels of members of the genus *Fragaria* range from diploid (2x) to decaploid (10x). Studying the stress tolerance of strawberry species, Na *et al.* (2014) reported that octoploid species have greater abiotic stress tolerance than diploid species according to analysis of their chlorophyll fluorescence characteristics. Indeed, strawberry plants are often exposed to extreme environmental conditions because of their year-round production cycle using various cultivation techniques. The major abiotic stress factors of salinity, temperature and water stress will be discussed in relation to strawberry cultivation in this chapter.

15.2. Salinity Stress

Salinity is a major environmental stress and is a substantial constraint to crop production. Excess salt, usually in the form of sodium chloride (NaCl), is the most widespread chemical condition inhibiting plant growth in nature (Munns and Termaat, 1986). In arid and semi-arid regions of the world, limited rainfall, high evapotranspiration, high temperature and inadequate water management contribute to increased soil salinity (Meloni *et al.*, 2003). Increased salinization of arable land is expected to have devastating global effects, resulting in 50% land loss by the middle of the 21st century (Wang *et al.*, 2003).

Strawberry represents an important commercial fruit crop with increasing areas of cultivation worldwide and rising consumption, but it is considered sensitive to NaCl salinity (Levitt, 1980; Schwarz, 1995), although the sensitivity criteria and causes of the injuries due to toxic Na^+ or Cl^- ions are important points of discussion (Barroso and Alvarez, 1997). Variations in salt tolerance among different strawberry cultivars have been reported (Awang *et al.*, 1993a; Kaya *et al.*, 2002; Turhan, 2002; Turhan and Eris, 2005a; Gulen *et al.*, 2006; Turhan *et al.*, 2008; Turhan and Eris, 2009). These differences were mainly in biomass production and leaf injury (Barroso and Alvarez, 1997), stomatal behaviour and antioxidant enzymes (Gulen *et al.*, 2006; Turhan *et al.*, 2008), membrane permeability (Kaya *et al.*, 2002; Gulen *et al.*, 2006) and yield (Awang *et al.*, 1993b; Keutgen and Pawelzik, 2008).

Below, we discuss the responses of strawberry plants to salinity stress at the whole-plant, cellular and subcellular levels, as well as tolerance mechanisms and strategies for genetic improvement of strawberry with salt-stress tolerance.

15.2.1. Salt injury

Salt stress with osmotic, nutritional and toxic effects prevents growth in many plant species (Cheeseman, 1988; Salisbury and Ross, 1992). High salt depositions in the soil generate a low water potential zone in the soil, making it increasingly difficult for the plant to acquire both water and nutrients. Therefore, salt stress essentially results in a water-deficit condition in the plant and takes the form of an osmotic stress or, as it is sometimes called, a physiological drought stress (Levitt, 1980; Mahajan and Tuteja, 2005). This osmotic dehydration leads rapidly to a decrease in the osmotic and water potential of cells and a decrease in cell volume (Levitt, 1980). It has been determined that salt treatments decrease the leaf water potential and osmotic potential (Awang *et al.*, 1993a; Awang and Atherton, 1994) and cause osmotic stress (Turhan, 2002; Turhan and Eris, 2005b, 2009) in strawberry plants.

The reduction in plant growth is explained by the lower osmotic potential in the soil, which leads to decreased water uptake, reduced transpiration and closure of stomata, which are associated with reduced growth (Levitt, 1980; Ben-Asher *et al.*, 2006). Transpiration rates generally tend to decline with increasing rhizospheric salinity in both halophytes and non-halophytes. Addition of salt to the growth medium causes a reduction in stomatal conduction (g_s) and the transpiration ratio (E) of strawberry (Turhan and Eris, 2007a).

Excessive uptake of Na^+ and Cl^- may result in limited assimilation, transport and distribution of mineral nutrients, as well as nutrient imbalances within the plant (Greenway and Munns, 1980; Levitt, 1980; Cheeseman, 1988). There is little information about the distribution of macronutrients and micronutrients in strawberry plants under NaCl salinity (Awang and Atherton, 1994; Saied *et al.*, 2005; Turhan and Eris, 2005a,b; Keutgen and Pawelzik, 2009; Khayyat *et al.*, 2009a). Different researchers have reported contradictory results. For example, it was determined that short-term salt application in the strawberry cultivar 'Rapella' did not affect potassium (K), sodium (Na), calcium (Ca) and magnesium (Mg) content on a dry-weight basis (Awang and Atherton, 1994). However, salt applications in the cultivar 'Camarosa' increased the amount of Na, Cl, Ca and Mg in its aerial parts, while

K and phosphorus (P) decreased. In the roots of plants, it was observed that, as Na and Cl levels increased, the amount of K and Mg decreased, but Ca and P were not affected by the salt application (Turhan and Eris, 2005b). High NaCl levels can also induce Ca deficiency in the leaves of strawberry (Kaya *et al.*, 2002). Keutgen and Pawelzik (2009) found that, under NaCl stress, plants of the cultivar 'Korona' achieved a significant increase in K content in the leaves and crowns, while 'Elsanta' plants showed an increase in K in the fruits and petioles. The accumulation of K under evaluated NaCl levels suggests an efficient K uptake system in strawberry plants. Concentrations of Ca were not significantly affected, with the exception of rising levels in roots of 'Elsanta' plants. Concentrations of Mg, manganese (Mn) and iron (Fe) significantly decreased in the leaves, while those of Mg and Mn rose remarkably in the crowns of both cultivars. Nitrogen (N) content in the leaves, petioles and roots of both cultivars increased. In addition, N levels rose in the fruits and crowns in 'Elsanta'. A significant limitation of N uptake by competition with Cl did not occur in these plants. Concentrations of P increased in the roots and petioles of both cultivars, and in the fruits of 'Elsanta'. These results were in line with those of Saied *et al.* (2005).

It is unclear how salinity affects the micronutrient composition of strawberry plants. El-Fouly and Salama (1999) suggested that micronutrient levels are generally less affected by salt stress than macronutrient levels. Turhan and Eris (2005a) and Khayyat *et al.* (2009a) observed that Fe, Zn and Mn concentrations in the shoots of salt-stressed strawberry plants increased. Moreover, they found that the copper (Cu) concentration of roots increased, whereas the Cu concentration of shoots remained unaffected under salt stress. Keutgen and Pawelzik (2009) also found that Zn and Cu content did not change with salt treatments in strawberry plants.

Direct toxicity due to Na^+ and Cl^- of saline water is common, especially in plants like strawberries and roses. These toxic effects occur below the osmotic levels, and both ions accumulate in relatively high amounts in plant tissues. The first toxicity symptom is leaf burn (Schwarz, 1995). Chlorine mainly causes the impairment of strawberry plant growth and carbohydrate production, because Na is effectively excluded from the plants, especially from the leaf tissue. Na exclusion is the most usual salinity adaptation mechanism found in glycophytes (Greenway and Munns, 1980). In contrast to Na, Cl tends to move more rapidly to the leaves. It has been determined that NaCl treatments increase Cl content in strawberry plants (Barroso and Alvarez, 1997; Turhan, 2002; Turhan and Eris, 2005b, 2009). According to Levitt (1980), NaCl treatments increased Na and Cl accumulation, and the toxic effects related to these ions caused necrosis and moulding in leaves. NaCl application causes necrosis at various levels in strawberry plants, and the severity of this injury varies depending on the NaCl concentration (Barroso and Alvarez, 1997; Turhan and Eris, 2005b; Keutgen and Pawelzik, 2009). High NaCl additions into nutrient solutions induced significant increases in electrolyte leakage for the strawberry cultivars 'Oso Grande', 'Camarosa' (Kaya *et al.*, 2002) and 'Chandler' (Gulen *et al.*, 2006).

Salinity is a factor that affects the whole plant metabolism, including its morphology and anatomy (Levitt, 1980). The decline in leaf growth is the earliest response of glycophytes exposed to salt stress. Reduced expansion could be due to NaCl increasing the osmotic potential of the solution and to the activity of Na^+ and Cl^- in the root zone (Greenway and Munns, 1980). Leaf number, leaf area, plant dry weight, and root and shoot growth also decrease with increasing salt concentration in strawberry plants (Awang *et al.*, 1993a; Turhan and Eris, 2007a; Keutgen and Pawelzik, 2009). Besides death of the plant, salt stress results in the development of leaf necrosis and accelerated leaf senescence, thus reducing the photosynthetic capability of the plants. In consequence, assimilation of carbohydrates available for fruit production is also reduced (Giuffrida *et al.*, 2001; Keutgen and Keutgen, 2003; Saied *et al.*, 2005).

The effects of increasing salinity in the soil on the growth of strawberry plants and on the yield of fruit have been well documented (Awang *et al.*, 1993a; Awang and Atherton, 1995a,b; Kaya *et al.*, 2002; Turhan, 2002; Saied *et al.*, 2005; Keutgen and Pawelzik, 2007, 2008). Negative influences of salinity on strawberry plant growth and fruit productivity have been reported by Awang *et al.* (1993a), Awang and Atherton (1995a,b), Turhan (2002) and Turhan and Eris (2005a,b, 2007a,b, 2009). NaCl salinity impairs leaf metabolism in sensitive species, photosynthesis is reduced and carbohydrate production is limited. This results in lower strawberry fruit yield and quality. Awang *et al.* (1993b) and Keutgen and Pawelzik (2008) reported a reduction in fruit yield, but fruit quality was improved at moderate salinity stress, because the concentrations of reducing sugars and acids increased on a fresh weight basis due to decreasing fruit water content. Salt stress in fruits of the cultivars 'Elsanta' and 'Korona' increased the antioxidant capacity, antioxidant pools (ascorbate (ASA), anthocyanins and SOD) and selected minerals such as Na, Cl, K, N, P and Zn, as well as lipid peroxidation. Furthermore, salt stress increased the contents of free and essential amino acids, especially in 'Elsanta'. The more tolerant 'Korona' was characterized by an increase in reduced glutathione (GSH) and a better fruit taste. In salt-stressed fruits of 'Elsanta', the taste was significantly impaired (Keutgen and Pawelzik, 2008). Fruit fresh weight is reduced markedly by increased salinity (Awang *et al.*, 1993b; Awang and Atherton, 1995a,b; Kaya *et al.*, 2002) but fruit dry matter content is less affected (Awang *et al.*, 1993b).

Salinity may affect various metabolic processes such as photosynthesis, protein synthesis, respiration, nitrogen assimilation and phytohormone turnover (Arshi *et al.*, 2002). Increasing salinity in the growth medium decreased the content of chlorophyll in the cultivars 'Camarosa' (Kaya *et al.*, 2002; Rahimi *et al.*, 2011) and 'Oso Grande' (Kaya *et al.*, 2002). Garriga *et al.* (2014) also showed reduced chlorophyll and anthocyanin content related to salinity stress in *Fragaria chiloensis*. However, total chlorophyll content was unaffected by salt applications in the cultivars 'Camarosa' and 'Tioga' (Turhan and Eris, 2005a). Different results can occur depending on the salt concentration and exposure time within a cultivar. Net photosynthesis was found to be reduced by high salinity, but only in 'Rapella' strawberry plants grown in unshaded conditions (Awang and Atherton, 1994).

One of the biochemical changes occurring when plants are subjected to biotic or abiotic stresses is the production of ROS. ROS are highly reactive and, in the absence of any protective mechanism, they can disrupt normal metabolism through oxidative damage to lipids, protein and nucleic acids (Allen, 1995). The chloroplasts and mitochondria of plant cells are important intracellular generators of activated oxygen species. Stomatal closure resulting from osmotic stress limits carbon dioxide availability for photosynthetic carbon assimilation, hence causing high accumulation of superoxide ($O_2^{\bullet-}$) in chloroplasts, which can cause oxidative damage (Ashraf, 2009). It was found that NaCl induces oxidative stress in strawberry leaves, as evidenced by hydrogen peroxide (H_2O_2)/$O_2^{\bullet-}$ anion accumulation, an increase in lipid peroxidation and carbonyl-group content. NaCl causes an accumulation in putrescine and spermidine, oxidation of ASA and GSH redox pairs, and inhibition of the activities of CAT, APX and GR (Tanou *et al.*, 2009).

15.2.2. Salt tolerance

Salt tolerance in plants is a complex phenomenon and depends on a number of interrelated factors based on morphological, biochemical and physiological processes (Greenway and Munns, 1980; Arshi *et al.*, 2002; Qasim *et al.*, 2003).

The maintenance of turgor by osmotic adjustment is an important physiological adaptation for minimizing the detrimental effects of salt stress (Munns, 2002). Turhan and Eris (2007a) suggested that the reductions in stomatal conduction (g_s) and the transpiration ratio (E) represent adaptive mechanisms to cope with excessive salt, especially in the strawberry cultivar 'Camarosa'. Moreover, control of the stomatal response to salinity has been indicated as a physiological trait that may differentiate salt-tolerant and salt-sensitive cultivars (Turhan *et al.*, 2008). Orsini *et al.* (2012) also demonstrate that low stomatal density may be beneficial for strawberry cultivars that are cultivated in marginal environments, in terms of salinity.

Cell membrane stability is widely used to express stress tolerance; and higher membrane stability is correlated with abiotic stress tolerance (Premachandra *et al.*, 1992). In this respect, leaf tissue of the cultivars 'Camarosa' and 'Tioga' exhibited salt tolerance, having less electrolyte leakage, which is symptomatic of cellular damage (Gulen *et al.*, 2006). Reduced electrolyte leakage has been correlated with greater membrane integrity under stressful conditions and these plants are characterized as salt stress-tolerant genotypes (Sudhakar *et al.*, 2001).

Plant species adjust to high salt concentrations by lowering their tissue osmotic potential by the accumulation of inorganic ions (e.g. Na^+, K^+ and Ca^{2+}), as well as organic solutes (e.g. sugars, organic acids, free amino acids and proline), depending on the species (Levitt, 1980; Hasegawa *et al.*, 1986). For example, in strawberry, Na content in fresh tissue was found to increase with salinity, suggesting that strawberry has adopted the mechanism of Na exclusion to tolerate high Na salinity in the nutrient solution (Awang and Atherton, 1994). The exclusion or maintenance of low concentrations of toxic Na^+ and Cl^-, especially in meristematic tissues and reproductive organs, is regarded as an essential phytophysiological mechanism for salinity tolerance (Munns, 2002). Rahimi *et al.* (2011) found that NaCl treatment increased proline content and soluble sugar levels in 'Camarosa' plants. The authors suggested that strawberry may uses osmoregulation by increasing proline and soluble sugar levels to tolerate salinity conditions.

Salt stress is reported to be responsible for some changes in nitrogen metabolism (Levitt, 1980). Amino acids and amides have been reported to accumulate in higher plants under salinity stress (Mansour, 2000). Keutgen and Pawelzik (2008) found that total amino acid levels, especially of essential amino acids, rose significantly in fruits of the cultivars 'Korona' and 'Elsanta'. Salt stress also increased the levels of free proline, asparagine and glutamine, and the authors suggested that these increases might contribute to osmotic adjustment. Turhan and Eris (2009) reported that the levels of total amino acids decreased with higher NaCl concentrations. Martínez-Ballesta *et al.* (2004) explained that this could be due to a toxic effect on cytoplasmic enzymes involved in amino acid synthesis. Turhan (2002) and Turhan and Eris (2009) reported that aspartic acid, glutamic acid, arginine, proline, serine and alanine were the most variable amino acids under salt stress in strawberry plants. Despite the presence of a strong correlation between stress tolerance and accumulation of proline in higher plants, this relationship may not be universal. For example, at low salt concentrations, proline accumulation is an indicator of salt tolerance, whereas at high salt concentrations, its accumulation is an indicator of salt damage, especially in more salt-sensitive strawberry cultivars such as 'Chandler'. It was determined that proline accumulation is not an indicator of salt tolerance or salt damage in more tolerant cultivars such as 'Camarosa' (Turhan, 2002; Turhan and Eris, 2009).

Salt-tolerant plants, besides being able to regulate ion and water movement, should also have a better antioxidative system for effective removal of ROS (Rout and Shaw, 2001). Antioxidative enzymes such as SOD, CAT and PRX are the most important components in the scavenging system of ROS. An association of antioxidant activity or production with salt tolerance at the cellular level has been observed in some studies. For example, Gulen *et al.* (2006) showed that total and specific PRX activities were significantly increased by salinity in strawberry plants. The same authors also reported that one basic isoperoxidase band (retardation factor (Rf) = 0.27) was commonly observed with different band intensities in all salt treatments. Salt stress decreased the total soluble protein content in leaf tissue of the cultivars 'Tioga' (salt tolerant) and

'Chandler' (salt sensitive). However, no effect of salt stress was observed on leaf soluble proteins in 'Camarosa', another salt-tolerant cultivar (Gulen *et al.*, 2006). In strawberry, although a higher constitutive activity of the antioxidant enzymes CAT, APX and GR was found to be associated with the salt tolerance of two salt-tolerant cultivars ('Camarosa' and 'Tioga'), these enzymes showed variable expression levels in the cultivars (Turhan *et al.*, 2008).

Exogenous application of some chemicals or plant growth regulators has been also studied as a means of improving salt-stress tolerance in strawberry. Supplementary Ca can ameliorate the parameters affected by high salinity (e.g. plant growth, water use and membrane permeability) and can also correct Ca deficiency (Kaya *et al.*, 2002). Kaya *et al.* (2001) and Khayyat *et al.* (2009b) showed that supplementary K also reduces some of the negative effects of NaCl stress in strawberry. In addition, salicylic acid (SA) treatments induced increases in leaf relative water content and decreases in electrolyte leakage under salt-stress conditions in strawberry. SA treatments increased the levels of almost all nutrients in the leaves and roots of strawberry plants under salt stress (Karlidag *et al.*, 2009). The authors suggested that the SA treatments could reduce the negative effects of salinity on the growth of strawberries (Karlidag *et al.*, 2009). It was also shown that arbuscular mycorrhizae fungi (AMF) not only stimulated plant growth but also contributed towards enhancing the tolerance to salinity stress. AMF were able to alleviate the damage caused by salinity stress on strawberry plants and promoted their growth (Fan *et al.*, 2012). Christou *et al.* (2013) described treatment with hydrogen sulphide (H_2S) as a potential priming agent in strawberry plants under salt stress. Plants pretreated with H_2S showed tolerance to 100 mM NaCl or 10% polyethylene glycol 6000 for 7 days. Data from quantitative real-time reverse transcription polymerase chain reaction gene expression analysis of antioxidative defence agents and salt overly sensitive (SOS) pathway genes revealed the ameliorative effects of H_2S. Christou *et al.* (2014) also showed that the priming effects of nitric oxide and H_2O_2 root pre-treatment correlated with increased transcript levels of enzymatic antioxidants, as well as ASA and GSH biosynthesis in leaves of the cultivar 'Camarosa'.

Investigations have become focused on biotechnology, transgenic plants, improvement of breeding and screening methodologies, and modification of the genetic structure of existing crops, aiming at enhanced adaptation to salinity conditions (Dajic, 2006). In spite of the complexity of salt tolerance, transfer of a single or a few genes can increase the tolerance of plants to saline conditions (Flowers, 2004). Therefore, there has been much interest in transferring salt-tolerance genes from wild salt-tolerant relative species to the cultivated species by means of breeding programmes. Husaini and Abdin (2008) successfully developed transgenic strawberry lines in which the tobacco osmotin gene was introduced. Osmotin-induced proline accumulation in transgenic strawberry plants imparted salinity tolerance, and the plants showed increased biomass, protein content, and shoot and root lengths compared with the wild-type plants, under stress conditions. The transgenic strawberry plants overexpressing osmotin were able to maintain a higher chlorophyll content under salt stress. Thus, under salinity stress conditions, the transgenic strawberry plantlets overexpressing the osmotin gene performed better than their counterparts (wild-type and transgenic lines that failed to express osmotin), but under permissive conditions, the rate of their growth was somewhat slower, perhaps due to the use of a constitutive promoter, thereby continuously causing diversion of basal metabolites, even in the absence of stress (Husaini and Abdin, 2008). Studies of gene expression in strawberry require suitable reference genes. Recently, Galli *et al.* (2015) reported that the candidate reference gene *HISTH4* (encoding histone H4) was the most stable of the genes tested under osmotic and salt stress. The traditional genes encoding glyceraldehyde 3-phosphate dehydrogenase and 18S ribosomal RNA were considered the most unstable genes in strawberry plants in all conditions tested.

15.3. Temperature Stress

One of the major environmental factors affecting plant growth and productivity is temperature (Havaux, 1993). Strawberry plants are known to be highly sensitive to variations in environmental conditions (Wang and Camp, 2000). However, field and greenhouse-grown strawberries are often subjected to extreme temperatures during cultivation. Wang and Camp (2000) reported the optimum day/night temperatures for leaf and petiole growth as 25/12°C and for roots and fruits as 18/12°C. For growth of the whole plant and the highest starch and total carbohydrate contents, 25/12°C was also stated as the optimum temperature. In general, studies related to temperature effects on strawberry have mostly considered the effects of low temperatures or freezing, rather than high temperatures. The effects of extreme temperature stress are discussed in the following sections.

15.3.1. Cold stress

Strawberry is a moderately cold-hardy species. It is one of the most important small fruit crops in several northern parts of the world (Ndong *et al.*, 1997). Barney *et al.* (1992) reported the cold hardiness of strawberry cultivars over a wide temperature range from –3.8 to –45.0°C. However, chilling injury of strawberry plants is a serious problem, particularly in subtropical cultivated areas (Zhang *et al.*, 2008a). Freezing tolerance is expressed as the temperature that kills 50% of the whole plant or tissues (LT_{50}), and is calculated by an ion leakage test of the cell membrane (Arora *et al.*, 1992). The LT_{50} level thus indicates the cold hardiness of a cultivar. For strawberry cultivars 'Chamly', 'Red Coat' and 'Oka', the LT_{50} was reported as –12°C, while for the freezing-sensitive 'Chandler' it was was –2°C (Ndong *et al.*, 1997). The wild strawberry is known as the most cold hardy, with the highest LT_{50} (–16°C). Turhan *et al.* (2012a) recently demonstrated seasonal changes in cold hardiness of eight strawberry cultivars. The cold hardiness of leaf tissues in non-acclimated-stage plants varied from –6 to –12°C, whereas it was reported as –9 to –22°C in acclimated plants, depending on the cultivar. 'Ventana' and 'Kabarla' were reported as the most and least cold-hardy cultivars, respectively, among the eight cultivars evaluated.

Studies about mechanisms of cold stress in strawberry have mostly focused on physiological and metabolic changes during cold acclimation or under cold-stress conditions, such as changes in protein content, antioxidative enzyme activities, alterations in lipid composition and soluble sugar concentrations. Freezing tolerance is induced in perennial and winter annual plants by low temperature and a short photoperiod, which are characteristics of autumn (McKersie and Leshem, 1994). Plant species differ greatly in their ability to develop cold or freezing tolerance through a process known as cold acclimation, which is the main factor in the plant's survival. Thus, understanding the mechanism involved in cold acclimation has great importance in the enhancement of cold tolerance of species, and is of considerable interest in terms of preventing cold damage. Some hormonal mechanisms are also important in plant cold acclimation. Rajashekar *et al.* (1999) showed that exogenous applications of abscisic acid (ABA) and glycine betaine to unhardened plants increased the cold tolerance of leaves. To elucidate the molecular basis of cold acclimation in strawberry, Ndong *et al.* (1997) conducted a study to identify genes associated with cold acclimation. The level of *Fcor1* (*Fragaria Cold-Regulated1*) transcript accumulation was correlated with the freezing tolerance of the strawberry cultivars 'Champly', 'Red Coat', 'Oka' and 'Chandler' and with wild strawberry (*Fragaria virginiana*), suggesting that *Fcor1* might be useful as a molecular marker.

According to our current knowledge, biochemical changes that have been associated with cold acclimation in various plant species include alterations in lipid composition, increased sugar and soluble protein content, expression of specific proteins, the appearance of new isozymes and alterations in the activity of antioxidative enzymes

(Thomashow, 1999; Sarnighausen *et al.*, 2004; Eris *et al.*, 2007; Gulen *et al.*, 2008, 2009; Cansev *et al.*, 2009). These metabolically active processes are regulated at the gene expression level and are associated with the induction of specific proteins that lead to an increased capacity to withstand cold temperatures (Guy, 1990; Hughes and Dunn, 1996; Danyluk *et al.*, 1998; Thomashow, 1999). Cold-induced genes and their products have been isolated and characterized in many species (Houde *et al.*, 1995; Hughes and Dunn, 1996; Limin *et al.*, 1997; Sarhan *et al.*, 1997; Breton *et al.*, 2000). Many of the cold tolerance-responsive genes have been predicted to encode proteins with characteristics of the dehydrin family of late embryogenesis abundant (LEA) proteins (Hughes and Dunn, 1996). The dehydrin family of proteins has been shown in many studies to be related to cold stress (Close, 1996). Dehydrin proteins accumulate at high levels under cold stress and are different from functionally known enzymes or proteins. Davik *et al.* (2013) demonstrated the variation in cold/freezing tolerance among 22 diploid *Fragaria* genotypes and a correlation of plant survival with expression of dehydrin and alcohol dehydrogenase proteins in the crown during acclimation.

The WCOR410 protein has been shown to be associated with plasma membrane cold tolerance in wheat (Danyluk *et al.*, 1998), and was employed by Houde *et al.* (2004) in strawberry to increase freezing tolerance using transformation of the WCOR410 acidic dehydrin gene. Data from the study demonstrated that the WCOR410 protein prevents membrane injury and greatly improves freezing tolerance in the leaves of transgenic strawberry. Koehler *et al.* (2012) also reported stress-responsive proteins in the crown tissue of strawberry plants exposed to cold treatments. Recently, Gu *et al.* (2013) investigated protein changes in non-transgenic and *rd29A:RdreB1B1*-transgenic strawberry leaves following low-temperature treatment. Twenty-one differentially displayed proteins that potentially participate in photosynthesis, carbohydrate metabolism, protein biosynthesis and the response to stress were identified. Proteins were identified as RuBisCO activase, RuBisCo large subunit, Cu/Zn-SOD, Lea14-A, elF5A and cold-stress-related expressed sequence tags (ESTs), all of which were sensitive to low-temperature stress and were differentially regulated between non-transgenic and *rd29A:RdreB1B1*-transgenic strawberries. Thus, they concluded that the *RdreB1B1* gene may elevate photosynthesis- and defence-related protein accumulation to increase plant cold tolerance. More recently, Rajashekar and Panda (2014) reported expression of the cold-responsive genes *COR47* and *COR78*, which are responsible for the cold acclimation process and freezing tolerance in strawberry plants.

Regarding the antioxidative defence mechanism, Gulen *et al.* (2008) studied alterations in the activity of PRX isozyme and malondialdehyde content (an indication of cellular damage through lipid peroxidation) during the process of cold acclimation in the strawberry cultivar 'Camarosa'. The possible role of the acidic and basic PRX bands with Rf = 0.23 and Rf = 0.17, respectively, was emphasized in cold acclimation of strawberry leaf tissues. Zhang *et al.* (2008a) reported that chilling acclimation at 0°C resulted in the enhancement of SOD, CAT, PRX, dehydroascorbate reductase, monodehydroascorbate reductase, APX and GR activities, and increased the levels of reduced ASA, dehydroascorbate and reduced GSH in the strawberry cultivar 'Toyonaka'. Thus, it was concluded that the ASA–GSH cycle plays an important role in the detoxification of H_2O_2, which is associated with the enhancement of chilling resistance in strawberry. In a study on the strawberry cultivars 'Toyonaka' and 'Zoji', the fast increase and maintenance at a high level of production of $O_2^{\bullet-}$ and the H_2O_2 in strawberry leaves treated at a low temperature suggested that $O_2^{\bullet-}$ and H_2O_2 might serve as signal molecules (Zhang *et al.*, 2008b). More recently, Turhan *et al.* (2012a) reported considerable increases in APX and GR enzyme activity in cold-acclimated plants and indicated a marker potential of these two enzymes in cold hardiness of strawberry plants. As ice formation occurs primarily in the intercellular spaces during freezing injury, Turhan *et al.*

(2012b) studied apoplastic antioxidant enzymes in the leaves of two strawberry cultivars, 'Aromas' and 'Diamante', to investigate their role in cold hardiness. It was found that the activities of apoplastic CAT, PRX and APX varied significantly depending on the cold-acclimation stage and cold-hardiness level of the cultivar. A lower malondialdehyde content and higher total carotenoid and enzyme activities were also reported during the hardening stage.

Carbohydrate metabolism is also very important in cold tolerance of plants. It is known that sugars reduce the freezing point and increase the intracellular osmotic potential (Levitt, 1980). Paquin *et al.* (1989) showed a direct effect of total soluble sugars, reducing sugars and sucrose in the cold hardiness of two strawberry cultivars, 'Redcoat' and 'Bounty'. Seasonal alterations in apoplastic and symplastic sugar metabolism were also explained clearly by Turhan (2012). Total soluble sugar, reducing sugar and sucrose content were higher in the cold-acclimated stage than in the non-acclimated stage in the symplast. Thus, symplastic sugar metabolism was found to be more effective in cold tolerance than apoplastic sugar metabolism. In addition, the regulatory role of acid invertase and sucrose synthase enzyme activities were reported in cold acclimation of strawberry plants (Turhan, 2012).

15.3.2. Heat stress

Heat stress is often defined as the increase in temperature above a threshold level for a period of time sufficient to cause irreversible damage to plant growth and development (Wahid *et al.*, 2007). According to Bray *et al.* (2000), a heat-stress response is observed in plants exposed to the temperatures at least 5°C above their optimal conditions. As an abiotic stress factor, high temperatures have a negative effect on plant growth and development, and limit crop production around the world. Exposure to high temperatures (≥35°C) causes a reduction in plant growth and yield (Renquist *et al.*, 1982; Hellman and Travis, 1988). It was also reported that cells of the strawberry cultivar 'Shikinari' grew very slowly and did not proliferate normally at 30°C in suspension cultures (Zhang *et al.*, 1997). However, thermotolerance can be developed as plant acclimation to a high temperature. High-temperature tolerance is usually expressed as the temperature required to produce a defined level of injury or a defined level of dead plants (usually 50%). Tolerance to high-temperature stress results from the ability to maintain high rates of photosynthesis and stability of proteins (which must resist conformational changes at high temperatures), the ability to repair or synthesize structural proteins and the possession of substances that protect proteins from change. There is evidence that the lower the hydration of protoplasm, the greater the hydrogen bonding and the greater the tolerance to high temperatures (Hale and Orcutt, 1987). One hypothesis on high-temperature acclimation is that there may be several mechanisms of adaptation to high temperature involving induction of protein synthesis or altered protein function (Sun *et al.*, 2002). Those plants that do not adapt may be incapable of carrying out the structural or functional change in the proteins.

Acclimation of plants to higher temperatures is minimal (a few degrees) compared with acclimation to drought or to freezing temperatures (Salisbury and Ross, 1992). Heat-shock proteins (HSPs) also appear when the increase in temperature is more gradual, as might occur under natural conditions, in order to acclimate the plants to high temperature (Salisbury and Ross, 1992). Gulen and Eris (2004) studied heat acclimation in strawberry plants under gradual heat stress. Gradually heat-stressed plants showed less membrane injury and exhibited greater heat-stress tolerance than shock heat-stressed plants. Thus, gradual high-temperature stress with long exposure increased heat-stress tolerance in strawberry leaves. Increased heat-stress tolerance was correlated with the accumulation of several heat-stable proteins in gradually heat-stressed plants. In addition significant differences between gradually and shock heat-stressed plants were reported in leaf relative water content, loss of turgidity, and total protein and DNA content.

Gulen and Eris (2003) also found that heat stress increased the chlorophyll content in plants of the strawberry cultivar 'Camarosa'. Similarly, Kesici *et al.* (2013) reported an increase in chlorophyll content of strawberry plants under heat stress.

Studies on the mechanism of stress tolerance in different strawberry species and cultivars are important for the development of heat-tolerant cultivars, but there is limited knowledge about the tolerance to heat stress of the commonly grown cultivars. Recently, Kesici *et al.* (2013) studied the heat tolerance of 15 commonly grown strawberry cultivars. The LT_{50} ranged from 51.8 and 52.9°C. The cultivars 'Elsanta', 'R. Hope' and 'Camarosa' were determined as relatively heat tolerant, while 'Whitney', 'Fern', 'Festival' and 'CG3' were relatively heat sensitive.

Regarding the role of antioxidant enzymes, similar to other stress factors, high temperature also increased the activities APX, PRX and CAT in strawberry plants (Ergin, 2012). However no effect was reported on GR activity under heat stress. The activity of PRX was closely related to cell membrane stability and lignification; thus, studies have mostly focused on PRX as a potential enzyme marker. Gulen and Eris (2004) reported one basic PRX band (Rf = 0.22) associated with cell membrane recovery and heat acclimation in plants of the strawberry cultivar 'Camarosa' under high-temperature stress. Similarly, Ergin *et al.* (2012) showed an increase in activity of basic PRX band (Rf = 0.55) under heat-stress treatments in two strawberry cultivars known to be heat tolerant ('R. Hope') and heat sensitive ('CG3'), and Ergin *et al.* (2014) reported a correlation between ASA application and high temperature. Plants to which ASA was applied showed less loss of turgidity and a higher relative water content under heat stress, and ASA content and PRX activity were higher, especially in the heat-tolerant 'R. Hope', compared with plants in which ASA was not applied, Thus, exogenously applied ASA may be effective in ameliorating the adverse effects of heat stress in strawberry (Ergin *et al.*, 2014).

Although a reduction in the contents of total soluble proteins and nucleic acids (DNA and RNA) has been observed under high-temperature stress in some plants (Waters and Schaal, 1996; Chaitanya *et al.*, 2001), the changes in protein that occurred and the adaptation to high temperature remain poorly understood. The potential role of HSPs in heat tolerance of strawberry has been noted in a limited number of studies. Ledesma *et al.* (2004) reported 19–29 kDa HSPs in leaves and 16–26 kDa HSPs in flowers of the strawberry cultivars 'Nyoho' and 'Toyonaka' under high-temperature conditions. More recently, Ergin (2012) detected a 23 kDa HSP in leaf tissues during heat acclimation of strawberry cultivars. Accumulation of the 23 kDa HSP was reported to be higher in heat-tolerant cultivars than in heat-sensitive cultivars, and thus correlated with heat tolerance in strawberry. The HSPs in these two studies could be considered as markers for future research.

In order to understand the molecular responses of strawberry to abiotic stresses, *Fragaria vesca* has been used, rather than cultivated hybrid strawberry species. ESTs were determined for this species for abiotic stresses including a combination of heat and osmotic stresses (Rivarola *et al.*, 2011). This EST collection includes useful data to improve our knowledge on the molecular responses of plants to abiotic stresses.

15.4. Water Stress

Water stress can be defined as either water excess (flooding) or water deficit (drought) (Mahajan and Tuteja, 2005). Drought stress decreases strawberry leaf water potential, transpiration rate and water channel activity, despite an increase in ABA. However, leaf water potentials and water channel activity were found to remain unchanged in flooded strawberry plants (Blanke and Cooke, 2004, 2006). These authors also reported that stomata closed more rapidly in flooded strawberry leaves than in drought-stressed ones, in spite of ABA delivery from the roots being significantly suppressed. Thus, strawberry plants appear to be more adversely affected by drought than by flooding.

Strawberry plants are known for their high water demand because of their shallow root system (almost all of the roots are in 0–15 cm zone), large leaf area and the high water content of the fruit (Chandler and Ferree, 1990; Klamkowski and Treder, 2006). They are therefore very susceptible to water-deficit stress. According to Sruamsiri and Lenz (1986), photosynthesis and transpiration decreased when the leaf water potential reached –1.0 MPa. Thus, it is important to optimize the irrigation strategies in strawberry cultivation. A partial root-zone drying (PRD) irrigation technique was developed to stimulate root growth and increase water-use efficiency because of general opinion about the better performance of PRD in comparison with deficit irrigation. However, Liu *et al.* (2007) demonstrated no significant difference between PRD-grown and deficit-irrigated strawberry plants when considering water-use efficiency and other physiological measurements. According to Jensen *et al.* (2009), PRD and deficit-irrigation strategies can be applied in short-term strawberry production in relation to water-use efficiency and root xylem ABA induction. However, PRD irrigation is not suggested in commercial strawberry production due to its cost and management complexity.

There are strong relationships among solar irradiance, temperature, field topography, soil-water content, water management, and fruit yield and quality in horticultural crops. According to Li *et al.* (2010), in strawberry plant, water stress occurs on the slope and shoulder areas, where there is high irradiance and temperature and low soil-water content, resulting in significantly lower fruit yield. Drought stress has an inhibitory effect on plant growth and yield in strawberry (Chandler and Ferree, 1990). Significant reductions in leaf water potential, fresh and dry masses, and leaf area and leaf number, and a gradual reduction in photochemical quenching and quantum efficiency were reported in strawberry plants under water stress (Razavi *et al.*, 2008). Moreover, it is known that the yield and fruit size of strawberries are reduced when plants are subjected to drought stress during flowering and fruit development (Liu *et al.*, 2007).

There are limited studies on the genotypic differences among members of the genus *Fragaria* and among genotypes in terms of drought tolerance, and more knowledge on cultivar variation for drought tolerance is needed for effective breeding programmes. Variation in drought tolerance has been indicated among *Fragaria* spp., and *Fragaria chiloensis* has been reported as the most drought-tolerant species (Zhang and Archbold, 1993). Regarding strawberry genotypes, Klamkowski and Treder (2008) reported that the strawberry cultivar 'Elsanta' was the most drought tolerant among the three cultivars tested, based on morphological and physiological measurements. Razavi *et al.* (2009) determined an ecophysiologically relevant grouping among 23 strawberry genotypes based on amplified fragment length polymorphism markers linked to the leaf relative water content and water-loss rate. Johnson *et al.* (2009) screened ten short-day strawberry cultivars to investigate the effect of water deficits on the vegetative and generative growth of the plants to determine their physical adaptations to drought stress. They reported that water-use efficiency (fruit yield) of the cultivars differed in their response to water deficit (66%), as an indication of the capability of the cultivars in developing tolerance to drought. Grant *et al.* (2010) also compared the performance of ten strawberry cultivars under water deficit (70%) and reported some drought-tolerant characteristics such as osmotic adjustment, small leaf area and transpiration rate as useful data in a breeding programme. However, studies on the combined effects of heat and water stress are needed to fully understand this scenario because of the close relationship between high temperature and water demand in plants. Recently, Çetinkaya (2013) showed a direct relationship between heat and drought tolerance of strawberry cultivars using physiological and molecular tools. The heat-tolerant cultivars 'R. Hope' and 'Camarosa' showed drought tolerance, while the heat-sensitive 'Festival' and 'CG3' showed drought sensitivity during drought and recovery treatments.

Stress avoidance and tolerance mechanisms have been described in drought stress, similar to the effect of other stress factors (Levitt, 1980). A drought-induced series of morphological and physiological responses in addition to lower fruit production were shown during stress tolerance and adaptation mechanisms in strawberry (O'Neill, 1983; Savé *et al.*, 1993; Klamkowski and Treder, 2006). The physiological basis underlying differences in sensitivity of different-aged leaves to water stress in strawberry was correlated with changes in cell size and other anatomical alterations in the leaf that affected the cellular osmotic volume and cellular water relationships (O'Neill, 1983). In addition, Savé *et al.* (1993) demonstrated changes in leaf osmotic and elastic properties and in the canopy structure of the strawberry cultivar 'Chandler' under mild water stress following an irrigation regime at a soil matric potential of –70 kPa. Strawberry plants exposed to water stress showed a tolerance mechanism based on high elasticity of the tissues as a morphological adaptation. The changes in whole-plant morphology and canopy architecture, from monolayer to polylayer leaf distribution and leaf orientation from south to north, were also reported as an indication of stress avoidance mechanisms in water-stressed plants. Klamkowski and Treder (2006) reported a reduction in growth of the above-ground part of a strawberry plant and an increase in root system development under water-deficit conditions. Determination of modifications in stomatal density and the size of stomata were used as an indication of water-use efficiency in the periods of decreased water.

External application of some plant growth regulators or similar chemicals has also been studied as a means of reducing water stress in strawberry plants. ABA and jasmonic acid or its methyl ester, methyl jasmonate, are the most effective plant growth regulators under osmotic-stress conditions due to their effects on a series of physiological and biochemical events in plant metabolism. ABA and jasmonic acid/methyl jasmonate are known as stress hormones (growth inhibitors), and when used as a foliar spray can inhibit stomatal opening, cell division, plant growth, photosynthesis activities, flower-bud formation, seed germination and embryogenesis, while they limit water loss and ameliorate injuries related to osmotic stress (Salisbury and Ross, 1992; Wang, 1999). The effect of methyl jasmonate on changes in transpiration, antioxidant enzyme activities and membrane lipid composition was studied in strawberry leaves under water stress by Wang (1999). Plants treated with methyl jasmonate showed a better ability to withstand water stress through alterations in their metabolism, such as increased CAT and SOD activity and reductions in transpiration, membrane lipid peroxidation, membrane lipids, glycolipids and phospholipids. In addition to plant hormones, some chemical applications such as Vapor Gard® and prohexadione-calcium (ProCa) have been reported to reduce water stress during first 2 weeks of strawberry transplantation, which is a critical stage of growth, when transplants are highly susceptible to water stress (Duval *et al.*, 2001; Reekie *et al.*, 2007). ProCa is known to be a gibberellin biosynthesis inhibitor, and has been shown to regulate leaf morphology in tolerance to water stress in the strawberry cultivars 'Camarosa' and 'Sweet Charlie' when used as a foliar spray (Reekie *et al.*, 2007). It was also reported that application of ProCa prevented the excessive water usage required to ensure the survival of strawberry transplants. Recently, Caulet *et al.* (2014) demonstrated the effect of external application of two furostanol glycosides on the capacity for acclimation to drought of strawberry plants by increased root development and water-use efficiency. Another external application that can be used to alleviate the effects of drought stress is inoculation of AMF. In general, mycorrhizal fungi stimulate root development, increase the uptake of plant nutrients from growth media and improve stress tolerance in plants. Borkowska (2002) showed that application of AMF strongly affected the growth and tolerance to drought stress of the strawberry cultivar 'Senga Sengana' cultivated in the greenhouse.

The development of transgenic strawberry plants has been well described and a protocol for *Agrobacterium*-mediated transformation of strawberry leaf discs has been reported by Husaini (2010) and Husaini *et al.* (2011). Husaini *et al.* (2012) also reported the role of overexpression of the osmotin gene (encoding a PR-5c protein) in tolerance to drought stress in three transgenic strawberry lines developed using *Agrobacterium*-mediated transformation. Transgenic strawberry plants have been produced with tolerance to drought stress by Chalavi and Raeini-Sarjaz (2012) using the coding region of a chitinase gene from *Lycopersicon chilense*. Chitinase is a 25–35 kDa protein and is expressed under biotic and abiotic stress conditions (Hong and Hwang, 2006). Expression of the chitinase gene is also related to the growth and development of strawberry plants exposed to water stress. Chalavi and Raeini-Sarjaz (2012) demonstrated that chitinase expression increased stomatal conductance and the photosynthetic rate in both well-watered and moderately watered transgenic strawberry lines in comparison with non-transgenic plants.

15.5. Conclusions and Future Prospects

All plants are subjected to various abiotic stresses in natural environments during their lives. Drought, salinity, temperature and heavy-metal stresses are among the major abiotic stresses that adversely affect plant growth and productivity. Almost all farmable land over the world is under sub-optimal conditions for plant growth and development as a result of various abiotic stresses. Serious yield reductions are recorded due to unfavourable physiochemical environments, even in high-tech agriculture. Alongside these losses, because of the possible effects of global climate changes it is thought that yield reductions in the future will be greater than expected. However, the world population is increasing and is estimated to reach more than 9 billion in 2050, so a considerable increase in world food production will be necessary. One of the most important ways of increasing productivity under stressful conditions is the development of new cultivars that can tolerate or avoid stress. Generally, breeding programmes, cross-breeding and more recently marker-assisted selection have been considered in relation to varietal improvement. As plant responses to stress are highly complex and depend on a number of interrelated factors based on morphological, biochemical and physiological processes, multidisciplinary work is needed for varietal improvement. In this respect, specific physiological and molecular data are required, as well as basic knowledge about a particular plant under a certain and/or multiple stress factor(s). Thus, physiological and molecular research into the mechanisms of abiotic stress in different species and cultivars has great importance for future strawberry breeding programmes.

Plants respond to extreme levels of environmental conditions in a wide variety of ways. The severity of the plant response also depends on other environmental conditions, such as plant species/cultivar, age of the plant and its physiological stage. Most studies have focused on salt, temperature and water stresses in strawberry, and there are very few data available about the response of strawberry plants to other abiotic factors such as heavy metals, radiation and nutrients. Thus, studies investigating the effects of these factors should be designed for future expectations. In addition, multidisciplinary studies considering the effects of more than one stress factor are necessary to understand the impact of stressful conditions on plant growth and development. As an example, when water is limiting, the effects of high temperature are increased.

A large number of researchers have devoted their efforts to investigating the possible effects of a stress factor on a particular tissue and physiological stage of a plant, and these results are summarized in Tables 15.1 and 15.2. Some studies have focused on investigating stress-associated genes in order to understand the molecular mechanism of abiotic stress. The primary objective of breeding programmes is generally to reduce the gap between yield potential and actual yield under unfavourable (stressful) conditions for sustainable agriculture. Conventional breeding or marker-assisted selection using molecular techniques,

Table 15.1. Morphological and physiological effects of abiotic stresses in strawberry plants.

Stress type	Response	Parameter	Reference(s)
Salinity	Negative	Leaf water potential and osmotic potential	Awang *et al.* (1993a); Awang and Atherton (1994)
	Negative	Stomatal conductance (g_s) and transpiration ratio (*E*)	Turhan and Eris (2007a)
	No effect	K, Na, Ca, and Mg content in 'Rapella'	Awang and Atherton (1994)
	Positive	Increased Ca, and Mg in its aerial parts 'Camarosa'	Turhan and Eris (2005b)
	Negative	K and P decreased in 'Camarosa'	Turhan and Eris (2005b)
	Increase	Na and Cl contents	Barroso and Alvarez (1997); Turhan (2002); Turhan and Eris (2005b, 2009)
	No effect	P content in roots	Turhan and Eris (2005b)
	Positive	P content in roots, petioles and fruits	Keutgen and Pawelzik (2009)
	No effect	Ca content	Turhan and Eris (2005b); Keutgen and Pawelzik (2009)
	Negative	Mg, Mn and Fe in leaves	Keutgen and Pawelzik (2009)
	Positive	Mg and Mn in crowns, N content in leaves, petioles, and roots and in fruits and crowns, P content in roots and petioles and in fruits	Keutgen and Pawelzik (2009)
	Positive	Fe, Zn, and Mn concentrations in shoot	Turhan and Eris (2005a); Khayyat *et al.* (2009a)
	Positive	Cu content in root	Turhan and Eris (2005a); Khayyat *et al.* (2009a)
	No effect	Cu content in shoot	Turhan and Eris (2005a); Khayyat *et al.* (2009a); Keutgen and Pawelzik (2009)
	No effect	Zn content	Keutgen and Pawelzik (2009)
	Negative	Causes necrosis	Barroso and Alvarez (1997); Turhan and Eris (2005b); Keutgen and Pawelzik (2009)
	Negative	Increased electrolyte leakage	Kaya *et al.* (2002); Gulen *et al.* (2006)
	Negative	Leaf number, leaf area, plant dry weight, root and shoot growth	Awang *et al.* (1993a); Turhan and Eris (2007a); Keutgen and Pawelzik (2009)
	Negative	Fruit productivity	Awang *et al.* (1993a), Awang and Atherton (1995a,b), Turhan (2002)

Continued

Table 15.1. Continued.

Stress type	Response	Parameter	Reference(s)
	Positive	Improves fruit quality (moderate salinity)	Awang *et al.* (1993b) Awang and Atherton (1995b); Kaya *et al.* (2002); Keutgen and Pawelzik (2008)
	Positive	Antioxidant capacity, and some minerals and lipid peroxidation in fruits, contents of free and essential amino acids, reduced GSH	Keutgen and Pawelzik (2008)
	No effect	Fruit dry matter content	Awang *et al.* (1993b)
	Negative/no effect (depending on cultivar, exposure time and salt concentration)	Chlorophyll content	Kaya *et al.* (2002); Turhan and Eris (2005a); Rahimi *et al.* (2011); Garriga *et al.* (2014)
	Negative	Net photosynthesis	Awang and Atherton (1994)
	Positive	Increase in H_2O_2 and $O_2^{\bullet -}$ accumulation, lipid peroxidation and carbonyl groups, therefore induces oxidative stress	Tanou *et al.* (2009)
	Positive	Proline content and soluble sugars	Turhan and Eris (2009); Rahimi *et al.* (2011)
	Negative	Total amino acids in leaves	Turhan and Eris (2009)
	Positive	Total amino acid levels, especially contents of essential amino acids in fruits	Keutgen and Pawelzik (2008)
	Positive	Specific PRX activities	Gulen *et al.* (2006)
	Negative/ no effect	Total soluble protein content in leaf tissue depending on cultivar	Gulen *et al.* (2006)
	Positive	CAT, APX and GR activities, but variable expression in salt-tolerant cultivars	Turhan *et al.* (2008)
Temperature			
Cold	Positive	*Fcor1* transcript accumulation was correlated with the freezing tolerance	NDong *et al.* (1997)
	Positive	Stress-responsive proteins in crown tissue	Koehler *et al.* (2012)
	Positive	PRX isozyme activity and MDA content	Gulen *et al.* (2008)
	Positive	SOD, CAT, PRX, DHAR, MDHAR, reduced ASA, DHA	Zhang *et al.* (2008a)
	Negative	GSH contents	Zhang *et al.* (2008a)
	Negative	Increased $O_2^{\bullet}$– generation and content of H_2O_2	Zhang *et al.* (2008b)
	Positive	APX and GR	Zhang *et al.* (2008a); Turhan *et al.* (2012a)
	Positive	CAT, PRX and APX in apoplast	Turhan *et al.* (2012b)
	Positive	Total soluble sugars, reducing sugars and sucrose	Paquin *et al.* (1989); Turhan (2012)
Heat	Positive	Causes less membrane injury (during gradual heat stress)	Gulen and Eris (2004)
	Positive	Chlorophyll content	Gulen and Eris (2003); Kesici *et al.* (2013)

Continued

Table 15.1. Continued.

Stress type	Response	Parameter	Reference(s)
	Positive	Increased heat-stress tolerance correlated with the accumulation of several heat-stable proteins in gradually heat-stressed plants	Gulen and Eris (2004)
	Positive	APX, PRX and CAT activities	Ergin (2012)
	No effect	GR activity	Ergin (2012)
	Positive	Accumulation of HSPs	Ledesma *et al.* (2004); Ergin (2012)
	Positive	Produced ESTs in *F. vesca*	Rivarola *et al.* (2011)
Water	Negative	Leaf water potential, transpiration rate and water channel activity	Blanke and Cooke (2004, 2006)
	Positive	ABA content	Blanke and Cooke (2004, 2006)
	Negative	Photosynthesis and transpiration	Sruamsiri and Lenz (1986)
	Negative	Growth and yield	O'Neill (1983); Chandler and Ferree (1990); Savé *et al.* (1993); Klamkowski and Treder (2006); Liu *et al.* (2007)
	Positive	Root system development	Klamkowski and Treder (2006)
	Negative	Leaf water potential, fresh and dry masses, leaf area and leaf number, gradual reduction of photochemical quenching and quantum efficiency	Razavi *et al.* (2008)
	Neutral	Leaf water potentials and water channel activity (during flooding)	Blanke and Cooke (2004, 2006)

GSH, glutathione; PRX, peroxidase; CAT, catalase; APX, ascorbate peroxidase; GR, glutathione reductase; MDA, malondialdehyde; SOD, superoxide dismutase; DHAR, dehydroascorbate reductase; MDHAR, monodehydroascorbate reductase; ASA, ascorbate; DHA, dehydroascorbate; HSP, heat-shock protein; ABA, abscisic acid.

or a combination of both, are being used to develop stress-tolerant cultivars. However, abiotic stresses are highly complex phenomena. Furthermore, cultivated strawberry is a hybrid species and an octoploid, making some studies difficult. Therefore, the diploid strawberry *F. vesca* has been an attractive model plant for genomic analysis because of its small genome size. Recently, an *F. vesca* EST collection has been established, and data from proteomic studies on abiotic stresses are proving useful and promising tools for further studies. A similar set of genes and stress-induced proteins have been described for tolerance to abiotic stresses. Transgenic strawberry plants that are tolerant to certain stress factors have been developed using genetic engineering approaches. However, results are still at the laboratory level rather than in field conditions where the whole cycle of plant growth and development can be considered. A detailed understanding of the complex mechanisms of stress and interactions with other related factors in the plant response is still needed. In this respect, variation among genotypes in response to stress factors should be analysed as a first step of any breeding programme to improve stress tolerance in plants. Thus, screening studies should be established to investigate the stress tolerance of different genotypes for each abiotic stress factor. A more complete understanding of the physiological and molecular mechanisms of the stress response in tolerant plants may help in the development of new cultivars and may prevent severe injury caused by stress factors. Research and breeding studies need not only to be based on present worldwide conditions but also planned to meet potential future conditions.

Table 15.2. Some exogenous applications to enhance stress tolerance in strawberry plants

Treatment	Stress type	Function	Reference
Calcium	Salinity	Ameliorates plant growth, water use and membrane permeability; prevents calcium deficiency	Kaya *et al.* (2002)
Potassium	Salinity	Positively influences leaf area development, chlorophyll content and reproductive parameters	Kaya *et al.* (2001); Khayyat *et al.* (2009b)
Arbuscular mycorrhizae fungi	Salinity	Promotes growth and increases plant dry weight	Fan *et al.* (2012)
Salicylic acid	Salinity	Induces increases in leaf water relative content and decreases in electrolyte leakage; increases contents of almost all nutrients in leaves and roots	Karlidag *et al.* (2009)
H_2S	Salinity	Antioxidative defence agents	Christou *et al.* (2013)
NO, H_2O_2	Salinity	Increases transcript levels of enzymatic antioxidants, ascorbate and glutathione biosynthesis	Christou *et al.* (2014)
Abscisic acid, glycine betaine	Cold	Increases the cold tolerance of leaves	Rajashekar *et al.* (1999)
ASA	Heat	Reduces loss of turgidity; higher RWC and ASA content, and higher PRX activity, especially in heat-tolerant cultivars	Ergin *et al.* (2014)
Methyl jasmonate	Water	Increases CAT and SOD activities and reducts transpiration, membrane lipid peroxidation, membrane lipids, glycolipids and phospholipids	Wang (1999)
Vapor Gard®, ProCa	Water	Regulates leaf morphology and prevents excessive water usage	Reekie *et al.* (2007)
Two furostonal glycosides	Water	Increases root development and water-use efficiency	Caulet *et al.* (2014)
Arbuscular mycorrhizae fungi	Water	Stimulates root development; increases uptake of plant nutrients	Borkowska (2002)

RWC, relative water content; ASA, ascorbate; CAT, catalase; SOD, superoxide dismutase.

References

Allen, R.D. (1995) Dissection of oxidative stress tolerance using transgenic plants. *Plant Physiology* 107, 1049–1054.

Arora, R., Wisniewski, M.E. and Scorza, R. (1992) Cold acclimation in genetically related (sibling) deciduous and evergreen peach (*Prunus persica* [L.] Batsch): I. Seasonal changes in cold hardiness and polypeptides of bark and xylem tissues. *Plant Physiology* 99, 1562–1568.

Arshi, A., Abdin, M.Z. and Iqbal, M. (2002) Growth and metabolism of senna as affected by salt stress. *Biologia Plantarum* 45, 295–298.

Ashraf, M. (2009) Biotechnological approach of improving plant salt tolerance using antioxidants as markers. *Biotechnology Advances* 27, 84–93.

Awang, Y.B. and Atherton, J.G. (1994) Salinity and shading effects on leaf water relations and ionic composition of strawberry plants grown on rockwool. *Journal of Horticultural Science* 69, 377–383.

Awang, Y.B. and Atherton, J.G. (1995a) Effect of plant size and salinity on the growth and fruiting of glasshouse strawberry. *Journal of Horticultural Science* 70, 257–262.

Awang, Y.B. and Atherton, J.G. (1995b) Growth and fruiting responses of strawberry plants grown on rockwool to shading and salinity. *Scientia Horticulturae* 62, 25–31.

Awang, Y.B., Atherton, J.G. and Taylor, A.J. (1993a) Salinity effects on strawberry plants grown in rockwool: I. Growth and leaf water relations. *Journal of Horticultural Science* 68, 783–790.

Awang, Y.B., Atherton, J.G. and Taylor, A.J. (1993b) Salinity effects on strawberry plants grown in rockwool: II. Fruit quality. *Journal of Horticultural Science* 68, 791–795.

Barney, D.L., Davis, B.B. and Fellman, J.K. (1992) *Strawberry Production: Overview.* College of Agriculture, University of Idaho, Moscow, Idaho.

Barroso, M.C.M. and Alvarez, C.E. (1997) Toxicity symptoms and tolerance of strawberry to salinity in the irrigation water. *Scientia Horticulturae* 71, 177–188.

Ben-Asher, J., Tsuyuki, I., Bravdo, B.-A. and Sagih, M. (2006) Irrigation of grapevines with saline water: I. Leaf area index, stomatal conductance, transpiration and photosynthesis. *Agricultural Water Management* 83, 13–21.

Blanke, M. and Cooke, D. (2004) Effects of flooding and drought on stomatal activity, transpiration, photosynthesis, water potential and water channel activity in strawberry stolons and leaves. *Plant Growth Regulation* 42, 153–160.

Blanke, M.M. and Cooke, D.T. (2006) Water channels in strawberry, and their role in the plant's response to water stress. *Acta Horticulturae* 708, 65–68.

Borkowska, B. (2002) Growth and photosynthetic activity of micropropagated strawberry plants inoculated with endomycorrhizal fungi (AMF) and growing under drought stress. *Acta Physiologiae Plantarum* 24, 365–370.

Bray, E.A., Bailey-Serres, J. and Weretilnyk, E. (2000) Responses to abiotic stresses. In: Gruissem, W., Buchnnan, B. and Jones, R. (eds) *Biochemistry and Molecular Biology of Plants.* American Society of Plant Physiologists, Rockville, Maryland, pp. 1158–1249.

Breton, G., Danyluk, J., Ouellet, F.O. and Sarhan, F. (2000) Biotechnological applications of plant freezing associated proteins. *Biotechnology Annual Review* 6, 59–101.

Cansev, A., Gulen, H. and Eris, A. (2009) Cold-hardiness of olive (*Olea europaea* L.) cultivars in cold-acclimated and non-acclimated stages: seasonal alteration of antioxidative enzymes and dehydrin-like proteins. *Journal of Agricultural Science* 147, 51–61.

Caulet, R.-P., Gradinariu, G., Iurea, D. and Morariu, A. (2014) Influence of furostanol glycosides treatments on strawberry (*Fragaria* × *ananassa* Duch.) growth and photosynthetic characteristics under drought condition. *Scientia Horticulturae* 169, 179–188.

Çetinkaya, C. (2013) Physiological effects of drought and recovery treatments in some strawberry cultivars. MSc thesis, Uludag University, Bursa, Turkey.

Chaitanya, K.V., Sundar, D. and Reddy, A.R. (2001) Mulberry leaf metabolism under high temperature stress. *Biologia Plantarum* 44, 379–384.

Chalavi, V. and Raeini-Sarjaz, M. (2012) The response of transgenic strawberry plants overexpressing a drought induced gene to water stress. *Journal of Applied Botany and Food Quality* 85, 182–187.

Chandler, C.K. and Ferree, D.C. (1990) Response of 'Raritan' and 'Surecrop' strawberry plants to drought stress. *Fruit Varieties Journal* 44, 183–184.

Cheeseman, J.M. (1988) Mechanisms of salinity tolerance in plants. *Plant Physiology* 87, 547–550.

Christou, A., Manganaris, G.A., Papadopoulos, I. and Fotopoulos, V. (2013) Hydrogen sulfide induces systemic tolerance to salinity and non-ionic osmotic stress in strawberry plants through modification of reactive species biosynthesis and transcriptional regulation of multiple defence pathways. *Journal of Experimental Botany* 64, 1953–1966.

Christou, A., Manganaris, G.A. and Fotopoulos, V. (2014) Systemic mitigation of salt stress by hydrogen peroxide and sodium nitroprusside in strawberry plants via transcriptional regulation of enzymatic and non-enzymatic antioxidants. *Environmental and Experimental Botany* 107, 46–54.

Close, T.J. (1996) Dehydrins: emergence of a biochemical role of a family of plant dehydration proteins. *Physiologia Plantarum* 97, 795–803.

Dajic, Z. (2006) Salt stress. In: Rao, K.V.M., Raghavendra, A.S. and Reddy, K.J. (ed.) *Physiology and Molecular Biology of Stress Tolerance in Plants*. Springer, Dordrecht, The Netherlands, pp. 41–99.

Danyluk, J., Perron, A., Houde, M., Limin, A., Fowler, B., Benhamou, N. and Sarhan, F. (1998) Accumulation of an acidic dehydrin in the vicinity of the plasma membrane during cold acclimation of wheat. *The Plant Cell* 10, 623–638.

Davik, J., Koehler, G., From, B., Torp, T., Rohloff, J., Eidem, P., Wilson, R.C., Sønsteby, A., Randall, S.K. and Alsheikh, M. (2013) Dehydrin, alcohol dehydrogenase, and central metabolite levels are associated with cold tolerance in diploid strawberry (*Fragaria* spp.). *Planta* 237, 265–277.

Duval, J.R., Stanley, C.D., Hicklenton, P.R. and Waldo, E. (2001) *Alternative Treatment of Bare-Rooted Strawberry Transplants to Reduce Plant Establishment Irrigation Requirements*. Gulf Coast Research and Education Center, University of Florida, Dover, Florida.

El-Fouly, M.M. and Salama, Z.H. (1999) Can foliar fertilization increase plant tolerance to salinity. In: *Dahlia Greidinger International Symposium: Nutrient Management under Salinity and Water Stress*, 28 February–4 March. Technion-ITT Haifa, Israel, pp. 113–125.

Ergin, S. (2012) Effects of high temperature stress on enzymatic and non-enzymatic antioxidants and protein metabolism in strawberry plants. PhD thesis, Uludag University, Bursa, Turkey.

Ergin, S., Kesici, M. and Gulen, H. (2012) Changes in H_2O_2 and peroxidase activities in strawberry plants under heat stress. *Harran University Journal of the Faculty of Agriculture* 16, 25–35.

Ergin, S., Aydogan, C., Ozturk, N. and Turhan, E. (2014) Effects of ascorbic acid application in strawberry plants during heat stress. *Turkish Journal of Agricultural and Natural Sciences* Special Issue 2, 1486–1491.

Eris, A., Gulen, H., Barut, E. and Cansev, A. (2007) Annual patterns of total soluble sugars and proteins related to cold-hardiness in olive (*Olea europaea* L. 'Gemlik'). *Journal of Horticultural Science and Biotechnology* 82, 597–604.

Fan, L., Fang, C., Dubé, C., Deschênes, M., Dalpé, Y., Tao, S. and Khanizadeh, S. (2012) Arbuscular mycorrhiza alleviates salinity stress of strawberry cultivars under salinity condition. *Acta Horticulturae* 926, 491–496.

FAO (2015a) *Agri-environmental Statistics*. UN Food and Agriculture Organization, Rome, Italy.

FAO (2015b) *Agricultural Productions by Crops*. UN Food and Agriculture Organization, Rome, Italy.

Flowers, T.J. (2004) Improving crop salt tolerance. *Journal of Experimental Botany* 55, 307–319.

Galli, V., Borowski, J.M., Perin, E.C., Messias, R.d.S., Labonde, J., Pereira, I.d.S., Silva, S.D.d.A. and Rombaldi, C.V. (2015) Validation of reference genes for accurate normalization of gene expression for real time-quantitative PCR in strawberry fruits using different cultivars and osmotic stresses. *Gene* 554, 205–214.

Garriga, M., Retamales, J.B., Romero-Bravo, S., Caligari, P.D.S. and Lobos, G.A. (2014) Chlorophyll, anthocyanin, and gas exchange changes assessed by spectroradiometry in *Fragaria chiloensis* under salt stress. *Journal of Integrative Plant Biology* 56, 505–515.

Giuffrida, F., Leonardi, C. and Noto, G. (2001) Response of soilless grown strawberry to different salinity levels in the nutrient solution. *Acta Horticulturae* 559, 675–680.

Grant, O.M., Johnson, A.W., Davies, M.J., James, C.M. and Simpson, D.W. (2010) Physiological and morphological diversity of cultivated strawberry (*Fragaria* × *ananassa*) in response to water deficit. *Environmental and Experimental Botany* 68, 264–272.

Greenway, H. and Munns, R. (1980) Mechanisms of salt tolerance in nonhalophytes. *Annual Review of Plant Physiology* 31, 149–190.

Gu, X., Gao, Z., Zhuang, W., Qiao, Y., Wang, X., Mi, L., Zhang, Z. and Lin, Z. (2013) Comparative proteomic analysis of *rd29A:RdreB1BI* transgenic and non-transgenic strawberries exposed to low temperature. *Journal of Plant Physiology* 170, 696–706.

Gulen, H. and Eris, A. (2003) Some physiological changes in strawberry (*Fragaria* × *ananassa* 'Camarosa') plants under heat stress. *Journal of Horticultural Science and Biotechnology* 78, 894–898.

Gulen, H. and Eris, A. (2004) Effect of heat stress on peroxidase activity and total protein content in strawberry plants. *Plant Science* 166, 739–744.

Gulen, H., Turhan, E. and Eris, A. (2006) Changes in peroxidase activities and soluble proteins in strawberry varieties under salt-stress. *Acta Physiologiae Plantarum* 28, 109–116.

Gulen, H., Çetinkaya, C., Kadıoğlu, M., Kesici, M., Cansev, A. and Eris, A. (2008) Peroxidase activity and lipid peroxidation in strawberry (*Fragaria* × *ananassa*) plants under low temperature. *Journal of Biological and Environmental Science* 2, 95–100.

Gulen, H., Cansev, A. and Eris, A. (2009) Cold hardiness of olive (*Olea europaea* L.) cultivars in cold-acclimated and non-acclimated stages: seasonal alteration of soluble sugars and phospholipids. *Journal of Agricultural Science* 147, 459–467.

Guy, C.L. (1990) Cold acclimation and freezing stress tolerance: role of protein metabolism. *Annual Review of Plant Physiology and Plant Molecular Biology* 41, 187–223.

Hale, M.G. and Orcutt, D.M. (1987) *The Physiology of Plants under Stress*. John Wiley & Sons, New York.

Hasegawa, P.M., Bressan, R.A. and Handa, A.K. (1986) Cellular mechanisms of salinity tolerance. *HortScience* 21, 1317–1324.

Havaux, M. (1993) Rapid photosynthetic adaptation to heat stress triggered in potato leaves by moderately elevated temperatures. *Plant, Cell and Environment* 16, 461–467.

Hellman, E.W. and Travis, J.D. (1988) Growth inhibition of strawberry at high temperatures. *Advances in Strawberry Production* 7, 36–38.
Hong, J.K. and Hwang, B.K. (2006) Promoter activation of pepper class II basic chitinase gene, *CAChi2*, and enhanced bacterial disease resistance and osmotic stress tolerance in the *CAChi2*-overexpressing *Arabidopsis*. *Planta* 223, 433–448.
Houde, M., Daniel, C., Lachapelle, M., Allard, F., Laliberté, S. and Sarhan, F. (1995) Immunolocalization of freezing-tolerance-associated proteins in the cytoplasm and nucleoplasm of wheat crown tissues. *The Plant Journal* 8, 583–593.
Houde, M., Dallaire, S., N'Dong, D. and Sarhan, F. (2004) Overexpression of the acidic dehydrin WCOR410 improves freezing tolerance in transgenic strawberry leaves. *Plant Biotechnology Journal* 2, 381–387.
Hughes, M.A. and Dunn, M.A. (1996) The molecular biology of plant acclimation to low temperature. *Journal of Experimental Botany* 47, 291–305.
Husaini, A.M. (2010) Pre- and post-agroinfection strategies for efficient leaf disk transformation and regeneration of transgenic strawberry plants. *Plant Cell Reports* 29, 97–110.
Husaini, A.M. and Abdin, M.Z. (2008) Development of transgenic strawberry (*Fragaria* × *ananassa* Duch.) plants tolerant to salt stress. *Plant Science* 174, 446–455.
Husaini, A.M., Mercado, J.A., Schaart, J.G. and Teixeira da Silva, J.A. (2011) Review of factors affecting organogenesis, somatic embryogenesis and *Agrobacterium tumefaciens*-mediated transformation of strawberry. In: Husaini, A.M. and Mercado, J.A. (eds) *Genomics, Transgenics, Molecular Breeding and Biotechnology of Strawberry*. Global Science Books, UK, pp. 1–11.
Husaini, A.M., Abdin, M.Z., Khan, S., Xu, Y.W., Aquil, S. and Anis, M. (2012) Modifying strawberry for better adaptability to adverse impact of climate change. *Current Science* 102, 1660–1673.
Jensen, N.L., Jensen, C.R., Liu, F. and Petersen, K.K. (2009) Water relations and abscisic acid in pot-grown strawberry plants under limited irrigation. *Journal of the American Society for Horticultural Science* 134, 574–580.
Johnson, A.W., Grant, O.M., Davies, M.J., James, C.M., Passey, A.J. and Simpson, D.W. (2009) Determining the response of ten strawberry cultivars to water-limited conditions. *Acta Horticulturae* 842, 769–772.
Karlidag, H., Yildirim, E. and Turan, M. (2009) Salicylic acid ameliorates the adverse effect of salt stress on strawberry. *Scientia Agricola* 66, 180–187.
Kaya, C., Kirnak, H. and Higgs, D. (2001) An experiment to investigate the ameliorative effects of foliar potassium phosphate sprays on salt-stressed strawberry plants. *Australian Journal of Agricultural Research* 52, 995–1000.
Kaya, C., Kirnak, H., Higgs, D. and Saltali, K. (2002) Supplementary calcium enhances plant growth and fruit yield in strawberry cultivars grown at high (NaCl) salinity. *Scientia Horticulturae* 93, 65–74.
Kesici, M., Gulen, H., Ergin, S., Turhan, E., Ipek, A. and Koksal, N. (2013) Heat-stress tolerance of some strawberry (*Fragaria* × *ananassa*) cultivars. *Notulae Botanicae Horti Agrobotanici Cluj-Napoca* 41, 238–243.
Keutgen, A.J. and Keutgen, N. (2003) Influence of NaCl salinity stress on fruit quality in strawberry. *Acta Horticulturae* 609, 155–157.
Keutgen, A.J. and Pawelzik, E. (2007) Modifications of strawberry fruit antioxidant pools and fruit quality under NaCl stress. *Journal of Agricultural and Food Chemistry* 55, 4066–4072.
Keutgen, A.J. and Pawelzik, E. (2008) Quality and nutritional value of strawberry fruit under long term salt stress. *Food Chemistry* 107, 1413–1420.
Keutgen, A.J. and Pawelzik, E. (2009) Impacts of NaCl stress on plant growth and mineral nutrient assimilation in two cultivars of strawberry. *Environmental and Experimental Botany* 65, 170–176.
Khayyat, M., Rajaee, S., Sajjadinia, A., Eshghi, S. and Tafazoli, E. (2009a) Calcium effects on changes in chlorophyll contents, dry weight and micronutrients of strawberry (*Fragaria* × *ananassa* Duch.) plants under salt-stress conditions. *Fruits* 64, 53–59.
Khayyat, M., Vazifeshenas, M.R., Rajaee, S. and Jamalian, S. (2009b) Potassium effect on ion leakage, water usage, fruit yield and biomass production by strawberry plants grown under NaCl stress. *Journal of Fruit and Ornamental Plant Research* 17, 79–88.
Klamkowski, K. and Treder, W. (2006) Morphological and physiological responses of strawberry plants to water stress. *Agriculturae Conspectus Scientificus* 71, 159–165.
Klamkowski, K. and Treder, W. (2008) Response to drought stress of three strawberry cultivars grown under greenhouse conditions. *Journal of Fruit and Ornamental Plant Research* 16, 179–188.

Koehler, G., Wilson, R.C., Goodpaster, J.V., Sønsteby, A., Lai, X., Witzmann, F.A., You, J.-S., Rohloff, J., Randall, S.K. and Alsheikh, M. (2012) Proteomic study of low-temperature responses in strawberry cultivars (*Fragaria × ananassa*) that differ in cold tolerance. *Plant Physiology* 159, 1787–1805.
Ledesma, N.A., Kawabata, S. and Sugiyama, N. (2004) Effect of high temperature on protein expression in strawberry plants. *Biologia Plantarum* 48, 73–79.
Levitt, J. (1980) *Responses of Plants to Environmental Stresses*, 2nd edn. Academic Press, New York.
Li, H., Li, T., Gordon, R.J., Asiedu, S.K. and Hu, K. (2010) Strawberry plant fruiting efficiency and its correlation with solar irradiance, temperature and reflectance water index variation. *Environmental and Experimental Botany* 68, 165–174.
Limin, A.E., Danyluk, J., Chauvin, L.P., Fowler, D.B. and Sarhan, F. (1997) Chromosome mapping of low-temperature induced *Wcs120* family genes and regulation of cold-tolerance expression in wheat. *Molecular and General Genetics* 253, 720–727.
Liu, F., Savić, S., Jensen, C.R., Shahnazari, A., Jacobsen, S.E., Stikić, R. and Andersen, M.N. (2007) Water relations and yield of lysimeter-grown strawberries under limited irrigation. *Scientia Horticulturae* 111, 128–132.
Mahajan, S. and Tuteja, N. (2005) Cold, salinity and drought stresses: an overview. *Archives of Biochemistry and Biophysics* 444, 139–158.
Mansour, M.M.F. (2000) Nitrogen containing compounds and adaptation of plants to salinity stress. *Biologia Plantarum* 43, 491–500.
Martınez-Ballesta, M.C., Martınez, V. and Carvajal, M. (2004) Osmotic adjustment, water relations and gas exchange in pepper plants grown under NaCl or KCl. *Environmental and Experimental Botany* 52, 161–174.
McKersie, B.D. and Leshem, Y.Y. (1994) *Stress and Stress Coping in Cultivated Plants*. Kluwer Academic Publishers, Dordrecht, The Netherlands.
Meloni, D.A., Oliva, M.A., Martinez, C.A. and Cambraia, J. (2003) Photosynthesis and activity of superoxide dismutase, peroxidase and glutathione reductase in cotton under salt stress. *Environmental and Experimental Botany* 49, 69–76.
Munns, R. (2002) Comparative physiology of salt and water stress. *Plant, Cell and Environment* 25, 239–250.
Munns, R. and Termaat, A. (1986) Whole-plant responses to salinity. *Functional Plant Biology* 13, 143–160.
Na, Y.-W., Jeong, H., Lee, S.-Y., Choi, H., Kim, S.-H. and Rho, I. (2014) Chlorophyll fluorescence as a diagnostic tool for abiotic stress tolerance in wild and cultivated strawberry species. *Horticulture, Environment, and Biotechnology* 55, 280–286.
Ndong, C., Ouellet, F., Houde, M. and Sarhan, F. (1997) Gene expression during cold acclimation in strawberry. *Plant and Cell Physiology* 38, 863–870.
O'Neill, S.D. (1983) Role of osmotic potential gradients during water stress and leaf senescence in *Fragaria virginiana*. *Plant Physiology* 72, 931–937.
Orsini, F., Alnayef, M., Bona, S., Maggio, A. and Gianquinto, G. (2012) Low stomatal density and reduced transpiration facilitate strawberry adaptation to salinity. *Environmental and Experimental Botany* 81, 1–10.
Paquin, R., Bolduc, R., Zizka, J., Pelletier, G. and Lechausser, P. (1989) Frost tolerance, sugar and proline content of strawberry (*Fragaria ananassa* Duch) during winter. *Canadian Journal of Plant Science* 69, 945–954.
Premachandra, G.S., Saneoka, H., Fujita, K. and Ogata, S. (1992) Leaf water relations, osmotic adjustment, cell membrane stability, epicuticular wax load and growth as affected by increasing water deficits in sorghum. *Journal of Experimental Botany* 43, 1569–1576.
Qasim, M., Ashraf, M., Ashraf, M.Y., Rehman, S.U. and Rha, E.S. (2003) Salt-induced changes in two canola cultivars differing in salt tolerance. *Biologia Plantarum* 46, 629–632.
Rahimi, A., Biglarifard, A. and Firozabadi, M. (2011) Influence of NaCl salinity on some physiological aspect of strawberry cv. Camarosa. *Russian Agricultural Sciences* 37, 378–384.
Rajashekar, C.B. and Panda, M. (2014) Water stress is a component of cold acclimation process essential for inducing full freezing tolerance in strawberry. *Scientia Horticulturae* 174, 54–59.
Rajashekar, C.B., Zhou, H., Marcum, K.B. and Prakash, O. (1999) Glycine betaine accumulation and induction of cold tolerance in strawberry (*Fragaria × ananassa* Duch.) plants. *Plant Science* 148, 175–183.
Razavi, F., Pollet, B., Steppe, K. and van Labeke, M.C. (2008) Chlorophyll fluorescence as a tool for evaluation of drought stress in strawberry. *Photosynthetica* 46, 631–633.
Razavi, F., van Labeke, M.C. and de Riek, J. (2009) Functional diversity for response to water deficit among different strawberry genotypes compared to their genetic structure as assessed by AFLP markers. *Acta Horticulturae* 839, 491–497.
Reekie, J.Y., Struik, P.C., Hicklenton, P.R. and Duval, J.R. (2007) Prohexadione-calcium changes morphological and physiological traits in strawberry plants and preconditions transplants to water stress. *European Journal of Horticultural Science* 72, 158–163.

Renquist, A.R., Breen, P.J. and Martin, L.W. (1982) Influences of water status and temperature on leaf elongation in strawberry. *Scientia Horticulturae* 18, 77–85.

Rivarola, M., Chan, A.P., Liebke, D.E., Melake-Berhan, A., Quan, H., Cheung, F., Ouyang, S., Folta, K.M., Slovin, J.P. and Rabinowicz, P.D. (2011) Abiotic stress-related expressed sequence tags from the diploid strawberry *Fragaria vesca* f. *semperflorens*. *Plant Genome* 4, 12–23.

Rout, N.P. and Shaw, B.P. (2001) Salt tolerance in aquatic macrophytes: possible involvement of the antioxidative enzymes. *Plant Science* 160, 415–423.

Saied, A.S., Keutgen, A.J. and Noga, G. (2005) The influence of NaCl salinity on growth, yield and fruit quality of strawberry cvs. 'Elsanta' and 'Korona'. *Scientia Horticulturae* 103, 289–303.

Salisbury, F.B. and Ross, C.W. (1992) *Plant Physiology*, 4th edn. Wadsworth Publishing, Belmont, California.

Sarhan, F., Ouellet, F. and Vazquez-Tello, A. (1997) The wheat *wcs120* gene family. A useful model to understand the molecular genetics of freezing tolerance in cereals. *Physiologia Plantarum* 101, 439–445.

Sarnighausen, E., Karlson, D.T., Zeng, Y., Goldsbrough, P.B., Raghothama, K.G. and Ashworth, E.N. (2004) Characterization of a novel Y_nSK_n class of dehydrin-like cDNAs from cold acclimated red-osier dogwood (*Cornus sericea* L.) xylem. In: Arora, R. (ed.) *Adaptations and Responses of Woody Plants to Environmental Stresses*. The Haworth Press, New York, pp. 17–36.

Savé, R., Peñuelas, J., Marfà, O. and Serrano, L. (1993) Changes in leaf osmotic and elastic properties and canopy structure of strawberries under mild water stress. *HortScience* 28, 925–927.

Scebba, F., Sebastiani, L. and Vitagliano, C. (1998) Changes in activity of antioxidative enzymes in wheat (*Triticum aestivum*) seedlings under cold acclimation. *Physiologia Plantarum* 104, 747–752.

Schwarz, M. (1995) *Soilless Culture Management*. Advanced Series in Agricultural Sciences, Springer, Heidelberg, Germany.

Sruamsiri, P. and Lenz, F. (1986) Photosynthesis and stomatal behaviour of strawberries VI. Effect of water deficiency. *Die Gartenbauwissenschaft* 51, 84–92.

Sudhakar, C., Lakshmi, A. and Giridarakumar, S. (2001) Changes in the antioxidant enzyme efficacy in two high yielding genotypes of mulberry (*Morus alba* L.) under NaCl salinity. *Plant Science* 161, 613–619.

Sun, W., Van Montagu, M. and Verbruggen, N. (2002) Small heat shock proteins and stress tolerance in plants. *Biochimica et Biophysica Acta* 1577, 1–9.

Tan, W., Meng, Q.W., Brestic, M., Olsovska, K. and Yang, X. (2011) Photosynthesis is improved by exogenous calcium in heat-stressed tobacco plants. *Journal of Plant Physiology* 168, 2063–2071.

Tanou, G., Molassiotis, A. and Diamantidis, G. (2009) Induction of reactive oxygen species and necrotic death-like destruction in strawberry leaves by salinity. *Environmental and Experimental Botany* 65, 270–281.

Thomashow, M.F. (1999) Plant cold acclimation: freezing tolerance genes and regulatory mechanisms. *Annual Review of Plant Physiology and Plant Molecular Biology* 50, 571–599.

Turhan, E. (2002) Researches on salt resistance physiology of strawberries growth in different media. PhD thesis, Uludag University, Bursa, Turkey.

Turhan, E. (2012) Seasonal alteration of sugar metabolism in strawberry (*Fragaria* × *ananassa*) plants during cold-acclimated and non-acclimated stages. *African Journal of Biotechnology* 11, 4558–4565.

Turhan, E. and Eris, A. (2005a) Changes of micronutrients, dry weight, and chlorophyll contents in strawberry plants under salt stress conditions. *Communications in Soil Science and Plant Analysis* 36, 1021–1028.

Turhan, E. and Eris, A. (2005b) Effects of sodium chloride applications and different growth media on ionic composition in strawberry plant. *Journal of Plant Nutrition* 27, 1653–1665.

Turhan, E. and Eris, A. (2007a) Growth and stomatal behaviour of two strawberry cultivars under long-term salinity stress. *Turkish Journal of Agriculture and Forestry* 31, 55–61.

Turhan, E. and Eris, A. (2007b) Determination of salt tolerance of 'Camarosa' and 'Chandler' strawberries grown in perlite and perlite:zeolite media. *International Journal of Fruit Science* 7, 87–96.

Turhan, E. and Eris, A. (2009) Changes of growth, amino acids, and ionic composition in strawberry plants under salt stress conditions. *Communications in Soil Science and Plant Analysis* 40, 3308–3322.

Turhan, E., Gulen, H. and Eris, A. (2008) The activity of antioxidative enzymes in three strawberry cultivars related to salt-stress tolerance. *Acta Physiologiae Plantarum* 30, 201–208.

Turhan, E., Aydogan, C., Akoglu, A., Baykul, A. and Evrenosoğlu, Y. (2012a) Relationship of seasonal changes in antioxidative enzymes and cold-hardiness in strawberry plant. *Journal of Food, Agriculture and Environment* 10, 445–450.

Turhan, E., Aydogan, C., Baykul, A., Akoglu, A., Evrenosoğlu, Y. and Ergin, S. (2012b) Apoplastic antioxidant enzymes in the leaves of two strawberry cultivars and their relationship to cold-hardiness. *Notulae Botanicae Horti Agrobotanici Cluj-Napoca* 40, 114–122.

Wahid, A., Gelani, S., Ashraf, M. and Foolad, M.R. (2007) Heat tolerance in plants: an overview. *Environmental and Experimental Botany* 61, 199–223.

Wang, S.Y. (1999) Methyl jasmonate reduces water stress in strawberry. *Journal of Plant Growth Regulation* 18, 127–134.

Wang, S.Y. and Camp, M.J. (2000) Temperatures after bloom affect plant growth and fruit quality of strawberry. *Scientia Horticulturae* 85, 183–199.

Wang, W., Vinocur, B. and Altman, A. (2003) Plant responses to drought, salinity and extreme temperatures: towards genetic engineering for stress tolerance. *Planta* 218, 1–14.

Waters, E.R. and Schaal, B.A. (1996) Heat shock induces a loss of rRNA-encoding DNA repeats in *Brassica nigra*. *Proceedings of the National Academy of Sciences USA* 93, 1449–1452.

Zhang, B. and Archbold, D.D. (1993) Solute accumulation in leaves of a *Fragaria chiloensis* and a *F. virginiana* selection responds to water deficit stress. *Journal of the American Society for Horticultural Science* 118, 280–285.

Zhang, W., Seki, M. and Furusaki, S. (1997) Effect of temperature and its shift on growth and anthocyanin production in suspension cultures of strawberry cells. *Plant Science* 127, 207–214.

Zhang, Y., Luo, Y., Hou, Y.-X., Jiang, H., Chen, Q. and Tang, H.-R. (2008a) Chilling acclimation induced changes in the distribution of H_2O_2 and antioxidant system of strawberry leaves. *Agriculture Journal* 3, 286–291.

Zhang, Y., Tang, H.-R. and Luo, Y. (2008b) Variation in antioxidant enzyme activities of two strawberry cultivars with short-term low temperature stress. *World Journal of Agricultural Sciences* 4, 458–462.

Index

Page numbers in **bold** type refer to figures and tables.